Financial Mathematics and Fintech

Series Editors

Zhiyong Zheng, Renmin University of China, Beijing, Beijing, China

Alan Peng, University of Toronto, Toronto, ON, Canada

This series addresses the emerging advances in mathematical theory related to finance and application research from all the fintech perspectives. It is a series of monographs and contributed volumes focusing on the in-depth exploration of financial mathematics such as applied mathematics, statistics, optimization, and scientific computation, and fintech applications such as artificial intelligence, block chain, cloud computing, and big data. This series is featured by the comprehensive understanding and practical application of financial mathematics and fintech. This book series involves cutting-edge applications of financial mathematics and fintech in practical programs and companies.

The Financial Mathematics and Fintech book series promotes the exchange of emerging theory and technology of financial mathematics and fintech between academia and financial practitioner. It aims to provide a timely reflection of the state of art in mathematics and computer science facing to the application of finance. As a collection, this book series provides valuable resources to a wide audience in academia, the finance community, government employees related to finance and anyone else looking to expand their knowledge in financial mathematics and fintech.

The key words in this series include but are not limited to:
a) Financial mathematics
b) Fintech
c) Computer science
d) Artificial intelligence
e) Big data

Yuanyuan Ke · Jing Li · Yifu Wang

Analysis of Reaction-Diffusion Models with the Taxis Mechanism

Yuanyuan Ke
School of Mathematics
Renmin University of China
Beijing, China

Jing Li
College of Science
Minzu University of China
Beijing, China

Yifu Wang
School of Mathematics and Statistics
Beijing Institute of Technology
Beijing, China

NNSF Project
NNSF Project 12071030, 12171498;
Beijing Natural Science Foundation Z210002

ISSN 2662-7167   ISSN 2662-7175 (electronic)
Financial Mathematics and Fintech
ISBN 978-981-19-3765-1   ISBN 978-981-19-3763-7 (eBook)
https://doi.org/10.1007/978-981-19-3763-7

This Springer imprint is published by the registered company Springer Nature Singapore Pte Ltd.
The registered company address is: 152 Beach Road, #21-01/04 Gateway East, Singapore 189721, Singapore

Preface

Taxis is an orientation mechanism under which the migration of the population is regulated by light, temperature, electric field, chemicals and many more in its environment. Among these, chemotaxis is an important sensory phenomenon in which cellular organisms direct their movements up or away the concentration gradient of stimulating chemical.

As a prototypical macroscopic model for self-enhanced chemotaxis, the mathematical feature of the Keller–Segel system has been the subject of intensive study over the past few decades, inter alia its ability to display cell aggregation in the utmost sense of finite-time blow-up of some solutions in two-even higher-dimensional settings. Motivated by numerical and modeling issues, suppressing taxis-driven blow-up in theory and numerics is a considerable challenging problem. There are some possible ways to avoid blow-up such as bounded chemotaxis sensibilities, nonlinear cell diffusion, logistic-type proliferation and death, and additional cross-diffusion term in the equation for the chemical signal.

In this book, we refrain from attempting to show that our results encompass all that have been done in the numerous relevant contributions on global classical solvability, boundedness and large time behavior in various types of chemotaxis systems, and rather put our recent research studies together in one place, and try to present in a somewhat systematic way some of the progress on these issues for more involved taxis-type cross-diffusive equations capable of adequately describing more complex biological systems.

The book is organized as follows. The first chapter focuses on global boundedness to a three-dimensional chemotaxis–Stokes system with nonlinear diffusion and rotation, and asymptotic profile of a two-dimensional chemotaxis–Navier–Stokes system with singular sensitivity and logistic source. The second chapter is concerned with Keller-Segel-fluid system where the chemoattractant is produced by bacteria rather than being consumed in the previous chapter. Relying on a variant of the natural gradient-like energy functional, the first part thereof shows that blow-up can be prevented by the slow diffusion of the cells in a two-dimensional Keller–Segel–Navier–Stokes system with rotational flux. In comparison with that the second part demonstrates that the suitable saturation of sensitivity is sufficient to guarantee global existence in a three-dimensional Keller–Segel–Navier–Stokes system involving tensor-valued sensitivity. The third chapter is divided into three parts. The first part investigates the logistic damping on Chaplain–Lolas model of cancer invasion with remodeling of tissue remodeling in two-dimensional spaces, while the second part of this chapter is devoted to the integrative interactions of chemotaxis, haptotaxis, logistic growth and remodeling mechanisms, and proves the global boundedness of solutions thereof rather comprehensively, as well as the global classical solutions under some smallness conditions in the three-dimensional setting. In the third part, we consider the long-time behavior of solutions to the evolution equations modeling tumor angiogenesis in a bounded smooth domain $\Omega \subset \mathbb{R}^N$ ($N = 1, 2$). In particular, in the one-dimensional case, it is shown that the corresponding solution converges to a steady state thereof with an explicit exponential rate. The fourth chapter is devoted to Keller–Segel–(Navier)–Stokes system modeling coral fertilization. The fifth chapter is concerned with the density-suppressed motility model. In the first part, by introducing an auxiliary parabolic problem to which the comparison principle applies and constructing relaxed super- and sub-solutions with spatially inhomogeneous decay rates, it is proved that the density-suppressed motility model admits traveling wave solutions in $\mathbb{R}^N$; In the second part, based on the duality argument, it is shown that for suitable fast diffusion of chemical signals the problem under consideration admits at least one global weak solution which will asymptotically converge to the spatially uniform equilibrium. The sixth chapter is devoted to a haptotactic cross-diffusion system modeling oncolytic virotherapy. In the first part, the corresponding solutions of the model with suitably small initial data is globally bounded and approach some constant profiles asymptotically. In the second part apart from the haptotaxis of uninfected cancer cells, the inclusion of two further haptotaxis mechanisms, both of infected tumor cells and virions, is considered with respect to aspects of classical solvability and boundedness in the presence of certain suitably strong further zero-order degradation.

It is our great pleasure to thank our collaborators, former students who were involved in this research. In particular, we would like to express our deep thanks to Professors Jingxue Yin, Peter Y. H. Pang, Zhian Wang, Li Chen and Jiashan Zheng, not only for our joint research but for the warm hospitality we enjoyed when visiting them as well. We would also like to acknowledge the financial support from the NNSF Project 12071030, 12171498, and Beijing Natural Science Foundation Z210002.

Beijing, China   Yuanyuan Ke
March 2022   Jing Li
                                                                                          Yifu Wang

Contents

Chapter 1
Chemotaxis–Fluid System

1.1 Introduction

In the early 1970s, Keller and Segel proposed the cross-diffusion system to describe the phenomenon of spatial structures in biological system through chemical induced processes (Keller and Segel 1970, 1971a). In particular, they looked at situations where cells partially orient their movement along gradients of a signal secreted by themselves, or instead, cells direct their movement in response to a substance which they consume. A prototypical example of the former is the Dictyostelium discoideum colony, while the latter is an E. coli population. The model in which the biased migration is induced by the consumed nutrient is usually called chemotaxis–consumption system. Often such chemotactic movements take place in a fluid environment, and experimental findings and analytical studies have revealed the remarkable effects of chemotaxis–fluid interaction on the overall behavior of the respective chemotaxis systems, such as the prevention of blow-up and improvement of efficiency of mixing (Chertock et al. 2012; Kiselev and Ryzhik 2012a; Kiselev and Xu 2016; Lorz 2012; Tuval et al. 2005). It should be noted that the derivation of chemotaxis models interacting with a fluid can be obtained by asymptotic methods inspired by Hilbert's sixth problem (Bellomo et al. 2016).

This chapter is concerned with a convective chemotaxis system for the oxygen-consuming and oxy-tactic bacteria, coupled with the incompressible Navier–Stokes equations. Section 1.3 is concerned with the following system

$$
\begin{cases}
n_t + u \cdot \nabla n = \Delta n^m - \nabla \cdot (n S(x, n, c) \cdot \nabla c), & x \in \Omega, t > 0, \\
c_t + u \cdot \nabla c = \Delta c - nc, & x \in \Omega, t > 0, \\
u_t + \nabla P = \Delta u + n \nabla \phi, & x \in \Omega, t > 0, \\
\nabla \cdot u = 0, & x \in \Omega, t > 0, \\
(\nabla n^m - n S(x, n, c) \cdot \nabla c) \cdot v = \partial_v c = 0, u = 0, & x \in \partial\Omega, t > 0, \\
n(x, 0) = n_0(x), c(x, 0) = c_0(x), u(x, 0) = u_0(x), & x \in \Omega
\end{cases}
\tag{1.1.1}
$$

© The Author(s) 2022
Y. Ke et al., *Analysis of Reaction-Diffusion Models with the Taxis Mechanism*,
Financial Mathematics and Fintech, https://doi.org/10.1007/978-981-19-3763-7_1

with $m > 0$, where Ω is a bounded domain in $\mathbb{R}^3$, $S(x, n, c)$ is a chemotactic sensitivity tensor satisfying

$$S \in C^2(\bar{\Omega} \times [0, \infty)^2; \mathbb{R}^{3\times 3}) \tag{1.1.2}$$

and

$$|S(x, n, c)| \le (1 + n)^{-\alpha} S_0(c) \quad \text{for all } (x, n, c) \in \Omega \times [0, \infty)^2 \tag{1.1.3}$$

with $\alpha \ge 0$ and some non-decreasing $S_0 : [0, \infty) \to \mathbb{R}$.

In the two-dimensional analogue of (1.1.1), the condition of $m = 1$, $\alpha = 0$ is sufficient to ensure global existence of some generalized solution thereof (see Winkler 2018d), which eventually becomes smooth (Winkler 2021a). Nevertheless, a new difficulty arises in the analytical studies of the three-dimensional version of (1.1.1). It is well known that, compared with the case $m = 1$ (see Ke and Zheng 2019, Wang et al. 2018, Winkler 2018e), the nonlinear diffusion mechanism $m \ne 1$ may inhibit the occurrence of blow-up phenomena (see Tao and Winkler 2012b, Winkler 2013). Up to now, system (1.1.1) with nonlinear diffusion has been studied systematically. Indeed, in **three** space dimensions ($N = 3$), many authors considered the global existence and boundedness of the solutions, and the restriction on m is weakened bit by bit. For example, when the chemotactic sensitivity function $S(x, n, c)$ is scalar-value, in 2010, the range of m can be belong to $[\frac{7+\sqrt{217}}{12}, 2]$ (Francesco et al. 2010); in 2013, for locally bounded solution, m can be greater than $\frac{8}{7}$ (Tao and Winkler 2013); in 2018, the result is pushed to $m > \frac{9}{8}$ (Winkler 2018c). If we only consider the global existence of the solutions, rather than its boundedness, the value of m can be even smaller, such as $m \ge 1$ in Duan and Xiang (2014) and $m \ge \frac{2}{3}$ in Zhang and Li (2015a). When the chemotactic sensitivity function $S(x, n, c)$ is **tensor**-value (S is a matrix), in 2015, Winkler (Winkler 2015b) established the uniform-in-time boundedness of global weak solutions in bounded and convex domains Ω for $m > \frac{7}{6}$. Zheng (2022) extended the previous global boundedness result to $m > \frac{10}{9}$. As an extension of this result, when S fulfills (1.1.3), the corresponding results are constantly updated. In 2017, it was shown in Wang and Li (2017) that $m \ge 1$ and $m + \alpha > \frac{7}{6}$ insures the global existence of bounded weak solution; in 2020, Wang (2020) extended the previous global boundedness result to $m + \frac{5}{4}\alpha > \frac{9}{8}$, $\alpha > 0$ and $m + \alpha > \frac{10}{9}$. The first section shows how far the porous medium type diffusion of bacteria and saturation of tensor-valued sensitivity ensure the global boundedness of the weak solutions to (1.1.1) in the standard sense by the method different from those in Tao and Winkler (2013), Wang (2020), Winkler (2015b), Winkler (2018c).

In order to prepare a precise statement of our main results in these respects, let us assume that the initial data satisfy

$$\begin{cases} n_0 \in C^\kappa(\bar{\Omega}) \text{ for certain } \kappa > 0 \text{ with } n_0 \geq 0 \text{ and } n_0 \not\equiv 0 \text{ in } \Omega, \\ c_0 \in W^{1,\infty}(\Omega) \text{ with } c_0 > 0 \text{ in } \bar{\Omega}, \\ u_0 \in D(A^\gamma) \text{ for some } \gamma \in \left(\dfrac{3}{4}, 1\right), \end{cases} \qquad (1.1.4)$$

where A denotes the Stokes operator with domain $D(A) := W^{2,2}(\Omega) \cap W_0^{1,2}(\Omega) \cap L_\sigma^2(\Omega)$, and $L_\sigma^2(\Omega) := \{\varphi \in L^2(\Omega) | \nabla \cdot \varphi = 0\}$ (see Sohr 2001). As for the time-independent gravitational potential function ϕ, we assume for simplicity that $\phi \in W^{2,\infty}(\Omega)$.

Within this framework, our main result can be stated as follows (Zheng and Ke 2021):

Theorem 1.1 *Let* (1.1.4) *hold and suppose that S satisfies* (1.1.2)–(1.1.3). *If $m + \alpha > \frac{10}{9}$ with $m > 0$ and $\alpha \geq 0$, then there exists at least one global weak solution (in the sense of Definition 1.1 below) of problem* (1.1.1). *Also, this solution is bounded in $\Omega \times (0, \infty)$ in the sense that for all $t > 0$*

$$\|n(\cdot, t)\|_{L^\infty(\Omega)} + \|c(\cdot, t)\|_{W^{1,\infty}(\Omega)} + \|u(\cdot, t)\|_{L^\infty(\Omega)} \leq C$$

with some positive constant C independent of t. Moreover, c and u are continuous in $\bar{\Omega} \times [0, \infty)$ and

$$n \in C_{\omega-*}^0([0, \infty); L^\infty(\Omega)).$$

The proof of Theorem 1.1 focuses on the derivation of regularity estimates for the component n_ε properly by means of a new bootstrap iteration in the case of $\frac{10}{9} < m + \alpha < \frac{3}{2}$, which seems to be quite different from those in Tao and Winkler (2013), Wang (2020), Winkler (2015b, 2018c), Zheng (2022). More precisely, based on the basic a priori estimates, we can establish the $L^p(\Omega)$-estimates on n_ε for some $p > \frac{3}{2}$ in the case $\frac{10}{9} < m + \alpha \leq 2, \alpha > \frac{7}{18}$ or $m + \alpha > 2$ by using some carefully analysis. Whereas for $\frac{10}{9} < m + \alpha \leq 2$ and smaller $\alpha \in [0, \frac{7}{18}]$, the derivation of the $L^{p*}(\Omega)$-estimates on n_ε with some $p* > \frac{3}{2}$ needs a new iteration. In fact, on the basis of the spatio-temporal estimate $\int_t^{t+1} \int_\Omega \frac{n_\varepsilon}{c_\varepsilon} |\nabla c_\varepsilon|^2$ provided by the quasi-energy functional, one can establish the boundedness of n_ε in $L^{p_1}(\Omega)$ (see Lemma 1.20) and $L^{p_n}(\Omega)$ (see Lemma 1.23), where $p_1 = \frac{16}{3}(m + \alpha)^2 - \frac{25}{3}(m + \alpha) + 4 + \frac{1}{3}\alpha[4(m + \alpha) - 1]$ and $p_n = \frac{2}{3}p_n^2 + \frac{2}{3}(4m - 5 + 3\alpha)p_n + (2m + 2\alpha - 3)(m - 1) + 1$. Based on the L^{p_n}-boundedness of n_ε, one can then archive the uniform bounds of n_ε in $L^p(\Omega)$ for any $p > 1$. With the aid of a standard Morse-type technique and the maximal Sobolev regularity, one can derive the boundedness of n_ε, ∇c_ε and u_ε in $L^\infty(\Omega)$, inter alia the further regularity properties thereof which seem necessary to obtain the global weak solution to system (1.1.1).

In the second part of this chapter, we are concerned with the chemotaxis–consumption system coupled with the incompressible Navier–Stokes equations

$$
\begin{cases}
n_t + u \cdot \nabla n = \Delta n - \chi \nabla \cdot \left(\dfrac{n}{c} \nabla c \right) + f(n), & x \in \Omega, t > 0, \\[2mm]
c_t + u \cdot \nabla c = \Delta c - nc, & x \in \Omega, t > 0, \\[2mm]
u_t + (u \cdot \nabla) u = \Delta u + \nabla P + n \nabla \phi, & x \in \Omega, t > 0, \\[2mm]
\nabla \cdot u = 0, & x \in \Omega, t > 0,
\end{cases}
\tag{1.1.5}
$$

describing the biological population density n, the chemical signal concentration c, the incompressible fluid velocity u and the associated pressure P of the fluid flow in the physical domain $\Omega \subset \mathbb{R}^N$. It is assumed that n and c diffuse randomly as well as are transported by the fluid, with a buoyancy effect on n through the presence of a given gravitational potential ϕ. Further, it is assumed that the chemotactic stimulus is perceived in accordance with the Weber–Fechner Law (Short et al. 2010; Wang 2013; Winkler 2019b) which states that subjective sensation is proportional to the logarithm of the stimulus intensity, in other words, the population n partially direct their movement toward increasing concentrations of the chemical nutrient c that they consume with the logarithmic sensitivity. In addition, on the considered time scales of cell migration, we allow for population growth to take place, through the term $f(n) = rn - \mu n^2$ with the effective growth rate $r \in \mathbb{R}$, which accounts for the mortality or population renewal, and strength of the overcrowding effect $\mu > 0$; we note that $r = 0$ is allowed and has indeed been argued for in certain models (Hillen and Painter 2009; Kiselev and Ryzhik 2012a).

The system (1.1.5) appears to generate interesting, nontrivial dynamics. However, to the best of our knowledge, no analytical result is available yet which rigorously describes the qualitative behavior of such solutions. This may be due to the circumstance that (1.1.5) joins two subsystems which are far from being fully understood even when decoupled from each other. Indeed, (1.1.5) contains the Navier–Stokes equations which themselves do not admit a complete existence and regularity theory (Wiegner 1999).

At the same time, by setting $u \equiv 0$ in (1.1.5), we arrive at the following chemotaxis–consumption model

$$
\begin{cases}
n_t = \Delta n - \chi \nabla \cdot \left(\dfrac{n}{c} \nabla c \right), \\[2mm]
c_t = \Delta c - nc,
\end{cases}
\tag{1.1.6}
$$

where population growth has been ignored, which was introduced by Keller and Segel (1971a) to describe the collective behavior of the bacteria E. coli set in one end of a capillary tube featuring a gradient of nutrient concentration observed in the celebrated experiment of Adler (1966). Later, this model was also employed to describe the dynamical interactions between vascular endothelial cells and vascular endothelial growth factor (VEGF) during the initiation of tumor angiogenesis (see Corrias et al. 2003; Levine et al. 2000). It has already been demonstrated that the logarithmic sensitivity featured in (1.1.6) renders a significant degree of complexity in the system; in particular, it plays an indispensable role in generating wave-like solutions without any type of cell kinetics (Hillen and Painter 2009; Keller and Segel 1970; Rosen 1978; Schwetlick 2003; Wang 2013), which is a prominent feature in the Fisher equation (Kolmogorov et al. 1937).

In comparison with (1.1.6), the related chemotaxis system

$$\begin{cases} n_t = \Delta n - \chi \nabla \cdot (\dfrac{n}{c} \nabla c) + f(n), \\ c_t = \Delta c - c + n, \end{cases} \tag{1.1.7}$$

where the chemical signal c is actively secreted by the bacteria rather than consumed (see Bellomo et al. 2015; Hillen and Painter 2009), has been more extensively studied. It is observed that the chemical signal production mechanism in the $c-$equation inhibits the tendency of c to take on small values, and thereby the singularity in the sensitivity function is mitigated. Accordingly, for such higher dimensional systems with reasonably smooth but arbitrarily large data, the global existence of bounded smooth solutions can be achieved. Indeed, global existence and boundedness of classical solutions to (1.1.7) without source terms is guaranteed if $\chi \in (0, \sqrt{\frac{2}{N}})$ (Fujie 2015; Winkler 2011a), or if $N = 2$, $\chi \in (0, \chi_0)$ with some $\chi_0 > 1.015$ (Lankeit 2016b), while certain generalized solutions have been constructed for general $\chi > 0$ in the two-dimensional radially symmetric case (Stinner and Winkler 2011; Winkler 2011a). Moreover, without any symmetry hypothesis, Winkler and Lankeit established the global solvability of generalized solutions for the cases $\chi < \infty, N = 2$; $\chi < \sqrt{8}, N = 3$; and $\chi < \frac{N}{N-2}, N \geq 4$ (Lankeit and Winkler 2017).

Furthermore, in accordance with known results for the classical Keller–Segel chemotaxis model (see Lankeit 2015; Winkler 2010a, 2014a for example), the presence of the logistic source term $f(n) = n(r - \mu n)$ in (1.1.7) can inhibit the tendency toward explosions of cells at least under some restrictions on certain parameters. Indeed, it is known that (1.1.7) with $N = 2$ possesses a global classical solution (n, c) for any $r \in \mathbb{R}$, $\chi, \mu > 0$, and (n, c) is globally bounded if $r > \frac{\chi^2}{4}$ for $0 < \chi \leq 2$ or $r > \chi - 1$ for $\chi > 2$ (Zhao and Zheng 2017). Moreover, (n, c) exponentially converges to $(\frac{r}{\mu}, \frac{r}{\mu})$ in $L^\infty(\Omega)$ provided that $\mu > 0$ is sufficiently large (Zheng et al. 2018). As for the higher dimensional cases ($N \geq 2$), the global very weak solution of (1.1.7) with $f(n) = rn - \mu n^k$ is constructed when k, χ and r fulfill a certain condition. In addition, when $N = 2$ or 3, this solution is global bounded provided $\frac{r}{\mu}$ and the initial data $\|n_0\|_{L^2}$, $\|\nabla c_0\|_{L^4}$ are suitably small (Zhao and Zheng 2019).

In contrast to (1.1.7), system (1.1.6) is more challenging due to the combination of the consumption of c with the singular chemotaxis sensitivity of n. Intuitively, the absorption mechanism in the $c-$equation of (1.1.6), which induces the preference for small values of c, considerably intensifies the destabilizing potential of singular sensitivity in the $n-$equation. Up to now, it seems that only limited results on global classical solvability in the spatial two-dimensional case are available. In fact, only recently have certain global generalized solutions to (1.1.6) been constructed for general initial data in Lankeit and Lankeit (2019b), Winkler (2016a), Winkler (2018a), whereas with respect to global classical solvability, it has only been shown for some small initial data (see Wang et al. 2016; Winkler 2016c). In particular, Winkler (2016c) showed that the global classical solutions to (1.1.6) in bounded convex two-dimensional domains exist and converge to the homogeneous steady state under an essentially explicit smallness condition on n_0 in $L \log L(\Omega)$ and $\nabla \ln c_0$ in $L^2(\Omega)$. We would, however, like to note that numerous variants of (1.1.6), such as those involving nonlinear diffusion, logistic-type cell kinetics and saturating signal production (Ding and Zhao 2018; Jia and Yang 2019; Lankeit and Lankeit 2019a; Lankeit 2017; Winkler 2022; Lankeit and Viglialoro 2020; Liu 2018; Viglialoro 2019; Zhao and Zheng 2018), have been studied. For example, the authors of Zhao and Zheng (2018) proved that the particular version of (1.1.6) by adding $f(n) = rn - \mu n^k$ ($r > 0$, $\mu > 0$, $k > 1$) into the n-equation admits a global classical solution (n, c) in the bounded domain $\Omega \subset \mathbb{R}^N$ if $k > 1 + \frac{N}{2}$, and in the two-dimensional setting, $(n, c, \frac{|\nabla c|}{c}) \to ((\frac{r}{\mu})^{\frac{1}{k-1}}, 0, 0)$ for sufficiently large μ. In particular, it is shown in the recent paper Lankeit and Lankeit (2019a) that (1.1.6) with logistic source $f(n) = rn - \mu n^2$ ($r \in \mathbb{R}$, $\mu > 0$) possesses a unique global classical solution if $0 < \chi < \sqrt{\frac{2}{N}}$, $\mu > \frac{N-2}{2N}$, and a globally bounded solution only in **one** dimension for any $\chi > 0$, $\mu > 0$. Also, the author of Wang (2019) showed that if $\mu > \mu_0$ with some $\mu_0 = \mu_0(\Omega, \chi) > 0$ then the corresponding classical solution is globally bounded, and $(n, c, \frac{|\nabla c|}{c}) \to (\frac{r_+}{\mu}, \lambda, 0)$ with $\lambda \in [0, \frac{1}{|\Omega|} \int_\Omega c_0)$ in $(L^\infty(\Omega))^3$ as $t \to \infty$. Of course, this leaves open the possibility of blow-up of solutions when μ is positive but small. Anyhow, it has been shown in Winkler (2017a) that when $\mu > 0$ is suitably small, the strongly destablizating action of chemotactic cross-diffusion may lead to the occurrence of solutions which attain possibly finite but arbitrarily large values.

Coming back to our chemotaxis–consumption–fluid model (1.1.5), as we have already pointed out, very little seems to be known regarding the qualitative behavior of solutions (Black 2018; Black et al. 2018, 2019). In fact, we are aware of one result only which is concerned with the asymptotic behavior and eventual regularity of solutions to the Stokes variant of (1.1.5). Namely, it is shown in Black (2018) that for small initial mass $\int_\Omega n_0$, the corresponding system upon neglection of $u \cdot \nabla u$ and $f(n)$ in (1.1.5) possesses at least one global generalized solutions, which will become smooth after some waiting time and stabilize toward the steady state $(\frac{1}{|\Omega|} \int_\Omega n_0, 0, 0)$ with respect to the topology of $(L^\infty(\Omega))^3$. Since the presence of the fluid interaction does not have any regularizing effect on the large time behavior, it is expected that instead of the small restriction on the initial data, the quadratic degradation may have a substantial regularizing effect on the dynamic behavior of solutions to (1.1.5).

The second part of this chapter focuses on the asymptotic profile in time of solutions to (1.1.5) in the two-dimensional case. In order to state our main results, we shall impose on (1.1.5) the boundary conditions

$$\nabla n \cdot \nu = \nabla c \cdot \nu = 0 \quad \text{and} \quad u = 0 \text{ for } x \in \partial\Omega, \tag{1.1.8}$$

and initial conditions

$$n(x, 0) = n_0(x), \ c(x, 0) = c_0(x), \ u(x, 0) = u_0(x) \ \text{ for } \ x \in \Omega. \tag{1.1.9}$$

Throughout this part, it is assumed that

$$\begin{cases} n_0 \in C^0(\bar{\Omega}), \ n_0 \geq 0 \text{ and } n_0 \not\equiv 0 \text{ in } \Omega, \\ c_0 \in W^{1,\infty}(\Omega), c_0 > 0 \text{ in } \bar{\Omega} \text{ as well as} \\ u_0 \in D(A^\beta) \text{ for all } \beta \in (\frac{1}{2}, 1) \end{cases} \tag{1.1.10}$$

with A denoting the Stokes operator $A = -\mathscr{P}\Delta$ with domain $D(A) := W^{2,2}(\Omega; \mathbb{R}^2) \cap W_0^{1,2}(\Omega; \mathbb{R}^2) \cap L_\sigma^2(\Omega)$, where $L_\sigma^2(\Omega) := \{\varphi \in L^2(\Omega; \mathbb{R}^2) | \nabla \cdot \varphi = 0\}$ and $\mathscr{P}$ stands for the Helmholtz projection of $L^2(\Omega)$ onto $L_\sigma^2(\Omega)$.

Within this framework, by straightforward adaptation of arguments in Lankeit and Lankeit (2019a) with only some necessary modifications, one can see that the problem (1.1.5), (1.1.8), (1.1.9) admits a global classical solution (n, c, u, P) whenever $\chi \in (0, 1), r \in \mathbb{R}$ and $\mu > 0$, which is unique up to addition of constants in the pressure variable P, and satisfies $n > 0, c > 0$ in $\Omega \times [0, \infty)$. The first of our main results is concerned with the global boundedness of the solution as well as its asymptotic behavior (Pang et al. 2021).

Theorem 1.2 *Let $f(n) = rn - \mu n^2, r \in \mathbb{R}, \mu > 0$ and $\phi \in W^{2,\infty}(\Omega)$, and suppose that (n_0, c_0, u_0) satisfy (1.1.10). If (n, c, u, P) denotes the corresponding global classical solution to (1.1.5), (1.1.8), (1.1.9), then there exists a value $\mu_0 = \mu_0(\Omega, \chi, r) \geq 0$ with $\mu_0(\Omega, \chi, 0) = 0$ such that whenever $\mu > \mu_0$, (n, c, u) is global bounded,*

$$\|n(\cdot, t) - \frac{r_+}{\mu}\|_{L^\infty(\Omega)} \to 0, \ \ \|\frac{\nabla c}{c}(\cdot, t)\|_{L^\infty(\Omega)} \to 0, \ \ \|u(\cdot, t)\|_{L^\infty(\Omega)} \to 0$$

and when $r > 0$, $\|c(\cdot, t)\|_{L^\infty(\Omega)} \to 0$ as $t \to \infty$.

As indicated in the above discussion, we need to introduce new ideas to show how the regularizing effect of the quadratic degradation in the chemotaxis–fluid model (1.1.5) can counterbalance the strongly destabilizing action of chemotactic cross-diffusion caused by the combination of the consumption of c with the singular chemotaxis sensitivity of n. Specifically, we develop the conditional energy functional method in Winkler (2016c) to show the global boundedness of solutions in the case of $r > 0$, in which the key point is to verify that

$$\mathscr{F}(n, w) := \int_\Omega H(n) + \frac{\chi}{2} \int_\Omega |\nabla w|^2, \ \ w := -\ln(\frac{c}{\|c_0\|_{L^\infty(\Omega)}}) \tag{1.1.11}$$

with $H(s) := s \ln \frac{\mu s}{er} + \frac{r}{\mu}$ constitutes an energy functional in the sense that $\mathscr{F}(n, w)$ is non-increasing in time whenever μ is appropriately large relative to r (see Lemma 1.41). Indeed, from (1.4.29), one can obtain the global bound of $\int_\Omega n|\ln n| dx$ and $\int_\Omega |\nabla w|^2 dx$, which then serves as a starting point to derive the uniform bound of

$\|n(\cdot, t)\|_{L^\infty(\Omega)}$ via the Neumann heat semigroup estimates. Furthermore, by making appropriate use of the dissipative information expressed in (1.4.29), we can establish the convergence result asserted in Theorem 1.2. It is noted that compared to that of the case $r > 0, \mu > 0$, the proof of Theorem 1.2 in the case of $r \leq 0, \mu > 0$ involves a more delicate analysis. In fact, unlike in the case $r = \mu = 0$ or $r > 0, \mu > 0$, (1.1.5) with $r \leq 0, \mu > 0$ seems to lack the favorable structure that facilitates such conditional energy-type inequalities. Taking full advantage of the decay information on n in L^1-norm expressed in (1.4.2), our approach toward Theorem 1.2 is to construct the quantity

$$\mathscr{F}(n, w) := \int_\Omega n(\ln n + a) + \frac{\chi}{2} \int_\Omega |\nabla w|^2 \qquad (1.1.12)$$

with parameter $a > 0$ determined below (see (1.4.64)). Unlike in the case of $r > 0$, $\mathscr{F}(n, w)$ does not enjoy monotonicity property, it however satisfies a favorable non-homogeneous differential inequality (1.4.71) in the sense that it can provide us a priori information on solution such as the global bound of $\int_\Omega n|\ln n|dx$ and $\int_\Omega |\nabla w|^2 dx$ (see Lemma 1.43), as well as $\lim_{t\to\infty} \int_\Omega |\nabla w(\cdot, t)|^2 = 0$ (see (1.4.77)).

As an important step to understand the model (1.1.5) more comprehensively, we shall consider the convergence rate of its classical solutions in the form of the following result:

Theorem 1.3 *Let the assumptions of Theorem 1.2 hold and $r > 0$. Then one can find $\mu_*(\chi, \Omega, r) > 0$ such that if $\mu > \mu_*(\chi, \Omega, r)$, the classical solution of (1.1.5), (1.1.8), (1.1.9) presented in Theorem 1.2 satisfies*

$$\left\|n(\cdot, t) - \frac{r}{\mu}\right\|_{L^\infty(\Omega)} \to 0, \quad \|c(\cdot, t)\|_{L^\infty(\Omega)} \to 0, \quad \|u(\cdot, t)\|_{L^\infty(\Omega)} \to 0$$

as well as $\left\|\frac{\nabla c}{c}(\cdot, t)\right\|_{L^p(\Omega)} \to 0$ for all $p > 1$ exponentially as $t \to \infty$.

This implies that suitably large μ relative to r enforces asymptotic stability of the corresponding constant equilibria of (1.1.5); however, the optimal lower bound on $\frac{\mu}{r}$ seems yet lacking. The main ingredient of our approach toward Theorem 1.3 involves a so-called self-map-type reasoning. More precisely, making use of the convergence properties of $(n, \frac{|\nabla c|}{c})$ asserted in Theorem 1.3, we prove by a self-map-type reasoning that whenever μ is suitably large compared with r,

$$\left(n(\cdot, t) - \frac{r}{\mu}, c(\cdot, t), u\right) \longrightarrow (0, 0, 0) \quad \text{and} \quad \frac{|\nabla c|}{c}(\cdot, t) \longrightarrow 0$$

in $(L^\infty(\Omega))^3$ and $L^6(\Omega)$ exponentially as $t \to \infty$, respectively (see Lemma 1.45).

As aforementioned, the limit case $r = 0$ becomes relevant in several applications. In this limiting situation, the total cell population can readily be seen to decay in the large time limit (cf. Lemma 1.36 below). As a consequence, we can obtain the decay properties of solutions, namely that the decay on n in L^1 actually occurs in L^∞, and also for c. More precisely, our result reads as follows:

Theorem 1.4 *Let the assumptions of Theorem 1.2 hold and $r = 0$. Then the classical solution of (1.1.5), (1.1.8), (1.1.9) from Theorem 1.2 satisfies $(n, c, \frac{|\nabla c|}{c}, u) \longrightarrow (0, 0, 0, 0)$ in $(L^\infty(\Omega))^4$ algebraically as $t \to \infty$.*

The result indicates that structure generating dynamics in the spatially two-dimensional version of (1.1.5), (1.1.8) and (1.1.9), if at all, occur on intermediate time scales rather than in the sense of a stable large time pattern formation process. Apparently, it leaves open the questions whether the more colorful large time behavior can appear in the three-dimensional version of (1.1.5).

The approach toward Theorem 1.4 uses an alternative method, which, at its core, is based on the argument that the L^∞-norm of n can be controlled from above by appropriate multiples of $\frac{1}{t+1}$. This results from a suitable variation-of-constants representation of n, by which and in view of the decay information on $|\nabla w|$ in $L^\infty(\Omega)$, the L^1 decay information on u from (1.4.2) can be turned into the L^∞-norm of n (see Lemma 1.46). As a consequence, by comparison argument, we have a pointwise upper estimate for w as well as a lower estimate for v (see Lemma 1.47). Using $L^p - L^q$ estimates for the Neumann heat semigroup $(e^{t\Delta})_{t>0}$, we then successively show that $\|\nabla w\|_{L^\infty}$ and $\|n\|_{L^\infty(\Omega)}$ can be controlled by appropriate multiples of $\frac{1}{t+1}$ from above and below, respectively (see Lemma 1.48). These a priori estimates allow us to get the pointwise lower estimate for w as well as the upper estimate for c, which complement the lower bound for c previously obtained, and thereby prove that c actually decays algebraically.

1.2 Preliminaries

Firstly let us recall the important $L^p - L^q$ estimates for the Neumann heat semigroup $(e^{t\Delta})_{t>0}$ on bounded domains, which plays an important role not only in Chap. 1, but also in Chaps. 3, 4 and 6.

Lemma 1.1 (Lemma 1.3 of Winkler 2010 and Lemma 2.1 of Cao 2015) *Let $(e^{t\Delta})_{t>0}$ denote the Neumann heat semigroup in the domain Ω and $\lambda_1 > 0$ denote the first nonzero eigenvalue of $-\Delta$ in $\Omega \subset \mathbb{R}^N$ under the Neumann boundary condition. There exists c_i, $i = 1, 2, 3, 4$, such that for all $t > 0$,*

(i) If $1 \le q \le p \le \infty$, then for all $\omega \in L^q(\Omega)$ with $\int_\Omega \omega = 0$,

$$\|e^{t\Delta}\omega\|_{L^p(\Omega)} \le c_1 \left(1 + t^{-\frac{N}{2}(\frac{1}{q} - \frac{1}{p})}\right) e^{-\lambda_1 t} \|\omega\|_{L^q(\Omega)};$$

(ii) If $1 \le q \le p \le \infty$, then for all $\omega \in L^q(\Omega)$,

$$\|\nabla e^{t\Delta}\omega\|_{L^p(\Omega)} \le c_2 \left(1 + t^{-\frac{1}{2} - \frac{N}{2}(\frac{1}{q} - \frac{1}{p})}\right) e^{-\lambda_1 t} \|\omega\|_{L^q(\Omega)};$$

(iii) If $2 \leq q \leq p \leq \infty$, *then for all* $\omega \in W^{1,q}(\Omega)$,

$$\|\nabla e^{t\Delta}\omega\|_{L^p(\Omega)} \leq c_3 \left(1 + t^{-\frac{N}{2}(\frac{1}{q}-\frac{1}{p})}\right) e^{-\lambda_1 t} \|\nabla\omega\|_{L^q(\Omega)};$$

(iv) If $1 \leq q \leq p < \infty$ *or* $1 < q < \infty$ *and* $p = \infty$, *then for all* $\omega \in (L^q(\Omega))^N$,

$$\|e^{t\Delta}\nabla\cdot\omega\|_{L^p(\Omega)} \leq c_4 \left(1 + t^{-\frac{1}{2}-\frac{N}{2}(\frac{1}{q}-\frac{1}{p})}\right) e^{-\lambda_1 t} \|\omega\|_{L^q(\Omega)}.$$

In order to obtain the solution of system (1.1.1) through a suitable approximation procedure, we follow the well-established approaches to regularize both the chemotactic sensitivity and nonlinear diffusion in the first equation in (1.1.1) (see Cao and Lankeit 2016; Li et al. 2015; Winkler 2015a,b; Ke and Zheng 2019). Let $(\rho_\varepsilon)_{\varepsilon\in(0,1)} \in C_0^\infty(\Omega)$ be a family of standard cut-off functions, which satisfying $0 \leq \rho_\varepsilon \leq 1$ in Ω and $\rho_\varepsilon \nearrow 1$ in Ω as $\varepsilon \searrow 0$, and $\chi_\varepsilon \in C_0^\infty([0,\infty))$ satisfying $0 \leq \chi_\varepsilon \leq 1$ in $[0,\infty)$ and $\chi_\varepsilon \nearrow 1$ as $\varepsilon \searrow 0$. Define

$$S_\varepsilon(x,n,c) := \rho_\varepsilon(x)\chi_\varepsilon(n)S(x,n,c), \quad x \in \bar{\Omega}, \ n \geq 0, \ c \geq 0$$

for $\varepsilon \in (0,1)$, which implies that $S_\varepsilon(x,n,c) = 0$ on $\partial\Omega$. As an approximation function of the sensitivity tensor S, S_ε also satisfies the condition (1.1.3), that is,

$$|S_\varepsilon(x,n,c)| \leq (1+n)^{-\alpha}S_0(c) \quad \text{for all } (x,n,c) \in \Omega \times [0,\infty)^2. \qquad (1.2.1)$$

The regularized problem of (1.1.1) can be presented as follows

$$\begin{cases} n_{\varepsilon t} + u_\varepsilon \cdot \nabla n_\varepsilon = \Delta(n_\varepsilon + \varepsilon)^m - \nabla\cdot(n_\varepsilon F_\varepsilon(n_\varepsilon)S_\varepsilon(x,n_\varepsilon,c_\varepsilon)\cdot\nabla c_\varepsilon), & x \in \Omega, t > 0, \\ c_{\varepsilon t} + u_\varepsilon \cdot \nabla c_\varepsilon = \Delta c_\varepsilon - n_\varepsilon c_\varepsilon, & x \in \Omega, t > 0, \\ u_{\varepsilon t} + \nabla P_\varepsilon = \Delta u_\varepsilon + n_\varepsilon \nabla\phi, & x \in \Omega, t > 0, \\ \nabla\cdot u_\varepsilon = 0, & x \in \Omega, t > 0, \\ \nabla n_\varepsilon \cdot \nu = \nabla c_\varepsilon \cdot \nu = 0, u_\varepsilon = 0, & x \in \partial\Omega, t > 0, \\ n_\varepsilon(x,0) = n_0(x), c_\varepsilon(x,0) = c_0(x), u_\varepsilon(x,0) = u_0(x), & x \in \Omega, \end{cases} \qquad (1.2.2)$$

where $F_\varepsilon(s) = \frac{1}{1+\varepsilon s}$ for $s \geq 0$.

Let us recall the local well-posedness of (1.2.2).

Lemma 1.2 (Winkler 2012, 2015b) *Let* $\Omega \subseteq \mathbb{R}^3$ *be a bounded domain with smooth boundary. Suppose that* (1.1.2)–(1.1.3) *hold. Assume that the initial data* (n_0, c_0, u_0) *fulfills* (1.1.4). *Then for each* $\varepsilon \in (0,1)$, *there exist functions*

$$\begin{cases} n_\varepsilon \in C^0(\bar{\Omega} \times [0, \infty)) \cap C^{2,1}(\bar{\Omega} \times (0, \infty)), \\ c_\varepsilon \in C^0(\bar{\Omega} \times [0, \infty)) \cap C^{2,1}(\bar{\Omega} \times (0, \infty)) \cap_{q>3} C^0([0, \infty); W^{1,q}(\Omega)), \\ u_\varepsilon \in C^0(\bar{\Omega} \times [0, \infty)) \cap C^{2,1}(\bar{\Omega} \times (0, \infty)), \\ P_\varepsilon \in C^{1,0}(\bar{\Omega} \times (0, \infty)), \end{cases}$$

$$(1.2.3)$$

such that $(n_\varepsilon, c_\varepsilon, u_\varepsilon, P_\varepsilon)$ solves (1.2.2) classically in $\Omega \times (0, \infty)$, and such that $n_\varepsilon \geq 0$ and $c_\varepsilon > 0$ in $\bar{\Omega} \times (0, \infty)$.

The following lemma reveals the relationship between the regularity of u_ε and n_ε.

Lemma 1.3 (Winkler 2015b; Zheng 2022, 2019) *Let $(n_\varepsilon, c_\varepsilon, u_\varepsilon, P_\varepsilon)$ be the solution of (1.2.2) in $\Omega \times (0, T)$ as well as $p \in [1, +\infty)$ and $q \in [1, +\infty)$, such that*

$$\begin{cases} q < \dfrac{3p}{3-p} & \text{if } p \leq 3, \\ q \leq \infty & \text{if } p > 3. \end{cases}$$

Then for all $K > 0$, there exists $C = C(p, q, K)$ such that if $\|n_\varepsilon(\cdot, t)\|_{L^p(\Omega)} \leq K$ for all $t \in (0, T)$, then $\|Du_\varepsilon(\cdot, t)\|_{L^q(\Omega)} \leq C$ for all $t \in (0, T)$.

The following lemmas will be used in the sequel.

Lemma 1.4 *Let $T > 0$, $\tau \in (0, T)$, $A > 0, \alpha > 0$ and $B > 0$, and suppose that $y : [0, T) \to [0, \infty)$ is absolutely continuous fulfilling $y'(t) + Ay^\alpha(t) \leq h(t)$ for a.e. $t \in (0, T)$ with some nonnegative function $h \in L^1_{loc}([0, T))$ satisfying $\int_t^{t+\tau} h(s)ds \leq B$ for all $t \in (0, T - \tau)$. Then*

$$y(t) \leq \max\left\{ y_0 + B, \frac{1}{\tau^{\frac{1}{\alpha}}}\left(\frac{B}{A}\right)^{\frac{1}{\alpha}} + 2B \right\} \quad \text{for all } t \in (0, T). \qquad (1.2.4)$$

For its elementary proof, we refer to Lemma 3.4 of Stinner et al. (2014) where the particular case $\tau = \alpha = 1$ is detailed.

As a crucial tool for analyzing the key term $\int_\Omega \frac{|\nabla c_\varepsilon|^2}{c_\varepsilon}$ below, we will use the following inequality established by Lemma 2.2.4 in Lankeit (2016a).

Lemma 1.5 (Lankeit 2016a) *There are $C_0 > 0$ and $\mu_0 > 0$ such that every positive $w \in C^2(\bar{\Omega})$ fulfilling $\nabla w \cdot \nu = 0$ on $\partial\Omega$ satisfies*

$$-2\int_\Omega \frac{|\Delta w|^2}{w} + \int_\Omega \frac{|\nabla w|^2 \Delta w}{w^2}$$
$$\leq -\mu_0 \int_\Omega w|D^2 \ln w|^2 - \mu_0 \int_\Omega \frac{|\nabla w|^4}{w^3} + C_0 \int_\Omega w. \qquad (1.2.5)$$

Now, we display an important auxiliary interpolation lemma in Winkler (2015b), Zheng and Wang (2017).

Lemma 1.6 (Winkler 2015b; Zheng and Wang 2017) *Let* $q \geq 1$,

$$\lambda \in [2q + 2, 4q + 1] \tag{1.2.6}$$

and $\Omega \subset \mathbb{R}^3$ *be a bounded domain with smooth boundary. Then there exists* $C > 0$ *such that for all* $\varphi \in C^2(\bar{\Omega})$ *fulfilling* $\varphi \cdot \frac{\partial \varphi}{\partial \nu} = 0$ *on* $\partial \Omega$, *we have*

$$\|\nabla\varphi\|_{L^\lambda(\Omega)} \leq C\||\nabla\varphi|^{q-1}D^2\varphi\|_{L^2(\Omega)}^{\frac{2(\lambda-3)}{(2q-1)\lambda}}\|\varphi\|_{L^\infty(\Omega)}^{\frac{6q-\lambda}{(2q-1)\lambda}} + C\|\varphi\|_{L^\infty(\Omega)}. \tag{1.2.7}$$

As an application of Lemma 1.6, (1.2.10) immediately leads to

Lemma 1.7 *Let* $\beta \in [1, \infty)$. *Then there exists a positive constant* $\lambda_{0,\beta}$ *such that*

$$\|\nabla c_\varepsilon\|_{L^{2\beta+2}(\Omega)}^{2\beta+2} \leq \lambda_{0,\beta}(\||\nabla c_\varepsilon|^{\beta-1}D^2 c_\varepsilon\|_{L^2(\Omega)}^2 + 1). \tag{1.2.8}$$

The basic boundedness information of solutions to (1.2.2) is stated as follows.

Lemma 1.8 *The solution* $(n_\varepsilon, c_\varepsilon, u_\varepsilon, P_\varepsilon)$ *of* (1.2.2) *satisfies*

$$\|n_\varepsilon(\cdot, t)\|_{L^1(\Omega)} = \|n_0\|_{L^1(\Omega)} \quad \text{for all } t > 0 \tag{1.2.9}$$

and

$$\|c_\varepsilon(\cdot, t)\|_{L^\infty(\Omega)} \leq \|c_0\|_{L^\infty(\Omega)} \quad \text{for all } t > 0. \tag{1.2.10}$$

Proof The identity (1.2.9) directly follows by integrating the first equation in (1.2.2). Moreover (1.2.10) is readily derived by applying the maximum principle to the second equation.

The following Gagliardo–Nirenberg inequality will be used several times in Sect. 1.4.

Lemma 1.9 *Let* $\Omega \subset \mathbb{R}^2$ *be a bounded Lipschitz domain. Then i) there is* $C > 0$ *such that* $\|\nabla\varphi\|_{L^4(\Omega)}^4 \leq C\|\Delta\varphi\|_{L^2(\Omega)}^2\|\nabla\varphi\|_{L^2(\Omega)}^2$ *for all* $\varphi \in W^{2,2}(\Omega)$ *fulfilling* $\frac{\partial\varphi}{\partial\nu}|_{\partial\Omega} = 0$; *(ii) there is* $C > 0$ *such that* $\|\varphi\|_{L^3(\Omega)}^3 \leq C\|\varphi\|_{W^{1,2}(\Omega)}^2\|\varphi\|_{L^1(\Omega)}$ *for all* $\varphi \in W^{1,2}(\Omega)$.

1.3 Global Boundedness of Solution to a Chemotaxis–Fluid System with Nonlinear Diffusion

1.3.1 A Quasi-energy Functional

Since some first regularity properties beyond those from Lemma 1.8 can be obtained by making use of a quasi-energy functional. Indeed it is a starting point of the derivation of further estimates for solutions to the approximate problems (1.2.2).

Lemma 1.10 *For any $\varepsilon \in (0, 1)$, the solution $(n_\varepsilon, c_\varepsilon, u_\varepsilon, P_\varepsilon)$ of (1.2.2) satisfies*

$$\frac{d}{dt}\int_\Omega \frac{|\nabla c_\varepsilon|^2}{c_\varepsilon} + \mu_0 \int_\Omega c_\varepsilon |D^2 \ln c_\varepsilon|^2 + \frac{\mu_0}{2}\int_\Omega \frac{|\nabla c_\varepsilon|^4}{c_\varepsilon^3} + \int_\Omega \frac{n_\varepsilon |\nabla c_\varepsilon|^2}{c_\varepsilon}$$
$$\leq 2\int_\Omega |\nabla n_\varepsilon||\nabla c_\varepsilon| + \frac{2\|c_0\|_{L^\infty(\Omega)}}{\mu_0}\int_\Omega |\nabla u_\varepsilon|^2 + C \quad \text{for all } t > 0 \tag{1.3.1}$$

for some $C > 0$, where μ_0 is the same as (1.2.5).

Proof Thanks to $c_\varepsilon > 0$, we integrate by parts and deduce from c_ε-equation in (1.2.2) that

$$\frac{d}{dt}\int_\Omega \frac{|\nabla c_\varepsilon|^2}{c_\varepsilon} = -2\int_\Omega \frac{\Delta c_\varepsilon c_{\varepsilon t}}{c_\varepsilon} + \int_\Omega \frac{|\nabla c_\varepsilon|^2 c_{\varepsilon t}}{c_\varepsilon^2}$$
$$= -2\int_\Omega \frac{|\Delta c_\varepsilon|^2}{c_\varepsilon} + 2\int_\Omega \frac{\Delta c_\varepsilon n_\varepsilon c_\varepsilon}{c_\varepsilon} + 2\int_\Omega \frac{\Delta c_\varepsilon}{c_\varepsilon}u_\varepsilon \cdot \nabla c_\varepsilon$$
$$\quad + \int_\Omega \frac{|\nabla c_\varepsilon|^2 \Delta c_\varepsilon}{c_\varepsilon^2} - \int_\Omega \frac{|\nabla c_\varepsilon|^2 n_\varepsilon c_\varepsilon}{c_\varepsilon^2} - \int_\Omega \frac{|\nabla c_\varepsilon|^2 u_\varepsilon \cdot \nabla c_\varepsilon}{c_\varepsilon^2}$$
$$= -2\int_\Omega \frac{|\Delta c_\varepsilon|^2}{c_\varepsilon} + 2\int_\Omega \Delta c_\varepsilon n_\varepsilon + 2\int_\Omega \frac{\Delta c_\varepsilon}{c_\varepsilon}u_\varepsilon \cdot \nabla c_\varepsilon$$
$$\quad + \int_\Omega \frac{|\nabla c_\varepsilon|^2 \Delta c_\varepsilon}{c_\varepsilon^2} - \int_\Omega \frac{|\nabla c_\varepsilon|^2 n_\varepsilon}{c_\varepsilon} - \int_\Omega \frac{|\nabla c_\varepsilon|^2 u_\varepsilon \cdot \nabla c_\varepsilon}{c_\varepsilon^2}. \tag{1.3.2}$$

Together with (1.2.10), an application of Lemma 1.5 yields that for some positive constants μ_0 and $C(\mu_0)$, it has

$$-2\int_\Omega \frac{|\Delta c_\varepsilon|^2}{c_\varepsilon} + \int_\Omega \frac{|\nabla c_\varepsilon|^2 \Delta c_\varepsilon}{c_\varepsilon^2}$$
$$\leq -\mu_0 \int_\Omega (c_\varepsilon |D^2 \ln c_\varepsilon|^2 + \frac{|\nabla c_\varepsilon|^4}{c_\varepsilon^3}) + C(\mu_0)\|c_0\|_{L^\infty(\Omega)}|\Omega| \quad \text{for } t > 0.$$

In addition, integrating by parts again, we have

$$2\int_\Omega \frac{\Delta c_\varepsilon}{c_\varepsilon}(u_\varepsilon \cdot \nabla c_\varepsilon)$$
$$= 2\int_\Omega \frac{|\nabla c_\varepsilon|^2}{c_\varepsilon^2}u_\varepsilon \cdot \nabla c_\varepsilon - 2\int_\Omega \frac{1}{c_\varepsilon}\nabla c_\varepsilon \cdot (\nabla u_\varepsilon \cdot \nabla c_\varepsilon)$$
$$\quad - 2\int_\Omega \frac{1}{c_\varepsilon}(u_\varepsilon \cdot D^2 c_\varepsilon) \cdot \nabla c_\varepsilon \quad \text{for all } t > 0$$

and

$$\int_\Omega \frac{|\nabla c_\varepsilon|^2}{c_\varepsilon^2}u_\varepsilon \cdot \nabla c_\varepsilon = 2\int_\Omega \frac{1}{c_\varepsilon}u_\varepsilon \cdot D^2 c_\varepsilon \cdot \nabla c_\varepsilon \quad \text{for all } t > 0.$$

So combining the above two inequalities, we get

$$2 \int_{\Omega} \frac{\Delta c_{\varepsilon}}{c_{\varepsilon}} (u_{\varepsilon} \cdot \nabla c_{\varepsilon}) - \int_{\Omega} \frac{|\nabla c_{\varepsilon}|^2}{c_{\varepsilon}^2} u_{\varepsilon} \cdot \nabla c_{\varepsilon}$$

$$\leq 2 \int_{\Omega} \frac{|\nabla c_{\varepsilon}|^2}{c_{\varepsilon}} |\nabla u_{\varepsilon}| \leq \frac{\mu_0}{2} \int_{\Omega} \frac{|\nabla c_{\varepsilon}|^4}{c_{\varepsilon}^3} + \frac{2 \|c_0\|_{L^{\infty}(\Omega)}}{\mu_0} \int_{\Omega} |\nabla u_{\varepsilon}|^2.$$

Therefore inequality (1.3.1) readily results from above inequalities.

In order to deal with the term $\int_{\Omega} |\nabla u_{\varepsilon}|^2$ on the right of (1.3.1), we recall the following standard energy inequality for the fluid component of solutions of (1.2.2).

Lemma 1.11 *Let $m + \alpha > \frac{2}{3}$. Then for any $\eta \in (0, 1)$, there exists $C(\eta) > 0$ such that*

$$\frac{d}{dt} \int_{\Omega} |u_{\varepsilon}|^2 + \int_{\Omega} |\nabla u_{\varepsilon}|^2 \leq \eta \int_{\Omega} (n_{\varepsilon} + \varepsilon)^{m+\alpha-2} |\nabla n_{\varepsilon}|^2 + C(\eta) \text{ for all } t > 0.$$

$$(1.3.3)$$

Proof Testing the third equation in (1.2.2) by u_{ε} and using $\nabla \cdot u_{\varepsilon} = 0$, we get

$$\frac{1}{2} \frac{d}{dt} \int_{\Omega} |u_{\varepsilon}|^2 + \int_{\Omega} |\nabla u_{\varepsilon}|^2 = \int_{\Omega} n_{\varepsilon} u_{\varepsilon} \cdot \nabla \phi \text{ for all } t > 0. \qquad (1.3.4)$$

By the Young inequality and the continuity of the embedding $W^{1,2}(\Omega) \hookrightarrow L^6(\Omega)$, we obtain that there is $C_1 > 0$ such that

$$\int_{\Omega} n_{\varepsilon} u_{\varepsilon} \cdot \nabla \phi \leq \|\nabla \phi\|_{L^{\infty}(\Omega)} \|n_{\varepsilon}\|_{L^{\frac{6}{5}}(\Omega)} \|\nabla u_{\varepsilon}\|_{L^2(\Omega)}$$

$$\leq C_1 \|n_{\varepsilon} + \varepsilon\|_{L^{\frac{6}{5}}(\Omega)} \|\nabla u_{\varepsilon}\|_{L^2(\Omega)} \text{ for all } t > 0.$$

$$(1.3.5)$$

Further, by the Gagliardo–Nirenberg inequality, we have

$$\|n_{\varepsilon} + \varepsilon\|_{L^{\frac{6}{5}}(\Omega)}$$

$$\leq C_2 (\|\nabla(n_{\varepsilon} + \varepsilon)^{\frac{m+\alpha}{2}}\|_{L^2(\Omega)}^{\frac{1}{3(m+\alpha)-1}} \|(n_{\varepsilon} + \varepsilon)^{\frac{m+\alpha}{2}}\|_{L^{\frac{2}{m+\alpha}}(\Omega)}^{\frac{2}{m+\alpha} - \frac{1}{3(m+\alpha)-1}} + \|(n_{\varepsilon} + \varepsilon)^{\frac{m+\alpha}{2}}\|_{L^{\frac{2}{m+\alpha}}(\Omega)}^{\frac{2}{m+\alpha}})$$

$$\leq C_3 (\|\nabla(n_{\varepsilon} + \varepsilon)^{\frac{m+\alpha}{2}}\|_{L^2(\Omega)}^{\frac{1}{3(m+\alpha)-1}} + 1) \text{ for all } t > 0,$$

$$(1.3.6)$$

for some $C_2 > 0$ and $C_3 > 0$ independent of ε. Combining (1.3.6) with (1.3.5) and noticing $m + \alpha > \frac{2}{3}$, we can see that for any $\eta \in (0, 1)$,

$$\int_\Omega n_\varepsilon u_\varepsilon \cdot \nabla \phi$$

$$\leq \frac{1}{2}\|\nabla u_\varepsilon\|_{L^2(\Omega)}^2 + C_4(\|\nabla(n_\varepsilon + \varepsilon)^{\frac{m+\alpha}{2}}\|_{L^2(\Omega)}^{\frac{2}{3(m+\alpha)-1}} + 1) \tag{1.3.7}$$

$$\leq \frac{1}{2}\|\nabla u_\varepsilon\|_{L^2(\Omega)}^2 + \eta \int_\Omega (n_\varepsilon + \varepsilon)^{m+\alpha-2}|\nabla n_\varepsilon|^2 + C(\eta) \quad \text{for all } t > 0.$$

This together with (1.3.4) arrives at (1.3.3).

Now, we turn to analyze $\int_\Omega (n_\varepsilon + \varepsilon)\ln(n_\varepsilon + \varepsilon)$ or $\|n_\varepsilon + \varepsilon\|_{L^{1+\alpha}(\Omega)}^{1+\alpha}$, which contributes to absorbing $\int_\Omega (n_\varepsilon + \varepsilon)^{m+\alpha-2}|\nabla n_\varepsilon|^2$ on the right-hand side of (1.3.3).

Lemma 1.12 *The solution* $(n_\varepsilon, c_\varepsilon, u_\varepsilon, P_\varepsilon)$ *of* (1.2.2) *satisfies*

$$\begin{cases} \dfrac{d}{dt}\displaystyle\int_\Omega (n_\varepsilon + \varepsilon)\ln(n_\varepsilon + \varepsilon) + m \int_\Omega (n_\varepsilon + \varepsilon)^{m-2}|\nabla n_\varepsilon|^2 \\[2mm] \leq C_S \displaystyle\int_\Omega |\nabla n_\varepsilon||\nabla c_\varepsilon| \quad \text{if } \alpha = 0, \\[3mm] \dfrac{1}{\alpha(1+\alpha)}\dfrac{d}{dt}\|n_\varepsilon + \varepsilon\|_{L^{1+\alpha}(\Omega)}^{1+\alpha} + m \int_\Omega (n_\varepsilon + \varepsilon)^{m+\alpha-2}|\nabla n_\varepsilon|^2 \\[2mm] \leq C_S \displaystyle\int_\Omega |\nabla n_\varepsilon||\nabla c_\varepsilon| \quad \text{if } \alpha > 0 \end{cases} \tag{1.3.8}$$

for all $t > 0$, *where* $C_S = \displaystyle\sup_{0 \leq s \leq \|c_0\|_{L^\infty(\Omega)}} S_0(s)$.

Proof The proof of the lemma is given separately for two cases.

(1) For the case $\alpha = 0$. Integration by parts, we deduce from n_ε-equation as well as $\nabla \cdot u_\varepsilon = 0$ and (1.2.1) that

$$\frac{d}{dt}\int_\Omega (n_\varepsilon + \varepsilon)\ln(n_\varepsilon + \varepsilon)$$

$$= \int_\Omega \Delta(n_\varepsilon + \varepsilon)^m \ln(n_\varepsilon + \varepsilon) - \int_\Omega \ln(n_\varepsilon + \varepsilon)\nabla \cdot (n_\varepsilon F_\varepsilon(n_\varepsilon)S_\varepsilon(x, n_\varepsilon, c_\varepsilon) \cdot \nabla c_\varepsilon)$$

$$- \int_\Omega \ln(n_\varepsilon + \varepsilon)u_\varepsilon \cdot \nabla n_\varepsilon \tag{1.3.9}$$

$$\leq -m \int_\Omega (n_\varepsilon + \varepsilon)^{m-2}|\nabla n_\varepsilon|^2 + \int_\Omega S_0(c_\varepsilon)|\nabla n_\varepsilon||\nabla c_\varepsilon|$$

$$\leq -m \int_\Omega (n_\varepsilon + \varepsilon)^{m-2}|\nabla n_\varepsilon|^2 + C_S \int_\Omega |\nabla n_\varepsilon||\nabla c_\varepsilon|.$$

(2) For the case $\alpha > 0$. Multiplying the first equation in (1.2.2) by $(n_\varepsilon + \varepsilon)^\alpha$, and noticing the hypothesis (1.1.3), we then have

$$\frac{1}{1+\alpha}\frac{d}{dt}\|n_\varepsilon + \varepsilon\|_{L^{1+\alpha}(\Omega)}^{1+\alpha} + m\alpha \int_\Omega (n_\varepsilon + \varepsilon)^{m+\alpha-2}|\nabla n_\varepsilon|^2$$

$$=\alpha \int_\Omega (n_\varepsilon + \varepsilon)^{\alpha-1} n_\varepsilon \nabla n_\varepsilon \cdot (F_\varepsilon(n_\varepsilon) S_\varepsilon(x, n_\varepsilon, c_\varepsilon) \cdot \nabla c_\varepsilon) \qquad (1.3.10)$$

$$\leq \alpha \int_\Omega (n_\varepsilon + \varepsilon)^\alpha (1 + n_\varepsilon)^{-\alpha} S_0(c_\varepsilon)|\nabla n_\varepsilon||\nabla c_\varepsilon|$$

$$\leq \alpha C_S \int_\Omega |\nabla n_\varepsilon||\nabla c_\varepsilon| \quad \text{for all } t > 0.$$

Hence, (1.3.8) readily follows from (1.3.9) and (1.3.10).

Remark 1.1 Note that when $S(x, n, c)$ is scalar-value, one can make use of the corresponding flavor thereof to neutralize $2 \int_\Omega |\nabla n_\varepsilon||\nabla c_\varepsilon|$ on the right-hand side of (1.3.1) and $C_S \int_\Omega |\nabla n_\varepsilon||\nabla c_\varepsilon|$ on the right side of (1.3.8).

In the sequel we shall derive an energy-type inequality under the assumption $\frac{10}{9} < m + \alpha \leq 2$, from which the regularity of solutions of (1.2.2) beyond that of Lemma 1.8 is achieved.

Lemma 1.13 *Let $\frac{10}{9} < m + \alpha \leq 2$ and S satisfy (1.1.2)–(1.1.3). Suppose that (1.1.4) hold. Then there exists $C > 0$ independent of ε such that the solution of (1.2.2) satisfies*

$$\int_\Omega \frac{|\nabla c_\varepsilon|^2}{c_\varepsilon} + \int_\Omega |u_\varepsilon|^2 \leq C \quad \text{for all } t > 0, \qquad (1.3.11)$$

$$\begin{cases} \int_\Omega (n_\varepsilon + \varepsilon)^{1+\alpha} \leq C & \text{if } \alpha > 0, \\ \int_\Omega (n_\varepsilon + \varepsilon) \ln(n_\varepsilon + \varepsilon) \leq C & \text{if } \alpha = 0, \end{cases} \qquad (1.3.12)$$

$$\int_t^{t+1} \int_\Omega \left(\frac{n_\varepsilon}{c_\varepsilon}|\nabla c_\varepsilon|^2 + \frac{|\nabla c_\varepsilon|^4}{c_\varepsilon^3} + (n_\varepsilon + \varepsilon)^{m+\alpha+\frac{2}{3}} + (n_\varepsilon + \varepsilon)^{m+\alpha-2}|\nabla n_\varepsilon|^2\right) \leq C, \qquad (1.3.13)$$

$$\int_t^{t+1} \int_\Omega |\nabla u_\varepsilon|^2 \leq C \qquad (1.3.14)$$

as well as

$$\int_t^{t+1} \int_\Omega c_\varepsilon |D^2 \ln c_\varepsilon|^2 \leq C. \qquad (1.3.15)$$

Proof Adding an suitable multiples of the inequalities in Lemmas 1.10–1.12, one can conclude that there exist positive constants c_i, $(i = 1, 2, 3)$, such that

$$\frac{d}{dt}\left(\int_\Omega \frac{|\nabla c_\varepsilon|^2}{c_\varepsilon} + \int_\Omega (n_\varepsilon + \varepsilon)\ln(n_\varepsilon + \varepsilon) + k_1 \int_\Omega |u_\varepsilon|^2\right) + k_2 \int_\Omega |\nabla u_\varepsilon|^2$$

$$+ k_2 \Omega c_\varepsilon |D^2 \ln c_\varepsilon|^2 + \int_\Omega \frac{n_\varepsilon |\nabla c_\varepsilon|^2}{c_\varepsilon} \tag{1.3.16}$$

$$+ k_2 \int_\Omega \frac{|\nabla c_\varepsilon|^4}{c_\varepsilon^3} + k_2 \int_\Omega (n_\varepsilon + \varepsilon)^{m-2} |\nabla n_\varepsilon|^2$$

$$\leq k_3 \int_\Omega |\nabla n_\varepsilon| |\nabla c_\varepsilon| + k_3 \quad \text{for all } t > 0 \quad \text{when } \alpha = 0$$

and

$$\frac{d}{dt} \left(\int_\Omega \frac{|\nabla c_\varepsilon|^2}{c_\varepsilon} + \frac{1}{\alpha(1+\alpha)} \|n_\varepsilon + \varepsilon\|^{1+\alpha}_{L^{1+\alpha}(\Omega)} + k_1 \int_\Omega |u_\varepsilon|^2 \right)$$

$$+ k_2 \int_\Omega |\nabla u_\varepsilon|^2 + k_2 \int_\Omega c_\varepsilon |D^2 \ln c_\varepsilon|^2 + \int_\Omega \frac{n_\varepsilon |\nabla c_\varepsilon|^2}{c_\varepsilon}$$

$$+ k_2 \int_\Omega \frac{|\nabla c_\varepsilon|^4}{c_\varepsilon^3} + k_2 \int_\Omega (n_\varepsilon + \varepsilon)^{m+\alpha-2} |\nabla n_\varepsilon|^2 \tag{1.3.17}$$

$$\leq k_3 \int_\Omega |\nabla n_\varepsilon| |\nabla c_\varepsilon| + k_3 \quad \text{for all } t > 0 \quad \text{when } \alpha > 0.$$

Next, we will estimate $\int_\Omega |\nabla n_\varepsilon| |\nabla c_\varepsilon|$ by the Gagliardo–Nirenberg inequality along with the basic priori information provided by Lemma 1.8. Indeed, making use of (1.2.10) and the Young inequality, we thereby find $k_5 > 0$ such that

$$k_3 \int_\Omega |\nabla n_\varepsilon| |\nabla c_\varepsilon|$$

$$\leq \frac{k_2}{2} \int_\Omega \frac{|\nabla c_\varepsilon|^4}{c_\varepsilon^3} + \frac{k_2}{4} \int_\Omega (n_\varepsilon + \varepsilon)^{m+\alpha-2} |\nabla n_\varepsilon|^2 + k_5 \int_\Omega (n_\varepsilon + \varepsilon)^{4-2m-2\alpha}. \tag{1.3.18}$$

Therefore for $\alpha > 0$, we insert (1.3.18) into (1.3.17) to get

$$\frac{d}{dt} \left(\int_\Omega \frac{|\nabla c_\varepsilon|^2}{c_\varepsilon} + \frac{1}{\alpha(1+\alpha)} \|n_\varepsilon + \varepsilon\|^{1+\alpha}_{L^{1+\alpha}(\Omega)} + k_1 \int_\Omega |u_\varepsilon|^2 \right)$$

$$+ k_2 \int_\Omega |\nabla u_\varepsilon|^2 + k_2 \int_\Omega c_\varepsilon |D^2 \ln c_\varepsilon|^2 + \int_\Omega \frac{n_\varepsilon |\nabla c_\varepsilon|^2}{c_\varepsilon}$$

$$+ \frac{k_2}{2} \int_\Omega \frac{|\nabla c_\varepsilon|^4}{c_\varepsilon^3} + \frac{3k_2}{4} \int_\Omega (n_\varepsilon + \varepsilon)^{m+\alpha-2} |\nabla n_\varepsilon|^2 \tag{1.3.19}$$

$$\leq k_5 \int_\Omega (n_\varepsilon + \varepsilon)^{4-2m-2\alpha} + k_3 \quad \text{for all } t > 0.$$

Next, we deal with $\int_\Omega (n_\varepsilon + \varepsilon)^{4-2m-2\alpha}$ separately for two cases. Indeed, in the case $\frac{10}{9} < m + \alpha < \frac{3}{2}$, by the Gagliardo–Nirenberg inequality, we get

$$k_5 \int_\Omega (n_\varepsilon + \varepsilon)^{4-2m-2\alpha}$$

$$= k_5 \| (n_\varepsilon + \varepsilon)^{\frac{m+\alpha}{2}} \|_{L^{\frac{2(4-2m-2\alpha)}{m+\alpha}}(\Omega)}^{\frac{2(4-2m-2\alpha)}{m+\alpha}}$$

$$\leq k_6 \| \nabla (n_\varepsilon + \varepsilon)^{\frac{m+\alpha}{2}} \|_{L^2(\Omega)}^{\frac{2(3-2m-2\alpha)}{3(m+\alpha)-1}} \| (n_\varepsilon + \varepsilon)^{\frac{m+\alpha}{2}} \|_{L^{\frac{2}{m+\alpha}}(\Omega)}^{\frac{2(4-2m-2\alpha)}{m+\alpha} - \frac{2(3-2m-2\alpha)}{3(m+\alpha)-1}} \qquad (1.3.20)$$

$$+ \| (n_\varepsilon + \varepsilon)^{\frac{m+\alpha}{2}} \|_{L^{\frac{2}{m+\alpha}}(\Omega)}^{\frac{2(4-2m-2\alpha)}{m+\alpha}}$$

$$= k_7 (\| \nabla (n_\varepsilon + \varepsilon)^{\frac{m+\alpha}{2}} \|_{L^2(\Omega)}^{\frac{6(3-2m-2\alpha)}{3(m+\alpha)-1}} + 1) \quad \text{for all } t > 0,$$

where k_6 and k_7 are positive constants. Hence, if $\frac{10}{9} < m + \alpha < \frac{3}{2}$, we have $\frac{6(3-2m-2\alpha)}{3(m+\alpha)-1} \in (0, 2)$, and then get

$$k_5 \int_\Omega (n_\varepsilon + \varepsilon)^{4-2m-2\alpha} \leq \frac{k_2}{4} \int_\Omega (n_\varepsilon + \varepsilon)^{m+\alpha-2} |\nabla n_\varepsilon|^2 + k_8 \quad \text{for all } t > 0 \quad (1.3.21)$$

with some $k_8 > 0$ by the Young inequality. While in the case $\frac{3}{2} \leq m + \alpha \leq 2$, we have $4 - 2m - 2\alpha \in (0, 1)$ and thereby immediately get

$$k_5 \int_\Omega (n_\varepsilon + \varepsilon)^{4-2m-2\alpha} \leq k_5 \int_\Omega n_\varepsilon + k_9 \qquad (1.3.22)$$

with some $k_9 > 0$. Therefore, (1.3.19) together with (1.3.20)–(1.3.22) leads to

$$\frac{d}{dt} \left(\int_\Omega \frac{|\nabla c_\varepsilon|^2}{c_\varepsilon} + \frac{1}{\alpha(1+\alpha)} \| n_\varepsilon + \varepsilon \|_{L^{1+\alpha}(\Omega)}^{1+\alpha} + k_1 \int_\Omega |u_\varepsilon|^2 \right)$$

$$+ k_2 \int_\Omega |\nabla u_\varepsilon|^2 + k_2 \int_\Omega c_\varepsilon |D^2 \ln c_\varepsilon|^2 + \int_\Omega \frac{n_\varepsilon |\nabla c_\varepsilon|^2}{c_\varepsilon} \qquad (1.3.23)$$

$$+ \frac{k_2}{2} \int_\Omega \frac{|\nabla c_\varepsilon|^4}{c_\varepsilon^3} + \frac{k_2}{2} \int_\Omega (n_\varepsilon + \varepsilon)^{m+\alpha-2} |\nabla n_\varepsilon|^2$$

$$\leq k_{10} \quad \text{for all } t > 0.$$

Since $m + \alpha > \frac{10}{9}$, we utilize the Gagliardo–Nirenberg inequality to see that there exists a positive constant k_{11} such that

$$\frac{1}{\alpha(1+\alpha)} \| n_\varepsilon + \varepsilon \|_{L^{1+\alpha}(\Omega)}^{1+\alpha}$$

$$\leq k_{11} \left(\int_\Omega (n_\varepsilon + \varepsilon)^{m+\alpha-2} |\nabla n_\varepsilon|^2 \right)^{\frac{3\alpha}{3(m+\alpha)-1}} + k_{11} \quad \text{for all } t > 0.$$

Hence recalling (1.2.10) and according to the Poincaré inequality, we can see that for all $t > 0$,

$$\int_{\Omega} \frac{|\nabla c_\varepsilon|^2}{c_\varepsilon} + \frac{1}{\alpha(1+\alpha)} \|n_\varepsilon + \varepsilon\|_{L^{1+\alpha}(\Omega)}^{1+\alpha} + k_1 \int_{\Omega} |u_\varepsilon|^2$$

$$\leq k_{12} \left(\int_{\Omega} |\nabla u_\varepsilon|^2 + \int_{\Omega} \frac{|\nabla c_\varepsilon|^4}{c_\varepsilon^3} + \int_{\Omega} (n_\varepsilon + \varepsilon)^{m+\alpha-2} |\nabla n_\varepsilon|^2 \right)^\zeta + k_{12} \tag{1.3.24}$$

with some $k_{12} > 0$ and $\zeta = \max\{\frac{3\alpha}{3(m+\alpha)-1}, 1\}$. Thus, we infer from (1.3.23) and (1.3.24) that there exist $k_{13} > 0$ and $k_{14} > 0$ such that for all $\varepsilon \in (0, 1)$,

$$\frac{d}{dt} \left(\int_{\Omega} \frac{|\nabla c_\varepsilon|^2}{c_\varepsilon} + \frac{1}{\alpha(1+\alpha)} \|n_\varepsilon + \varepsilon\|_{L^{1+\alpha}(\Omega)}^{1+\alpha} + \frac{8}{\mu_0} \|c_0\|_{L^\infty(\Omega)} \int_{\Omega} |u_\varepsilon|^2 \right)$$

$$+ k_{13} \left(\int_{\Omega} \frac{|\nabla c_\varepsilon|^2}{c_\varepsilon} + \frac{1}{\alpha(1+\alpha)} \|n_\varepsilon + \varepsilon\|_{L^{1+\alpha}(\Omega)}^{1+\alpha} + \frac{8}{\mu_0} \|c_0\|_{L^\infty(\Omega)} \int_{\Omega} |u_\varepsilon|^2 \right)^{\frac{1}{\zeta}}$$

$$+ k_{13} \left(\int_{\Omega} |\nabla u_\varepsilon|^2 + \int_{\Omega} c_\varepsilon |D^2 \ln c_\varepsilon|^2 + \int_{\Omega} \frac{n_\varepsilon |\nabla c_\varepsilon|^2}{c_\varepsilon} \right)$$

$$+ k_{13} \left(\int_{\Omega} \frac{|\nabla c_\varepsilon|^4}{c_\varepsilon^3} + \frac{5m\alpha}{8} \int_{\Omega} (n_\varepsilon + \varepsilon)^{m+\alpha-2} |\nabla n_\varepsilon|^2 \right)$$

$$\leq k_{14} \quad \text{for all } t > 0 \quad \text{if } \alpha > 0$$

$$\tag{1.3.25}$$

which along with Lemma 1.4, implies that (1.3.11)–(1.3.12) are valid. Further, (1.3.13)–(1.3.15) result from integrating the inequality (1.3.25). The proof for the case $\alpha = 0$ can be proved similarly, and is thus omitted here.

1.3.2 $L^\infty((0, \infty); L^p(\Omega))$ Estimate of n_ε for Some $p > \frac{3}{2}$

The further regularity properties of solutions can be obtained by means of a bootstrap iteration in the case of $\frac{10}{9} < m + \alpha < \frac{3}{2}$. In this direction, we first shall make use of results in Lemma 1.13 to improve the regularities, in particular for n_ε.

Lemma 1.14 *Let $p > 1$. Then the solution $(n_\varepsilon, c_\varepsilon, u_\varepsilon, P_\varepsilon)$ of (1.2.2) satisfies*

$$\frac{1}{p} \frac{d}{dt} \|n_\varepsilon + \varepsilon\|_{L^p(\Omega)}^p + \frac{m(p-1)}{2} \int_{\Omega} (n_\varepsilon + \varepsilon)^{m+p-3} |\nabla n_\varepsilon|^2$$

$$\leq \frac{(p-1)C_S^2}{2m} \int_{\Omega} (n_\varepsilon + \varepsilon)^{p+1-m} (1 + n_\varepsilon)^{-2\alpha} |\nabla c_\varepsilon|^2 \quad \text{for all } t > 0. \tag{1.3.26}$$

Proof Multiplying the first equation in (1.2.2) by $(n_\varepsilon + \varepsilon)^{p-1}$, using $\nabla \cdot u_\varepsilon = 0$ as well as (1.1.3), we get

$$\frac{1}{p}\frac{d}{dt}\|n_\varepsilon + \varepsilon\|^p_{L^p(\Omega)} + m(p-1)\int_\Omega (n_\varepsilon + \varepsilon)^{m+p-3}|\nabla n_\varepsilon|^2$$

$$=(p-1)\int_\Omega (n_\varepsilon + \varepsilon)^{p-2}n_\varepsilon \nabla n_\varepsilon \cdot (F_\varepsilon(n_\varepsilon)S_\varepsilon(x,n_\varepsilon,c_\varepsilon)\cdot \nabla c_\varepsilon)$$

$$\leq (p-1)\int_\Omega (n_\varepsilon + \varepsilon)^{p-1}(1+n_\varepsilon)^{-\alpha}S_0(c_\varepsilon)|\nabla n_\varepsilon||\nabla c_\varepsilon|$$ (1.3.27)

$$\leq (p-1)C_S \int_\Omega (n_\varepsilon + \varepsilon)^{p-1}(1+n_\varepsilon)^{-\alpha}|\nabla n_\varepsilon||\nabla c_\varepsilon| \quad \text{for all } t > 0.$$

Hence (1.3.26) follows from (1.3.27) and Young's inequality.

As a consequence of Lemma 1.13, we have

Lemma 1.15 *Under the assumptions of Lemma 1.13, there exists a positive constant C independent of ε such that*

$$\int_\Omega |\nabla c_\varepsilon|^2 \leq C \text{ for all } t > 0.$$ (1.3.28)

Proof Noticing that $|\nabla c_\varepsilon|^2 \leq \frac{|\nabla c_\varepsilon|^2}{c_\varepsilon}\|c_\varepsilon(\cdot,t)\|_{L^\infty(\Omega)}$, (1.3.28) results from (1.3.11) and (1.2.10).

Combining Lemma 1.14 and estimate (1.3.28) immediately leads to

Lemma 1.16 *Let $\frac{10}{9} < m + \alpha \leq 2$, S satisfy (1.1.2)–(1.1.3) and $(n_\varepsilon, c_\varepsilon, u_\varepsilon, P_\varepsilon)$ be the solution of (1.2.2). Then there exists $C > 0$ independent of ε such that*

$$\sup_{t\in(0,\infty)} \int_\Omega (n_\varepsilon + \varepsilon)^{m+2\alpha} + \sup_{t\in(0,\infty)}\int_t^{t+1}\int_\Omega (n_\varepsilon + \varepsilon)^{2m+2\alpha-3}|\nabla n_\varepsilon|^2 \leq C \quad (1.3.29)$$

for all $t > 0$.

Proof Taking $p = m + 2\alpha$ in (1.3.26), we get

$$\frac{1}{m+2\alpha}\frac{d}{dt}\|n_\varepsilon + \varepsilon\|^{m+2\alpha}_{L^{m+2\alpha}(\Omega)} + \frac{m(m+2\alpha-1)}{2}\int_\Omega (n_\varepsilon + \varepsilon)^{2m+2\alpha-3}|\nabla n_\varepsilon|^2$$

$$\leq k_1 \int_\Omega (n_\varepsilon + \varepsilon)^{1+2\alpha}(1+n_\varepsilon)^{-2\alpha}|\nabla c_\varepsilon|^2$$

$$\leq \|c_0\|_{L^\infty(\Omega)}k_1\int_\Omega \frac{n_\varepsilon}{c_\varepsilon}|\nabla c_\varepsilon|^2 + k_1\int_\Omega |\nabla c_\varepsilon|^2$$ (1.3.30)

for some positive constant $k_1 > 0$. Now, applying the Gagliardo–Nirenberg inequality and (1.2.9), one can find constants $k_2 > 0$, $k_3 > 0$ and $k_4 > 0$ independent of $\varepsilon \in (0,1)$ such that

$$\int_\Omega (n_\varepsilon + \varepsilon)^{m+2\alpha}$$

$$= \| (n_\varepsilon + \varepsilon)^{\frac{2m+2\alpha-1}{2}} \|_{L^{\frac{2(m+2\alpha)}{2m+2\alpha-1}}(\Omega)}^{\frac{2(m+2\alpha)}{2m+2\alpha-1}}$$

$$\leq c_2 \left(\| \nabla (n_\varepsilon + \varepsilon)^{\frac{2m+2\alpha-1}{2}} \|_{L^2(\Omega)}^{\frac{3m+6\alpha-3}{3m+3\alpha-2}} \| (n_\varepsilon + \varepsilon)^{\frac{2m+2\alpha-1}{2}} \|_{L^{\frac{2}{2m+2\alpha-1}}(\Omega)}^{\frac{2(m+2\alpha)}{2m+2\alpha-1} - \frac{3m+6\alpha-3}{3m+3\alpha-2}} \right.$$

$$\left. + \| (n_\varepsilon + \varepsilon)^{\frac{2m+2\alpha-1}{2}} \|_{L^{\frac{2}{2m+2\alpha-1}}(\Omega)}^{\frac{2(m+2\alpha)}{2m+2\alpha-1}} \right)$$

$$\leq c_3 (\| \nabla (n_\varepsilon + \varepsilon)^{\frac{2m+2\alpha-1}{2}} \|_{L^2(\Omega)}^{\frac{3m+6\alpha-3}{3m+3\alpha-2}} + 1)$$

$$\leq \frac{m(m-1)}{(2m-1)^2} \| \nabla (n_\varepsilon + \varepsilon)^{\frac{2m+2\alpha-1}{2}} \|_{L^2(\Omega)}^2 + k_4.$$

Inserting the above inequality into (1.3.30), one then has

$$\frac{1}{m+2\alpha} \frac{d}{dt} \| n_\varepsilon + \varepsilon \|_{L^{m+2\alpha}(\Omega)}^{m+2\alpha} + \frac{m(m+2\alpha-1)}{4} \int_\Omega (n_\varepsilon + \varepsilon)^{2m+2\alpha-3} |\nabla n_\varepsilon|^2$$

$$+ \int_\Omega (n_\varepsilon + \varepsilon)^{m+2\alpha}$$

$$\leq \| c_0 \|_{L^\infty(\Omega)} k_1 \int_\Omega \frac{n_\varepsilon}{c_\varepsilon} |\nabla c_\varepsilon|^2 + k_1 \int_\Omega |\nabla c_\varepsilon|^2 + k_4.$$

As the application of Lemma 1.4, this together with (1.3.28) and (1.3.13) then arrives at (1.3.29).

According to Lemma 1.3, the bound of $L^p(\Omega)$ for Du_ε can be suitably enlarge upon the result of Lemma 1.16 in asserting the following.

Lemma 1.17 *Let* $\frac{10}{9} < m + \alpha \leq \frac{3}{2}$. *Then for* $r < \frac{3(m+\alpha)}{3-(m+\alpha)}$, *there exists* $K := K(r, m)$ *such that*

$$\| Du_\varepsilon(\cdot, t) \|_{L^r(\Omega)} \leq K \quad for \ all \ t > 0. \tag{1.3.31}$$

Proof In light of (1.3.29), (1.3.31) is the consequence of an application of Lemma 1.3 with $p = m + \alpha$.

In order to obtain the further regularity of c_ε, one can establish the time evolution of ∇c_ε in $L^{2\beta}(\Omega)$, similar to that of Lemma 3.6 in Winkler (2015b).

Lemma 1.18 *For any* $\beta > 1$, *the solution of* (1.2.2) *satisfies*

$$\frac{1}{2\beta} \frac{d}{dt} \| \nabla c_\varepsilon \|_{L^{2\beta}(\Omega)}^{2\beta} + \frac{2(\beta-1)}{\beta^2} \int_\Omega |\nabla |\nabla c_\varepsilon|^\beta|^2$$

$$+ \frac{1}{2} \int_\Omega |\nabla c_\varepsilon|^{2\beta-2} |D^2 c_\varepsilon|^2 + \int_\Omega n_\varepsilon |\nabla c_\varepsilon|^{2\beta} \tag{1.3.32}$$

$$\leq - \int_\Omega c_\varepsilon |\nabla c_\varepsilon|^{2\beta-2} \nabla n_\varepsilon \cdot \nabla c_\varepsilon + \int_\Omega |Du_\varepsilon| |\nabla c_\varepsilon|^{2\beta} + C$$

for all $t > 0$, where $C > 0$ is a positive constant independent of ε.

Proof Noticing the boundedness of $\|\nabla c_\varepsilon(\cdot, t)\|_{L^2(\Omega)}$ obtained in Lemma 1.15, and applying the arguments as those in the proof (3.10) of Ishida et al. (2014) (see also Wang and Xiang 2016; Zheng 2016, 2017a), one can find a positive constant k_1 such that

$$\int_{\partial\Omega} \frac{\partial|\nabla c_\varepsilon|^2}{\partial \nu}|\nabla c_\varepsilon|^{2\beta-2} \leq \frac{(\beta-1)}{\beta^2}\int_\Omega \left|\nabla|\nabla c_\varepsilon|^\beta\right|^2 + k_1.$$

Hence by pursuing quite a similar strategy in the proof of Lemma 3.6 in Winkler (2015b), one can derive (1.3.32).

Now, we address the question how far the regularity information such as provided by Lemma 1.17 is convenient to estimate the term $\int_\Omega |Du_\varepsilon||\nabla c_\varepsilon|^{2\beta}$ on the right of (1.3.32).

Lemma 1.19 *Let $r > \frac{3}{2}$ and $\beta \in [r-1, \frac{r-1}{(4-2r)_+}]$. Then for any $\eta > 0$ and $K > 0$ there exists $C = C(\beta, r, K) > 0$ such that if $\|Du_\varepsilon\|_{L^r(\Omega)} \leq K$, then*

$$\int_\Omega |\nabla c_\varepsilon|^{2\beta}|Du_\varepsilon| \leq \eta \int_\Omega |\nabla c_\varepsilon|^{2\beta-2}|D^2 c_\varepsilon|^2 + C \text{ for all } t > 0. \qquad (1.3.33)$$

Proof We invoke the Hölder inequality with exponents $\frac{r}{r-1}$ and r to see that

$$\int_\Omega |\nabla c_\varepsilon|^{2\beta}|Du_\varepsilon| \leq \left(\int_\Omega |\nabla c_\varepsilon|^{\frac{2\beta r}{r-1}}\right)^{\frac{r-1}{r}}\left(\int_\Omega |Du_\varepsilon|^r\right)^{\frac{1}{r}}$$

$$\leq K\left(\int_\Omega |\nabla c_\varepsilon|^{\frac{2\beta r}{r-1}}\right)^{\frac{r-1}{r}}$$

$$\leq K\|\nabla c_\varepsilon\|_{L^{\frac{2\beta r}{r-1}}(\Omega)}^{2\beta} \quad \text{for all } t > 0.$$

Since $\beta \in [r-1, \frac{r-1}{(4-2r)_+}]$ ensures that $\lambda := \frac{2\beta r}{r-1} \in [2\beta+2, 4\beta+1]$. Therefore, we may apply Lemma 1.7 and (1.2.10) to see that for some $k_1 = k_1(K, r, \beta) > 0$ and $k_2 = k_2(K, \beta, r) > 0$, it has

$$K\|\nabla c_\varepsilon\|_{L^{\frac{2\beta r}{r-1}}(\Omega)}^{2\beta} \leq k_1 \||\nabla c_\varepsilon|^{\beta-1}D^2 c_\varepsilon\|_{L^2(\Omega)}^{2\beta\frac{4\beta(\lambda-3)}{(2\beta-1)\lambda}}\|c_\varepsilon\|_{L^\infty(\Omega)}^{2\beta\frac{6\beta-\lambda}{(2\beta-1)\lambda}} + k_1\|c_\varepsilon\|_{L^\infty(\Omega)}^{2\beta}$$

$$\leq k_2(\||\nabla c_\varepsilon|^{\beta-1}D^2 c_\varepsilon\|_{L^2(\Omega)}^{\frac{4\beta(\lambda-3)}{(2\beta-1)\lambda}} + 1) \text{ for all } t > 0.$$

Thanks to the assumption $r > \frac{3}{2}$ and $\beta \in [r-1, \frac{r-1}{(4-2r)_+}]$, we have

$$\frac{4\beta(\lambda-3)}{(2\beta-1)\lambda} = \frac{4\beta(\frac{2\beta r}{r-1}-3)}{(2\frac{2\beta r}{r-1}-1)\frac{2\beta r}{r-1}} < 2,$$

and thus arrive at (1.3.33) by means of the Young inequality.

At this position, on the basis of space–time regularity property of n_ε provided by Lemmas 1.16, 1.14 can be exploited so as to derive the further regularity features of n_ε.

Lemma 1.20 Let $(n_\varepsilon, c_\varepsilon, u_\varepsilon, P_\varepsilon)$ be the solution of (1.2.2) as well as $\frac{10}{9} < m + \alpha \le \frac{3}{2}$. Then there exists $C > 0$ such that

$$\sup_{t \in (0,\infty)} \|n_\varepsilon(\cdot, t) + \varepsilon\|_{L^{p_1}(\Omega)}^{p_1} + \sup_{t \in (0,\infty)} \int_t^{t+1} \int_\Omega (n_\varepsilon + \varepsilon)^{p_1 + m - 3} |\nabla n_\varepsilon|^2 \le C,$$

(1.3.34)

where $p_1 = \frac{16}{3}(m + \alpha)^2 - \frac{25(m+\alpha)}{3} + 4 + \frac{\alpha}{3}(4(m + \alpha) - 1)$.

Proof Let $\beta_1 = 2m + 2\alpha - 1$. Then in view of (1.3.32) and (1.2.10), we obtain that for some $C_1 > 0$ and all $t > 0$,

$$\frac{1}{2\beta_1} \frac{d}{dt} \|\nabla c_\varepsilon\|_{L^{2\beta_1}(\Omega)}^{2\beta_1} + \frac{2(\beta_1 - 1)}{\beta_1^2} \int_\Omega |\nabla |\nabla c_\varepsilon|^{\beta_1}|^2 + \int_\Omega |\nabla c_\varepsilon|^{2\beta_1 - 2} |D^2 c_\varepsilon|^2$$

$$\le \|c_0\|_{L^\infty(\Omega)} \int_\Omega |\nabla c_\varepsilon|^{2\beta_1 - 1} |\nabla n_\varepsilon| - \int_\Omega n_\varepsilon |\nabla c_\varepsilon|^{2\beta_1} + \int_\Omega |\nabla c_\varepsilon|^{2\beta_1} |Du_\varepsilon| + C_1.$$

(1.3.35)

By Lemma 1.7 and the Young inequality twice, we can conclude that for some $C_2 > 0$,

$$\|c_0\|_{L^\infty(\Omega)} \int_\Omega |\nabla c_\varepsilon|^{2\beta_1 - 1} |\nabla n_\varepsilon|$$

$$= \|c_0\|_{L^\infty(\Omega)} \int_\Omega \left(n_\varepsilon^{\frac{2m+2\alpha-3}{2}} |\nabla n_\varepsilon| \right) \left(n_\varepsilon^{\frac{3-2m-2\alpha}{2}} |\nabla c_\varepsilon|^{\frac{3-2m-2\alpha}{2} 2\beta_1} \right) |\nabla c_\varepsilon|^{2\beta_1 - 1 - \frac{3-2m-2\alpha}{2} 2\beta_1}$$

$$\le \frac{1}{2} \int_\Omega n_\varepsilon |\nabla c_\varepsilon|^{2\beta_1} + \frac{1}{2\lambda_{0,\beta_1}} \int_\Omega |\nabla c_\varepsilon|^{[2\beta_1 - 1 - \frac{3-2m-2\alpha}{2} 2\beta_1] \frac{1}{m+\alpha-1}}$$

$$+ C_2 \int_\Omega (n_\varepsilon + \varepsilon)^{2m+2\alpha-3} |\nabla n_\varepsilon|^2$$

$$= \frac{1}{2} \int_\Omega n_\varepsilon |\nabla c_\varepsilon|^{2\beta_1} + \frac{1}{2\lambda_{0,\beta_1}} \int_\Omega |\nabla c_\varepsilon|^{2\beta_1 + 2} + C_2 \int_\Omega (n_\varepsilon + \varepsilon)^{2m+2\alpha-3} |\nabla n_\varepsilon|^2 + \frac{1}{2}$$

$$\le \frac{1}{2} \int_\Omega n_\varepsilon |\nabla c_\varepsilon|^{2\beta_1} + \frac{1}{2} \||\nabla c_\varepsilon|^{\beta_1 - 1} D^2 c_\varepsilon\|_{L^2(\Omega)}^2 + C_2 \int_\Omega (n_\varepsilon + \varepsilon)^{2m+2\alpha-3} |\nabla n_\varepsilon|^2 + \frac{1}{2}.$$

(1.3.36)

Here we have used the fact that $\frac{1}{2} + \frac{2m+2\alpha-2}{2} + \frac{3-2m-2\alpha}{2} = 1$. Inserting (1.3.36) into (1.3.35), we then have

$$\frac{1}{2\beta_1}\frac{d}{dt}\|\nabla c_\varepsilon\|_{L^{2\beta_1}(\Omega)}^{2\beta_1} + \frac{2(\beta_1-1)}{\beta_1^2}\int_\Omega |\nabla|\nabla c_\varepsilon|^{\beta_1}|^2$$

$$+ \frac{1}{2}\int_\Omega |\nabla c_\varepsilon|^{2\beta_1-2}|D^2 c_\varepsilon|^2 + \frac{1}{2}\int_\Omega n_\varepsilon |\nabla c_\varepsilon|^{2\beta_1} \qquad (1.3.37)$$

$$\leq C_2 \int_\Omega (n_\varepsilon + \varepsilon)^{2m+2\alpha-3}|\nabla n_\varepsilon|^2 + \int_\Omega |\nabla c_\varepsilon|^{2\beta_1}|Du_\varepsilon| + \frac{1}{2}.$$

In addition, it is observed that

$$\frac{3(m+\alpha)}{3-(m+\alpha)} - 1 < \beta_1 < \frac{\frac{3(m+\alpha)}{3-(m+\alpha)}-1}{(4-2\frac{3(m+\alpha)}{3-(m+\alpha)})_+}$$

can be warranted by $m + \alpha \in (\frac{10}{9}, \frac{3}{2}]$, and thereby by Lemmas 1.16 and 1.17, there exists a constant $C_3 > 0$ such that

$$\int_\Omega |\nabla c_\varepsilon|^{2\beta_1}|Du_\varepsilon| \leq \frac{1}{4}\int_\Omega |\nabla c_\varepsilon|^{2\beta_1-2}|D^2 c_\varepsilon|^2 + C_3 \quad \text{for all } t > 0.$$

Substituting it into (1.3.37), one immediately obtains that for some $C_4 > 0$

$$\frac{1}{2\beta_1}\frac{d}{dt}\|\nabla c_\varepsilon\|_{L^{2\beta_1}(\Omega)}^{2\beta_1} + \frac{2(\beta_1-1)}{\beta_1^2}\int_\Omega |\nabla|\nabla c_\varepsilon|^{\beta_1}|^2$$

$$+ \frac{1}{4}\int_\Omega |\nabla c_\varepsilon|^{2\beta_1-2}|D^2 c_\varepsilon|^2 + \frac{1}{2}\int_\Omega n_\varepsilon |\nabla c_\varepsilon|^{2\beta_1}$$

$$\leq C_2 \int_\Omega (n_\varepsilon + \varepsilon)^{2m+2\alpha-3}|\nabla n_\varepsilon|^2 + C_4,$$

which along with (1.3.29) leads to

$$\sup_{t\in(0,\infty)} \|\nabla c_\varepsilon\|_{L^{2\beta_1}(\Omega)}^{2\beta_1} + \sup_{t\in(0,\infty)}\int_t^{t+1}\int_\Omega (|\nabla c_\varepsilon|^{2\beta_1-2}|D^2 c_\varepsilon|^2 + n_\varepsilon |\nabla c_\varepsilon|^{2\beta_1}) \leq C_5$$

$$(1.3.38)$$

for some $C_5 > 0$.

Moreover, denoting $p_0 = m + 2\alpha$ and taking $p := p_1 = \frac{16}{3}(m+\alpha)^2 - \frac{25}{3}(m+\alpha) + 4 + \frac{1}{3}\alpha[4(m+\alpha)-1]$ in (1.3.26), and applying the Young inequality, we conclude that for any $\delta > 0$, there exists constant $C(\delta) > 0$ such that

$$\frac{1}{p_1}\frac{d}{dt}\|n_\varepsilon + \varepsilon\|_{L^{p_1}(\Omega)}^{p_1} + \frac{m(p_1-1)}{2}\int_\Omega (n_\varepsilon+\varepsilon)^{m+p_1-3}|\nabla n_\varepsilon|^2$$

$$\leq C_1 \int_\Omega (n_\varepsilon+\varepsilon)^{p_1+1-m-2\alpha}|\nabla c_\varepsilon|^2$$

$$\leq \delta \int_\Omega (n_\varepsilon+\varepsilon)^{[p_1+1-m-2\alpha-\frac{1}{\beta_1}]\frac{\beta_1}{\beta_1-1}} + C(\delta)\int_\Omega (n_\varepsilon+\varepsilon)|\nabla c_\varepsilon|^{2\beta_1}$$

$$\leq \delta \int_\Omega (n_\varepsilon+\varepsilon)^{\frac{5m}{3}+p_1-1+\frac{4\alpha}{3}} + C(\delta)\int_\Omega (n_\varepsilon+\varepsilon)|\nabla c_\varepsilon|^{2\beta_1},$$

$$(1.3.39)$$

thanks to $(p_1+1-m-2\alpha-\frac{1}{\beta_1})\frac{\beta_1}{\beta_1-1} = \frac{5m}{3}+p_1-1+\frac{4\alpha}{3} = m+p_1-1+\frac{2}{3}p_0$.
Further, by the Gagliardo–Nirenberg interpolation inequality, we infer from (1.3.29) that

$$\int_\Omega (n_\varepsilon+\varepsilon)^{m+p_1-1+\frac{2}{3}p_0}$$

$$=\|(n_\varepsilon+\varepsilon)^{\frac{m+p_1-1}{2}}\|_{L^{\frac{2(m+p_1-1+\frac{2}{3}p_0)}{m+p_1-1}}(\Omega)}^{\frac{2(m+p_1-1+\frac{2}{3}p_0)}{m+p_1-1}}$$

$$\leq C_6\left(\|\nabla(n_\varepsilon+\varepsilon)^{\frac{p_1+m-1}{2}}\|_{L^2(\Omega)}^2\|(n_\varepsilon+\varepsilon)^{\frac{p_1+m-1}{2}}\|_{L^{\frac{2p_0}{m+p_1-1}}(\Omega)}^{\frac{2(m+p_2-1+\frac{2}{3}p_0)}{m+p_1-1}-2}\right.$$

$$\left.+\|(n_\varepsilon+\varepsilon)^{\frac{p_1+m-1}{2}}\|_{L^{\frac{2p_0}{m+p_1-1}}(\Omega)}^{\frac{2(m+p_1-1+\frac{2}{3}p_0)}{m+p_1-1}}\right)$$

$$\leq C_7(\|\nabla(n_\varepsilon+\varepsilon)^{\frac{p_1+m-1}{2}}\|_{L^2(\Omega)}^2+1)$$

with constants $C_6 > 0$ and $C_7 > 0$. Inserting the above inequality into (1.3.29) and picking $\delta > 0$ appropriately small, one concludes that there exists a positive constant C_8 such that

$$\frac{1}{p_1}\frac{d}{dt}\|n_\varepsilon+\varepsilon\|_{L^{p_1}(\Omega)}^{p_1} + \frac{m(p_1-1)}{4}\int_\Omega (n_\varepsilon+\varepsilon)^{m+p_1-3}|\nabla n_\varepsilon|^2 + \int_\Omega (n_\varepsilon+\varepsilon)^{p_1}$$

$$\leq C_8\int_\Omega n_\varepsilon|\nabla c_\varepsilon|^{2\beta_1} + \int_\Omega |\nabla c_\varepsilon|^{2\beta_1+2} + C_8.$$

Now by (1.3.38) and (1.2.8), one can get

$$\sup_{t\in(0,\infty)}\|n_\varepsilon(\cdot,t)+\varepsilon\|_{L^{p_1}(\Omega)}^{p_1} + \sup_{t\in(0,\infty)}\int_t^{t+1}\int_\Omega (n_\varepsilon+\varepsilon)^{p_1+m-3}|\nabla n_\varepsilon|^2 \leq C_9$$

for some positive constant C_9.

Lemma 1.21 *Let $\frac{10}{9} < m+\alpha \leq \frac{3}{2}$ and $(n_\varepsilon, c_\varepsilon, u_\varepsilon)$ be the solution of (1.2.2). Then for any $\beta > 1$ and $\eta > 0$ there exists a constant $C = C(\beta,\eta) > 0$ such that*

$$\int_{\Omega} |\nabla c_{\varepsilon}|^{2\beta} + \int_{\Omega} |\nabla c_{\varepsilon}|^{2\beta} |Du_{\varepsilon}| \leq \eta \int_{\Omega} |\nabla c_{\varepsilon}|^{2\beta-2} |D^2 c_{\varepsilon}|^2 + C \ \text{for all} \ t > 0.$$

Proof It is observed that $m + \alpha > \frac{10}{9}$ ensures $p_1 = \frac{16}{3}(m+\alpha)^2 - \frac{25}{3}(m+\alpha) + 4 + \frac{1}{3}\alpha[4(m+\alpha) - 1] > \frac{322}{243}$, and thus from Lemma 1.20, we have

$$\sup_{t\in(0,\infty)} \|n_{\varepsilon}(\cdot, t)\|_{L^{\frac{322}{243}}(\Omega)} \leq C_1.$$

By Lemma 1.3, for any $2 < r < \frac{966}{407}$, $\|Du_{\varepsilon}(\cdot, t)\|_{L^r(\Omega)} \leq C_2$ for some positive constant C_2. Moreover, by Lemma 1.19, one can conclude that for any $\beta > 1$

$$\int_{\Omega} |\nabla c_{\varepsilon}|^{2\beta} |Du_{\varepsilon}| \leq \frac{\eta}{2} \int_{\Omega} |\nabla c_{\varepsilon}|^{2\beta-2} |D^2 c_{\varepsilon}|^2 + C_3 \quad \text{for all} \ t > 0 \qquad (1.3.40)$$

with some positive constant C_3. On the other hand, in view of Lemma 1.7, it follows from the Young inequality that there is $C_4 > 0$ satisfying
$$\int_{\Omega} |\nabla c_{\varepsilon}|^{2\beta} \leq \frac{\eta}{2} \int_{\Omega} |\nabla c_{\varepsilon}|^{2\beta-2} |D^2 c_{\varepsilon}|^2 + C_4 \quad \text{for all} \ t > 0,$$

which together with (1.3.40) leads to the desired inequality.

The regularity of ∇c_{ε} from Lemma 1.21 can be readily developed to the following basis for the iterative reason, which can elevate $L^{p_1}(\Omega)$ of n_{ε} from Lemma 1.20 to the $L^p(\Omega)$-boundedness of n_{ε} with some $p > \frac{3}{2}$. To this end, we consider the properties of the iteration sequence $\{p_n\}_{n\geq 1}$ stated in the following.

Lemma 1.22 *Let* $p_1 = \frac{16}{3}(m+\alpha)^2 - \frac{25}{3}(m+\alpha) + \frac{1}{3}\alpha(4(m+\alpha) - 1) + 4$, $\frac{10}{9} < m \leq \frac{3}{2}$ *and* $0 \leq \alpha \leq \frac{7}{18}$. *Assume that for any* $n = 1, 2, \cdots$,

$$p_{n+1} = \frac{2}{3}p_n^2 + \frac{2}{3}(4m - 5 + 3\alpha)p_n + (2m + 2\alpha - 3)(m - 1) + 1,$$

then p_1 *is monotonically non-decreasing functions with respect to m as well as p_n is monotonically non-decreasing functions with respect to n, inter alia* $\lim_{n\to\infty} p_n = +\infty$.

Proof A direct calculation shows that

$$\begin{aligned} p_{n+1} &= \frac{2}{3}p_n^2 + \frac{2}{3}(4m - 5 + 3\alpha)p_n + (2m + 2\alpha - 3)(m - 1) + 1 \\ &= \frac{2}{3}p_n[p_n + (4m - 5) + 3\alpha] + (2m + 2\alpha - 3)(m - 1) + 1. \end{aligned}$$

Due to $0 < \alpha \leq \frac{7}{18}$ and $\frac{10}{9} < m + \alpha \leq \frac{3}{2}$, the mathematical induction implies that for any $n \in \mathbb{N}^*$, $p_n \geq m + \alpha$. In addition, it is observed that

$$p_{n+1} - p_n = \frac{2}{3}(p_n^2 - p_{n-1}^2) + \frac{2}{3}(4m - 5)(p_n - p_{n-1})$$
$$= \frac{2}{3}(p_n - p_{n-1})(p_n + p_{n-1} + 4m - 5 + 3\alpha).$$

Hence in light of $p_2 > p_1$ and $m + \alpha > \frac{10}{9}$, one can see that $p_3 > p_2$, and thereby $p_{n+1} > p_n$ by the induction.

By the contradiction argument, one can show that $\lim_{n \to \infty} p_n = +\infty$. In fact, supposed that $\{p_n\}_{n \geq 1}$ is bounded, then $\lim_{n \to \infty} p_n = p_*$ with some positive constant $p_* < \infty$, which implies that $p_* = \frac{2}{3}p_*^2 + \frac{2}{3}(4m - 5 + 3\alpha)p_* + (2m + 2\alpha - 3)(m - 1) + 1$, that is

$$\frac{2}{3}p_*^2 + (\frac{8}{3}m - \frac{13}{3} + 2\alpha)p_* + (2m + 2\alpha - 3)(m - 1) + 1 = 0. \tag{1.3.41}$$

By Weda's Theorem, we have

$$0 \leq \Delta = (\frac{8}{3}m - \frac{13}{3} + 2\alpha)^2 - 4 \times \frac{2}{3} \times [(2m + 2\alpha - 3)(m - 1) + 1]$$
$$= \frac{16\rho^2 - 8(11 - 2\alpha)\rho + 4\alpha^2 - 20\alpha + 73}{9} \tag{1.3.42}$$
$$:= \frac{H(\rho, \alpha)}{9}$$

with $\rho = m + \alpha$. Note that for any $0 \leq \alpha \leq \frac{7}{18}$ and $\frac{10}{9} < m + \alpha \leq \frac{3}{2}$, $\frac{\partial H(\rho, \alpha)}{\partial \rho} = 32\rho - 8(11 - 2\alpha) < 0$. So for $0 \leq \alpha \leq \frac{7}{18}$ and $\frac{10}{9} < m + \alpha \leq \frac{3}{2}$,

$$H(\rho, \alpha) \leq H(\frac{10}{9}, \alpha) = 16(\frac{10}{9})^2 - 8(11 - 2\alpha)\frac{10}{9} + 4\alpha^2 - 20\alpha + 73 < 0,$$

which contradicts with (1.3.42).

Lemma 1.23 Let $\frac{10}{9} < m + \alpha \leq \frac{3}{2}$ as well as $0 \leq \alpha \leq \frac{7}{18}$. If

$$\sup_{t \in (0, \infty)} \|n_\varepsilon(\cdot, t) + \varepsilon\|_{L^{p_n}(\Omega)}^{p_n} + \sup_{t \in (0, \infty)} \int_t^{t+1} \int_\Omega (n_\varepsilon + \varepsilon)^{p_n + m - 3} |\nabla n_\varepsilon|^2 \leq K \tag{1.3.43}$$

with $p_n + m - 3 < 0$ for some $K > 0$, then there exists $C = C(K) > 0$ independent of ε, such that

$$\sup_{t \in (0, \infty)} \|n_\varepsilon(\cdot, t) + \varepsilon\|_{L^{p_{n+1}}(\Omega)}^{p_{n+1}} + \sup_{t \in (0, \infty)} \int_t^{t+1} \int_\Omega (n_\varepsilon + \varepsilon)^{p_{n+1} + m - 3} |\nabla n_\varepsilon|^2 \leq C,$$

where $p_{n+1} = \frac{2}{3}p_n^2 + 2(\frac{4}{3}m - \frac{5}{3} + \alpha)p_n + (2m + 2\alpha - 3)(m - 1) + 1$ for any $n = 1, 2, 3, \cdots$, and p_1 is taken from Lemma 1.20.

Proof Let $\beta_n = p_n + m - 1$ for any $n = 1, 2, 3, \cdots$. Recalling (1.3.32), there is $C_1 > 0$ such that for all $t > 0$,

$$\frac{1}{2\beta_n} \frac{d}{dt} \|\nabla c_\varepsilon\|_{L^{2\beta_n}(\Omega)}^{2\beta_n} + \frac{2(\beta_n - 1)}{\beta_n^2} \int_\Omega |\nabla |\nabla c_\varepsilon|^{\beta_n}|^2 + \int_\Omega |\nabla c_\varepsilon|^{2\beta_n - 2} |D^2 c_\varepsilon|^2$$
$$\leq \|c_0\|_{L^\infty(\Omega)} \int_\Omega |\nabla c_\varepsilon|^{2\beta_n - 1} |\nabla n_\varepsilon| - \int_\Omega n_\varepsilon |\nabla c_\varepsilon|^{2\beta_n} + \int_\Omega |\nabla c_\varepsilon|^{2\beta_n} |Du_\varepsilon| + C_1.$$
$$(1.3.44)$$

As done in (1.3.36), we can conclude that there exists a positive constant C_2 such that

$$\|c_0\|_{L^\infty(\Omega)} \int_\Omega |\nabla c_\varepsilon|^{2\beta_n - 1} |\nabla n_\varepsilon|$$
$$\leq \frac{1}{2} \int_\Omega n_\varepsilon |\nabla c_\varepsilon|^{2\beta_n} + \frac{1}{2} \||\nabla c_\varepsilon|^{\beta_n - 1} D^2 c_\varepsilon\|_{L^2(\Omega)}^2 \qquad (1.3.45)$$
$$+ C_2 \int_\Omega (n_\varepsilon + \varepsilon)^{m + p_n - 3} |\nabla n_\varepsilon|^2 + \frac{1}{2},$$

which along with (1.3.44) implies that

$$\frac{1}{2\beta_n} \frac{d}{dt} \|\nabla c_\varepsilon\|_{L^{2\beta_n}(\Omega)}^{2\beta_n} + \frac{2(\beta_n - 1)}{\beta_n^2} \int_\Omega |\nabla |\nabla c_\varepsilon|^{\beta_n}|^2$$
$$+ \frac{1}{2} \int_\Omega |\nabla c_\varepsilon|^{2\beta_n - 2} |D^2 c_\varepsilon|^2 + \frac{1}{2} \int_\Omega n_\varepsilon |\nabla c_\varepsilon|^{2\beta_n} \qquad (1.3.46)$$
$$\leq C_2 \int_\Omega (n_\varepsilon + \varepsilon)^{m + p_n - 3} |\nabla n_\varepsilon|^2 + \int_\Omega |\nabla c_\varepsilon|^{2\beta_n} |Du_\varepsilon| + \frac{1}{2}$$

and thereby together with Lemma 1.21 leads to

$$\frac{1}{2\beta_n} \frac{d}{dt} \|\nabla c_\varepsilon\|_{L^{2\beta_n}(\Omega)}^{2\beta_n} + \|\nabla c_\varepsilon\|_{L^{2\beta_n}(\Omega)}^{2\beta_n} + \frac{1}{8} \int_\Omega |\nabla c_\varepsilon|^{2\beta_n - 2} |D^2 c_\varepsilon|^2 + \frac{1}{2} \int_\Omega n_\varepsilon |\nabla c_\varepsilon|^{2\beta_n}$$
$$\leq C_2 \int_\Omega (n_\varepsilon + \varepsilon)^{m + p_n - 3} |\nabla n_\varepsilon|^2 + C_4$$

with some constant $C_4 > 0$. By (1.3.43), there exists some positive constant C_5 such that

$$\sup_{t \in (0, \infty)} \|\nabla c_\varepsilon\|_{L^{2\beta_n}(\Omega)}^{2\beta_n} + \sup_{t \in (0, \infty)} \int_t^{t+1} \int_\Omega [|\nabla c_\varepsilon|^{2\beta_n - 2} |D^2 c_\varepsilon|^2 + n_\varepsilon |\nabla c_\varepsilon|^{2\beta_n}] \leq C_5.$$
$$(1.3.47)$$

Furthermore, in view of $p_n > m + \alpha$, taking $p_{n+1} = \frac{2}{3} p_n^2 + 2(\frac{4}{3}m - \frac{5}{3} + \alpha)p_n + (2m + 2\alpha - 3)(m - 1) + 1$ in (1.3.26) and by the Young inequality, it follows that for any $\eta > 0$, there is $C(\eta) > 0$ such that

$$\frac{1}{p_{n+1}}\frac{d}{dt}\|n_\varepsilon + \varepsilon\|_{L^{p_{n+1}}(\Omega)}^{p_{n+1}} + \frac{m(p_{n+1}-1)}{2}\int_\Omega (n_\varepsilon + \varepsilon)^{m+p_{n+1}-3}(1+n_\varepsilon)^{-2\alpha}|\nabla n_\varepsilon|^2$$

$$\leq \frac{(p_{n+1}-1)C_S^2}{2m}\int_\Omega (n_\varepsilon + \varepsilon)^{p_{n+1}+1-m-2\alpha}|\nabla c_\varepsilon|^2$$

$$= \frac{(p_{n+1}-1)C_S^2}{2m}\int_\Omega (n_\varepsilon + \varepsilon)^{\frac{1}{\beta_n}}|\nabla c_\varepsilon|^2 (n_\varepsilon + \varepsilon)^{p_{n+1}+1-m-2\alpha-\frac{1}{\beta_n}}$$

$$\leq C(\eta)\int_\Omega (n_\varepsilon + \varepsilon)|\nabla c_\varepsilon|^{2\beta_n} + \eta\int_\Omega (n_\varepsilon + \varepsilon)^{[p_{n+1}+1-m-2\alpha-\frac{1}{\beta_n}]\frac{\beta_n}{\beta_n-1}}.$$

$$(1.3.48)$$

Thanks to $p_{n+1} = \frac{2}{3}p_n^2 + 2(\frac{4}{3}m - \frac{5}{3} + \alpha)p_n + (2m + 2\alpha - 3)(m-1) + 1$, we can see that

$$(p_{n+1}+1-m-2\alpha-\frac{1}{\beta_n})\frac{\beta_n}{\beta_n-1} = (p_{n+1}+1-m-2\alpha-\frac{1}{p_n+m-1})\frac{p_n+m-1}{p_n+m-2}$$

$$= m + p_{n+1} - 1 + \frac{2}{3}p_n$$

and thereby

$$\int_\Omega (n_\varepsilon + \varepsilon)^{m+p_{n+1}-1+\frac{2}{3}p_n}$$

$$= \|(n_\varepsilon + \varepsilon)^{\frac{m+p_{n+1}-1}{2}}\|_{L^{\frac{2(m+p_{n+1}-1+\frac{2}{3}p_n)}{m+p_{n+1}-1}}(\Omega)}^{\frac{2(m+p_{n+1}-1+\frac{2}{3}p_n)}{m+p_{n+1}-1}}$$

$$\leq C_6\|\nabla(n_\varepsilon + \varepsilon)^{\frac{p_{n+1}+m-1}{2}}\|_{L^2(\Omega)}^2\|(n_\varepsilon + \varepsilon)^{\frac{p_{n+1}+m-1}{2}}\|_{L^{\frac{2p_n}{m+p_{n+1}-1}}(\Omega)}^{\frac{2(m+p_{n+1}-1+\frac{2}{3}p_n)}{m+p_{n+1}-1}-2}$$

$$+ C_6\|(n_\varepsilon + \varepsilon)^{\frac{p_{n+1}+m-1}{2}}\|_{L^{\frac{2p_n}{m+p_{n+1}-1}}(\Omega)}^{\frac{2(m+p_n-1+\frac{2}{3}p_n)}{m+p_n-1}}$$

$$\leq C_7(\|\nabla(n_\varepsilon + \varepsilon)^{\frac{p_{n+1}+m-1}{2}}\|_{L^2(\Omega)}^2 + 1).$$

Substituting the above inequality into (1.3.48) and taking $\eta > 0$ appropriately small, one may derive that there is $C_8 > 0$ such that for any $\varepsilon \in (0,1)$,

$$\frac{1}{p_{n+1}}\frac{d}{dt}\|n_\varepsilon + \varepsilon\|_{L^{p_{n+1}}(\Omega)}^{p_{n+1}} + \frac{m(p_{n+1}-1)}{4}\int_\Omega (n_\varepsilon + \varepsilon)^{m+p_{n+1}-3}|\nabla n_\varepsilon|^2$$

$$+ \int_\Omega (n_\varepsilon + \varepsilon)^{m+p_{n+1}-1+\frac{2}{3}p_n}$$

$$\leq C(\eta)\int_\Omega n_\varepsilon|\nabla c_\varepsilon|^{2\beta_n} + \int_\Omega |\nabla c_\varepsilon|^{2\beta_n} + C_8.$$

This together with (1.3.47) implies that for some positive constant C_9,

$$\sup_{t\in(0,\infty)} \|n_\varepsilon(\cdot, t) + \varepsilon\|_{L^{p_{n+1}}(\Omega)}^{p_{n+1}} + \sup_{t\in(0,\infty)} \int_t^{t+1} \int_\Omega (n_\varepsilon + \varepsilon)^{p_{n+1}+m-3}|\nabla n_\varepsilon|^2 \leq C_9$$

and thus completes the proof of Lemma 1.23.

Combining Lemma 1.3.34 with Lemma 1.23, we immediately have

Lemma 1.24 *Let* $0 \leq \alpha \leq \frac{7}{18}$ *and* $\frac{10}{9} < m + \alpha \leq \frac{3}{2}$. *Then there exist constants* $p^* > \frac{3}{2}$ *and* $C = C(p^*) > 0$ *such that*

$$\int_\Omega n_\varepsilon^{p^*}(x, t)dx \leq C$$

for all $t > 0$.

By the similar strategy as above, one can also derive the boundedness of $\int_\Omega n_\varepsilon^p$ with some $p > \frac{3}{2}$ in the case $m + \alpha > 2$.

Lemma 1.25 *Let* $m + \alpha > 2$. *There exists* $C > 0$ *independent of* ε *such that the solution of* (1.2.2) *satisfies*

$$\int_\Omega (n_\varepsilon + \varepsilon)^{m+2\alpha-1} \leq C \quad v\,for\,all \quad t > 0 \tag{1.3.49}$$

as well as

$$\int_t^{t+1} \int_\Omega \left[(n_\varepsilon + \varepsilon)^{2(m+\alpha-\frac{2}{3})} + (n_\varepsilon + \varepsilon)^{2m+2\alpha-4}|\nabla n_\varepsilon|^2 \right] \leq C. \tag{1.3.50}$$

Proof Taking c_ε as the test function for the second equation of (1.2.2) and using $\nabla \cdot u_\varepsilon = 0$, it yields that

$$\frac{1}{2}\frac{d}{dt}\|c_\varepsilon\|_{L^2(\Omega)}^2 + \int_\Omega |\nabla c_\varepsilon|^2 = -\int_\Omega n_\varepsilon c_\varepsilon^2,$$

which together with $n_\varepsilon \geq 0$ and $c_\varepsilon \geq 0$ implies that for some positive constant C_1,

$$\int_\Omega c_\varepsilon^2 + \int_t^{t+1} \int_\Omega |\nabla c_\varepsilon|^2 \leq C_1 \quad \text{for all } t > 0.$$

Now, choosing $p = m + 2\alpha - 1$ in (1.3.26), we get

$$\frac{1}{m + 2\alpha - 1}\frac{d}{dt}\|n_\varepsilon + \varepsilon\|_{L^{m+2\alpha-1}(\Omega)}^{m+2\alpha-1} + \frac{m(m + 2\alpha - 2)}{2}\int_\Omega (n_\varepsilon + \varepsilon)^{2m+2\alpha-4}|\nabla n_\varepsilon|^2$$

$$\leq \frac{(m + 2\alpha - 2)C_S^2}{2m}\int_\Omega (n_\varepsilon + \varepsilon)^{2\alpha}(1 + n_\varepsilon)^{-2\alpha}|\nabla c_\varepsilon|^2$$

$$\leq \frac{(m + 2\alpha - 2)C_S^2}{2m}\int_\Omega |\nabla c_\varepsilon|^2 \quad \text{for all } t > 0.$$

(1.3.51)

Furthermore, applying the Gagliardo–Nirenberg inequality, we obtain that there are $C_i > 0$, $(i = 1, 2, 3)$, such that

$$\int_\Omega (n_\varepsilon + \varepsilon)^{m+2\alpha-1}$$

$$= \|(n_\varepsilon + \varepsilon)^{m+\alpha-1}\|_{L^{\frac{m+2\alpha-1}{m+\alpha-1}}(\Omega)}^{\frac{m+2\alpha-1}{m+\alpha-1}}$$

$$\leq C_1 \|\nabla(n_\varepsilon + \varepsilon)^{m+\alpha-1}\|_{L^2(\Omega)}^{2\frac{3(m+2\alpha-2)}{6m+6\alpha-7}} \|(n_\varepsilon + \varepsilon)^{m+\alpha-1}\|_{L^{\frac{1}{m+\alpha-1}}(\Omega)}^{\frac{m+2\alpha-1}{m+\alpha-1} - 2\frac{3(m+2\alpha-2)}{6m+6\alpha-7}}$$

$$+ C_1 \|(n_\varepsilon + \varepsilon)^{m+\alpha-1}\|_{L^{\frac{1}{m+\alpha-1}}(\Omega)}^{\frac{m+2\alpha-1}{m+\alpha-1}}$$

$$\leq C_2(\|\nabla(n_\varepsilon + \varepsilon)^{m+\alpha-1}\|_{L^2(\Omega)}^{\frac{6(m+2\alpha-2)}{6m+6\alpha-7}} + 1)$$

$$\leq \frac{m(m + 2\alpha - 2)}{8}\int_\Omega (n_\varepsilon + \varepsilon)^{2m+2\alpha-4}|\nabla n_\varepsilon|^2 + C_3$$

thanks to $\frac{6(m+2\alpha-2)}{6m+6\alpha-7} < 2$, and

$$\int_\Omega (n_\varepsilon + \varepsilon)^{2(m+\alpha-\frac{2}{3})} = \|(n_\varepsilon + \varepsilon)^{m+\alpha-1}\|_{L^{\frac{2(m+\alpha-\frac{2}{3})}{m+\alpha-1}}(\Omega)}^{\frac{2(m+\alpha-\frac{2}{3})}{m+\alpha-1}}$$

$$\leq C_3(\|\nabla(n_\varepsilon + \varepsilon)^{m+\alpha-1}\|_{L^2(\Omega)}^2 + 1).$$

Inserting above two inequalities into (1.3.51), we derive

$$\frac{1}{m + 2\alpha - 1}\frac{d}{dt}\|n_\varepsilon + \varepsilon\|_{L^{m+2\alpha-1}(\Omega)}^{m+2\alpha-1} + \int_\Omega (n_\varepsilon + \varepsilon)^{m+2\alpha-1}$$

$$+ \frac{m(m + 2\alpha - 2)}{2}\int_\Omega (n_\varepsilon + \varepsilon)^{2m+2\alpha-4}|\nabla n_\varepsilon|^2 + (n_\varepsilon + \varepsilon)^{2(m+\alpha-\frac{2}{3})}$$

(1.3.52)

$$\leq \frac{(m + 2\alpha - 2)C_S^2}{2m}\int_\Omega |\nabla c_\varepsilon|^2 + C_4 \quad \text{for all } t > 0$$

with some positive constant C_4. Now, we define $y_\varepsilon(t) := \|n_\varepsilon(\cdot, t) + \varepsilon\|_{L^{m+2\alpha-1}(\Omega)}^{m+2\alpha-1}$ and

$$h_\varepsilon(t) := \frac{m(m + 2\alpha - 2)}{2}\int_\Omega (n_\varepsilon + \varepsilon)^{2m+2\alpha-4}|\nabla n_\varepsilon|^2 + (n_\varepsilon + \varepsilon)^{2(m+\alpha-\frac{2}{3})} \quad \text{for all } t > 0.$$

As an application of Lemma 1.4, this together with (1.3.51) readily yields (1.3.49) and (1.3.50).

With the space–time regularity property of n_ε in (1.3.50), we can improve the regularity of ∇u_ε beyond (1.3.14) through following lemma.

Lemma 1.26 *Let $m + \alpha > 2$. There exists constant $C > 0$ such that for all $t > 0$,*

$$\int_\Omega |\nabla u_\varepsilon(\cdot, t)|^2 + \int_t^{t+1} \int_\Omega |\Delta u_\varepsilon|^2 \leq C. \tag{1.3.53}$$

Proof Multiplying the projected Stokes equation $u_{\varepsilon t} + A u_\varepsilon = \mathcal{P}[n_\varepsilon \nabla \phi]$ by $A u_\varepsilon$, we derive

$$\frac{1}{2} \frac{d}{dt} \|A^{\frac{1}{2}} u_\varepsilon\|_{L^2(\Omega)}^2 + \int_\Omega |A u_\varepsilon|^2 = \int_\Omega \mathcal{P}(n_\varepsilon \nabla \phi) A u_\varepsilon$$
$$\leq \frac{1}{2} \int_\Omega |A u_\varepsilon|^2 + \frac{1}{2} \|\nabla \phi\|_{L^\infty(\Omega)}^2 \int_\Omega n_\varepsilon^2 \text{ for all } t > 0. \tag{1.3.54}$$

Recalling that $\|A^{\frac{1}{2}} u_\varepsilon\|_{L^2(\Omega)}^2 = \|\nabla u_\varepsilon\|_{L^2(\Omega)}^2$ (see p. 133 of Sohr 2001), and with some $C_1 > 0$, we have

$$\int_\Omega |\nabla u_\varepsilon(\cdot, t)|^2 \leq C_1 \int_\Omega |A u_\varepsilon|^2 \text{ for all } t > 0.$$

Thanks to the fact that $\| \cdot \|_{W^{2,2}(\Omega)}$ and $\|A(\cdot)\|_{L^2(\Omega)}$ are equivalent on $D(A)$ (see p. 129 of Sohr 2001), we see that for

$$y(t) := \int_\Omega |\nabla u_\varepsilon(\cdot, t)|^2, t > 0$$

and

$$h(t) := \frac{1}{2} \|\nabla \phi\|_{L^\infty(\Omega)}^2 \int_\Omega n_\varepsilon^2, t > 0.$$

Equation (1.3.54) implies the inequality

$$y'(t) + \frac{1}{2C_1} y(t) + \frac{1}{2} \int_\Omega |A u_\varepsilon|^2 \leq h(t) \text{ for all } t > 0.$$

As an application of Lemma 1.4, this yields (1.3.53) thanks to (1.3.50). $\qquad \blacksquare$

At this position, we can achieve the regularity of c_ε in the case $m + \alpha > 2$ just as that in Lemma 1.13.

Lemma 1.27 *Let $m + \alpha > 2$. There exists $C > 0$ independent of ε such that*

$$\int_\Omega |\nabla c_\varepsilon(\cdot, t)|^2 + \int_t^{t+1} \int_\Omega |\Delta c_\varepsilon|^2 \leq C \text{ for all } t > 0. \tag{1.3.55}$$

Proof Similar to the proof (1.3.42), we can conclude that

$$
\frac{1}{2}\frac{d}{dt}\|\nabla c_\varepsilon\|_{L^2(\Omega)}^2
$$
$$
\leq -\frac{1}{2}\int_\Omega |\Delta c_\varepsilon|^2 + \|c_0\|_{L^\infty(\Omega)}\int_\Omega n_\varepsilon^2 - \int_\Omega \nabla c_\varepsilon (\nabla u_\varepsilon \cdot \nabla c_\varepsilon)
$$
$$
\leq -\frac{1}{2}\int_\Omega |\Delta c_\varepsilon|^2 + \|c_0\|_{L^\infty(\Omega)}\int_\Omega n_\varepsilon^2 + \|\nabla u_\varepsilon\|_{L^2(\Omega)}\|\nabla c_\varepsilon\|_{L^4(\Omega)}^2 \quad \text{for all } t > 0.
$$

(1.3.56)

Recalling (1.2.10), the Gagliardo–Nirenberg inequality entails that there exist $C_1 > 0$ and $C_2 > 0$ such that

$$
\|\nabla c_\varepsilon\|_{L^4(\Omega)}^2 \leq C_1 \|\Delta c_\varepsilon\|_{L^2(\Omega)}\|c_\varepsilon\|_{L^\infty(\Omega)}^2 + C_1 \|c_\varepsilon\|_{L^\infty(\Omega)}^2
$$
$$
\leq C_2 \|\Delta c_\varepsilon\|_{L^2(\Omega)} + C_2 \quad \text{for all } t > 0.
$$

(1.3.57)

Substituting (1.3.57) into (1.3.56) and by the Young inequality, we obtain that for some positive constant C_3,

$$
\frac{1}{2}\frac{d}{dt}\|\nabla c_\varepsilon\|_{L^2(\Omega)}^2
$$
$$
\leq -\frac{1}{2}\int_\Omega |\Delta c_\varepsilon|^2 + \|c_0\|_{L^\infty(\Omega)}\int_\Omega n_\varepsilon^2 + \|\nabla u_\varepsilon\|_{L^2(\Omega)}[C_2\|\Delta c_\varepsilon\|_{L^2(\Omega)}^{\frac{1}{2}} + C_2]^2
$$
$$
\leq -\frac{1}{4}\int_\Omega |\Delta c_\varepsilon|^2 + \|c_0\|_{L^\infty(\Omega)}\int_\Omega n_\varepsilon^2 + C_3\|\nabla u_\varepsilon\|_{L^2(\Omega)}^2 + C_3 \quad \text{for all } t > 0,
$$

(1.3.58)

which together with (1.3.57) implies that for some positive constants C_4, C_5,

$$
\frac{1}{2}\frac{d}{dt}\|\nabla c_\varepsilon\|_{L^2(\Omega)}^2 + \frac{1}{8}\|\Delta c_\varepsilon\|_{L^2(\Omega)}^2 + C_4\|\nabla c_\varepsilon\|_{L^2(\Omega)}^2
$$
$$
\leq \|c_0\|_{L^\infty(\Omega)}\int_\Omega n_\varepsilon^2 + C_3\|\nabla u_\varepsilon\|_{L^2(\Omega)}^2 + C_5 \quad \text{for all } t > 0.
$$

(1.3.59)

Now, we define $g_\varepsilon(t) := \|\nabla c_\varepsilon(\cdot, t)\|_{L^2(\Omega)}^2$ and

$$
h_\varepsilon(t) := \|c_0\|_{L^\infty(\Omega)}\int_\Omega n_\varepsilon^2(\cdot, t) + C_3\|\nabla u_\varepsilon(\cdot, t)\|_{L^2(\Omega)}^2 + C_5.
$$

As an application of Lemma 1.4, this in conjunction with (1.3.53) entails (1.3.55).

Proceeding as the proof of Lemma 1.16, we can arrive at

Lemma 1.28 *Let $m + \alpha > 2$. Then there exists $C > 0$ independent of ε such that the solution of (1.2.2) satisfies*

$$
\int_\Omega (n_\varepsilon + \varepsilon)^{m+2\alpha-\frac{1}{2}} \leq C \quad \text{for all } t > 0.
$$

(1.3.60)

Proof Choosing $p = m + 2\alpha - \frac{1}{2}$ in (1.3.26), we obtain that for some $C_1 > 0$,

$$
\frac{1}{m + 2\alpha - \frac{1}{2}} \frac{d}{dt} \| n_\varepsilon + \varepsilon \|_{L^{m+2\alpha-\frac{1}{2}}(\Omega)}^{m+2\alpha-\frac{1}{2}} + \frac{m(m + 2\alpha - \frac{3}{2})}{2} \int_\Omega (n_\varepsilon + \varepsilon)^{2m+2\alpha-\frac{1}{2}-3} |\nabla n_\varepsilon|^2
$$

$$
\leq C_1 \int_\Omega (n_\varepsilon + \varepsilon)^{\frac{1}{2}+2\alpha}(1 + n_\varepsilon)^{-2\alpha} |\nabla c_\varepsilon|^2
$$

$$
\leq C_1 \int_\Omega (n_\varepsilon + \varepsilon)^{\frac{1}{2}} |\nabla c_\varepsilon|^2
$$

$$
\leq C_1^2 [\| n_0 \|_{L^1(\Omega)} + |\Omega|] + \frac{1}{4} \int_\Omega |\nabla c_\varepsilon|^4 \quad \text{for all } t > 0.
$$

$$(1.3.61)$$

On the other hand, we employ the Gagliardo–Nirenberg inequality to derive that there exists positive constants C_2, C_3 and C_4 such that

$$
\int_\Omega (n_\varepsilon + \varepsilon)^{m+2\alpha-\frac{1}{2}}
$$

$$
= \| (n_\varepsilon + \varepsilon)^{m+\alpha-\frac{3}{4}} \|_{L^{\frac{m+2\alpha-\frac{1}{2}}{m+\alpha-\frac{3}{4}}}(\Omega)}^{\frac{m+2\alpha-\frac{1}{2}}{m+\alpha-\frac{3}{4}}}
$$

$$
\leq C_2 \| \nabla (n_\varepsilon + \varepsilon)^{m+\alpha-\frac{3}{4}} \|_{L^2(\Omega)}^{2\frac{3(2m+4\alpha-3)}{3(4m+4\alpha-3)-2}} \| (n_\varepsilon + \varepsilon)^{m+\alpha-\frac{3}{4}} \|_{L^1(\Omega)}^{\frac{m+2\alpha-\frac{1}{2}}{m+\alpha-\frac{3}{4}}-2\frac{3(2m+4\alpha-3)}{3(4m+4\alpha-3)-2}}
$$

$$
+ C_2 \| (n_\varepsilon + \varepsilon)^{m+\alpha-\frac{3}{4}} \|_{L^1(\Omega)}^{\frac{m+2\alpha-\frac{1}{2}}{m+\alpha-\frac{3}{4}}}
$$

$$
\leq C_3 (\| \nabla (n_\varepsilon + \varepsilon)^{m+\alpha-\frac{3}{4}} \|_{L^2(\Omega)}^{2\frac{3(2m+4\alpha-3)}{3(4m+4\alpha-3)-2}} + 1)
$$

$$
= C_4 \left(\int_\Omega (n_\varepsilon + \varepsilon)^{2m+2\alpha-\frac{1}{2}-3} |\nabla n_\varepsilon|^2 \right)^{\frac{3(2m+4\alpha-3)}{3(4m+4\alpha-3)-2}} + C_3 \quad \text{for all } t > 0.
$$

Inserting the above inequality into (1.3.61), one has

$$
\frac{1}{m + 2\alpha - \frac{1}{2}} \frac{d}{dt} \| n_\varepsilon + \varepsilon \|_{L^{m+2\alpha-\frac{1}{2}}(\Omega)}^{m+2\alpha-\frac{1}{2}} + C_5 \left(\int_\Omega (n_\varepsilon + \varepsilon)^{m+2\alpha-\frac{1}{2}} \right)^{\frac{3(4m+4\alpha-3)-2}{3(2m+4\alpha-3)}}
$$

$$
\leq C_6 + \frac{1}{4} \int_\Omega |\nabla c_\varepsilon|^4 \quad \text{for all } t > 0.
$$

With the help of (1.3.55) and (1.3.57), we derive that (1.3.60) by Lemma 1.4.

At this position, by the result stated in Lemmas 1.28, 1.24 and 1.16, we have

Lemma 1.29 *Let* $m + \alpha > \frac{10}{9}$. *Then there exist positive constants* $q_0 > \frac{3}{2}$ *and* $C > 0$ *such that the solution of* (1.2.2) *satisfies*

$$
\int_\Omega n_\varepsilon^{q_0}(x, t)dx \leq C \quad \text{for all } t > 0.
$$

$$(1.3.62)$$

Proof Let

$$
q_0 = \begin{cases}
p^* & \text{if } \dfrac{10}{9} < m + \alpha \le \dfrac{3}{2} \text{ and } 0 \le \alpha \le \dfrac{7}{18}, \\[2mm]
m + 2\alpha & \text{if } \dfrac{10}{9} < m + \alpha \le \dfrac{3}{2} \text{ and } \alpha > \dfrac{7}{18}, \\[2mm]
m + 2\alpha & \text{if } \dfrac{3}{2} < m + \alpha \le 2, \\[2mm]
m + 2\alpha - \dfrac{1}{2} & \text{if } m + \alpha > 2,
\end{cases}
$$

with p^* given in Lemma 1.24. Then it is easy to see that $q_0 > \frac{3}{2}$ and thereby (1.3.62) readily follows from Lemmas 1.28, 1.24 and 1.16.

1.3.3 Uniform L^∞-Boundedness of n_ε as Well as ∇c_ε and u_ε

With Lemma 1.29 at hand, further regularity properties of n_ε, c_ε and u_ε can now be obtained by essentially rather standard arguments (see the proof of Corollary 3.4 in Winkler 2015b or Lemma 6.1 in Winkler 2018c for example). We firstly use the heat semigroup to obtain the $L^\infty(\Omega)$-bound for u_ε.

Lemma 1.30 *Let $m + \alpha > \frac{10}{9}$ and assume that the hypothesis of Theorem 1.1 holds. Then there exists a positive constant C independent of ε such that, the solution of (1.2.2) satisfies*

$$\|u_\varepsilon(\cdot, t)\|_{L^\infty(\Omega)} \le C \quad \text{for all } t > 0. \tag{1.3.63}$$

Proof Let $h_\varepsilon(x, t) = \mathscr{P}[n_\varepsilon \nabla \phi]$. Then by Lemma 1.29, there is $C_1 > 0$ such that for all $t > 0$

$$\|h_\varepsilon(\cdot, t)\|_{L^{q_0}(\Omega)} \le C_1. \tag{1.3.64}$$

Fixing r_0 and δ with $r_0 \in (\frac{3}{2q_0}, 1)$ and $\delta \in (0, 1 - r_0)$, one can chooses $r_1 > \frac{3}{\delta}$ such that $W^{\delta, r_1}(\Omega) \hookrightarrow L^\infty(\Omega)$. It then follows from the variation-of-constants representation, the Young inequality, the Sobolev embedding theorem and (1.3.64) that

$$
\begin{aligned}
&\|u_\varepsilon(\cdot, t)\|_{L^\infty(\Omega)} \\
&\le \|e^{-tA} u_0\|_{L^\infty(\Omega)} + \int_0^t \|A^{r_0} e^{-(t-\tau)A} A^{-r_0} h_\varepsilon(\cdot, \tau) d\tau\|_{L^\infty(\Omega)} d\tau \\
&\le \|e^{-tA} u_0\|_{L^\infty(\Omega)} + \int_0^t \|A^{r_0+\delta} e^{-(t-\tau)A} A^{-r_0} h_\varepsilon(\cdot, \tau) d\tau\|_{L^{r_1}(\Omega)} d\tau \\
&\le \|e^{-tA} u_0\|_{L^\infty(\Omega)} + C_1 \int_0^t (t-\tau)^{-r_0-\delta} e^{-\lambda(t-\tau)} \|e^{-(t-\tau)A} A^{-r_0} h_\varepsilon(\cdot, \tau) d\tau\|_{L^{r_1}(\Omega)} d\tau \\
&\le \|A^\gamma u_0\|_{L^2(\Omega)} + C_2 \int_0^t (t-\tau)^{-r_0-\delta} e^{-\lambda(t-\tau)} \|h_\varepsilon(\cdot, \tau)\|_{L^{q_0}(\Omega)} d\tau \\
&\le C_3 \quad \text{for all } t > 0.
\end{aligned}
$$

Here, we have used the fact that $r_0 > \frac{3}{2q_0} > \frac{3}{2}(\frac{1}{q_0} - \frac{1}{r_1})$ and

$$\int_0^t (t-\tau)^{-r_0-\delta} e^{-\lambda(t-\tau)} \leq \int_0^\infty t^{-r_0-\delta} e^{-\lambda\tau} d\tau < \infty.$$

Lemma 1.31 *Assume that the hypothesis of Theorem 1.1 holds. Then there exists a positive constant C independent of ε such that the solution of (1.2.2) satisfies*

$$\|\nabla c_\varepsilon(\cdot, t)\|_{L^{r_0}(\Omega)} \leq C \quad \text{for all } t > 0 \tag{1.3.65}$$

with $3 < r_0 < \min\{\frac{3q_0}{(3-q_0)_+}, 4\}$, where $q_0 > \frac{3}{2}$ is given by Lemma 1.29.

Proof Involving the variation-of-constants formula for c_ε and applying $\nabla \cdot u_\varepsilon = 0$ in $x \in \Omega, t > 0$, we have

$$c_\varepsilon(t) = e^{t(\Delta-1)} c_0 - \int_0^t e^{(t-s)(\Delta-1)}(n_\varepsilon(s)c_\varepsilon(s) - c_\varepsilon(s) - \nabla \cdot (u_\varepsilon(s)c_\varepsilon(s))ds,$$

and thus

$$\begin{aligned}
&\|\nabla c_\varepsilon(\cdot, t)\|_{L^{r_0}(\Omega)} \\
&\leq \|\nabla e^{t\Delta} c_0\|_{L^{r_0}(\Omega)} + \int_0^t \|\nabla e^{(t-s)\Delta}[n_\varepsilon(s) - 1]c_\varepsilon(s)\|_{L^{r_0}(\Omega)} ds \\
&\quad + \int_0^t \|\nabla e^{(t-s)\Delta} \nabla \cdot (u_\varepsilon(s)c_\varepsilon(s))\|_{L^{r_0}(\Omega)} ds \quad \text{for all } t > 0,
\end{aligned} \tag{1.3.66}$$

where $r_0 \in (3, \min\{\frac{3q_0}{(3-q_0)_+}, 4\})$.

Now, we will estimate the terms on the right of (1.3.66) one by one. In view of (1.1.4), there is $C_1 > 0$ such that

$$\|\nabla e^{t(\Delta-1)} c_0\|_{L^{r_0}(\Omega)} \leq C_1 \quad \text{for all } t > 0. \tag{1.3.67}$$

Since $q_0 > \frac{3}{2}$, it yields

$$-\frac{1}{2} - \frac{3}{2}\left(\frac{1}{q_0} - \frac{1}{r_0}\right) > -1,$$

which together with Lemmas 1.29 and 1.8 implies that for some positive constants C_2 and C_3,

$$\begin{aligned}
&\int_0^t \|\nabla e^{(t-s)(\Delta-1)}[(n_\varepsilon(s) - 1)c_\varepsilon(s)]\|_{L^{r_0}(\Omega)} ds \\
&\leq C_2 \int_0^t [1 + (t-s)^{-\frac{1}{2} - \frac{3}{2}(\frac{1}{q_0} - \frac{1}{r_0})}] e^{-\lambda(t-s)}[\|n_\varepsilon(s)\|_{L^{q_0}(\Omega)} + 1]\|c_\varepsilon(s)\|_{L^\infty(\Omega)} ds \\
&\leq C_3 \quad \text{for all } t > 0.
\end{aligned} \tag{1.3.68}$$

Finally, we choose $\iota = \frac{1}{3}$ satisfying $\frac{1}{2} + \frac{3}{2}(\frac{1}{\infty} - \frac{1}{6}) < \frac{1}{3}$ and $\tilde{\kappa} = \frac{1}{12} \in (0, \frac{1}{6})$. In view of Hölder's inequality, we derive from Lemma 1.30 that there exist positive constants C_i, $i = 4, \cdots, 8$ such that

$$
\int_0^t \|\nabla e^{(t-s)(\Delta-1)} \nabla \cdot (u_\varepsilon(s)c_\varepsilon(s))\|_{L^{r_0}(\Omega)} ds
$$

$$
\leq C_4 \int_0^t \|e^{(t-s)(\Delta-1)} \nabla \cdot (u_\varepsilon(s)c_\varepsilon(s))\|_{W^{1,r_0}(\Omega)} ds
$$

$$
\leq C_5 \int_0^t \|(-\Delta+1)^\iota e^{(t-s)(\Delta-1)} \nabla \cdot (u_\varepsilon(s)c_\varepsilon(s))\|_{L^6(\Omega)} ds \tag{1.3.69}
$$

$$
\leq C_6 \int_0^t (t-s)^{-\iota-\frac{1}{2}-\tilde{\kappa}} e^{-\tilde{\lambda}(t-s)} \|u_\varepsilon(s)c_\varepsilon(s)\|_{L^\infty(\Omega)} ds
$$

$$
\leq C_7 \int_0^t (t-s)^{-\iota-\frac{1}{2}-\tilde{\kappa}} e^{-\tilde{\lambda}(t-s)} \|u_\varepsilon(s)\|_{L^\infty(\Omega)} \|c_\varepsilon(s)\|_{L^\infty(\Omega)} ds
$$

$$
\leq C_8 \quad \text{for all } t > 0.
$$

Here, we have used the fact that

$$
\int_0^t (t-s)^{-\iota-\frac{1}{2}-\tilde{\kappa}} e^{-\lambda(t-s)} ds \leq \int_0^\infty \sigma^{-\iota-\frac{1}{2}-\tilde{\kappa}} e^{-\lambda\sigma} d\sigma < +\infty.
$$

Combining with (1.3.66)–(1.3.69), we arrive at (1.3.65).

From the regularity property of solutions obtained above, we can infer the higher regularity about n_ε.

Lemma 1.32 *Assuming that $m + \alpha > \frac{10}{9}$. Then for all $p > 2$, there exists $C > 0$ such that*

$$
\|n_\varepsilon(\cdot, t) + \varepsilon\|_{L^p(\Omega)} \leq C \text{ for all } t > 0. \tag{1.3.70}
$$

Proof Recalling Lemmas 1.14 and by (1.31), we have

$$
\frac{1}{p}\frac{d}{dt}\|n_\varepsilon + \varepsilon\|_{L^p(\Omega)}^p + \frac{m(p-1)}{2}\int_\Omega (n_\varepsilon + \varepsilon)^{m+p-3} |\nabla n_\varepsilon|^2
$$

$$
\leq C_1 \int_\Omega (n_\varepsilon + \varepsilon)^{p+1-m}(1 + n_\varepsilon)^{-2\alpha} |\nabla c_\varepsilon|^2
$$

$$
\leq C_1 \left(\int_\Omega (n_\varepsilon + \varepsilon)^{3(p+1-m-2\alpha)} \right)^{\frac{1}{3}} \left(\int_\Omega |\nabla c_\varepsilon|^3 \right)^{\frac{2}{3}}
$$

$$
\leq C_2 \left(\int_\Omega (n_\varepsilon + \varepsilon)^{3(p+1-m-2\alpha)} \right)^{\frac{1}{3}} \quad \text{for all } t > 0
$$

for constants $C_1 > 0$ and $k_2 > 0$. By the Gagliardo–Nirenberg inequality, there exist positive constants $C_i > 0$, $(i = 3, 4, 5, 6)$ fulfilling

$$\left(\int_\Omega (n_\varepsilon + \varepsilon)^{3(p+1-m-2\alpha)}\right)^{\frac{1}{3}}$$

$$=\|(n_\varepsilon + \varepsilon)^{\frac{m+p-1}{2}}\|_{L^{\frac{6(p+1-m)}{m+p-1}}(\Omega)}^{\frac{2(p+1-m-2\alpha)}{m+p-1}}$$

$$\leq C_3 \|\nabla(n_\varepsilon + \varepsilon)^{\frac{m+p-1}{2}}\|_{L^2(\Omega)}^{2\frac{3p-3m+2-6\alpha}{3m+3p-4}} \|(n_\varepsilon + \varepsilon)^{\frac{m+p-1}{2}}\|_{L^{\frac{2}{m+p-1}}(\Omega)}^{\frac{2(p+1-m)}{m+p-1}-2\frac{3p-3m+2-6\alpha}{3m+3p-4}}$$

$$+ C_3 \|(n_\varepsilon + \varepsilon)^{\frac{m+p-1}{2}}\|_{L^{\frac{2}{m+p-1}}(\Omega)}^{\frac{2(p+1-m)}{m+p-1}}$$

$$\leq C_4 (\|\nabla(n_\varepsilon + \varepsilon)^{\frac{m+p-1}{2}}\|_{L^2(\Omega)}^{2\frac{3p-3m+2-6\alpha}{3m+3p-4}} + 1) \quad \text{for all } t > 0$$

and

$$\int_\Omega (n_\varepsilon + \varepsilon)^p$$

$$=\|(n_\varepsilon + \varepsilon)^{\frac{m+p-1}{2}}\|_{L^{\frac{2p}{m+p-1}}(\Omega)}^{\frac{2p}{m+p-1}}$$

$$\leq C_5 \|\nabla(n_\varepsilon + \varepsilon)^{\frac{m+p-1}{2}}\|_{L^2(\Omega)}^{2\frac{3p-3}{3m+3p-4}} \|(n_\varepsilon + \varepsilon)^{\frac{m+p-1}{2}}\|_{L^{\frac{2}{m+p-1}}(\Omega)}^{\frac{2p}{m+p-1}-2\frac{3p-3}{3m+3p-4}}$$

$$+ C_4 \|(n_\varepsilon + \varepsilon)^{\frac{m+p-1}{2}}\|_{L^{\frac{2}{m+p-1}}(\Omega)}^{\frac{2p}{m+p-1}}$$

$$\leq C_6 (\|\nabla(n_\varepsilon + \varepsilon)^{\frac{m+p-1}{2}}\|_{L^2(\Omega)}^{2\frac{3p-3}{3m+3p-4}} + 1) \quad \text{for all } t > 0.$$

With the help of $m + \alpha > \frac{10}{9}$, we have $\frac{3p-3m+2-6\alpha}{3m+3p-4} < 1$ and hence obtain that for some constant $C_7 > 0$

$$\frac{1}{p}\frac{d}{dt}\|n_\varepsilon + \varepsilon\|_{L^p(\Omega)}^p + C_7 \left(\int_\Omega (n_\varepsilon + \varepsilon)^p\right)^{\frac{3m+3p-4}{3p-3}} \leq C_7 \quad \text{for all } t > 0.$$

Therefore, (1.3.70) follows from the application of Lemma 1.4.

By applying the general semigroup estimates, the standard parabolic regularity arguments and a Moser-type iteration (see, e.g., Lemma A.1 of Tao and Winkler 2012a), we can now establish the existence of global bounded classical solutions to the regularized system (1.2.2).

Proposition 1.1 *Let $m + \alpha > \frac{10}{9}$. Then there exists $C > 0$ independent of $\varepsilon \in (0, 1)$ such that*

$$\|n_\varepsilon(\cdot, t)\|_{L^\infty(\Omega)} + \|c_\varepsilon(\cdot, t)\|_{W^{1,\infty}(\Omega)} + \|A^\gamma u_\varepsilon(\cdot, t)\|_{L^2(\Omega)} \leq C \quad \text{for all } t > 0. \tag{1.3.71}$$

Proof Let $h_\varepsilon(x, t) = \mathscr{P}[n_\varepsilon \nabla \phi]$. Then by (1.3.70), there is $C_1 > 0$ such that

$$\|h_\varepsilon(\cdot, t)\|_{L^2(\Omega)} \leq C_1 \text{ for all } t > 0.$$

So combining the known smoothing properties of the Stokes semigroup (see Giga 1986) with (1.1.4), there are positive constants C_2 and C_3 such that

$$\|A^\gamma u_\varepsilon(\cdot, t)\|_{L^2(\Omega)} \leq \|A^\gamma e^{-tA} u_0\|_{L^2(\Omega)} + \int_0^t \|A^\gamma e^{-(t-\tau)A} h_\varepsilon(\cdot, \tau) d\tau\|_{L^2(\Omega)} d\tau$$

$$\leq \|A^\gamma u_0\|_{L^2(\Omega)} + C_2 \int_0^t (t-\tau)^{-\gamma} e^{-\lambda(t-\tau)} \|h_\varepsilon(\cdot, \tau)\|_{L^2(\Omega)} d\tau$$

$$\leq C_3 \quad \text{for all } t > 0.$$

Next, we rewrite the variation-of-constants formula for c_ε in the form

$$c_\varepsilon(\cdot, t) = e^{t(\Delta-1)} c_0 + \int_0^t e^{(t-s)(\Delta-1)} (c_\varepsilon - n_\varepsilon c_\varepsilon - u_\varepsilon \cdot \nabla c_\varepsilon)(\cdot, s) ds \quad \text{for all } t > 0.$$

Due to $3 < r_0 < \min\{\frac{3q_0}{(3-q_0)_+}, 4\}$ (see Lemma 1.31), one can pick $\theta \in (\frac{1}{2} + \frac{3}{2r_0}, 1)$ and thereby the domain of the fractional power $D((-\Delta+1)^\theta) \hookrightarrow W^{1,\infty}(\Omega)$ (see Winkler 2010). Hence, in view of L^p-L^q estimates associated heat semigroup, Lemma 1.31 as well as (1.1.4), we conclude that there exist positive constants λ_1, C_4 as well as C_5 and C_6 such that

$$\|\nabla c_\varepsilon(\cdot, t)\|_{W^{1,\infty}(\Omega)}$$

$$\leq C_4 e^{-\lambda_1 t} \|\nabla c_0\|_{L^\infty(\Omega)}$$

$$+ \int_0^t (t-s)^{-\theta} e^{-\lambda_1(t-s)} \|(c_\varepsilon - n_\varepsilon c_\varepsilon - u_\varepsilon \cdot \nabla c_\varepsilon)(s)\|_{L^{r_0}(\Omega)} ds$$

$$\leq C_5 + C_5 \int_0^t (t-s)^{-\theta} e^{-\lambda_1(t-s)} ds \tag{1.3.72}$$

$$+ C_5 \int_0^t (t-s)^{-\theta} e^{-\lambda_1(t-s)} [\|n_\varepsilon(\cdot, s)\|_{L^{r_0}(\Omega)} + \|\nabla c_\varepsilon(\cdot, s)\|_{L^{r_0}(\Omega)}] ds$$

$$\leq C_6 \quad \text{for all } t \in (0, \infty).$$

Finally, we rewrite the first equation of (1.2.2) as

$$n_{\varepsilon t} = \Delta(n_\varepsilon + \varepsilon)^m - \nabla \cdot (n_\varepsilon u_\varepsilon + n_\varepsilon F_\varepsilon(n_\varepsilon) S_\varepsilon(x, n_\varepsilon, c_\varepsilon) \cdot \nabla c_\varepsilon), \tag{1.3.73}$$

Hence, in view of (1.3.72) and using the outcome of Lemma 1.32 with suitably large p as a starting point, we may invoke Lemma A.1 in Tao and Winkler (2012a) which by means of a Moser-type iteration applied to (1.3.73) and establish

$$\|n_\varepsilon(\cdot, t)\|_{L^\infty(\Omega)} \leq C_7 \quad \text{for all } t > 0 \tag{1.3.74}$$

with some positive constant C_7 independent of ε.

To achieve the convergence result, we still need the following further regularity estimate. With the help of Proposition 1.1, we can straightforwardly deduce the uniform Hölder properties of c_ε as well as ∇c_ε and u_ε by using the standard parabolic regularity property and the standard semigroup estimation techniques.

Lemma 1.33 *Let $m + \alpha > \frac{10}{9}$. Then one can find $\mu \in (0, 1)$ such that for some $C > 0$,*

$$\|c_\varepsilon(\cdot, t)\|_{C^{\mu, \frac{\mu}{2}}(\Omega \times [t, t+1])} \leq C \text{ for all } t \in (0, \infty)$$

as well as

$$\|u_\varepsilon(\cdot, t)\|_{C^{\mu, \frac{\mu}{2}}(\Omega \times [t, t+1])} \leq C \text{ for all } t \in (0, \infty),$$

and for any $\tau > 0$, there exists $C(\tau) > 0$ fulfilling

$$\|\nabla c_\varepsilon(\cdot, t)\|_{C^{\mu, \frac{\mu}{2}}(\Omega \times [t, t+1])} \leq C \text{ for all } t \in (\tau, \infty).$$

Proof Based on the uniform boundedness of $\{(n_\varepsilon, c_\varepsilon, u_\varepsilon)\}_{\varepsilon \in (0,1)}$ as claimed in Proposition 1.1 and the assumptions on ϕ, we conclude the desired estimates by applying the standard parabolic regularity theory (see, e.g., Ladyzenskaja et al. 1968) and some standard semigroup estimation techniques, which is omitted here.

Unlike c_ε and u_ε, we are not able to attain the Hölder regularity for n_ε due to the presence of nonlinear diffusion. We now make full use of the a priori bounds derived so far to obtain the boundedness property of the time derivatives of certain powers of n_ε and spatio-temporal integrability property of $\int_0^\infty \int_\Omega (n_\varepsilon + \varepsilon)^{m+p-3}|\nabla n_\varepsilon|^2$, which plays a key role in deriving strong compactness properties for n_ε. Let us provide the following spatio-temporal estimates at first.

Lemma 1.34 *Let $m + \alpha > \frac{10}{9}$. Then there exists a positive constant C such that for any $\varepsilon \in (0, 1)$*

$$\int_0^\infty \int_\Omega (n_\varepsilon + \varepsilon)^{m+p-3}|\nabla n_\varepsilon|^2 \leq C \text{ for all } p > 1 \text{ and } p \geq m + 2\alpha - 1. \quad (1.3.75)$$

Proof In light of Proposition 1.1, there exists $C_1 > 0$ such that for all $\varepsilon \in (0, 1)$, $n_\varepsilon \leq C_1$ in $\Omega \times (0, \infty)$. For any $p \geq m + 2\alpha - 1$ and $p > 1$, using Proposition 1.1, we can thereby estimate the integral on the right of (1.3.26) according to

$$\begin{aligned}
&\frac{1}{p}\|n_\varepsilon(\cdot, t) + \varepsilon\|_{L^p(\Omega)}^p + \frac{m(p-1)}{2}\int_0^t \int_\Omega (n_\varepsilon + \varepsilon)^{m+p-3}|\nabla n_\varepsilon|^2 \\
&\leq \frac{(p-1)C_S^2}{2m}\int_0^t \int_\Omega (n_\varepsilon + \varepsilon)^{p+1-m-2\alpha}|\nabla c_\varepsilon|^2 + \frac{1}{p}\|n_0 + \varepsilon\|_{L^p(\Omega)}^p \\
&\leq \frac{(p-1)C_S^2}{2m}(C_1 + 1)^{p+1-m-2\alpha}\int_0^\infty \int_\Omega |\nabla c_\varepsilon|^2 + \frac{1}{p}\|n_0 + 1\|_{L^p(\Omega)}^p \\
&\leq \frac{(p-1)C_S^2}{4m}(C_1 + 1)^{p+1-m-2\alpha}\int_\Omega c_0^2 + \frac{1}{p}\|n_0 + 1\|_{L^p(\Omega)}^p \text{ for all } t \in (0, \infty),
\end{aligned}$$

which immediately leads to our conclusion.

In order to pass to the limit in system (1.2.2) by compactness argument, we intend to supplement Proposition 1.1 with an appropriate boundedness property of the time derivatives of n_ε.

Lemma 1.35 *Let $m + \alpha > \frac{10}{9}$. Then one can find $C > 0$ such that for any $\varepsilon \in (0, 1)$*

$$\|\partial_t n_\varepsilon(\cdot, t)\|_{(W_0^{2,2}(\Omega))^*} \leq C \quad \text{for all } t \in (0, \infty). \tag{1.3.76}$$

In particular,

$$\|n_\varepsilon(\cdot, t) - n_\varepsilon(\cdot, s)\|_{(W_0^{2,2}(\Omega))^*} \leq C|t - s| \quad \text{for all } t \geq 0, s \geq 0 \quad \text{and } \varepsilon \in (0, 1). \tag{1.3.77}$$

Moreover, let $\varsigma > m$ and $\varsigma \geq 2(m - 1)$. Then for all $T > 0$ and $\varepsilon \in (0, 1)$, there exists a positive constant $C(T)$ such that

$$\int_0^T \|\partial_t (n_\varepsilon + \varepsilon)^\varsigma(\cdot, t)\|_{(W_0^{3,2}(\Omega))^*} dt \leq C(T) \quad \text{for all } \varepsilon \in (0, 1). \tag{1.3.78}$$

Proof To estimate the integrals on the right of (1.3.80) below appropriately, we first apply Proposition 1.1 to find C_1 such that

$$(n_\varepsilon + \varepsilon)^m \leq C_1, \quad n_\varepsilon \leq C_1 \quad \text{as well as} \quad |\nabla c_\varepsilon| \leq C_1 \quad \text{and} \quad |u_\varepsilon| \leq C_1 \quad \text{in } \Omega \times (0, \infty). \tag{1.3.79}$$

For any fixed $\psi \in C_0^\infty(\Omega)$, we multiply the first equation in (1.2.2) by $(n_\varepsilon + \varepsilon)^{\varsigma-1}\psi$ and then get

$$\frac{1}{\varsigma} \int_\Omega \partial_t (n_\varepsilon + \varepsilon)^\varsigma(\cdot, t) \cdot \psi$$

$$= \int_\Omega (n_\varepsilon + \varepsilon)^{\varsigma-1} \left[\Delta(n_\varepsilon + \varepsilon)^m - \nabla \cdot (n_\varepsilon S_\varepsilon(x, n_\varepsilon, c_\varepsilon)\nabla c_\varepsilon) - u_\varepsilon \cdot \nabla n_\varepsilon \right] \cdot \psi$$

$$= -(\varsigma - 1)m \int_\Omega (n_\varepsilon + \varepsilon)^{\varsigma-2}(n_\varepsilon + \varepsilon)^{m-1} |\nabla n_\varepsilon|^2 \psi$$

$$-m \int_\Omega (n_\varepsilon + \varepsilon)^{\varsigma-1}(n_\varepsilon + \varepsilon)^{m-1} \nabla n_\varepsilon \cdot \nabla \psi$$

$$+ (\varsigma - 1) \int_\Omega (n_\varepsilon + \varepsilon)^{\varsigma-1} \nabla n_\varepsilon \cdot (S_\varepsilon(x, n_\varepsilon, c_\varepsilon) \cdot \nabla c_\varepsilon)\psi$$

$$+ \int_\Omega (n_\varepsilon + \varepsilon)^\varsigma S_\varepsilon(x, n_\varepsilon, c_\varepsilon)\nabla c_\varepsilon \cdot \nabla \psi$$

$$+ \frac{1}{\varsigma} \int_\Omega (n_\varepsilon + \varepsilon)^\varsigma u_\varepsilon \cdot \nabla \psi \quad \text{for all } t \in (0, \infty). \tag{1.3.80}$$

In what follows, we shall estimate the right of the above equality appropriately by (1.3.79). Indeed, since $\varsigma > m$ and $\varsigma \geq 2(m + \alpha - 1)$, the number $p := \varsigma - m + 1$ satisfies $p > 1$ and $p \geq m + 2\alpha - 1$, so that, (1.3.75) becomes applicable so as to yield $C_3 > 0$ fulfilling

$$\int_0^\infty \int_\Omega (n_\varepsilon + \varepsilon)^{\varsigma-2} |\nabla n_\varepsilon|^2 = \int_0^\infty \int_\Omega (n_\varepsilon + \varepsilon)^{m+p-3} |\nabla n_\varepsilon|^2 \le C_3$$

for some positive constant C_3. Now, applying (1.3.79), we conclude from the Young inequality that

$$
\begin{aligned}
&-(\varsigma - 1)m \int_\Omega (n_\varepsilon + \varepsilon)^{\varsigma-2} (n_\varepsilon + \varepsilon)^{m-1} |\nabla n_\varepsilon|^2 \psi \\
&-m \int_\Omega (n_\varepsilon + \varepsilon)^{\varsigma-1} (n_\varepsilon + \varepsilon)^{m-1} \nabla n_\varepsilon \cdot \nabla \psi \\
&+(\varsigma - 1) \int_\Omega (n_\varepsilon + \varepsilon)^{\varsigma-1} \nabla n_\varepsilon \cdot (S_\varepsilon(x, n_\varepsilon, c_\varepsilon) \cdot \nabla c_\varepsilon) \psi \\
&+ \int_\Omega (n_\varepsilon + \varepsilon)^\varsigma S_\varepsilon(x, n_\varepsilon, c_\varepsilon) \nabla c_\varepsilon \cdot \nabla \psi + \frac{1}{\varsigma} \int_\Omega (n_\varepsilon + \varepsilon)^\varsigma u_\varepsilon \cdot \nabla \psi \\
&\le C_4(\varsigma - 1) \int_\Omega (n_\varepsilon + \varepsilon)^{\varsigma-2} |\nabla n_\varepsilon|^2 \|\psi\|_{L^\infty(\Omega)} \\
&+ C_5 \int_\Omega [(n_\varepsilon + \varepsilon)^{\varsigma-2} |\nabla n_\varepsilon|^2 + (C_1 + 1)^{\varsigma-1}] \|\nabla \psi\|_{L^\infty(\Omega)} \\
&+ C_6 \int_\Omega [(n_\varepsilon + \varepsilon)^{\varsigma-2} |\nabla n_\varepsilon|^2 + C_1^\varsigma] \|\psi\|_{L^\infty(\Omega)} \\
&+ C_1^{\varsigma+1} C_S |\Omega| \|\nabla \psi\|_{L^\infty(\Omega)} + \frac{1}{\varsigma} C_1^{\varsigma+1} |\Omega| \|\psi\|_{L^\infty(\Omega)}
\end{aligned}
\tag{1.3.81}
$$

with some positive constants C_4 as well as C_5 and C_6. Inserting (1.3.81) into (1.3.80), we derive that there is $C_7 > 0$ such that for all $t > 0$ and any $\varepsilon \in (0, 1)$,

$$\left| \int_\Omega \partial_t (n_\varepsilon + \varepsilon)^\varsigma (\cdot, t) \cdot \psi \right| \le C_7 \left(\int_\Omega (n_\varepsilon + \varepsilon)^{\varsigma-2} |\nabla n_\varepsilon|^2 + 1 \right) \|\psi\|_{W^{1,\infty}(\Omega)}.$$

As in the three-dimensional space, we have $W_0^{3,2}(\Omega) \hookrightarrow W^{1,\infty}(\Omega)$. Collecting the above inequalities, we infer the existence of $C_8 > 0$ such that for any $\varepsilon \in (0, 1)$,

$$\|\partial_t (n_\varepsilon + \varepsilon)^\varsigma (\cdot, t)\|_{(W_0^{3,2}(\Omega))^*} \le C_8 \left(\int_\Omega (n_\varepsilon + \varepsilon)^{\varsigma-2} |\nabla n_\varepsilon|^2 + 1 \right) \text{ for all } t \in (0, \infty).$$

Therefore, we obtain the desired estimate (1.3.78).

Testing the first equation in (1.2.2) by an arbitrary $\varphi \in C_0^\infty(\Omega)$, we have

$$
\begin{aligned}
\int_\Omega n_{\varepsilon,t}(\cdot, t) \cdot \varphi &= \int_\Omega \left[\Delta(n_\varepsilon + \varepsilon)^m - \nabla \cdot (n_\varepsilon S_\varepsilon(x, n_\varepsilon, c_\varepsilon) \nabla c_\varepsilon) - u_\varepsilon \cdot \nabla n_\varepsilon \right] \cdot \varphi \\
&= \int_\Omega (n_\varepsilon + \varepsilon)^m \Delta \varphi + \int_\Omega n_\varepsilon S_\varepsilon(x, n_\varepsilon, c_\varepsilon) \nabla c_\varepsilon \cdot \nabla \varphi + \int_\Omega n_\varepsilon u_\varepsilon \cdot \nabla \varphi
\end{aligned}
$$

for all $t \in (0, \infty)$. Then combining this with (1.3.79) as well as (1.1.3), we get

$$| \int_\Omega n_{\varepsilon,t}(\cdot, t) \cdot \varphi | \le C_9 [\int_\Omega |\Delta\varphi| + \int_\Omega |\nabla\varphi|] \text{ in } \Omega \times (0, \infty)$$

for all $\varepsilon \in (0, 1)$ with some positive constant C_9, which establishes implies (1.3.76) and thus also (1.3.77).

1.3.4 Global Boundedness of Weak Solutions

The a-priori estimates achieved so far allow us to construct weak solutions by compactness arguments. To this end, let us define what a weak solution is supposed to be.

Definition 1.1 (*Weak solutions*) By a global weak solution of (1.1.1), we mean a triple (n, c, u) of functions

$$\begin{cases} n \in L^1_{loc}(\bar{\Omega} \times [0, \infty)), \\ c \in L^1_{loc}([0, \infty); W^{1,1}(\Omega)), \\ u \in L^1_{loc}([0, \infty); W^{1,1}_0(\Omega; \mathbb{R}^3)), \end{cases}$$

such that $n \ge 0$ and $c \ge 0$ a.e. in $\Omega \times (0, \infty)$,

$$nc, n^m \in L^1_{loc}(\bar{\Omega} \times [0, \infty)), \quad u \otimes u \in L^1_{loc}(\bar{\Omega} \times [0, \infty); \mathbb{R}^{3\times3}), \quad \text{and}$$
$$nS(x, n, c)\nabla c, \quad cu \text{ and } nu \text{ belong to } L^1_{loc}(\bar{\Omega} \times [0, \infty); \mathbb{R}^3),$$

$\nabla \cdot u = 0$ a.e. in $\Omega \times (0, \infty)$, and

$$-\int_0^T \int_\Omega n\varphi_t - \int_\Omega n_0\varphi(\cdot, 0)$$
$$= \int_0^T \int_\Omega n^m \Delta\varphi + \int_0^T \int_\Omega n(S(x, n, c) \cdot \nabla c) \cdot \nabla\varphi + \int_0^T \int_\Omega nu \cdot \nabla\varphi$$

for any $\varphi \in C_0^\infty(\bar{\Omega} \times [0, \infty))$ as well as

$$-\int_0^T \int_\Omega c\varphi_t - \int_\Omega c_0\varphi(\cdot, 0) = -\int_0^T \int_\Omega \nabla c \cdot \nabla\varphi - \int_0^T \int_\Omega nc \cdot \varphi + \int_0^T \int_\Omega cu \cdot \nabla\varphi$$

for any $\varphi \in C_0^\infty(\bar{\Omega} \times [0, \infty))$ and

$$-\int_0^T \int_\Omega u\varphi_t - \int_\Omega u_0\varphi(\cdot, 0) = -\int_0^T \int_\Omega \nabla u \cdot \nabla\varphi - \int_0^T \int_\Omega n\nabla\phi \cdot \varphi$$

for any $\varphi \in C_0^\infty(\Omega \times [0, \infty); \mathbb{R}^3)$ fulfilling $\nabla \cdot \varphi \equiv 0$.

The Proof of Theorem 1.1 We first give a series of convergence results. According to Lemma 1.33, the Arzelà-Ascoli theorem and a standard extraction procedure, we can find a sequence $(\varepsilon_j)_{j \in \mathbb{N}} \subseteq (0, 1)$ with $\varepsilon_j \searrow 0$ as $j \to \infty$ such that

$$c_{\varepsilon_j} \to c \text{ in } C^0_{loc}(\bar{\Omega} \times [0, \infty)), \qquad (1.3.82)$$

$$\nabla c_{\varepsilon_j} \to \nabla c \text{ in } C^0_{loc}(\bar{\Omega} \times [0, \infty)), \qquad (1.3.83)$$

and

$$u_{\varepsilon_j} \to u \text{ in } C^0_{loc}(\bar{\Omega} \times (0, \infty)) \qquad (1.3.84)$$

hold with some limit functions c and u belonging to the indicated spaces. On the other hand, Proposition 1.1 ensures the existence of a subsequence such that

$$\nabla c_{\varepsilon_j} \rightharpoonup \nabla c \text{ weakly star in } L^\infty(\Omega \times (0, \infty)), \qquad (1.3.85)$$

$$Du_{\varepsilon_j} \rightharpoonup Du \text{ weakly star in } L^\infty(\Omega \times [0, \infty)), \qquad (1.3.86)$$

and

$$n_{\varepsilon_j} \rightharpoonup n \text{ weakly star in } L^\infty(\Omega \times (0, \infty)) \qquad (1.3.87)$$

hold for some $n \in L^\infty(\Omega \times (0, \infty))$.

Fix $\zeta > m - 1$. Then Lemmas 1.34 and 1.35 assert that for any $T > 0$,

$$((n_\varepsilon + \varepsilon)^\zeta)_{\varepsilon \in (0,1)} \text{ is bounded in } L^2((0, T); W^{1,2}(\Omega))$$

and

$$(\partial_t(n_\varepsilon + \varepsilon)^\zeta)_{\varepsilon \in (0,1)} \text{ is bounded in } L^1((0, T); (W^{3,2}_0(\Omega))^*)$$

respectively. So the embedding $W^{1,2}(\Omega) \hookrightarrow\hookrightarrow L^2(\Omega) \hookrightarrow (W^{3,2}_0(\Omega))^*$ and the Aubin–Lions compactness lemma yield that $(n_\varepsilon + \varepsilon)^\zeta_{\varepsilon \in (0,1)}$ is a relatively compact subset of the space $L^2(\Omega \times (0, T))$. This in conjunction with the Egorov theorem gives that for some subsequence of $\varepsilon = \varepsilon_j$,

$$(n_\varepsilon + \varepsilon)^\zeta \to n^\zeta \text{ strongly in } L^2(\Omega \times (0, T))$$

and hence

$$(n_\varepsilon + \varepsilon) \to z \text{ a.e. in } \Omega \times (0, T),$$

for some nonnegative measurable $z : \Omega \times (0, T) \to \mathbb{R}$. This combined with the Egorov theorem, then we can see that $z = n$, and thereby

$$n_\varepsilon \to n \text{ a.e. in } \Omega \text{ for all } (0, \infty) \setminus N. \qquad (1.3.88)$$

Next, noticing that $L^\infty(\Omega) \hookrightarrow (W_0^{2,2}(\Omega))^*$ is compact, in view of Proposition 1.1 and Lemma 1.35, we can use the Arzelà-Ascoli theorem again to assert

$$n_\varepsilon \to n \quad \text{in} \quad C_{loc}^0([0, \infty); (W_0^{2,2}(\Omega))^*). \tag{1.3.89}$$

With the help of (1.3.89) and the fact that $\|n\|_{L^\infty(\Omega \times (0,\infty))}$ is finite, we can derive

$$n \in C_{\omega-*}^0([0, \infty); L^\infty(\Omega)) \tag{1.3.90}$$

by using the similar methods in the proof of Lemma 4.1 in Winkler (2015b).

Combining (1.3.83) with (1.3.88), noticing the definition of S_ε, we may further infer that

$$n_\varepsilon S_\varepsilon(x, n_\varepsilon, c_\varepsilon) \cdot \nabla c_\varepsilon \to n S(x, n, c) \cdot \nabla c \quad \text{a.e. in} \quad \Omega \times (0, \infty) \quad \text{as} \quad \varepsilon := \varepsilon_j \searrow 0.$$

Then we may use the dominated convergence theorem, along with a subsequence (still denoted by $\{\varepsilon_j\}_{j=1}^\infty$), we derive that

$$n_\varepsilon S_\varepsilon(x, n_\varepsilon, c_\varepsilon) \cdot \nabla c_\varepsilon \to n S(x, n, c) \cdot \nabla c \quad \text{strongly in} \quad L_{loc}^2(\bar{\Omega} \times [0, \infty)) \quad \text{as} \quad \varepsilon := \varepsilon_j \searrow 0. \tag{1.3.91}$$

In the following, we shall show that the triple (n, c, u) is exactly a global weak solution to system (1.1.1). Indeed, multiplying the first equation in (1.2.2) by $\varphi \in C_0^\infty(\bar{\Omega} \times [0, \infty))$, integrating by parts, we obtain

$$
\begin{aligned}
&- \int_0^\infty \int_\Omega n_\varepsilon \varphi_t - \int_\Omega n_0 \varphi(\cdot, 0) \\
&= \int_0^\infty \int_\Omega (n_\varepsilon + \varepsilon)^m \Delta\varphi + \int_0^\infty \int_\Omega n_\varepsilon (S_\varepsilon(x, n_\varepsilon, c_\varepsilon) \cdot \nabla c_\varepsilon) \cdot \nabla\varphi \\
&\quad + \int_0^\infty \int_\Omega n_\varepsilon u_\varepsilon \cdot \nabla\varphi.
\end{aligned}
$$

In view of (1.3.89), (1.3.91) as well as (1.3.84), we conclude from the dominated convergence theorem that

$$
\begin{aligned}
&- \int_0^\infty \int_\Omega n \varphi_t - \int_\Omega n_0 \varphi(\cdot, 0) \\
&= \int_0^\infty \int_\Omega n^m \Delta\varphi + \int_0^\infty \int_\Omega n(S(x, n, c) \cdot \nabla c) \cdot \nabla\varphi \\
&\quad + \int_0^\infty \int_\Omega n u \cdot \nabla\varphi.
\end{aligned}
$$

Next, multiplying the second equation and the third equation in (1.2.2) by $\varphi \in C_0^\infty(\Omega \times [0, \infty))$ and $\psi \in C_0^\infty(\bar{\Omega} \times [0, \infty); \mathbb{R}^3)$, respectively, then with the help of (1.3.85)–(1.3.86) and by a limit procedure, we also derive that

$$-\int_0^\infty \int_\Omega c\varphi_t - \int_\Omega c_0\varphi(\cdot,0)$$
$$=-\int_0^\infty \int_\Omega \nabla c \cdot \nabla\varphi - \int_0^\infty \int_\Omega nc\varphi + \int_0^\infty \int_\Omega cu \cdot \nabla\varphi$$

and

$$-\int_0^\infty \int_\Omega u\varphi_t - \int_\Omega u_0\varphi(\cdot,0) = -\int_0^\infty \int_\Omega \nabla u \cdot \nabla\varphi - \int_0^\infty \int_\Omega n\nabla\phi \cdot \varphi$$

in a completed similar manner. This means that (n, c, u) is a weak solution of (1.1.1). The convergence properties in (1.3.82)–(1.3.89) lead to the stated boundedness of global weak solutions thereof, and thus complete the proof of Theorem 1.1.

1.4 Asymptotic Profile of Solution to a Chemotaxis–Fluid System with Singular Sensitivity

1.4.1 Basic a Priori Bounds

In order to derive some essential estimates, it would be more convenient to deal with a nonsingular chemotaxis term of the form $\nabla \cdot (n\nabla w)$ instead of $\nabla \cdot (\frac{n}{c}\nabla c)$ in (1.1.5). To this end, we employ the following transformation as in Lankeit and Lankeit (2019a), Lankeit (2017), Winkler (2016a): $w := -\ln(\frac{c}{\|c_0\|_{L^\infty(\Omega)}})$, whereupon $0 \le w \in C^0(\bar{\Omega} \times (0,\infty)) \cap C^{2,1}(\bar{\Omega} \times (0,\infty))$, and the problem (1.1.5), (1.1.8), (1.1.9) transforms to

$$
\begin{cases}
n_t + u \cdot \nabla n = \Delta n + \chi\nabla \cdot (n\nabla w) + n(r - \mu n), & x \in \Omega, t > 0, \\
w_t + u \cdot \nabla w = \Delta w - |\nabla w|^2 + n, & x \in \Omega, t > 0, \\
u_t + (u \cdot \nabla)u = \Delta u + \nabla P + n\nabla\phi, & x \in \Omega, t > 0, \\
\nabla \cdot u = 0, & x \in \Omega, t > 0, \\
\nabla n \cdot v = \nabla w \cdot v = 0, \quad u = 0, & x \in \partial\Omega, t > 0, \\
n(x,0) = n_0(x), \; w(x,0) = -\ln(\dfrac{c_0(x)}{\|c_0\|_{L^\infty(\Omega)}}), \; u = u_0(x), & x \in \Omega.
\end{cases}
$$
$$(1.4.1)$$

Let us first recall some basic but important information about (n, w) due to the presence of the quadratic degradation term in the first equation of (1.4.1).

Lemma 1.36 *The classical solution* (n, w, u, P) *of (1.4.1) satisfies*

$(i)\ \limsup\limits_{t\to\infty} \|n(\cdot,t)\|_{L^1(\Omega)} \le \dfrac{|\Omega|r_+}{\mu};$

(ii) $\int_{t_0}^t \|n(\cdot, s)\|_{L^2(\Omega)}^2 ds \leq \dfrac{r_+}{\mu} \int_{t_0}^t \|n(\cdot, s)\|_{L^1(\Omega)} ds + \dfrac{1}{\mu} \|n(\cdot, t_0)\|_{L^1(\Omega)}$ for all $t >$ t_0;

(iii) $\int_{t_0}^t \int_{\Omega} |\nabla w|^2 dx ds \leq \int_{\Omega} w(x, t_0) dx + \int_{t_0}^t \|n(\cdot, s)\|_{L^1(\Omega)} ds$ for all $t > t_0$.
In particular, if $r \leq 0$, then

$$\|n(\cdot, t)\|_{L^1(\Omega)} \leq \frac{|\Omega|}{\mu(t + \gamma)} \quad \text{for all } t > t_0 \tag{1.4.2}$$

with $\gamma = \dfrac{|\Omega|}{\mu \int_{\Omega} n_0(x) dx}$.

Proof Integrating the first equation in (1.4.1) and using the Cauchy–Schwarz inequality, we get

$$\frac{d}{dt} \int_{\Omega} n = r \int_{\Omega} n - \mu \int_{\Omega} n^2 \leq r_+ \int_{\Omega} n - \frac{\mu}{|\Omega|} \left(\int_{\Omega} n \right)^2 \tag{1.4.3}$$

which yields (i) readily. By the time integration of (1.4.3) over (t_0, t), we get (ii) immediately. In addition, from the second equation in (1.4.1), $\nabla \cdot u = 0$ and $u = 0$ on $\partial \Omega$, it follows that

$$\frac{d}{dt} \int_{\Omega} w = - \int_{\Omega} |\nabla w|^2 + \int_{\Omega} n, \tag{1.4.4}$$

and thus establishes (iii).

When $r \leq 0$, it follows from (1.4.3) that

$$\frac{d}{dt} \int_{\Omega} n \leq - \frac{\mu}{|\Omega|} \left(\int_{\Omega} n \right)^2 \tag{1.4.5}$$

which then yields (1.4.2) by the time integration.

In order to make use of the spatio-temporal properties provided by Lemma 1.36(ii) to estimate the ultimate bound of $\int_{\Omega} |\nabla u|^2$, we shall utilize the following elementary lemma (see Lemma 3.4 of Winkler 2019a):

Lemma 1.37 Let $t_0 \geq 0$, $T \in (t_0, \infty]$, $a > 0$ and $b > 0$, and suppose that the nonnegative function $h \in L^1_{loc}(\mathbb{R})$ satisfies $\int_t^{t+1} h(s) ds \leq b$ for all $t \in [t_0, T]$. If $y \in C^0([t_0, T)) \cap C^1([t_0, T))$ has the property that $y'(t) + a y(t) \leq h(t)$ for all $t \in (t_0, T)$, then $y(t) \leq e^{-a(t-t_0)} y(t_0) + \frac{b}{1-e^{-a}}$ for all $t \in [t_0, T)$.

With Lemmas 1.36 and 1.37 at hand, we can employ the standard energy inequality associated with the fluid evolution system in (1.4.1) to derive some boundedness results for u.

Lemma 1.38 *For the global classical solution* (n, w, u) *of* (1.4.1), *we have*
(i) if $r > 0$, *then*

$$\limsup_{t \to \infty} \|u(\cdot, t)\|_{L^2(\Omega)}^2 \leq \frac{3(1 + r)|\Omega|}{\mu} \frac{\|\nabla \phi\|_{L^\infty(\Omega)}^2}{C_p(1 - e^{-\frac{C_p}{2}})} \frac{r}{\mu} \qquad (1.4.6)$$

as well as

$$\limsup_{t \to \infty} \int_t^{t+1} \|\nabla u(\cdot, s)\|_{L^2(\Omega)}^2 ds \leq \frac{5(1 + r)|\Omega|}{\mu} \frac{\|\nabla \phi\|_{L^\infty(\Omega)}^2}{C_p(1 - e^{-\frac{C_p}{2}})} \frac{r}{\mu} \qquad (1.4.7)$$

with Poincaré constant $C_P > 0$.
(ii) if $r \leq 0$, *then*

$$\int_\Omega |u(\cdot, t)|^2 \leq \|u(\cdot, t_0)\|_{L^2(\Omega)}^2 e^{-\frac{C_p}{2}(t - t_0)} + \frac{2|\Omega|}{\mu^2} \frac{\|\nabla \phi\|_{L^\infty(\Omega)}^2}{C_p(1 - e^{-\frac{C_p}{2}})} \frac{1}{t_0 + \gamma} \quad \text{for all } t > t_0$$
$$(1.4.8)$$

as well as

$$\int_t^{t+1} \|\nabla u(\cdot, s)\|_{L^2(\Omega)}^2 ds$$
$$\leq \|u(\cdot, t_0)\|_{L^2(\Omega)}^2 e^{-\frac{C_p}{2}(t - t_0)} + \frac{4|\Omega|}{\mu^2} \frac{\|\nabla \phi\|_{L^\infty(\Omega)}^2}{C_p(1 - e^{-\frac{C_p}{2}})} \frac{1}{t_0 + \gamma} \quad \text{for all } t > t_0. \quad (1.4.9)$$

Proof (i) According to the Poincaré inequality, one can find some constant $C_p > 0$ such that $C_p \int_\Omega |u|^2 \leq \int_\Omega |\nabla u|^2$. Testing the third equation in (1.4.1) by u and using the Hölder inequality, we obtain

$$\frac{d}{dt} \int_\Omega |u|^2 + C_p \int_\Omega |u|^2 + \int_\Omega |\nabla u|^2$$
$$\leq 2 \int_\Omega n \nabla \phi \cdot u \leq 2 \|\nabla \phi\|_{L^\infty(\Omega)} \|n\|_{L^2(\Omega)} \|u\|_{L^2(\Omega)}$$
$$\leq \frac{C_p}{2} \|u\|_{L^2(\Omega)}^2 + \frac{2}{C_p} \|\nabla \phi\|_{L^\infty(\Omega)}^2 \|n\|_{L^2(\Omega)}^2,$$

due to $u|_{\partial \Omega} = 0$ and $\nabla \cdot u = 0$.
Writing $h(t) = \frac{2}{C_p} \|\nabla \phi\|_{L^\infty(\Omega)}^2 \|n(\cdot, t)\|_{L^2(\Omega)}^2$, we see that $y(t) := \int_\Omega |u(\cdot, t)|^2$ satisfies

$$y'(t) + \frac{C_p}{2} y(t) + \int_\Omega |\nabla u(\cdot, t)|^2 \leq h(t) \quad \text{for all } t > 0. \qquad (1.4.10)$$

In view of Lemma 1.36 (i) and (ii), we know that

$$\limsup_{t\to\infty} \int_t^{t+1} h(s)\,ds \le \frac{2}{C_p}\|\nabla\phi\|_{L^\infty(\Omega)}^2 \frac{(1+r)|\Omega|}{\mu}\frac{r}{\mu}. \tag{1.4.11}$$

An application of Lemma 1.37 thus shows that there exists positive $t_0 > 0$ such that

$$\int_\Omega |u(\cdot,t)|^2 \le \|u(\cdot,t_0)\|_{L^2(\Omega)}^2 e^{-\frac{C_p}{2}(t-t_0)} + \frac{3(1+r)|\Omega|}{\mu}\frac{\|\nabla\phi\|_{L^\infty(\Omega)}^2}{C_p(1-e^{-\frac{C_p}{2}})}\frac{r}{\mu} \quad \text{for all } t > t_0$$

and thereby verifies (1.4.6). Thereafter, again thanks to (1.4.11), an integration of (1.4.10) in time yields (1.4.7).

(ii) In view of (1.4.2), we have

$$\int_t^{t+1} h(s)\,ds \le \frac{2}{C_p}\|\nabla\phi\|_{L^\infty(\Omega)}^2 \frac{|\Omega|}{\mu^2}\frac{1}{t+\gamma}, \tag{1.4.12}$$

whereupon Lemma 1.37 guarantees that

$$\int_\Omega |u(\cdot,t)|^2 \le \|u(\cdot,t_0)\|_{L^2(\Omega)}^2 e^{-\frac{C_p}{2}(t-t_0)} + \frac{2|\Omega|}{\mu^2}\frac{\|\nabla\phi\|_{L^\infty(\Omega)}^2}{C_p(1-e^{-\frac{C_p}{2}})}\frac{1}{t_0+\gamma} \quad \text{for all } t > t_0.$$

This precisely warrants (1.4.8), and thereby in turn yields (1.4.9) after integrating (1.4.10) over $(t, t+1)$ and once more employing (1.4.12).

Now by a further testing procedure, we can turn the above information into the estimate of $\|\nabla u(\cdot,t)\|_{L^2(\Omega)}$, particularly its decay in the case of $r = 0$, on the basis of an interpolation argument, which is inspired by an approach illustrated in section 3.2 of Tao and Winkler (2016).

Lemma 1.39 *For the global classical solution (n, w, u, P) of (1.4.1), we have*

(i) if $r > 0$, then there exists $\mu_1 := \mu_1(\Omega, r) > 0$ such that for all $\mu > \mu_1$,

$$\limsup_{t\to\infty} \|\nabla u(\cdot,t)\|_{L^2(\Omega)} \le \frac{1}{17K_1|\Omega|} \tag{1.4.13}$$

(ii) if $r \le 0$, then for any $\mu > 0$,

$$\lim_{t\to\infty} \|\nabla u(\cdot,t)\|_{L^2(\Omega)} = 0. \tag{1.4.14}$$

Proof Applying the Helmholtz projector $\mathscr{P}$ to the third equation in (1.4.1), multiplying the resulting identity $u_t + Au = -\mathscr{P}[(u\cdot\nabla)u] + \mathscr{P}[n\nabla\phi]$ by Au, and using the Gagliardo–Nirenberg inequality, we can find $C_1 > 0$ such that

$$\frac{1}{2}\frac{d}{dt}\int_\Omega |\nabla u|^2 + \int_\Omega |Au|^2$$

$$= -\int_\Omega \mathscr{P}[(u\cdot\nabla)u]\cdot Au + \int_\Omega \mathscr{P}[n\nabla\phi]\cdot Au$$

$$\leq \frac{1}{2}\int_\Omega |Au|^2 + \int_\Omega |(u\cdot\nabla)u|^2 + \|\nabla\phi\|^2_{L^\infty(\Omega)}\int_\Omega n^2$$

$$\leq \frac{1}{2}\int_\Omega |Au|^2 + \|u\|^2_{L^\infty(\Omega)}\|\nabla u\|^2_{L^2(\Omega)} + \|\nabla\phi\|^2_{L^\infty(\Omega)}\int_\Omega n^2$$

$$\leq \frac{1}{2}\int_\Omega |Au|^2 + C_1\|Au\|_{L^2(\Omega)}\|u\|_{L^2(\Omega)}\|\nabla u\|^2_{L^2(\Omega)} + \|\nabla\phi\|^2_{L^\infty(\Omega)}\int_\Omega n^2$$

$$\leq \int_\Omega |Au|^2 + \frac{C_1^2}{2}\|u\|^2_{L^2(\Omega)}\|\nabla u\|^4_{L^2(\Omega)} + \|\nabla\phi\|^2_{L^\infty(\Omega)}\int_\Omega n^2,$$

which entails $y(t) := \int_\Omega |\nabla u(\cdot,t)|^2$ satisfies

$$y'(t) \leq h_1(t)y(t) + h_2(t) \quad \text{for all } t > 0 \tag{1.4.15}$$

with $h_1(t)=C_1^2\|u(\cdot,t)\|^2_{L^2(\Omega)}\|\nabla u(\cdot,t)\|^2_{L^2(\Omega)}$ and $h_2(t) = 2\|\nabla\phi\|^2_{L^\infty(\Omega)}\|n(\cdot,t)\|^2_{L^2(\Omega)}$.

(i) In order to prepare the integration of (1.4.15), we may use Lemma 1.38 (i) to find some $t_0 > 0$ such that

$$\|u(\cdot,t)\|^2_{L^2(\Omega)} \leq C_2 := \frac{3(1+r)|\Omega|}{\mu}\frac{\|\nabla\phi\|^2_{L^\infty(\Omega)}}{C_p(1-e^{-\frac{C_p}{2}})}\frac{r}{\mu}$$

and $\int_{t-1}^t \|\nabla u(\cdot,s)\|^2_{L^2(\Omega)}ds \leq 2C_2$ for all $t > t_0 + 1$.

Hence for any $t > t_0 + 1$, we can find $t_* = t_*(t) \in [t-1,t)$ such that

$$\|\nabla u(\cdot,t_*)\|^2_{L^2(\Omega)} \leq 2C_2, \tag{1.4.16}$$

and then integrating (1.4.15) over (t_*,t) yields

$$y(t) \leq y(t_*)e^{\int_{t_*}^t h_1(\sigma)d\sigma} + \int_{t_*}^t e^{\int_s^t h_1(\sigma)d\sigma}h_2(s)ds \leq (2+C_p)C_2e^{2C_1^2C_2^2}$$

and thereby verifies (1.4.13).

(ii) For any $t_0 > 1$ and $t > t_0 + 2$, we use Lemma 1.38 (ii) to pick $t_* = t_*(t) \in [t-1,t)$ fulfilling

$$\|\nabla u(\cdot,t_*)\|^2_{L^2(\Omega)} = \int_{t-1}^t \|\nabla u(\cdot,s)\|^2_{L^2(\Omega)}ds$$

$$\leq \|u(\cdot,t_0)\|^2_{L^2(\Omega)}e^{-\frac{C_p}{2}(t-1-t_0)} + \frac{4|\Omega|}{\mu^2}\frac{\|\nabla\phi\|^2_{L^\infty(\Omega)}}{C_p(1-e^{-\frac{C_p}{2}})}\frac{1}{t_0+\gamma},$$

as well as

$$\int_{t-1}^{t} h_1(\sigma)d\sigma \leq C_1^2 \max_{t-1 \leq s \leq t} \|u(\cdot, s)\|_{L^2(\Omega)}^2 \int_{t-1}^{t} \|\nabla u(\cdot, s)\|_{L^2(\Omega)}^2 ds$$

$$\leq C_1^2 (\|u(\cdot, t_0)\|_{L^2(\Omega)}^2 e^{-\frac{C_p}{2}(t-1-t_0)} + \frac{4|\Omega|}{\mu^2} \frac{\|\nabla\phi\|_{L^\infty(\Omega)}^2}{C_p(1 - e^{-\frac{C_p}{2}})} \frac{1}{t_0 + \gamma})^2.$$

In addition, by (1.4.12) we also have

$$\int_{t-1}^{t} h_2(\sigma)d\sigma = 2\|\nabla\phi\|_{L^\infty(\Omega)}^2 \int_{t-1}^{t} \|n(\cdot, s)\|_{L^2(\Omega)}^2 ds \leq 2\|\nabla\phi\|_{L^\infty(\Omega)}^2 \frac{|\Omega|}{\mu^2} \frac{1}{t-1+\gamma}.$$

Therefore combining the above inequalities, (1.4.15) implies that

$$y(t) \leq y(t_*) e^{\int_{t-1}^{t} h_1(\sigma)d\sigma} + e^{\int_{t-1}^{t} h_1(\sigma)d\sigma} \int_{t-1}^{t} h_2(s)ds$$

and thus (1.4.14) holds readily.

1.4.2 Global Boundedness of Solutions

In this party, we show that the classical solution of problem (1.4.1) is globally bounded in the cases of $r > 0$ and $r \leq 0$, respectively.

1. The Case $r > 0$

In this subsection, we derive the global boundedness of solutions to (1.4.1) whenever μ is suitably large compared with r. As in Winkler (2016c), the main idea is to examine the behavior of the functional

$$\mathscr{F}(n, w) := \int_{\Omega} H(n) + \frac{\chi}{2} \int_{\Omega} |\nabla w|^2 \tag{1.4.17}$$

where $H(s) := s \ln \frac{\mu s}{er} + \frac{r}{\mu}$, along trajectories of the boundary value problem (1.4.1).

The following elementary property of $H(n)$ will be used in the sequel.

Lemma 1.40 *For all nonnegative function $n \in C(\bar{\Omega})$, $H(n) \geq 0$.*

Proof It is easy to verify that $H(\frac{r}{\mu}) = 0$, $H'(\frac{r}{\mu}) = 0$ and $H''(s) = \frac{1}{s} \geq 0$, which implies $H(n) \geq 0$ for all $n \geq 0$.

Now we can describe the evolution of $\mathscr{F}(n, w)$ along the trajectories of (1.4.1) by the standard testing procedure.

Lemma 1.41 *Let $\Omega \subset \mathbb{R}^2$ be a smooth bounded domain and (n, w, u) be the global classical solution of (1.4.1) with $r > 0$, $\mu > 0$. Then whenever $\mu > \mu_2(\Omega, \chi, r) := \max\{\mu_1, \frac{K_1(36+16\chi)|\Omega|}{\chi} r\}$, there exists $t_* > 0$ such that*

$$\frac{d}{dt}\mathscr{F}(n, w) \le 0 \ \text{for all} \ t \ge t_*. \tag{1.4.18}$$

Proof Multiplying the first equation in (1.4.1) by $H'(n)$ and integrating by parts, we get

$$\begin{aligned}
\frac{d}{dt}\int_\Omega H(n) &= \int_\Omega H'(n)(\triangle n + \chi \nabla \cdot (n\nabla w) + rn - \mu n^2 - u \cdot \nabla n) \\
&= -\int_\Omega H''(n)(|\nabla n|^2 + \chi n\nabla n \cdot \nabla w) + \int_\Omega H'(n)(rn - \mu n^2) \\
&= -\int_\Omega \frac{|\nabla n|^2}{n} - \chi \int_\Omega \nabla n \cdot \nabla w + \int_\Omega (\ln n - \ln \frac{r}{\mu})(rn - \mu n^2) \\
&\le -\int_\Omega \frac{|\nabla n|^2}{n} - \chi \int_\Omega \nabla n \cdot \nabla w
\end{aligned}$$

$$\tag{1.4.19}$$

due to $(\ln n - \ln \frac{r}{\mu})(rn - \mu n^2) \le 0$, $\nabla \cdot u = 0$ and $u = 0$ on $\partial\Omega$.

On the other hand, testing the second equation in (1.4.1) by $-\triangle w$, using $\nabla \cdot u = 0$ and $u = 0$ on $\partial\Omega$ again, we can obtain

$$\begin{aligned}
&\frac{1}{2}\frac{d}{dt}\int_\Omega |\nabla w|^2 + \int_\Omega |\triangle w|^2 \\
&= \int_\Omega |\nabla w|^2 \triangle w + \int_\Omega \nabla n \cdot \nabla w + \int_\Omega (u \cdot \nabla w)\triangle w \\
&\le \frac{1}{2}\int_\Omega |\triangle w|^2 + \frac{1}{2}\int_\Omega |\nabla w|^4 + \int_\Omega \nabla n \cdot \nabla w + \int_\Omega (u \cdot \nabla w)\triangle w \\
&= \frac{1}{2}\int_\Omega |\triangle w|^2 + \frac{1}{2}\int_\Omega |\nabla w|^4 + \int_\Omega \nabla n \cdot \nabla w - \int_\Omega \nabla w \cdot (\nabla u \cdot \nabla w) - \int_\Omega u(D^2 w \cdot \nabla w) \\
&= \frac{1}{2}\int_\Omega |\triangle w|^2 + \frac{1}{2}\int_\Omega |\nabla w|^4 + \int_\Omega \nabla n \cdot \nabla w - \int_\Omega \nabla w \cdot (\nabla u \cdot \nabla w) - \frac{1}{2}\int_\Omega u \cdot \nabla|\nabla w|^2 \\
&= \frac{1}{2}\int_\Omega |\triangle w|^2 + \frac{1}{2}\int_\Omega |\nabla w|^4 + \int_\Omega \nabla n \cdot \nabla w - \int_\Omega \nabla w \cdot (\nabla u \cdot \nabla w).
\end{aligned}$$

Furthermore, by Lemma 1.9 (i) and the Cauchy–Schwarz inequality, we get

$$\begin{aligned}
&\frac{1}{2}\frac{d}{dt}\int_\Omega |\nabla w|^2 + \frac{1}{2}\int_\Omega |\triangle w|^2 \\
&\le \frac{K_1}{2}\|\nabla w\|_{L^2(\Omega)}^2 \int_\Omega |\triangle w|^2 + \int_\Omega \nabla n \cdot \nabla w + \int_\Omega |\nabla u||\nabla w|^2 \\
&\le (\frac{K_1}{2}\|\nabla w\|_{L^2(\Omega)}^2 + K_1|\Omega|^{\frac{1}{2}}\|\nabla u\|_{L^2(\Omega)}) \int_\Omega |\triangle w|^2 + \int_\Omega \nabla n \cdot \nabla w
\end{aligned}$$

and thus

$$\frac{1}{2}\frac{d}{dt}\int_{\Omega}|\nabla w|^2 + \frac{1}{2}(1 - K_1\|\nabla w\|^2_{L^2(\Omega)} - 2K_1|\Omega|^{\frac{1}{2}}\|\nabla u\|_{L^2(\Omega)})\int_{\Omega}|\triangle w|^2$$
$$\leq \int_{\Omega}\nabla n \cdot \nabla w.$$

(1.4.20)

Since $2\mathscr{F}(n, w) \geq \chi\|\nabla w\|^2_{L^2(\Omega)}$ due to $H(n) \geq 0$, combining (1.4.20) with (1.4.19) yields

$$\frac{d}{dt}\mathscr{F}(n, w) + \int_{\Omega}\frac{|\nabla n|^2}{n} + (\frac{\chi}{2} - K_1\mathscr{F}(n, w) - 2\chi K_1|\Omega|^{\frac{1}{2}}\|\nabla u\|_{L^2(\Omega)})\int_{\Omega}|\triangle w|^2 \leq 0 \quad (1.4.21)$$

for $t > 0$.

On the other hand, when $\mu > \mu_1$, it follows from (1.4.13) that it is possible to pick some $t_0 > 0$ such that $16K_1|\Omega|^{\frac{1}{2}}\|\nabla u(\cdot, t)\|_{L^2(\Omega)} < 1$ for all $t > t_0$, and thereby

$$\frac{d}{dt}\mathscr{F}(n, w) + \int_{\Omega}\frac{|\nabla n|^2}{n} + (\frac{3\chi}{8} - K_1\mathscr{F}(n, w))\int_{\Omega}|\triangle w|^2 \leq 0 \text{ for } t > t_0.$$

(1.4.22)

In what follows, we shall show that there exists $t_* > t_0$ such that $4K_1\mathscr{F}(n, w)(t_*) < \chi$.

Firstly by Lemma 1.36 (i), there exists $t_1 > t_0$ such that for all $t > t_1$

$$\|n(\cdot, t)\|_{L^1(\Omega)} \leq \frac{3|\Omega|r}{2\mu},$$

(1.4.23)

which along with Lemma 1.36 (iii) yields

$$\int_{t_1}^{t_2}\int_{\Omega}|\nabla w|^2 \leq \int_{\Omega}w(\cdot, t_1) + \int_{t_1}^{t_2}\|n(\cdot, s)\|_{L^1(\Omega)}ds$$
$$\leq \int_{\Omega}w_0(x) + \int_0^{t_1}\|n(\cdot, s)\|_{L^1(\Omega)}ds + \frac{3|\Omega|r}{2\mu}(t_2 - t_1).$$

Similarly invoking Lemma 1.36 (i) and (ii), we find that

$$\int_{t_1}^{t_2}\|n(\cdot, s)\|^2_{L^2(\Omega)}ds \leq \frac{3|\Omega|}{2}(\frac{r}{\mu})^2(t_2 - t_1) + \frac{1}{\mu}\|n(\cdot, t_1)\|_{L^1(\Omega)}.$$

Hence there exists $t^* > t_1$ suitably large such that whenever $t_2 \geq t^*$,

$$\int_{t_1}^{t_2}\int_{\Omega}|\nabla w|^2 \leq \frac{2|\Omega|r}{\mu}(t_2 - t_1)$$

(1.4.24)

and

$$\int_{t_1}^{t_2}\|n(\cdot, s)\|^2_{L^2(\Omega)}ds \leq 2|\Omega|(\frac{r}{\mu})^2(t_2 - t_1).$$

(1.4.25)

Let

$$\mathscr{S}_1 := \{t \in [t_1, t_2] | \int_\Omega |\nabla w(\cdot, t)|^2 \geq \frac{8|\Omega|r}{\mu}\}$$

and

$$\mathscr{S}_2 := \{t \in [t_1, t_2] | \|n(\cdot, t)\|^2_{L^2(\Omega)} \geq 8|\Omega|(\frac{r}{\mu})^2\}.$$

Then

$$|\mathscr{S}_1| \leq \frac{|t_2 - t_1|}{4}, \quad |\mathscr{S}_2| \leq \frac{|t_2 - t_1|}{4}. \tag{1.4.26}$$

In order to estimate the size of $\mathscr{S}_1$ and $\mathscr{S}_2$, we recall (1.4.24) to get

$$\frac{8|\Omega|r}{\mu}|\mathscr{S}_1| \leq \int_{t_1}^{t_2} \int_\Omega |\nabla w|^2 \leq \frac{2|\Omega|r}{\mu}(t_2 - t_1)$$

and thus $|\mathscr{S}_1| \leq \frac{|t_2-t_1|}{4}$ is valid. Similarly, one can verify that $|\mathscr{S}_2| \leq \frac{|t_2-t_1|}{4}$.

As (1.4.26) warrants that $|(t_1, t_2) \setminus (\mathscr{S}_1 \cup \mathscr{S}_2)| \geq \frac{|t_2-t_1|}{2}$, one can conclude that there exists $t_* \in (t_1, t_2)$ such that

$$\|n(\cdot, t_*)\|^2_{L^2(\Omega)} < 8|\Omega|(\frac{r}{\mu})^2 \tag{1.4.27}$$

and

$$\int_\Omega |\nabla w(\cdot, t_*)|^2 < \frac{8|\Omega|r}{\mu}. \tag{1.4.28}$$

Applying $\xi \ln \frac{\xi}{\sigma} \leq \eta \xi^2 + \ln \frac{1}{\eta\sigma} \cdot \xi$ for all $\xi > 0, \eta > 0, \sigma > 0$ with $\eta = \frac{\mu}{r}$ (see Lemma 5.5 of Winkler 2016c) and (1.4.27), we then arrive at

$$\int_\Omega H(n)(\cdot, t_*) \leq \frac{\mu}{r} \int_\Omega n^2(\cdot, t_*) - \int_\Omega n(\cdot, t_*) + \frac{r}{\mu}|\Omega| \leq \frac{9|\Omega|r}{\mu}.$$

Thereupon from (1.4.28) and the definition of $\mathscr{F}(n, w)$, it follows that $\mathscr{F}(n, w)(t_*) < (9 + 4\chi)|\Omega|\frac{r}{\mu}$, which entails that $4K_1\mathscr{F}(n, w)(t_*) < \chi$ provided $\mu > \frac{K_1(36+16\chi)|\Omega|r}{\chi}$.

As an immediate consequence of (1.4.22), we have

$$\frac{d}{dt}\mathscr{F}(n, w) + \int_\Omega \frac{|\nabla n|^2}{n} + \frac{\chi}{8} \int_\Omega |\Delta w|^2 \leq 0 \quad \text{for all } t > t_* \tag{1.4.29}$$

when $\mu > \mu_2(\Omega, \chi, r)$, and thus end the proof of this lemma.

Additionally from (1.4.29), one can also conclude that

Corollary 1.1 *Under the conditions of Lemma 1.41, we have*

$$\mathscr{F}(n,w)(t) + \int_{t_*}^{\infty}\int_{\Omega}\frac{|\nabla n|^2}{n} + \frac{\chi}{8}\int_{t_*}^{\infty}\int_{\Omega}|\Delta w|^2 \le (9+4\chi)|\Omega|\frac{r}{\mu} \quad \text{for all } t > t_*.$$

(1.4.30)

Next by a further testing procedure, we can turn the above information into the uniform-in-time boundedness of $\|n(\cdot,t)\|_{L^2(\Omega)}$ and $\|\nabla w(\cdot,t)\|_{L^4(\Omega)}$ if μ is appropriately large compared with r, which will serve as the foundation for the proof of global boundedness of $\|n(\cdot,t)\|_{L^{\infty}(\Omega)}$ and $\|\nabla w(\cdot,t)\|_{L^{\infty}(\Omega)}$.

Lemma 1.42 *If* $\mu > \mu_0(\chi,\Omega,r) := \max\{\mu_2(\chi,\Omega,r), \frac{208K_2|\Omega|r}{\chi^2}\}$, *then there exists* $C > 0$ *such that*

$$\|n(\cdot,t)\|_{L^2(\Omega)} + \|\nabla w(\cdot,t)\|_{L^4(\Omega)} \le C \quad \text{for all } t \ge t_*.$$

(1.4.31)

Proof Since $\mu > \mu_2(\chi,\Omega,r)$, it follows from (1.4.30) that

$$\int_{\Omega}|\nabla w|^2 \le \frac{r}{\mu}(\frac{18}{\chi} + \frac{8}{\chi^2})|\Omega| \quad \text{for all } t > t_*$$

and moreover due to $\frac{r}{\mu} < \frac{\chi^2}{208K_2|\Omega|}$,

$$K_2\int_{\Omega}|\nabla w|^2 \le \frac{1}{8} \quad \text{for all } t > t_*.$$

(1.4.32)

Multiplying the first equation in (1.4.1) by n and integrating the result over Ω, we get

$$\frac{1}{2}\frac{d}{dt}\int_{\Omega}n^2 = -\int_{\Omega}|\nabla n|^2 - \chi\int_{\Omega}n\nabla n\nabla w + r\int_{\Omega}n^2 - \mu\int_{\Omega}n^3$$
$$\le -\frac{1}{2}\int_{\Omega}|\nabla n|^2 + \frac{1}{2}\int_{\Omega}n^2|\nabla w|^2 + r\int_{\Omega}n^2 - \mu\int_{\Omega}n^3.$$

(1.4.33)

On the other hand, by the second equation in (1.4.1) and the identity $\nabla w \cdot \nabla\Delta w = \frac{1}{2}\Delta|\nabla w|^2 - |D^2w|^2$, we obtain

$$\frac{d}{dt}\int_{\Omega}|\nabla w|^4$$
$$= 2\int_{\Omega}|\nabla w|^2\Delta|\nabla w|^2 - 4\int_{\Omega}|\nabla w|^2|D^2w|^2 - 4\int_{\Omega}|\nabla w|^2\nabla w\cdot\nabla|\nabla w|^2$$
$$+ 4\int_{\Omega}|\nabla w|^2\nabla n\cdot\nabla w - 4\int_{\Omega}|\nabla w|^2\nabla w\cdot\nabla(u\cdot\nabla w)$$
$$= -2\int_{\Omega}|\nabla|\nabla w|^2|^2 - 4\int_{\Omega}|\nabla w|^2|D^2w|^2$$
$$- 4\int_{\Omega}|\nabla w|^2\nabla w\cdot\nabla|\nabla w|^2 - 4\int_{\Omega}n|\nabla w|^2\Delta w$$

$$-4\int_{\Omega} n\nabla|\nabla w|^2 \cdot \nabla w + 2\int_{\partial\Omega} |\nabla w|^2 \frac{\partial|\nabla w|^2}{\partial\nu} - 4\int_{\Omega} |\nabla w|^2 \nabla w \cdot (\nabla u \cdot \nabla w)$$

$$\tag{1.4.34}$$

due to $\nabla \cdot u = 0$ and $u = 0$ on $\partial\Omega$.

According to $\frac{\partial|\nabla w|^2}{\partial\nu} \leq C_1|\nabla w|^2$ on $\partial\Omega$ for some $C_1 > 0$ and

$$\||\nabla w|^2\|_{L^2(\partial\Omega)} \leq \eta\|\nabla|\nabla w|^2\|_{L^2(\Omega)} + C_2(\eta)\||\nabla w|^2\|_{L^1(\Omega)} \text{ for any } \eta \in (0, \frac{5}{4})$$

(see Lemma 4.2 of Mizoguchi and Souplet (2014) and Remark 52.9 in Quittner and Souplet 2007), one can conclude that

$$2\int_{\partial\Omega} |\nabla w|^2 \frac{\partial|\nabla w|^2}{\partial\nu} \leq \frac{1}{4}\int_{\Omega} |\nabla|\nabla w|^2|^2 + C_3(\int_{\Omega} |\nabla w|^2)^2 \tag{1.4.35}$$

for some $C_3 > 0$.

For the other integrals on the right side of (1.4.34), we use the Young inequality to estimate

$$-4\int_{\Omega} |\nabla w|^2 \nabla w \cdot \nabla|\nabla w|^2 \leq \frac{1}{3}\int_{\Omega} |\nabla|\nabla w|^2|^2 + 12\int_{\Omega} |\nabla w|^6 \tag{1.4.36}$$

$$-4\int_{\Omega} n\nabla|\nabla w|^2 \cdot \nabla w \leq \frac{1}{3}\int_{\Omega} |\nabla|\nabla w|^2|^2 + 12\int_{\Omega} n^2|\nabla w|^2 \tag{1.4.37}$$

as well as

$$-4\int_{\Omega} n|\nabla w|^2\triangle w \leq \frac{1}{6}\int_{\Omega} |\nabla w|^2|\triangle w|^2 + 24\int_{\Omega} n^2|\nabla w|^2$$
$$\leq \frac{1}{3}\int_{\Omega} |\nabla w|^2|D^2 w|^2 + 24\int_{\Omega} n^2|\nabla w|^2 \tag{1.4.38}$$

due to $|\triangle w|^2 \leq 2|D^2 w|^2$ on Ω.

Substituting (1.4.35)–(1.4.38) into (1.4.34), we readily get

$$\frac{d}{dt}\int_{\Omega} |\nabla w|^4 + \frac{13}{12}\int_{\Omega} |\nabla|\nabla w|^2|^2 + \frac{11}{3}\int_{\Omega} |\nabla w|^2|D^2 w|^2$$
$$\leq 12\int_{\Omega} |\nabla w|^6 + 36\int_{\Omega} n^2|\nabla w|^2 + C_3(\int_{\Omega} |\nabla w|^2)^2 + 4\int_{\Omega} |\nabla w|^4|\nabla u|$$

and thus

$$\frac{d}{dt}\int_\Omega |\nabla w|^4 + 2\int_\Omega |\nabla |\nabla w|^2|^2$$
$$\leq 12\int_\Omega |\nabla w|^6 + 36\int_\Omega n^2|\nabla w|^2 + C_3(\int_\Omega |\nabla w|^2)^2 + 4\int_\Omega |\nabla w|^4|\nabla u| \tag{1.4.39}$$

due to the fact $|\nabla |\nabla w|^2|^2 \leq 4|\nabla w|^2|D^2 w|^2$ on Ω.

Therefore combining (1.4.33) with (1.4.39) leads to

$$\frac{d}{dt}\int_\Omega (n^2 + |\nabla w|^4) + 2\int_\Omega |\nabla |\nabla w|^2|^2 + \int_\Omega |\nabla n|^2$$
$$\leq 12\int_\Omega |\nabla w|^6 + 37\int_\Omega n^2|\nabla w|^2 + C_3(\int_\Omega |\nabla w|^2)^2$$
$$+ 2r\int_\Omega n^2 - 2\mu\int_\Omega n^3 + 4\int_\Omega |\nabla w|^4|\nabla u| \tag{1.4.40}$$
$$\leq 13\int_\Omega |\nabla w|^6 + 37^2\int_\Omega n^3 + C_3(\int_\Omega |\nabla w|^2)^2$$
$$+ 2r\int_\Omega n^2 - 2\mu\int_\Omega n^3 + 4\int_\Omega |\nabla w|^4|\nabla u|.$$

Furthermore by Lemma 1.9 (ii), we get $\|\varphi\|_{L^3}^3 \leq K_2\|\nabla\varphi\|_{L^2}^2\|\varphi\|_{L^1} + C_4\|\varphi\|_{L^1}^3$ and thus

$$\int_\Omega |\nabla w|^6 \leq K_2(\int_\Omega |\nabla |\nabla w|^2|^2)\left(\int_\Omega |\nabla w|^2\right) + C_4(\int_\Omega |\nabla w|^2)^3.$$

Upon inserting this into (1.4.40) and (1.4.30), we obtain

$$\frac{d}{dt}\int_\Omega (n^2 + |\nabla w|^4) + (2 - 13K_2\int_\Omega |\nabla w|^2)\int_\Omega |\nabla |\nabla w|^2|^2$$
$$+ \int_\Omega |\nabla n|^2 + \int_\Omega (n^2 + |\nabla w|^4)$$
$$\leq 37^2\int_\Omega n^3 + (2r+1)\int_\Omega n^2 - 2\mu\int_\Omega n^3 + \int_\Omega |\nabla w|^4 + 4\int_\Omega |\nabla w|^4|\nabla u| + C_5,$$

which, along with

$$\int_\Omega |\nabla w|^4 \leq \frac{1}{7}\int_\Omega |\nabla |\nabla w|^2|^2 + C_6$$

and

$$4\int_\Omega |\nabla w|^4|\nabla u| \leq 4\||\nabla w|^2\|_{L^6(\Omega)}^2\|\nabla u\|_{L^{\frac{3}{2}}(\Omega)} \leq \frac{13}{56}\int_\Omega |\nabla |\nabla w|^2|^2 + C_7$$

by the Gagliardo–Nirenberg inequality and (1.4.13), implies that

$$\frac{d}{dt} \int_\Omega (n^2 + |\nabla w|^4) + (\frac{13}{8} - 13K_2 \int_\Omega |\nabla w|^2) \int_\Omega |\nabla |\nabla w|^2|^2$$
$$+ \int_\Omega |\nabla n|^2 + \int_\Omega (n^2 + |\nabla w|^4) \tag{1.4.41}$$
$$\leq 37^2 \int_\Omega n^3 + (2r + 1) \int_\Omega n^2 - 2\mu \int_\Omega n^3 + C_8.$$

On the other hand, according to an extended variant (Biler et al. 1994), (1.4.23) and (1.4.30), one can infer that

$$37^2 \int_\Omega n^3 \leq C_9 \left(\int_\Omega |\nabla n|^2 \right) \left(\int_\Omega n |\ln n| \right) + C_9 (\int_\Omega n)^3 + C_9 \leq \frac{1}{2} \int_\Omega |\nabla n|^2 + C_{10}.$$

Hence from (1.4.41) it follows that there exists $C_{11} > 0$ such that for all $t > t_*$

$$\frac{d}{dt} \int_\Omega (n^2 + |\nabla w|^4) + \int_\Omega (n^2 + |\nabla w|^4) + (\frac{13}{8} - 13K_2 \int_\Omega |\nabla w|^2) \int_\Omega |\nabla |\nabla w|^2|^2 \leq C_{11},$$
$$\tag{1.4.42}$$

which, along with (1.4.32), entails that

$$\frac{d}{dt} \int_\Omega (n^2 + |\nabla w|^4) + \int_\Omega (n^2 + |\nabla w|^4) \leq C_{11}$$

for all $t > t_*$ and thereby (1.4.31) is valid.

We are now ready to prove Theorem 1.2 in the case of $r > 0$.
Proof of Theorem 1.2 *in the case of* $r > 0$. From the above lemmas, it follows that there exists $C > 0$ such that

$$\|n(\cdot, t)\|_{L^2(\Omega)} + \|\nabla w(\cdot, t)\|_{L^4(\Omega)} + \|\nabla u(\cdot, t)\|_{L^2(\Omega)} \leq C$$

whenever $\mu > \mu_0(\chi, \Omega, r) := \max\{\mu_2(\chi, \Omega, r), \frac{208K_2|\Omega|r}{\chi^2}\}$. So, by the argument in, e.g., Lemma 4.4 of Black (2018), we can readily prove that $\|n(\cdot, t)\|_{L^\infty(\Omega)}$, $\|\nabla w(\cdot, t)\|_{L^\infty(\Omega)}$ and $\|A^\alpha u(\cdot, t)\|_{L^2(\Omega)}$ with some $\alpha \in (\frac{1}{2}, 1)$ are globally bounded; we refer the reader to the proof of Lemma 4.4 in Black (2018), Lemmas 3.12 and 3.11 in Tao and Winkler (2016) for the details.

Based on the global boundedness of solutions, we are able to derive the convergence result claimed in Theorem 1.2, namely,

$$\lim_{t \to \infty} \|n(\cdot, t) - \frac{r}{\mu}\|_{L^\infty(\Omega)} = 0, \tag{1.4.43}$$

$$\lim_{t \to \infty} \|\nabla w(\cdot, t)\|_{L^\infty(\Omega)} = 0, \tag{1.4.44}$$

$$\lim_{t \to \infty} \|u(\cdot, t)\|_{L^\infty(\Omega)} = 0 \tag{1.4.45}$$

as well as

$$\lim_{t \to \infty} \inf_{x \in \Omega} w(x, t) = \infty. \tag{1.4.46}$$

In fact, due to

$$\int_{t_*}^{\infty} \int_{\Omega} \frac{|\nabla n|^2}{n} + \int_{t_*}^{\infty} \int_{\Omega} |\Delta w|^2 \leq C$$

established in (1.4.30), we can show (1.4.44), (1.4.45) and

$$\lim_{t \to \infty} \|n(\cdot, t) - \overline{n}(t)\|_{L^\infty(\Omega)} = 0 \tag{1.4.47}$$

with $\overline{n}(t) = \frac{1}{|\Omega|} \int_{\Omega} n(\cdot, t)$ by the arguments in Proposition 4.15 of Black (2018), where we have used

$$\int_{t_*}^{\infty} \|n(\cdot, t) - \overline{n}(t)\|_{L^2(\Omega)}^2 \leq C \int_{t_*}^{\infty} \|\nabla n\|_{L^1(\Omega)}^2 \leq C \int_{t_*}^{\infty} \left(\int_{\Omega} \frac{|\nabla n|^2}{n} \int_{\Omega} n \right)$$

and the regularity of n. Therefore it suffices to show that

$$\lim_{t \to \infty} |\overline{n}(t) - \frac{r}{\mu}| = 0. \tag{1.4.48}$$

To this end, we adapt the idea of Lițcanu and Morales-Rodrigo (2010b) and give the details of the proof for the convenience of readers.

Integrating the first equation in (1.4.1) on the spatial variable over Ω, we obtain

$$\overline{n}_t = r\overline{n} - \frac{\mu}{|\Omega|} \int_{\Omega} n^2 = r\overline{n} - \mu \overline{n}^2 - \frac{\mu}{|\Omega|} \int_{\Omega} (n - \overline{n})^2.$$

Putting $a(t) := \frac{\mu}{|\Omega|} \int_{\Omega} (n(\cdot, t) - \overline{n})^2$, the above equation then becomes

$$\overline{n}_t = \mu \overline{n}(\frac{r}{\mu} - \overline{n}) - a(t). \tag{1.4.49}$$

Thereupon multiplying (1.4.49) by $\overline{n} - \frac{r}{\mu}$, we get

$$\frac{d}{dt}(\overline{n} - \frac{r}{\mu})^2 + 2\mu \overline{n}(\overline{n} - \frac{r}{\mu})^2 = -2a(t)(\overline{n} - \frac{r}{\mu})$$

and then

$$2\mu \int_1^{\infty} \overline{n}(\overline{n} - \frac{r}{\mu})^2 \leq (\overline{n}(1) - \frac{r}{\mu})^2 + 2 \sup_{t \geq 1} |\overline{n}(t) - \frac{r}{\mu}| \int_1^{\infty} a(t). \tag{1.4.50}$$

In addition, invoking the Poincaré–Wirtinger inequality

$$\int_{\Omega} |\varphi - \frac{1}{|\Omega|} \int_{\Omega} \varphi(y) dy|^2 \le C_p \int_{\Omega} |\varphi| \int_{\Omega} \frac{|\nabla\varphi|^2}{|\varphi|} \quad \text{for all } \varphi \in W^{1,2}(\Omega)$$

for some $C_p > 0$, one can find

$$\int_1^{\infty} a(s) ds \le C_p \sup_{t \ge 1} \|n(t)\|_{L^1(\Omega)} \int_1^{\infty} \int_{\Omega} \frac{|\nabla n(s)|^2}{n(s)} ds \le C \tag{1.4.51}$$

due to (1.4.30) and Lemma 1.38 (i). Hence combining (1.4.51) with (1.4.50) yields

$$\int_1^{\infty} \overline{n}(\overline{n} - \frac{r}{\mu})^2 \le C. \tag{1.4.52}$$

On the other hand, $\frac{d}{dt}\overline{n}(\overline{n} - \frac{r}{\mu})^2 = \overline{n}_t((\overline{n} - \frac{r}{\mu})^2 + 2\overline{n}(\overline{n} - \frac{r}{\mu}))$, which along with $|\overline{n}_t| \le r\overline{n} + \frac{\mu}{|\Omega|}\int_{\Omega} n^2 \le C$ implies that

$$\left| \frac{d}{dt}\overline{n}(\overline{n} - \frac{r}{\mu})^2 \right| \le C. \tag{1.4.53}$$

Therefore by Lemma 6.3 of Liţcanu and Morales-Rodrigo (2010b), (1.4.53) and (1.4.52) show that

$$\lim_{t \to \infty} \overline{n}(t)(\overline{n}(t) - \frac{r}{\mu})^2 = 0. \tag{1.4.54}$$

From (1.4.47), it follows that there exists $t_1 > t_*$ such that $\|n(\cdot, t) - \overline{n}(t)\|_{L^{\infty}(\Omega)} \le \frac{r}{2\mu}$ for all $t > t_1$, and thus

$$\begin{aligned}
\overline{n}_t &= r\overline{n} - \mu\overline{n}^2 - \frac{\mu}{|\Omega|} \int_{\Omega} n(n - \overline{n}) \\
&\ge \mu\overline{n}(\frac{r}{\mu} - \overline{n} - \sup_{t > t_1} \|n(\cdot, t) - \overline{n}(t)\|_{L^{\infty}(\Omega)}) \\
&\ge \mu\overline{n}(\frac{r}{2\mu} - \overline{n}).
\end{aligned} \tag{1.4.55}$$

On the other hand, noticing that the solution $y(t)$ of the ODE

$$y'(t) = \mu\overline{y}(\frac{r}{2\mu} - \overline{y}), \quad y(t_1) > 0$$

satisfies $\lim_{t \to \infty} y(t) = \frac{r}{2\mu}$, by the comparison principle, (1.4.55) implies that there exists $t_2 > t_1$ such that for all $t \ge t_2$, $\overline{n}(t) \ge \frac{r}{4\mu}$. This together with (1.4.54) yields (1.4.48).

Finally, in view of (1.4.43), one can find $t_3 > 1$ such that $n(x, t) \geq \frac{r}{2\mu}$ for all $x \in \Omega$ and $t \geq t_3$, and thereby $w(x, t)$ satisfies $w_t \geq \Delta w - |\nabla w|^2 + \frac{r}{2\mu} - u \cdot \nabla w$ for $t \geq t_3$. Hence if $y(t)$ denotes the solution of ODE: $y'(t) = \frac{r}{2\mu}$, $y(t_3) = \min_{x \in \Omega} w(\cdot, t_3)$, then

$$w(x, t) \geq \frac{r}{2\mu}(t - t_3) \tag{1.4.56}$$

by means of a straightforward parabolic comparison which warrants that (1.4.46) holds and thereby completes the proof.

2. The Case $r \leq 0$

In this subsection, we show the global boundedness of solutions to (1.1.5), (1.1.8), (1.1.9) in the case $r \leq 0$, $\mu > 0$. As mentioned in the introduction, due to the structure of (1.1.5) with $r \leq 0$, $\mu > 0$, it is difficult to find a decreasing energy functional compared with the situation when $r > 0$, $\mu > 0$ considered in the previous subsection or when $r = \mu = 0$ considered in Winkler (2016c). Indeed, the energy-type functional $\mathscr{F}(n, w)$ in (3.1) of Winkler (2016c) decreases along a solution in $\Omega \times (t_0, \infty)$ if $\mathscr{F}(n(\cdot, t_0), w(\cdot, t_0))$ is suitably small, namely

$$\frac{d}{dt}\mathscr{F}(n, w) \leq 0 \text{ for all } t \geq t_0.$$

The main idea underlying our approach is to make use of the quadratic degradation in the first equation of (1.1.5) which should enforce some suitable regularity properties. More precisely, on the basis of (1.4.2), we can show that the quantity of form

$$\mathscr{F}(n, w) := \int_\Omega n(\ln n + a)dx + \frac{\chi}{2}\int_\Omega |\nabla w|^2 dx, \tag{1.4.57}$$

with parameter $a > 0$ determined below (see (1.4.64)), satisfies a certain of differential inequality. Although unlike the case of $r > 0$ in which it enjoys the monotonicity property, $\mathscr{F}(n, w)$ also provides us the global boundedness of $\int_\Omega n|\ln n|dx$ and $\int_\Omega |\nabla w|^2 dx$. This is encapsulated in the following lemma.

Lemma 1.43 *Let $\Omega \subset \mathbb{R}^2$ be a smooth bounded domain and (n, w, u) be the global classical solution (1.4.1) with $r \leq 0$, $\mu > 0$. Then there exists $t_* > 0$ such that for all $t > t_*$*

$$\int_\Omega |\nabla w(\cdot, t)|^2 \leq \frac{1}{4K_1} \tag{1.4.58}$$

with K_1 given in Lemma 1.9 as well as

$$\int_\Omega n|\ln n| \leq C \tag{1.4.59}$$

for some $C > 0$.

Proof We test the first equation in (1.4.1) against $\ln n + a + 1$, and integrate by parts to see that

$$\frac{d}{dt} \int_\Omega n(\ln n + a)$$

$$\leq -\int_\Omega \frac{|\nabla n|^2}{n} - \chi \int_\Omega \nabla n \cdot \nabla w + \int_\Omega (n(r - \mu n) - u \cdot \nabla n)(\ln n + a + 1)$$

$$\leq -\int_\Omega \frac{|\nabla n|^2}{n} - \chi \int_\Omega \nabla n \cdot \nabla w + \int_\Omega n(r - \mu n)(\ln n + a) \qquad (1.4.60)$$

due to $r \leq 0$ and $\nabla \cdot u = 0$.

On the other hand, recalling (1.4.20) and (1.4.14), it is possible to fix $t_0 > 0$ such that for all $t \geq t_0$, we have

$$\frac{1}{2} \frac{d}{dt} \int_\Omega |\nabla w|^2 + \frac{1}{4} \int_\Omega |\Delta w|^2 + \frac{1}{4}(\frac{3}{4} - 2K_1 \|\nabla w\|_{L^2(\Omega)}^2) \int_\Omega |\Delta w|^2$$

$$\leq \int_\Omega \nabla u \cdot \nabla w. \qquad (1.4.61)$$

From Lemma 1.9 (i), there exists a constant $K_3 > 0$ such that

$$8K_3 \|\nabla w\|_{L^2(\Omega)}^2 \leq \|\Delta w\|_{L^2(\Omega)}^2. \qquad (1.4.62)$$

Hence combining (1.4.61) with (1.4.60), we get

$$\frac{d}{dt} \mathscr{F}(n, w) + \int_\Omega \frac{|\nabla n|^2}{n} + \frac{\chi}{4} \int_\Omega |\Delta w|^2 + K_3 \int_\Omega n(\ln n + a)$$

$$+ \frac{\chi}{4}(\frac{3}{4} - 2K_1 \|\nabla w\|_{L^2(\Omega)}^2) \int_\Omega |\Delta w|^2$$

$$\leq \int_\Omega n(K_3 - \mu n)(\ln n + a) + r \int_\Omega n(\ln n + a) \quad \text{for } t \geq t_0. \qquad (1.4.63)$$

Now for any fixed $\varepsilon < \min\{\frac{\chi}{24K_1}, \frac{\chi}{42K_2}\}$, we pick $a > 1$ sufficiently large such that

$$e^{-a} < \frac{K_3}{\mu}, \quad (1 - r)|\Omega| \max_{0 < n \leq e^{-a}} |n \ln n| < \varepsilon \min\{K_3, 1\}, \qquad (1.4.64)$$

due to $\lim_{n \to 0} n \ln n = 0$ and $n \ln n < 0$ for all $n \in (0, 1)$, and thereby fix $t_1 > \max\{1, t_0\}$ fulfilling

$$\frac{a|\Omega|}{\mu(t_1 + \gamma)} < \frac{\varepsilon}{4}, \quad \frac{|\Omega|}{\mu^2(t_1 + \gamma)} < \frac{\varepsilon}{16} \qquad (1.4.65)$$

as well as

$$\frac{(a + (\ln \frac{K_3}{\mu})_+)|\Omega|}{\mu(t_1 + \gamma)} + \frac{2\chi(\frac{|\Omega|}{\mu} + \|n_0\|_{L^1(\Omega)} + \|w_0\|_{L^1(\Omega)})}{t_1} < \frac{\varepsilon}{4}. \qquad (1.4.66)$$

Let $t_2 = t_1 + t_1^2$,

$$\mathscr{S}_1 \triangleq \{t \in [t_1, t_2]| \int_\Omega |\nabla w(\cdot, t)|^2 \geq \frac{\varepsilon}{2\chi}\}$$

and

$$\mathscr{S}_2 \triangleq \{t \in [t_1, t_2]| \|n(\cdot, t)\|_{L^2(\Omega)}^2 \geq \frac{\varepsilon}{4}\}.$$

Then

$$|\mathscr{S}_1| \leq \frac{|t_2 - t_1|}{4}, \quad |\mathscr{S}_2| \leq \frac{|t_2 - t_1|}{4}. \qquad (1.4.67)$$

By Lemma 1.36(iii), (1.4.2) and the second equation in (1.4.1), we obtain that

$$\int_{t_1}^{t_2} \int_\Omega |\nabla w|^2 \leq \int_{t_1}^{t_2} \int_\Omega n + \int_\Omega w(\cdot, t_1)$$

$$= \int_{t_1}^{t_2} \int_\Omega n + \int_\Omega w_0 + \int_0^{t_1} \int_\Omega n$$

$$\leq \frac{|\Omega|}{\mu(t_1 + \gamma)}(t_2 - t_1) + \int_\Omega w_0 + t_1 \int_\Omega n_0.$$

Furthermore, by (1.4.65) and (1.4.66)

$$\int_{t_1}^{t_2} \int_\Omega |\nabla w|^2 dx ds \leq (\frac{|\Omega|}{\mu(t_1 + \gamma)} + \frac{t_1 \|n_0\|_{L^1(\Omega)} + \|w_0\|_{L^1(\Omega)}}{t_2 - t_1})(t_2 - t_1)$$

$$\leq \frac{\frac{|\Omega|}{\mu} + \|n_0\|_{L^1(\Omega)} + \|w_0\|_{L^1(\Omega)}}{t_1}(t_2 - t_1)$$

$$< \frac{\varepsilon}{8\chi}(t_2 - t_1).$$

On the other hand, by the definition of $\mathscr{S}_1$, we see that $\frac{\varepsilon}{2\chi}|\mathscr{S}_1| \leq \int_{t_1}^{t_2} \int_\Omega |\nabla w|^2$ and thereby $|\mathscr{S}_1| \leq \frac{|t_2 - t_1|}{4}$.

In addition, by (1.4.2) and (1.4.65), we get

$$\int_{t_1}^{t_2} \int_\Omega n^2 \leq \frac{1}{\mu} \int_\Omega n(\cdot, t_1) \leq \frac{|\Omega|}{\mu^2(t_1 + \gamma)} < \frac{\varepsilon}{16},$$

which implies that $|\mathscr{S}_2| \leq \frac{|t_2 - t_1|}{4}$.

Therefore from (1.4.67), it follows that $|(t_1, t_2) \setminus (\mathscr{S}_1 \cup \mathscr{S}_2)| \geq \frac{|t_2 - t_1|}{2}$, and thereby there exists $t_* \in (t_1, t_2)$ such that

$$\|n(\cdot, t_*)\|_{L^2(\Omega)}^2 < \frac{\varepsilon}{4} \tag{1.4.68}$$

and

$$\int_\Omega |\nabla w(\cdot, t_*)|^2 < \frac{\varepsilon}{2\chi} < \frac{1}{6K_1}. \tag{1.4.69}$$

By (1.4.69), we can see that the set

$$\mathbf{S} \triangleq \{t \in (t_*, \infty) | \ K_1 \int_\Omega |\nabla w(\cdot, s)|^2 < \frac{1}{4} \quad \text{for all } s \in (t_*, t)\}$$

is not empty and hence $T_S = \sup \mathbf{S}$ is a well-defined element of $(t_*, \infty]$. In fact, we claim that $T_S = \infty$. To this end, supposing on the contrary that $T_S < \infty$, we then have $K_1 \int_\Omega |\nabla w(\cdot, t)|^2 < \frac{1}{4}$ for all $t \in [t_*, T_S)$, but

$$K_1 \int_\Omega |\nabla w(\cdot, T_S)|^2 = \frac{1}{4}. \tag{1.4.70}$$

Hence from (1.4.63) and (1.4.62), it follows that for all $t \in [t_*, T_S)$,

$$\begin{aligned}
&\frac{d}{dt} \mathscr{F}(n, w) + \int_\Omega \frac{|\nabla n|^2}{n} + \frac{\chi}{4} \int_\Omega |\Delta w|^2 + K_3 \int_\Omega n(\ln n + a) + \frac{K_3 \chi}{2} \int_\Omega |\nabla w|^2 \\
&\leq \int_\Omega n(K_3 - \mu n)(\ln n + a) + r \int_\Omega n(\ln n + a) \\
&\leq \int_{e^{-a} < n \leq \frac{K_3}{\mu}} n(K_3 - \mu n)(\ln n + a) + r \int_{0 < n \leq e^{-a}} n(\ln n + a) \\
&\leq K_3 \int_{e^{-a} < n \leq \frac{K_3}{\mu}} n(\ln n + a) + r \int_{0 < n \leq e^{-a}} n \ln n \tag{1.4.71} \\
&\leq a K_3 \int_\Omega n + K_3 \int_{e^{-a} < n \leq \frac{K_3}{\mu}} n \ln n - r|\Omega| \max_{0 < n \leq e^{-a}} |n \ln n| \\
&\leq K_3 (a + (\ln \frac{K_3}{\mu})_+) \int_\Omega n + \varepsilon K_3 \\
&\leq \frac{(a + (\ln \frac{K_3}{\mu})_+) K_3 |\Omega|}{\mu(t_1 + \gamma)} + \varepsilon K_3,
\end{aligned}$$

where we have made use of $t_* \geq t_1$, the decay estimate (1.4.2) and (1.4.65), and thus

$$\mathcal{F}(n, w)(T_s) + \int_{t_*}^{T_s} e^{-K_3(T_s - \sigma)} \Big(\int_\Omega \frac{|\nabla n|^2}{n}(\cdot, \sigma) + \frac{\chi}{4} \int_\Omega |\triangle w(\cdot, \sigma)|^2 \Big) d\sigma$$

$$\leq \mathcal{F}(n, w)(t_*) + \frac{(a + (\ln \frac{K_3}{\mu})_+)|\Omega|}{\mu(t_1 + \gamma)} + \varepsilon,$$

which implies that

$$\frac{\chi}{2} \int_\Omega |\nabla w(\cdot, T_s)|^2 \leq \mathcal{F}(n, w)(t_*) + \frac{(a + (\ln \frac{K_3}{\mu})_+)|\Omega|}{\mu(t_1 + \gamma)} - \int_\Omega n(\ln n + a)(\cdot, T_s) + \varepsilon$$

$$\leq \int_\Omega n(\ln n + a)(\cdot, t_*) + \frac{\chi}{2} \int_\Omega |\nabla w|^2(\cdot, t_*) + \varepsilon$$

$$+ \frac{(a + (\ln \frac{K_3}{\mu})_+)|\Omega|}{\mu(t_1 + \gamma)} - \int_\Omega n(\ln n + a)(\cdot, T_s)$$

$$\leq \int_\Omega (n^2 + an)(\cdot, t_*) + \frac{\chi}{2} \int_\Omega |\nabla w|^2(\cdot, t_*) + \varepsilon \qquad (1.4.72)$$

$$+ \frac{(a + (\ln \frac{K_3}{\mu})_+)|\Omega|}{\mu(t_1 + \gamma)} - \int_\Omega n(\ln n + a)(\cdot, T_s),$$

due to $n \geq \ln n$ for all $n > 0$.

In addition, by (1.4.65), we see that

$$\int_\Omega n(\ln n + a)(\cdot, T_s) \geq \int_{0 < n \leq e^{-a}} n(\ln n + a)(\cdot, T_s) \qquad (1.4.73)$$

$$\geq \int_{0 < n \leq e^{-a}} n \ln n(\cdot, T_s)$$

$$\geq - |\Omega| \max_{0 < n \leq e^{-a}} |n \ln n|$$

$$\geq - \varepsilon.$$

Upon inserting (1.4.73) into (1.4.72), we see that

$$\frac{\chi}{2} \int_\Omega |\nabla w(\cdot, T_s)|^2 \leq \int_\Omega (n^2 + an)(\cdot, t_*) + \frac{\chi}{2} \int_\Omega |\nabla w|^2(\cdot, t_*) c1.2 - 3.58$$

$$\qquad (1.4.74)$$

$$+ \frac{(a + (\ln \frac{K_3}{\mu})_+)|\Omega|}{\mu(t_1 + \gamma)} + 2\varepsilon,$$

which along with (1.4.68), (1.4.69), (1.4.2) and (1.4.65), establishes that

$$\frac{\chi}{2} \int_{\Omega} |\nabla w(\cdot, T_S)|^2 \leq \frac{5\varepsilon}{2} + a \int_{\Omega} n(\cdot, t_*) + \frac{(a + (\ln \frac{K_3}{\mu})_+)|\Omega|}{\mu(t_1 + \gamma)}$$

$$< 3\varepsilon \tag{1.4.75}$$

$$\leq \frac{\chi}{8K_1}.$$

This contradicts (1.4.70) and thereby $T_S = \infty$, which means that the differential inequality (1.4.71) is actually valid for all $t > t_*$.

Now revisiting the proof of (1.4.75), upon integration in time over (t_*, t), we have $\frac{\chi}{2} \int_{\Omega} |\nabla w(\cdot, t)|^2 \leq 3\varepsilon$ for all $t > t_*$ which implies that (1.4.58) is valid by the choice of ε, as well as

$$\int_{\Omega} n \ln n(\cdot, t) \leq C_1 \quad \text{for all } t > t_* \tag{1.4.76}$$

for some $C_1 > 0$.

Since $\xi \ln \xi \geq -\frac{1}{e}$ for all $\xi > 0$,

$$\int_{\Omega} n |\ln n|(\cdot, t) = \int_{\Omega} n \ln n(\cdot, t) - 2 \int_{0 < n < 1} n \ln n(\cdot, t) \leq \int_{\Omega} n \ln n(\cdot, t) + \frac{2|\Omega|}{e},$$

which along with (1.4.76) readily implies that (1.4.59) is actually valid with $C = C_1 + \frac{2|\Omega|}{e}$.

Furthermore, from (1.4.71), one can also conclude that

Corollary 1.2 *Under the conditions of Lemma 1.43, we have*

$$\lim_{t \to \infty} \int_t^{t+1} \int_{\Omega} \left(\frac{|\nabla n|^2}{n} + |\Delta w|^2\right) = 0, \quad \lim_{t \to \infty} \int_{\Omega} |\nabla w(\cdot, t)|^2 = 0. \tag{1.4.77}$$

Proof On the basis of the decay estimate (1.4.2) and revisiting the argument in the proof of Lemma 1.43, one can conclude that for any $\varepsilon \in (0, \min\{\frac{\chi}{24K_1}, \frac{\chi}{42K_2}\})$, there exists $t_\varepsilon > 1$ such that

$$\int_{\Omega} |\nabla w(\cdot, t)|^2 + \int_{t_\varepsilon}^t e^{-K_3(t-\sigma)} \left(\int_{\Omega} \frac{|\nabla n|^2}{n}(\cdot, \sigma) + \frac{\chi}{8} \int_{\Omega} |\Delta w(\cdot, \sigma)|^2\right) d\sigma \leq \varepsilon$$

for all $t > t_\varepsilon$. Furthermore, it follows from the above inequality that

$$\int_{t-1}^t \left(\int_{\Omega} \frac{|\nabla n|^2}{n}(\cdot, \sigma) + \frac{\chi}{8} \int_{\Omega} |\Delta w(\cdot, \sigma)|^2\right) d\sigma \leq \varepsilon e^{K_3}$$

for any $t > t_\varepsilon + 1$, which implies that (1.4.77) is indeed valid.

At this point, we can prove Theorem 1.2 in the case of $r \leq 0$.

Proof of Theorem 1.2 *in the case* $r \leq 0$. We can repeat the argument in the proof of Theorem 1.2 in the case $r > 0$. In fact, in view of (1.4.58) and (1.4.59), (1.4.31) is also valid for $r \leq 0$, $\mu > 0$, and thereby the global boundedness of solutions can be proven. In addition, similar to the case of $r > 0$, we can show

$$\lim_{t \to \infty} \|n(\cdot, t)\|_{L^\infty(\Omega)} = 0, \tag{1.4.78}$$

$$\lim_{t \to \infty} \|\nabla w(\cdot, t)\|_{L^\infty(\Omega)} = 0 \tag{1.4.79}$$

as well as

$$\lim_{t \to \infty} \|u(\cdot, t)\|_{L^\infty(\Omega)} = 0. \tag{1.4.80}$$

For the sake of completeness we shall only recount the main steps and refer to the mentioned sources for more details. Invoking standard parabolic regularity theory (see the proofs of Lemma 4.5 and Lemma 4.9 of Winkler 2016c for details), one can see that there exist $\theta \in (0, 1)$ and $\alpha \in (\frac{1}{2}, 1)$ and $C_1 > 0$ such that for all $t > 1$

$$\|n\|_{C^{\theta, \frac{\theta}{2}}(\overline{\Omega} \times [t, t+1])} + \|\nabla w(\cdot, t)\|_{C^\theta(\overline{\Omega})} + \|A^\alpha u(\cdot, t)\|_{L^2(\Omega)} \leq C_1. \tag{1.4.81}$$

If (1.4.78) were false, then there would be $C_2 > 0$, $(t_k)_{k \in \mathbb{N}}$ and $(x_k)_{k \in \mathbb{N}} \subseteq \Omega$ such that $t_k \to \infty$ as $k \to \infty$, and $n(x_k, t_k) > C_2$ for all $k \in \mathbb{N}$, which, along with the uniform continuity of n in $\overline{\Omega} \times [t, t+1]$ as shown by (1.4.81), entails that one can find $r > 0$ such that $B(x_k, r) \subseteq \Omega$ for all $k \in \mathbb{N}$ and $n(x, t_k) > \frac{C_2}{2}$ for all $x \in B(x_k, r)$. This shows

$$\int_\Omega n(\cdot, t_k) \geq \int_{B(x_k, r)} n(\cdot, t_k) \geq \frac{C_2}{2} \pi r^2$$

which contradicts (1.4.2) and thus proves (1.4.78). Similarly, on the basis of (1.4.77) and (1.4.81), (1.4.79) can be proved. Finally, (1.4.80) results from (1.4.14), (1.4.81) and a simple interpolation, and thereby completes the proof.

1.4.3   Asymptotic Profile of Solutions

It is observed that in the case $r < 0$, solutions to (1.1.5), (1.1.8), (1.1.9) enjoy the exponential decay property due to the exponential decay of $\|n(\cdot, t)\|_{L^1(\Omega)}$. Therefore, we pay our attention to the asymptotic profile of (1.1.5), (1.1.8), (1.1.9) in the cases $r > 0$ and $r = 0$, namely, we will give the proofs of Theorems 1.3 and 1.4 respectively.

1. The Case $r > 0$

Making use of the convergence properties of $(n, \frac{|\nabla c|}{c})$ asserted in Theorem 1.2, we apply $L^p - L^q$ estimates for the Neumann heat semigroup $(e^{t\Delta})_{t>0}$ to show

$(n, c, u) \to (\frac{r}{\mu}, 0, 0)$ in $L^\infty(\Omega)$ and $\frac{|\nabla c|}{c} \to 0$ in $L^p(\Omega)$ at some exponential rate as $t \to \infty$, respectively, whenever μ is suitably large compared with r. To this end, we first make an observation which will be used in the proof of the subsequent lemma:

Lemma 1.44 *For any fixed* $\alpha \in (0, \min\{\lambda_1, r\})$, *there exists* $\varepsilon_1 > 0$ *such that*

$$8\mu\varepsilon_1 < r - \alpha \qquad (1.4.82)$$

as well as

$$4c_1 I \varepsilon_2 < 1, \quad 8\chi c_4 I \varepsilon_2 < 1 \qquad (1.4.83)$$

where $c_i > 0$ *(i=1,4) is given in Lemma 1.1,* $I = \int_0^\infty (1 + \sigma^{-\frac{2}{3}} + \sigma^{-\frac{1}{2}}) e^{-(\lambda_1 - \alpha)\sigma} d\sigma$ *and* $\varepsilon_2 = 4c_1 |\Omega|^{\frac{1}{6}} I \varepsilon_1$.

Lemma 1.45 *Let* (n, w, u) *be the global bounded solution of* (1.4.1). *For fixed* $\alpha \in (0, \min\{\lambda_1, r\})$ *and* $\mu > 32\chi c_4 c_1 |\Omega|^{\frac{1}{6}} I^2 r$, *one can find constants* $C_i > 0$ *(i =* 1, 2, 3*) and* $\beta < \alpha$ *such that*

$$\|n(\cdot, t) - \frac{r}{\mu}\|_{L^\infty(\Omega)} \le C_1 e^{-\alpha t}, \qquad (1.4.84)$$

$$\|\nabla w(\cdot, t)\|_{L^6(\Omega)} \le C_2 e^{-\alpha t} \qquad (1.4.85)$$

as well as

$$\|u(\cdot, t)\|_{L^\infty(\Omega)} \le C_3 e^{-\beta t} \qquad (1.4.86)$$

for all $t \ge 1$.

Proof Let $\tilde{N}(x, t) = n(x, t) - \frac{r}{\mu}, \varepsilon_1 > 0$ and $\varepsilon_2 > 0$ be given by Lemma 1.44. Then from (1.4.43), (1.4.44) and (1.4.45), there exists $t_0 > 1$ suitably large such that for $t \ge t_0$

$$\|\tilde{N}(\cdot, t)\|_{L^\infty(\Omega)} \le \frac{\varepsilon_1}{8}, \quad (c_2 + 1)\|\nabla w(\cdot, t)\|_{L^\infty(\Omega)} \le \frac{\varepsilon_2}{8} \qquad (1.4.87)$$

and

$$8c_1 \|u(\cdot, t)\|_{L^\infty(\Omega)} \int_0^\infty (1 + \sigma^{-\frac{1}{2}}) e^{-(\lambda_1 - \alpha)\sigma} d\sigma \le 1. \qquad (1.4.88)$$

Now we consider

$$T \triangleq \sup\left\{\tilde{T} \in (t_0, \infty) \,\middle|\, \begin{array}{ll} \|\tilde{N}(\cdot, t)\|_{L^\infty(\Omega)} \le \varepsilon_1 e^{-\alpha(t-t_0)} & \text{for all } t \in [t_0, \tilde{T}), \\ \|\nabla w(\cdot, t)\|_{L^6(\Omega)} \le \varepsilon_2 e^{-\alpha(t-t_0)} & \text{for all } t \in [t_0, \tilde{T}). \end{array}\right\} \qquad (1.4.89)$$

By (1.4.87), T is well-defined. In what follows, we shall demonstrate that $T = \infty$.

To this end, we first invoke the variation-of-constants representation of w:

$$w(\cdot, t) = e^{(t-t_0)\Delta} w(\cdot, t_0) - \int_{t_0}^{t} e^{(t-s)\Delta} |\nabla w(\cdot, s)|^2 ds + \int_{t_0}^{t} e^{(t-s)\Delta} \tilde{N}(\cdot, s) ds$$
$$- \int_{t_0}^{t} e^{(t-s)\Delta} (u \cdot \nabla w)(\cdot, s) ds + \frac{r}{\mu}(t - t_0),$$

$$(1.4.90)$$

and use Lemma 1.1(i), (ii) to estimate

$$\|\nabla w(\cdot, t)\|_{L^6(\Omega)}$$
$$\leq \|\nabla e^{(t-t_0)\Delta} w(\cdot, t_0)\|_{L^6(\Omega)} + \int_{t_0}^{t} \|\nabla e^{(t-s)\Delta} |\nabla w(\cdot, s)|^2\|_{L^6(\Omega)} ds$$
$$+ \int_{t_0}^{t} \|\nabla e^{(t-s)\Delta} N(\cdot, s)\|_{L^6(\Omega)} ds + \int_{t_0}^{t} \|\nabla e^{(t-s)\Delta} (u \cdot \nabla w)(\cdot, s)\|_{L^6(\Omega)} ds$$
$$\leq 2c_2 e^{-\lambda_1(t-t_0)} \|\nabla w(\cdot, t_0)\|_{L^6(\Omega)}$$
$$+ c_1 \int_{t_0}^{t} (1 + (t-s)^{-\frac{2}{3}}) e^{-\lambda_1(t-s)} \|\nabla w(\cdot, s)\|_{L^6(\Omega)}^2 ds$$
$$+ c_1 |\Omega|^{\frac{1}{6}} \int_{t_0}^{t} (1 + (t-s)^{-\frac{1}{2}}) e^{-\lambda_1(t-s)} \|\tilde{N}(\cdot, s)\|_{L^\infty(\Omega)} ds$$
$$+ c_1 \int_{t_0}^{t} (1 + (t-s)^{-\frac{1}{2}}) e^{-\lambda_1(t-s)} \|u(\cdot, s)\|_{L^\infty(\Omega)} \|\nabla w(\cdot, s)\|_{L^6(\Omega)} ds$$
$$:= I_1 + I_2 + I_3$$

$$(1.4.91)$$

for all $t_0 < t < T$.

Now we estimate the terms I_i ($i = 1, 2, 3$), respectively. Firstly, from (1.4.87), we have $I_1 \leq \frac{\varepsilon_2}{4} e^{-\lambda_1(t-t_0)}$. By the definition of T and (1.4.83), we can see that

$$I_2 \leq c_1 \varepsilon_2^2 \int_{t_0}^{t} (1 + (t-s)^{-\frac{2}{3}}) e^{-\lambda_1(t-s)} e^{-2\alpha(s-t_0)} ds$$
$$\leq c_1 \varepsilon_2^2 \int_{t_0}^{t} (1 + (t-s)^{-\frac{2}{3}}) e^{-\lambda_1(t-s)} e^{-\alpha(s-t_0)} ds$$
$$\leq c_1 \varepsilon_2^2 \int_{0}^{\infty} (1 + \sigma^{-\frac{2}{3}}) e^{-(\lambda_1-\alpha)\sigma} d\sigma \cdot e^{-\alpha(t-t_0)}$$
$$\leq \frac{\varepsilon_2}{4} e^{-\alpha(t-t_0)}.$$

By the definition of T, (1.4.88) and $\varepsilon_2 = 4c_1 |\Omega|^{\frac{1}{6}} I \varepsilon_1$, we also have

$$I_3 \leq (c_1|\Omega|^{\frac{1}{6}}\varepsilon_1 + c_1 \sup_{t \geq t_0} \|u(\cdot,t)\|_{L^\infty(\Omega)}\varepsilon_2) \int_{t_0}^t (1 + (t-s)^{-\frac{1}{2}})e^{-\lambda_1(t-s)}e^{-\alpha(s-t_0)}ds$$

$$= (c_1|\Omega|^{\frac{1}{6}}\varepsilon_1 + c_1 \sup_{t \geq t_0} \|u(\cdot,t)\|_{L^\infty(\Omega)}\varepsilon_2) \int_{t_0}^t (1 + (t-s)^{-\frac{1}{2}})e^{-(\lambda_1-\alpha)(t-s)}e^{-\alpha(t-t_0)}ds$$

$$\leq (c_1|\Omega|^{\frac{1}{6}}\varepsilon_1 + c_1 \sup_{t \geq t_0} \|u(\cdot,t)\|_{L^\infty(\Omega)}\varepsilon_2) \int_0^\infty (1 + \sigma^{-\frac{1}{2}})e^{-(\lambda_1-\alpha)\sigma}d\sigma \cdot e^{-\alpha(t-t_0)}$$

$$\leq \frac{3\varepsilon_2}{8}e^{-\alpha(t-t_0)}.$$

Substituting these estimates into (1.4.91), we get

$$\|\nabla w(\cdot,t)\|_{L^6(\Omega)} \leq \frac{7\varepsilon_2}{8}e^{-\alpha(t-t_0)} < \varepsilon_2 e^{-\alpha(t-t_0)} \quad \text{for all } t \in [t_0, T). \qquad (1.4.92)$$

On the other hand, since $\tilde{N}_t = \Delta\tilde{N} + \chi\nabla\cdot(n\nabla w) - r\tilde{N} - \mu\tilde{N}^2 - u\cdot\nabla\tilde{N}$, the variation-of-constants representation of $\tilde{N}$ yields

$$\tilde{N}(\cdot,t) = e^{(t-t_0)(\Delta-r)}\tilde{N}(\cdot,t_0) + \chi\int_{t_0}^t e^{(t-s)(\Delta-r)}\nabla\cdot(n\nabla w)(\cdot,s)ds$$

$$- \mu\int_{t_0}^t e^{(t-s)(\Delta-r)}\tilde{N}^2(\cdot,s)ds - \int_{t_0}^t e^{(t-s)(\Delta-r)}(u\cdot\nabla\tilde{N})(\cdot,s)ds.$$

Then by $\nabla\cdot u = 0$ we can see that

$$\|\tilde{N}(\cdot,t)\|_{L^\infty(\Omega)}$$

$$\leq \|e^{(t-t_0)(\Delta-r)}\tilde{N}(\cdot,t_0)\|_{L^\infty(\Omega)} + \mu\int_{t_0}^t \|e^{(t-s)(\Delta-r)}\tilde{N}^2(\cdot,s)\|_{L^\infty(\Omega)}ds$$

$$+ \int_{t_0}^t \|e^{(t-s)(\Delta-r)}\nabla\cdot(u\tilde{N})(\cdot,s)\|_{L^\infty(\Omega)}ds$$

$$+ \chi\int_{t_0}^t \|e^{(t-s)(\Delta-r)}\nabla\cdot(n\nabla w)(\cdot,s)\|_{L^\infty(\Omega)}ds$$

$$:= J_1 + J_2 + J_3 + J_4.$$

Here the maximum principle together with (1.4.87) ensures that

$$J_1 \leq e^{-r(t-t_0)}\|\tilde{N}(\cdot,t_0)\|_{L^\infty(\Omega)} \leq \frac{\varepsilon_1}{8}e^{-\alpha(t-t_0)}.$$

By the definition of T and comparison principle, we infer that

$$J_2 \leq \mu\int_{t_0}^t e^{-r(t-s)}\|e^{(t-s)\Delta}\tilde{N}^2(\cdot,s)\|_{L^\infty(\Omega)}ds$$

$$\leq \mu\int_{t_0}^t e^{-r(t-s)}\|\tilde{N}(\cdot,s)\|_{L^\infty(\Omega)}^2 ds$$

$$\leq \quad \mu\varepsilon_1^2 \int_{t_0}^t e^{-r(t-s)}e^{-2\alpha(s-t_0)}ds$$

$$\leq \quad \mu\varepsilon_1^2 \int_{t_0}^t e^{-(r-\alpha)(t-s)}ds \cdot e^{-\alpha(t-t_0)}$$

$$\leq \quad \frac{\mu\varepsilon_1^2}{r-\alpha}e^{-\alpha(t-t_0)}$$

$$\leq \quad \frac{\varepsilon_1}{8}e^{-\alpha(t-t_0)}$$

due to (1.4.82) and $\alpha < r$. Similarly by (1.4.88), we have

$$J_3 \leq c_1 \sup_{t \geq t_0} \|u(\cdot,t)\|_{L^\infty(\Omega)} \int_{t_0}^t (1+(t-s)^{-\frac{1}{2}})e^{-(\lambda_1+r)(t-s)}\|\tilde{N}(\cdot,s)\|_{L^\infty(\Omega)}ds$$

$$\leq c_1 \sup_{t \geq t_0} \|u(\cdot,t)\|_{L^\infty(\Omega)}\varepsilon_1 \int_{t_0}^t (1+(t-s)^{-\frac{1}{2}})e^{-(\lambda_1+r)(t-s)}e^{-\alpha(s-t_0)}ds$$

$$\leq c_1 \sup_{t \geq t_0} \|u(\cdot,t)\|_{L^\infty(\Omega)}\varepsilon_1 \int_0^\infty (1+\sigma^{-\frac{1}{2}})e^{-(\lambda_1-\alpha)\sigma}d\sigma \cdot e^{-\alpha(t-t_0)}$$

$$\leq \frac{\varepsilon_1}{8}e^{-\alpha(t-t_0)}.$$

As for the term J_4, we recall (1.4.83), (1.4.89) and apply Lemma 1.1 (iv) to get

$$J_4 \leq \chi c_4 \int_{t_0}^t (1+(t-s)^{-\frac{2}{3}})e^{-(\lambda_1+r)(t-s)}\|(n\nabla w)(\cdot,s)\|_{L^6(\Omega)}ds$$

$$\leq \chi c_4\varepsilon_2 \int_{t_0}^t (1+(t-s)^{-\frac{2}{3}})e^{-(\lambda_1+r)(t-s)}(\frac{r}{\mu}+\varepsilon_1 e^{-\alpha(s-t_0)})e^{-\alpha(s-t_0)}ds$$

$$\leq \chi c_4\varepsilon_2(\frac{r}{\mu}+\varepsilon_1) \int_0^\infty (1+\sigma^{-\frac{2}{3}})e^{-(\lambda_1+r-\alpha)\sigma}d\sigma \cdot e^{-\alpha(t-t_0)}$$

$$\leq \frac{\varepsilon_1}{8}e^{-\alpha(t-t_0)} + \chi c_4\frac{r}{\mu}I\varepsilon_2 e^{-\alpha(t-t_0)}$$

$$= \frac{\varepsilon_1}{8}e^{-\alpha(t-t_0)} + 4\chi c_4 c_1 |\Omega|^{\frac{1}{6}}I^2\varepsilon_1\frac{r}{\mu}e^{-\alpha(t-t_0)}$$

$$\leq \frac{\varepsilon_1}{4}e^{-\alpha(t-t_0)}$$

due to $\mu > 32\chi c_4 c_1|\Omega|^{\frac{1}{6}}I^2 r$. Hence combining above inequalities, we arrive at

$$\|\tilde{N}(\cdot,t)\|_{L^\infty(\Omega)} \leq \frac{5\varepsilon_1}{8}e^{-\alpha(t-t_0)} \quad \text{for all } t \in [t_0, T).$$

This along with (1.4.92) readily shows that T cannot be finite. In combination with the decay property (1.4.84), a straightforward interpolation argument can be employed to prove (1.4.86).

Proof of Theorem 1.3.   According to (1.4.56) and $w = -\ln(\frac{c}{\|c_0\|_{L^\infty(\Omega)}})$, we have
$c(x,t) \leq \|c_0\|_{L^\infty(\Omega)} e^{-\frac{r}{2\mu}(t-t_3)}$ for all $t \geq t_3$. On the other hand, if $\mu_*(\chi, \Omega, r) :=$
$\max\{\mu_0, 32\chi c_4 c_1 |\Omega|^{\frac{1}{6}} I^2 r\}$, then as an immediate consequence of Theorem 1.3 and
Lemma 1.45, $n(\cdot, t) \to \frac{r}{\mu}$ and $\frac{|\nabla c|}{c}(\cdot, t) \to 0$ in $L^\infty(\Omega)$ and $L^6(\Omega)$, respectively, at
an exponential rate when $\mu > \mu_*(\chi, \Omega, r)$. Moreover, with the help of the uniform
boundedness of $\|\frac{|\nabla c|}{c}(\cdot, t)\|_{L^\infty(\Omega)}$ with respect to $t > 0$, one can show that $\frac{|\nabla c|}{c}(\cdot, t) \to$
0 in $L^p(\Omega)$ for any $p > 1$ exponentially by the interpolation argument. The proof
of this theorem is thus complete.

2. The Case $r = 0$

The proof of Theorem 1.4 proceeds on an alternative reasoning. To this end, making
use of the decay information on $|\nabla w|$ in $L^\infty(\Omega)$ in (1.4.77) and the quadratic degra-
dation in the n−equation, we first turn the decay property of $\|n(\cdot, t)\|_{L^1(\Omega)}$ from
(1.4.2) into an upper bound estimate of $\|n(\cdot, t)\|_{L^\infty(\Omega)}$.

Lemma 1.46 *Let (n, w, u) be the global bounded solution of* (1.4.1) *obtained in
Theorem 1.2 with $r = 0$, $\mu > 0$. Then one can find constant $C > 0$ such that*

$$\|n(\cdot, t)\|_{L^\infty(\Omega)} \leq \frac{C}{t+1} \quad \text{for all } t > 0. \tag{1.4.93}$$

Proof According to the known smoothing properties of the Neumann heat semigroup
$(e^{\tau\Delta})_{t>0}$ on $\Omega \subset \mathbb{R}^n$ (see Winkler 2010), one can pick $c_1 > 0$ and $c_2 > 0$ such that
for all $0 < \tau \leq 1$,

$$\|e^{\tau\Delta}\varphi\|_{L^\infty(\Omega)} \leq C_1\tau^{-\frac{n}{2}}\|\varphi\|_{L^1(\Omega)} \quad \text{for all } \varphi \in L^1(\Omega) \tag{1.4.94}$$

and

$$\|e^{\tau\Delta}\nabla \cdot \varphi\|_{L^\infty(\Omega)} \leq C_2\tau^{-\frac{1}{2}-\frac{n}{2p}}\|\varphi\|_{L^p(\Omega)} \quad \text{for all } \varphi \in C^1(\Omega; \mathbb{R}^n). \tag{1.4.95}$$

By (1.4.79) and (1.4.80), there exists $t_0 > 3$ such that

$$24C_2(\chi\|\nabla w(\cdot, t)\|_{L^3(\Omega)} + \|u(\cdot, t)\|_{L^3(\Omega)}) \leq 1 \quad \text{for all } t > t_0 - 1. \tag{1.4.96}$$

Now in order to prove the lemma, it is sufficient to derive a bound, independent of
$T \in (t_0, \infty)$, for $M(T) \triangleq \sup_{t_0-1<t<T} \{t\|n(\cdot, t)\|_{L^\infty(\Omega)}\}$.

By the variation-of-constants representation of n, we have

$$n(\cdot, t) = e^{\Delta}n(\cdot, t-1) + \chi\int_{t-1}^t e^{(t-s)\Delta}\nabla \cdot (n\nabla w)(\cdot, s)ds - \int_{t-1}^t e^{(t-s)\Delta}(u \cdot \nabla n)(\cdot, s)ds$$
$$- \mu\int_{t-1}^t e^{(t-s)\Delta}n^2(\cdot, s)ds.$$

$$\tag{1.4.97}$$

Since $e^{(t-s)\Delta}$ is nonnegative in Ω for all $0 < s < t$ due to the maximum principle, it
follows from the nonnegativity of n that for all $t \in (t_0, T)$

$$\|n(\cdot,t)\|_{L^\infty(\Omega)}$$

$$\leq \|e^\Delta n(\cdot,t-1)\|_{L^\infty(\Omega)} + \chi \int_{t-1}^t \|e^{(t-s)\Delta}\nabla\cdot(n\nabla w)(\cdot,s)\|_{L^\infty(\Omega)}ds$$

$$+ \int_{t-1}^t \|e^{(t-s)\Delta}(u\cdot\nabla n)(\cdot,s)\|_{L^\infty(\Omega)}ds$$

which along with (1.4.94)–(1.4.96) and (1.4.2) yields

$$\|n(\cdot,t)\|_{L^\infty(\Omega)}$$

$$\leq C_1\|n(\cdot,t-1)\|_{L^1(\Omega)} + C_2\chi \int_{t-1}^t (t-s)^{-\frac{5}{6}}\|(n\nabla w)(\cdot,s)\|_{L^3(\Omega)}ds$$

$$+ C_2 \int_{t-1}^t (t-s)^{-\frac{5}{6}}\|(un)(\cdot,s)\|_{L^3(\Omega)}ds$$

$$\leq \frac{C_1|\Omega|}{\mu(t-1+\gamma)} + \frac{6C_2}{t-1}(\chi \max_{t_0-1<s<T}\|\nabla w(\cdot,s)\|_{L^3(\Omega)} + \|u(\cdot,t)\|_{L^2(\Omega)})\cdot M(T)$$

$$\leq \frac{C_1|\Omega|}{\mu(t-1+\gamma)} + \frac{1}{4(t-1)}M(T).$$

Hence,

$$M(T) \leq \frac{4C_1|\Omega|}{\mu} + 2\sup_{t_0-1<s<t_0}\{s\|n(\cdot,s)\|_{L^\infty(\Omega)}\},$$

which readily yields (1.4.93) since $T > t_0$ is arbitrary, and thus ends the proof.

In light of Lemma 1.46, we can derive a pointwise estimate $c(x,t)$ from below.

Lemma 1.47 *Let (n,w,u) be the global classical solution of (1.4.1) obtained in Thenrem 1.2 with $r = 0$, $\mu > 0$. Then there exists $\kappa > 0$ fulfilling*

$$c(x,t) \geq \frac{\inf_{x\in\Omega} c_0(x)}{(t+1)^\kappa}. \tag{1.4.98}$$

Proof By the second equation of (1.4.1) and Lemma 1.46, we can see that

$$w_t \leq \Delta w - |\nabla w|^2 + \frac{C_1}{t+1} - u\cdot\nabla w$$

with some $C_1 > 0$ for all $t > 0$. Let $y \in C^1([0,\infty))$ denote the solution of the initial-value problem $y'(t) = \frac{C_1}{t+1}$, $y(0) = \|w_0\|_{L^\infty(\Omega)}$, then from the comparison principle, we infer that

$$w(x,t) \leq \|w_0\|_{L^\infty(\Omega)} + C_1\ln(t+1) \quad \text{for all } t > 0, \tag{1.4.99}$$

which along with $w = -\ln(\frac{c}{\|c_0\|_{L^\infty(\Omega)}})$, yields (1.4.98) with $\kappa = C_1$.

Now utilizing the decay information on $|\nabla w|$ in $L^\infty(\Omega)$ in (1.4.77) again, and thanks to the precise information on the decay of $\|n(\cdot, t)\|_{L^\infty(\Omega)}$ in Lemma 1.46, we can obtain the desired estimate for $\|n(\cdot, t)\|_{L^\infty(\Omega)}$ from below as well as the upper estimate for $\|\nabla w(\cdot, t)\|_{L^\infty(\Omega)}$.

Lemma 1.48 *Let (n, w, u) be the solution of (1.4.1) obtained in Thenrem 1.2 with $r = 0$, $\mu > 0$. Then one can find $C_1 > 0$ and $C_2 > 0$ fulfilling*

$$\|n(\cdot, t)\|_{L^\infty(\Omega)} \geq \frac{1}{|\Omega|}\|n(\cdot, t)\|_{L^1(\Omega)} \geq \frac{C_1}{t+1} \quad \text{for all } t > 0 \qquad (1.4.100)$$

as well as

$$\|\nabla w(\cdot, t)\|_{L^\infty(\Omega)} \leq \frac{C_2}{t+1} \quad \text{for all } t > 0. \qquad (1.4.101)$$

Proof We first adapt the method in Lemma 1.46 to derive the precise decay rate of $\|\nabla w(\cdot, t)\|_{L^\infty(\Omega)}$. By (1.4.79) and (1.4.80), one can choose some $t_0 > 2$ such that

$$4c_1 \int_0^\infty (1 + \sigma^{-\frac{1}{2}}) e^{-\lambda_1 \sigma} d\sigma (\|\nabla w(\cdot, t)\|_{L^\infty(\Omega)} + \|u(\cdot, t)\|_{L^\infty(\Omega)}) \leq 1 \quad (1.4.102)$$

for all $t > \frac{t_0}{2}$ and then let $M(T) \triangleq \sup_{\frac{t_0}{2} < s < T} \{s\|\nabla w(\cdot, s)\|_{L^\infty(\Omega)}\}$ for all $T > t_0$.

By the variation-of-constants representation of w, we have

$$w(\cdot, t) = e^{\frac{t}{2}\Delta} w\left(\cdot, \frac{t}{2}\right) - \int_{\frac{t}{2}}^t e^{(t-s)\Delta}|\nabla w|^2(\cdot, s)ds + \int_{\frac{t}{2}}^t e^{(t-s)\Delta}(n - u \cdot \nabla w)(\cdot, s)ds$$

for all $t_0 < t < T$. We then show that

$$\|\nabla w(\cdot, t)\|_{L^\infty(\Omega)}$$

$$\leq \|\nabla e^{\frac{t}{2}\Delta} w(\cdot, \frac{t}{2})\|_{L^\infty(\Omega)} + \int_{\frac{t}{2}}^t \|\nabla e^{(t-s)\Delta}|\nabla w|^2\|_{L^\infty(\Omega)} + \int_{\frac{t}{2}}^t \|\nabla e^{(t-s)\Delta} n\|_{L^\infty(\Omega)}$$

$$+ \int_{\frac{t}{2}}^t \|\nabla e^{(t-s)\Delta}(u \cdot \nabla w)\|_{L^\infty(\Omega)}$$

$$\leq c_1(1 + t^{-\frac{1}{2}}) e^{-\frac{\lambda_1 t}{2}} \|w(\cdot, \frac{t}{2})\|_{L^\infty(\Omega)} + c_1 \int_{\frac{t}{2}}^t (1 + (t-s)^{-\frac{1}{2}}) e^{-\lambda_1(t-s)} \|n(\cdot, s)\|_{L^\infty(\Omega)}$$

$$+ c_1 \int_{\frac{t}{2}}^t (1 + (t-s)^{-\frac{1}{2}}) e^{-\lambda_1(t-s)} \|\nabla w(\cdot, s)\|_{L^\infty(\Omega)}(\|\nabla w(\cdot, s)\|_{L^\infty(\Omega)} + \|u(\cdot, s)\|_{L^\infty(\Omega)})$$

$$\leq c_1(1 + t^{-\frac{1}{2}}) e^{-\frac{\lambda_1 t}{2}} (\|w_0\|_{L^\infty(\Omega)} + c_2 \ln(t+1)) + \frac{2c_1 c_2}{t} \int_0^\infty (1 + \sigma^{-\frac{1}{2}}) e^{-\lambda_1 \sigma} d\sigma$$

$$+ \frac{2c_1}{t} \int_0^\infty (1 + \sigma^{-\frac{1}{2}}) e^{-\lambda_1 \sigma} d\sigma \sup_{t \geq \frac{t_0}{2}} (\|\nabla w(\cdot, t)\|_{L^\infty(\Omega)} + \|u(\cdot, t)\|_{L^\infty(\Omega)}) \cdot M(T)$$

$$\leq c_1(1+t^{-\frac{1}{2}})e^{-\frac{\lambda_1 t}{2}}(\|w_0\|_{L^\infty(\Omega)} + c_2\ln(t+1)) + \frac{2c_1c_2}{t}\int_0^\infty (1+\sigma^{-\frac{1}{2}})e^{-\lambda_1\sigma}\,d\sigma$$
$$+ \frac{1}{2t}M(T)$$

by using Lemma 1.4.1(i), (1.4.99), (1.4.93) and (1.4.102). This along with the definition of $M(T)$ yields

$$M(T) \leq 2\sup_{\frac{t_0}{2}<s<t_0}\{s\|\nabla w(\cdot,s)\|_{L^\infty(\Omega)}\} + c_4$$

with some constant $c_4 > 0$ as $\lim_{t\to\infty} t\ln(t+1)e^{-\lambda_1 t} = 0$. Hence, upon the definition of $M(T)$, we arrive at (1.4.101) with an evident choice of C_2.

Continuing with the proof, we claim that there exists $c_4 > 0$ such that

$$\|n(\cdot,t)\|_{L^\infty(\Omega)} \geq \frac{1}{|\Omega|}\|n(\cdot,t)\|_{L^1(\Omega)} \geq \frac{c_4}{t+1} \quad \text{for all } t > 0. \tag{1.4.103}$$

Indeed, from the n-equation of (1.4.1) with $r = 0$ and Young's inequality, it follows that $\frac{d}{dt}\int_\Omega \ln n = \int_\Omega \frac{|\nabla n|^2}{n^2} + \chi\int_\Omega \frac{1}{n}\nabla\cdot(n\nabla w) - \mu\int_\Omega n \geq -\frac{\chi^2}{4}\int_\Omega |\nabla w|^2 - \mu\int_\Omega n$. Inserting (1.4.2) and (1.4.101) into the above inequality yields $\frac{d}{dt}\int_\Omega \ln n \geq -\frac{\chi^2}{4}\frac{C_2^2|\Omega|}{(t+1)^2} - \frac{|\Omega|}{t+\gamma}$ and thus

$$\int_\Omega \ln n(\cdot,t) \geq -|\Omega|\ln(t+\gamma) - c_5 \quad \text{for all } t > 1 \tag{1.4.104}$$

with some $c_5 > 0$. On the other hand, by the Jensen inequality, we have

$$|\Omega|\ln(\int_\Omega n(\cdot,t)) - |\Omega|\ln|\Omega| = |\Omega|\ln\{\frac{1}{|\Omega|}\int_\Omega n(\cdot,t))\} \geq \int_\Omega \ln n(\cdot,t).$$

This inequality together with (1.4.104) readily leads to (1.4.100).

With the above lemmas at hand, we can now complete the proof of Theorem 1.4.
Proof of Theorem 1.4. By $w = -\ln(\frac{c}{\|c_0\|_{L^\infty(\Omega)}})$, Lemma 1.46 and Lemma 1.48, one can see that $(n, \frac{|\nabla c|}{c}) \longrightarrow (0,0)$ in $L^\infty(\Omega)$ algebraically as $t \to \infty$. Hence, it suffices to show the decay property of $c(x,t)$. In view of the w-equation in (1.4.1), (1.4.103), (1.4.101) and $\nabla\cdot u = 0$, we can pick $C_i > 0$ $(i = 1,2,3)$ such that

$$\frac{d}{dt}\int_\Omega w = \int_\Omega n - \int_\Omega |\nabla w|^2 - \int_\Omega u\cdot\nabla w \geq \frac{C_1|\Omega|}{t+1} - \frac{C_2|\Omega|}{(t+1)^2},$$

and hence $\int_\Omega w(\cdot,t) \geq C_1|\Omega|\ln(t+1) - C_3$, which entails that for any $t > 0$ there exists $x_0(t) \in \Omega$ such that $w(x_0(t),t) \geq C_1\ln(t+1) - \frac{C_3}{|\Omega|}$. Since for each $\varphi \in W^{1,p}(\Omega)$ with $p > 2$, there exists $C_4 > 0$ such that

$$|\varphi(x) - \varphi(y)| \le C_4 |x-y|^{1-\frac{2}{p}} \|\nabla\varphi\|_{L^p(\Omega)} \quad \text{for all } x, y \in \Omega,$$

we therefore obtain from (1.4.101) that

$$w(x, t) \ge w(x_0(t), t) - |x - x_0(t)| \|\nabla w(\cdot, t)\|_{L^\infty(\Omega)} \tag{1.4.105}$$

$$\ge C_1 \ln(t+1) - \frac{C_3}{|\Omega|} - C_4 \text{diam}(\Omega),$$

and thereby $c(x, t) \le \dfrac{C_5}{(t+1)^{C_1}}$ for $x \in \Omega, t > 0$ for some $C_5 > 0$. This together with (1.4.98) shows that $c(x, t)$ actually converges to 0 in $L^\infty(\Omega)$ algebraically as $t \to \infty$, and thus ends the proof of Theorem 1.4.

Chapter 2
Keller–Segel–Navier–Stokes System Involving Tensor-Valued Sensitivity

2.1 Introduction

Chemotaxis, the biased movement of cells in response to chemical gradients, plays an important role in coordinating cell migration in many biological phenomena (see Hillen and Painter 2009). For example, the fruit fly Drosophila melanogaster navigates up gradients of attractive odors during food location, and male moths follow pheromone gradients released by the female during mate location. In 1970, Keller and Segel (1971b) proposed a mathematical model describing chemotactic aggregation of cellular slime molds

$$\begin{cases} n_t = \Delta n - \nabla \cdot (n \nabla c), & x \in \Omega, \ t > 0, \\ c_t = \Delta c - c + n & x \in \Omega, \ t > 0, \end{cases} \tag{2.1.1}$$

where $\Omega \subset \mathbb{R}^N$, and n and c denote the density of the cell population and the concentration of the attracting chemical substance, respectively. One of the most characteristic mathematical features of system (2.1.1) is the possibility of blow-up of solutions in a finite or infinite time. It is well known that solutions of system (2.1.1) may blow up when $N = 2$ with large total mass of cells and $N \geq 3$ with arbitrarily small prescribed total mass of cells (see Bellomo et al. 2015; Horstmann 2003; Nagai et al. 1997; Winkler 2013). In order to describe the nonlinear dependence on the cell density in cell movement, the following variant has also been widely studied:

$$\begin{cases} n_t = \Delta n^m - \nabla \cdot (n \nabla c), & x \in \Omega, \ t > 0, \\ c_t = \Delta c - c + n, & x \in \Omega, \ t > 0, \end{cases} \tag{2.1.2}$$

where $m > 0$. Recent results indicates that $m = 2 - \frac{2}{N}$ is the critical blow-up exponent of (2.1.2) in some sense. Indeed, all solutions are global and uniformly bounded if $m > 2 - \frac{2}{N}$ (see Tao and Winkler 2012a; Winkler 2010b); whereas if $m < 2 - \frac{2}{N}$,

© The Author(s) 2022
Y. Ke et al., *Analysis of Reaction-Diffusion Models with the Taxis Mechanism*,
Financial Mathematics and Fintech, https://doi.org/10.1007/978-981-19-3763-7_2

(2.1.2) has some solutions which blow up in a finite time (see Cieślak and Stinner 2012; Winkler 2010b).

Recent analytical findings show that already cell transport through a given fluid can substantially influence the solution behavior in certain Keller–Segel-type chemotaxis systems (Kiselev and Ryzhik 2012a, b); in fact, even complete suppression of blow-up may occur (Kiselev and Xu 2016). To the best of our knowledge, however, the full mutual coupling of equations from fluid dynamics to chemotaxis systems, including buoyancy-driven feedback on the fluid motion, has been considered in the analytical literature in cases when the signal substance is produced by the cells, which seems rather thin compared with that in cases for the oxygen-consumed though, such as the Keller–Segel–Navier–Stokes of the form

$$
\begin{cases}
n_t + u \cdot \nabla n = \Delta n - \nabla \cdot (n S(x, n, c) \nabla c), & x \in \Omega, t > 0, \\
c_t + u \cdot \nabla c = \Delta c - c + n, & x \in \Omega, t > 0, \\
u_t + \kappa (u \cdot \nabla) u + \nabla P = \Delta u + n \nabla \phi, & x \in \Omega, t > 0, \\
\nabla \cdot u = 0, & x \in \Omega, t > 0, \\
(\nabla n - n S(x, n, c)) \cdot v = \nabla c \cdot v = 0, \ u = 0, & x \in \partial \Omega, t > 0, \\
n(x, 0) = n_0(x), c(x, 0) = c_0(x), \ u(x, 0) = u_0(x), & x \in \Omega,
\end{cases}
\tag{2.1.3}
$$

where n and c are defined as before and $\Omega \subset \mathbb{R}^3$ is a bounded domain with a smooth boundary. Here, u, P, ϕ and $\kappa \in \mathbb{R}$ denote, respectively, the velocity field, the associated pressure of the fluid, the potential of the gravitational field and the strength of nonlinear fluid convection. $S(x, n, c)$ is a chemotactic sensitivity tensor satisfying

$$
S \in C^2(\bar{\Omega} \times [0, \infty)^2; \mathbb{R}^{3 \times 3})
\tag{2.1.4}
$$

and

$$
|S(x, n, c)| \leq C_S (1 + n)^{-\alpha} \quad \text{for all} \ (x, n, c) \in \Omega \times [0, \infty)^2
\tag{2.1.5}
$$

with some $C_S > 0$ and $\alpha > 0$. Problem (2.1.3) is proposed to describe the chemotaxis–fluid interaction in cases when the evolution of the chemoattractant is essentially dominated by production through cells (see Winkler et al. 2015 and Hillen and Painter 2009). For example, in two dimensions, if $S = S(x, n, c)$ is a tensor-valued sensitivity fulfilling (2.1.4) and (2.1.5), Wang and Xiang (2015) proved that the Stokes version ($\kappa = 0$ in the first equation of (2.1.3)) of system (2.1.3) admits a unique global classical solution that is bounded. Recently, Wang, Winkler and Xiang (2018) extended the above result Wang and Xiang (2015) to the Navier–Stokes version ($\kappa \neq 0$ in the first equation of (2.1.3)). In both papers Wang et al. (2018) and Wang and Xiang (2015), the condition $\alpha > 0$ is optimal for the existence of the solution. Furthermore, similar results are also valid for the three-dimensional Stokes version ($\kappa = 0$ in the first equation of (2.1.3)) of system (2.1.3) with $\alpha > \frac{1}{2}$ (see Wang and Xiang 2016). In the three-dimensional case, Wang and Liu (2017) showed that the Keller–Segel–Navier–Stokes ($\kappa \neq 0$ in the first equation of (2.1.3)) system (2.1.3) admits a global weak solution for tensor-valued sensitivity $S(x, n, c)$ satisfying (2.1.4) and (2.1.5) with $\alpha > \frac{3}{7}$. Recently, due to the lack of enough regularity and compactness properties for the first equation, by using the idea proposed by Winkler (2015a), Wang (2017) presented the existence of global **very weak** solutions for the system (2.1.3) under the assumption that S satisfies (2.1.4) and (2.1.5) with $\alpha > \frac{1}{3}$, which, in light of the known results for the fluid-free system mentioned above, is an optimal restriction on α. However, the existence of global (stronger than the result of Wang 2017) **weak** solutions is still open.

When taking the nonlinear diffusion of the cells into account, the system above may be reformed as

$$
\begin{cases}
n_t + u \cdot \nabla n = \Delta n^m - \nabla \cdot (nS(x, n, c)\nabla c), & x \in \Omega, t > 0, \\
c_t + u \cdot \nabla c = \Delta c - c + n, & x \in \Omega, t > 0, \\
u_t + \kappa(u \cdot \nabla)u + \nabla P = \Delta u + n\nabla\phi, & x \in \Omega, t > 0, \\
\nabla \cdot u = 0, & x \in \Omega, t > 0,
\end{cases}
\tag{2.1.6}
$$

with S fulfilling

$$|S(x, n, c)| \leq C_S$$

for some positive constant C_S. When $\kappa = 0$, Li et al. (2016) and Zheng (2019) considered the chemotaxis–Stokes system (2.1.6) for $N = 2$ and $N = 3$, respectively. They concluded that when $m > 2 - \frac{2}{N}$, the weak solutions of the simplified system (2.1.6) ($\kappa = 0$) are global existent and bounded. But till now, as far as we know, it is still not clear that in the case that $\kappa \neq 0$, whether the solution of the chemotaxis–Navier–Stokes system (2.1.6) is bounded or not.

The emergence of degenerate diffusion, full Navier–Stokes fluid ($\kappa \neq 0$) and rotational flux (tensor-valued sensitivity S) makes the system (2.1.6) contain a more complex cross-diffusion mechanism, which brings more mathematical difficulties to the problem. In fact, if $\kappa = 0$, by utilizing the L^1 estimate on n, one can invoke Lemma 2.4 in Wang and Xiang (2015) and the Sobolev embedding theorem (Theorem 5.6.6 in Evans 2010) to obtain the regularity of u in arbitrary L^p spaces (see Lemma 2.4 in Li et al. 2016). Then one can also obtain L^p estimate on c, by using the variation-of-constants representation for c (see the proof of Lemma 2.6 in Wang and Xiang 2015 and Lemma 2.6 in Li et al. 2016). By using the estimates on c and u, one can finally derive the entropy-like estimate involving the functional $\int_\Omega n^p + \int_\Omega |\nabla c|^{2q}$ (see Lemma 2.9 in Li et al. 2016 or Lemma 2.10 in Wang and Xiang 2015). Once the crucial step has been accomplished, the main results can be easily obtained by using the standard Alikakos–Moser iteration. However, when $\kappa \neq 0$, one cannot acquire the regularity of u in arbitrary L^p spaces directly. Here, we develop some L^p-estimate techniques to raise the a priori estimates of solutions from $L^1(\Omega) \to L^{m-1}(\Omega) \to L^m(\Omega) \to L^p(\Omega)$ (for any $p > 2$), which even seems a new method in the case of fluid-free system.

The first part of this chapter is concerned with system (2.1.6) along with the initial data

$$n(x, 0) = n_0(x), \quad c(x, 0) = c_0(x), \quad u(x, 0) = u_0(x), \qquad x \in \Omega, \qquad (2.1.7)$$

and under the boundary conditions

$$\left(\nabla n^m - nS(x, n, c)\nabla c\right) \cdot \nu = \nabla c \cdot \nu = 0, \quad u = 0, \qquad x \in \partial\Omega, t > 0 \qquad (2.1.8)$$

in a bounded domain $\Omega \subset \mathbb{R}^2$ with smooth boundary, where the chemotactic sensitivity tensor $S(x, n, c)$ satisfies

$$S \in C^2(\bar{\Omega} \times [0, \infty)^2; \mathbb{R}^{2\times2}) \qquad (2.1.9)$$

and

$$|S(x, n, c)| \leq C_S \quad \text{for all } (x, n, c) \in \Omega \times [0, \infty)^2 \qquad (2.1.10)$$

with some $C_S > 0$. Throughout this part $\phi \in W^{2,\infty}(\Omega)$ and the initial data (n_0, c_0, u_0) fulfills

$$\begin{cases} n_0 \in C^\kappa(\bar{\Omega}) \text{ for certain } \kappa > 0 \text{ with } n_0 \geq 0 \text{ in } \Omega, \\ c_0 \in W^{2,\infty}(\Omega) \text{ with } c_0 \geq 0 \text{ in } \bar{\Omega}, \\ u_0 \in D(A), \end{cases} \qquad (2.1.11)$$

where A denotes the Stokes operator with domain $D(A) := W^{2,2}(\Omega) \cap W_0^{1,2}(\Omega) \cap L_\sigma^2(\Omega)$, and $L_\sigma^2(\Omega) := \{\varphi \in L^2(\Omega) | \nabla \cdot \varphi = 0\}$ (see Sohr 2001).

Within the above frameworks, the main result on global existence and boundedness of solutions to (2.1.6)–(2.1.8) is stated as follows (Zheng and Ke 2020).

Theorem 2.1 *Let $m > 1$, $\Omega \subset \mathbb{R}^2$ be a bounded domain with smooth boundary, and assume (2.1.9)–(2.1.11) hold. Then the problem (2.1.6)–(2.1.8) admits a global-in-time weak solution (n, c, u, P), which is uniformly bounded in the sense that*

$$\|n(\cdot, t)\|_{L^\infty(\Omega)} + \|c(\cdot, t)\|_{W^{1,\infty}(\Omega)} + \|u(\cdot, t)\|_{L^\infty(\Omega)} \leq C \text{ for all } t > 0 \qquad (2.1.12)$$

with some positive constant C.

Remark 2.1 (i) If $u \equiv 0$, Theorem 2.1 coincides with Theorem 5.1 in Winkler (2010b), which seems to be optimal according to the two-dimensional fluid-free system.

(ii) Theorem 2.1 extends the results of Li et al. (2016), in which the authors discussed the chemotaxis–Stokes system ($\kappa = 0$) in a two-dimensional **convex** domain. As mentioned earlier, we not only extend the results to the chemotaxis–Navier–Stokes system ($\kappa \neq 0$), but also remove the convexity assumption of the domain. In Li et al. (2016), in order to get the regularity of ∇c, the assumption that the domain should be convex is required. Applying the boundedness of $\|\nabla c\|_{L^2(\Omega)}$ (see Lemma 2.8) and the fractional Gagliardo–Nirenberg inequality (see Lemma 2.5 in Ishida et al. 2014) to gain the regularity of ∇c in arbitrary L^p spaces, the hypothesis of convexity for Ω is removed herein.

The second part of this chapter considers the globally defined **weak** solution (see Definition 2.1) to system (2.1.3) with the initial data (n_0, c_0, u_0) fulfilling

$$
\begin{cases}
n_0 \in C^\kappa(\bar{\Omega}) \text{ for certain } \kappa > 0 \text{ with } n_0 \geq 0 \text{ in } \Omega, \\
c_0 \in W^{1,\infty}(\Omega) \text{ with } c_0 \geq 0 \text{ in } \bar{\Omega}, \\
u_0 \in D(A_r^\gamma) \text{ for some } \gamma \in (3/4, 1) \text{ and any } r \in (1, \infty),
\end{cases} \tag{2.1.13}
$$

where A_r denotes the Stokes operator with domain $D(A_r) := W^{2,r}(\Omega) \cap W_0^{1,r}(\Omega) \cap L_\sigma^r(\Omega)$ and $L_\sigma^r(\Omega) := \{\varphi \in L^r(\Omega) | \nabla \cdot \varphi = 0\}$ for $r \in (1, \infty)$ (similar to that in Sohr 2001).

Theorem 2.2 *Let $\Omega \subset \mathbb{R}^3$ be a bounded domain with a smooth boundary, and (2.1.13) hold. Suppose that S satisfies (2.1.4) and (2.1.5) with some $\alpha > \frac{1}{3}$. Then problem (2.1.3) possesses at least one global weak solution (n, c, u, P) in the sense of Definition 2.1.*

2.2 Preliminaries

In order to construct the weak solutions to (2.1.6)–(2.1.8) by an approximation procedure, we consider the approximate variant of (2.1.3) given by

$$
\begin{cases}
n_{\varepsilon t} + u_\varepsilon \cdot \nabla n_\varepsilon = \Delta(n_\varepsilon + \varepsilon)^m - \nabla \cdot (n_\varepsilon S_\varepsilon(x, n_\varepsilon, c_\varepsilon)\nabla c_\varepsilon), & x \in \Omega, \, t > 0, \\
c_{\varepsilon t} + u_\varepsilon \cdot \nabla c_\varepsilon = \Delta c_\varepsilon - c_\varepsilon + n_\varepsilon, & x \in \Omega, \, t > 0, \\
u_{\varepsilon t} + \nabla P_\varepsilon = \Delta u_\varepsilon - \kappa(u_\varepsilon \cdot \nabla)u_\varepsilon + n_\varepsilon \nabla \phi, & x \in \Omega, \, t > 0, \\
\nabla \cdot u_\varepsilon = 0, & x \in \Omega, \, t > 0, \\
\nabla n_\varepsilon \cdot \nu = \nabla c_\varepsilon \cdot \nu = 0, u_\varepsilon = 0, & x \in \partial\Omega, \, t > 0, \\
n_\varepsilon(x, 0) = n_0(x), c_\varepsilon(x, 0) = c_0(x), u_\varepsilon(x, 0) = u_0(x), & x \in \Omega,
\end{cases} \tag{2.2.1}
$$

where $S_\varepsilon(x, n, c) := \rho_\varepsilon(x)\chi_\varepsilon(n)S(x, n, c)$, $n \geq 0$, $c \geq 0$, $\rho_\varepsilon \in C_0^\infty(\Omega)$ such that $0 \leq \rho_\varepsilon \leq 1$ in Ω and $\rho_\varepsilon \nearrow 1$ in Ω as $\varepsilon \searrow 0$, $\chi_\varepsilon \in C_0^\infty([0, \infty))$ such that $0 \leq \chi_\varepsilon \leq 1$ in $[0, \infty)$ and $\chi_\varepsilon \nearrow 1$ in $[0, \infty)$ as $\varepsilon \searrow 0$.

By the well-established fixed-point arguments (see Lemma 2.1 in Winkler 2016v, Winkler 2015b and Lemma 2.1 in Painter and Hillen 2002), we could show the local solvability of system (2.2.1).

Lemma 2.1 *Let $\Omega \subset \mathbb{R}^2$ be a bounded domain with smooth boundary, and assume (2.1.9)–(2.1.11) hold. For any $\varepsilon \in (0, 1)$, there exist $T_{max,\varepsilon} \in (0, \infty]$ and a classical solution $(n_\varepsilon, c_\varepsilon, u_\varepsilon, P_\varepsilon)$ of system (2.2.1) in $\Omega \times [0, T_{max,\varepsilon})$. Here,*

$$\begin{cases} n_\varepsilon \in C^0(\bar{\Omega} \times [0, T_{max,\varepsilon})) \cap C^{2,1}(\bar{\Omega} \times (0, T_{max,\varepsilon})), \\ c_\varepsilon \in C^0(\bar{\Omega} \times [0, T_{max,\varepsilon})) \cap C^{2,1}(\bar{\Omega} \times (0, T_{max,\varepsilon})) \cap \bigcap_{p>1} L^\infty([0, T_{max,\varepsilon}); W^{1,p}(\Omega)), \\ u_\varepsilon \in C^0(\bar{\Omega} \times [0, T_{max,\varepsilon})) \cap C^{2,1}(\bar{\Omega} \times (0, T_{max,\varepsilon})) \cap \bigcap_{\gamma \in (0,1)} C^0([0, T_{max,\varepsilon}); D(A^\gamma)), \\ P_\varepsilon \in C^{1,0}(\bar{\Omega} \times (0, T_{max,\varepsilon})). \end{cases}$$

$$(2.2.2)$$

Moreover, n_ε and c_ε are nonnegative in $\Omega \times (0, T_{max,\varepsilon})$, and if $T_{max,\varepsilon} < +\infty$, then

$$\limsup_{t \nearrow T_{max,\varepsilon}} [\|n_\varepsilon(\cdot, t)\|_{L^\infty(\Omega)} + \|c_\varepsilon(\cdot, t)\|_{W^{1,\infty}(\Omega)} + \|A^\gamma u_\varepsilon(\cdot, t)\|_{L^2(\Omega)}] = \infty$$

for all $p > 2$ and $\gamma \in (\frac{1}{2}, 1)$.

Lemma 2.2 (Tao and Winkler 2015b) *Let $T \in (0, \infty]$, $\sigma \in (0, T)$, $A > 0$ and $B > 0$, and suppose that $y : [0, T) \to [0, \infty)$ is absolutely continuous such that $y'(t) + Ay(t) \leq h(t)$ for a.e. $t \in (0, T)$ with some nonnegative function $h \in L^1_{loc}([0, T))$ satisfying $\int_t^{t+\sigma} h(s)ds \leq B$ for all $t \in (0, T - \sigma)$. Then $y(t) \leq \max\{y_0 + B, \frac{B}{A\tau} + 2B\}$ for all $t \in (0, T)$.*

In light of the strong nonlinear term $(u \cdot \nabla)u$, problem (2.1.3) has no classical solutions in general, thus we consider its weak solutions.

Definition 2.1 Let $T > 0$ and assume that (n_0, c_0, u_0) fulfills (2.1.13). Then a triple of functions (n, c, u) is called a weak solution of (2.1.3) if the following conditions are satisfied:

$$\begin{cases} n \in L^1_{loc}(\bar{\Omega} \times [0, T)), \\ c \in L^1_{loc}([0, T); W^{1,1}(\Omega)), \\ u \in L^1_{loc}([0, T); W^{1,1}(\Omega); \mathbb{R}^3), \end{cases} \quad (2.2.3)$$

where $n \geq 0$ and $c \geq 0$ in $\Omega \times (0, T)$ as well as $\nabla \cdot u = 0$ in the distributional sense in $\Omega \times (0, T)$. Moreover,

$$\begin{aligned} &u \otimes u \in L^1_{loc}(\bar{\Omega} \times [0, \infty); \mathbb{R}^{3\times 3}) \text{ and } n \text{ belongs to } L^1_{loc}(\bar{\Omega} \times [0, \infty)), \\ &cu, \ nu, \text{ and } nS(x, n, c)\nabla c \text{ belong to } L^1_{loc}(\bar{\Omega} \times [0, \infty); \mathbb{R}^3) \end{aligned} \quad (2.2.4)$$

and

$$\begin{aligned} &-\int_0^T \int_\Omega n\varphi_t - \int_\Omega n_0\varphi(\cdot, 0) \\ &= -\int_0^T \int_\Omega \nabla n \cdot \nabla\varphi + \int_0^T \int_\Omega nS(x, n, c)\nabla c \cdot \nabla\varphi + \int_0^T \int_\Omega nu \cdot \nabla\varphi \end{aligned} \quad (2.2.5)$$

for any $\varphi \in C_0^\infty(\bar{\Omega} \times [0, T))$ satisfying $\frac{\partial\varphi}{\partial\nu} = 0$ on $\partial\Omega \times (0, T)$, as well as

$$-\int_0^T \int_\Omega c\varphi_t - \int_\Omega c_0\varphi(\cdot, 0)$$
$$= -\int_0^T \int_\Omega \nabla c \cdot \nabla\varphi - \int_0^T \int_\Omega c\varphi + \int_0^T \int_\Omega n\varphi + \int_0^T \int_\Omega cu \cdot \nabla\varphi$$

(2.2.6)

for any $\varphi \in C_0^\infty(\bar{\Omega} \times [0, T))$ and

$$-\int_0^T \int_\Omega u\varphi_t - \int_\Omega u_0\varphi(\cdot, 0) - \kappa \int_0^T \int_\Omega u \otimes u \cdot \nabla\varphi$$
$$= -\int_0^T \int_\Omega \nabla u \cdot \nabla\varphi - \int_0^T \int_\Omega n\nabla\phi \cdot \varphi$$

(2.2.7)

for any $\varphi \in C_0^\infty(\bar{\Omega} \times [0, T); \mathbb{R}^3)$ fulfilling $\nabla\varphi \equiv 0$ in $\Omega \times (0, T)$.

If $(n, c, u) : \Omega \times (0, \infty) \longrightarrow \mathbb{R}^5$ is a weak solution of (2.1.3) in $\Omega \times (0, T)$ for all $T > 0$, then (n, c, u) is called a global weak solution of (2.1.3).

To obtain the solution of system (2.1.3), we first consider the following approximate system of (2.1.3):

$$\begin{cases} n_{\varepsilon t} + u_\varepsilon \cdot \nabla n_\varepsilon = \Delta n_\varepsilon - \nabla \cdot (n_\varepsilon F_\varepsilon'(n_\varepsilon) S_\varepsilon(x, n_\varepsilon, c_\varepsilon)\nabla c_\varepsilon), & x \in \Omega, \ t > 0, \\ c_{\varepsilon t} + u_\varepsilon \cdot \nabla c_\varepsilon = \Delta c_\varepsilon - c_\varepsilon + F_\varepsilon(n_\varepsilon), & x \in \Omega, \ t > 0, \\ u_{\varepsilon t} + \nabla P_\varepsilon = \Delta u_\varepsilon - \kappa(Y_\varepsilon u_\varepsilon \cdot \nabla)u_\varepsilon + n_\varepsilon \nabla\phi, & x \in \Omega, \ t > 0, \\ \nabla \cdot u_\varepsilon = 0, & x \in \Omega, \ t > 0, \\ \nabla n_\varepsilon \cdot \nu = \nabla c_\varepsilon \cdot \nu = 0, u_\varepsilon = 0, & x \in \partial\Omega, \ t > 0, \\ n_\varepsilon(x, 0) = n_0(x), c_\varepsilon(x, 0) = c_0(x), \ u_\varepsilon(x, 0) = u_0(x), & x \in \Omega, \end{cases}$$

(2.2.8)

where

$$F_\varepsilon(s) := \frac{1}{\varepsilon}\ln(1 + \varepsilon s) \quad \text{for all} \ s \geq 0 \ \text{and} \ \varepsilon > 0,$$

(2.2.9)

as well as

$$S_\varepsilon(x, n, c) := \rho_\varepsilon(x)S(x, n, c), \quad x \in \bar{\Omega}, \ n \geq 0, \ c \geq 0$$

(2.2.10)

and

$$Y_\varepsilon w := (1 + \varepsilon A)^{-1} w \quad \text{for all} \ w \in L_\sigma^2(\Omega)$$

is a standard Yosida approximation and A is the realization of the Stokes operator (see Sohr 2001). Here, $(\rho_\varepsilon)_{\varepsilon\in(0,1)} \in C_0^\infty(\Omega)$ is a family of standard cutoff functions satisfying $0 \leq \rho_\varepsilon \leq 1$ in Ω and $\rho_\varepsilon \nearrow 1$ in Ω as $\varepsilon \searrow 0$.

The local solvability of (2.2.8) can be derived by a suitable extensibility criterion and a slight modification of the well-established fixed-point arguments in Lemma 2.1 of Winkler (2016v) (see also Winkler 2015b and Lemma 2.1 of Painter and Hillen 2002), so here we omit the proof.

Lemma 2.3 *For each $\varepsilon \in (0, 1)$, there exist $T_{max,\varepsilon} \in (0, \infty]$ and a classical solution $(n_\varepsilon, c_\varepsilon, u_\varepsilon, P_\varepsilon)$ of (2.2.8) in $\Omega \times (0, T_{max,\varepsilon})$ such that*

$$
\begin{cases}
n_\varepsilon \in C^0(\bar{\Omega} \times [0, T_{max,\varepsilon})) \cap C^{2,1}(\bar{\Omega} \times (0, T_{max,\varepsilon})), \\
c_\varepsilon \in C^0(\bar{\Omega} \times [0, T_{max,\varepsilon})) \cap C^{2,1}(\bar{\Omega} \times (0, T_{max,\varepsilon})), \\
u_\varepsilon \in C^0(\bar{\Omega} \times [0, T_{max,\varepsilon}); \mathbb{R}^3) \cap C^{2,1}(\bar{\Omega} \times (0, T_{max,\varepsilon}); \mathbb{R}^3), \\
P_\varepsilon \in C^{1,0}(\bar{\Omega} \times (0, T_{max,\varepsilon})),
\end{cases}
$$

classically solving (2.2.8) in $\Omega \times [0, T_{max,\varepsilon})$. Moreover, n_ε and c_ε are nonnegative in $\Omega \times (0, T_{max,\varepsilon})$, and

$$\|n_\varepsilon(\cdot, t)\|_{L^\infty(\Omega)} + \|c_\varepsilon(\cdot, t)\|_{W^{1,\infty}(\Omega)} + \|A^\gamma u_\varepsilon(\cdot, t)\|_{L^2(\Omega)} \to \infty \quad as \quad t \to T_{max,\varepsilon},$$

where γ is given by (2.1.13).

Lemma 2.4 (Winkler 2010; Zheng 2017c) *Let $(e^{\tau\Delta})_{\tau\geq 0}$ be the Neumann heat semigroup in Ω and $p > 3$. Then there exist positive constants $k_1 := k_1(\Omega)$, $k_2 := k_2(\Omega)$ and $k_3 := k_3(\Omega)$ such that for all $\tau > 0$ and any $\varphi \in W^{1,p}(\Omega)$,*

$$\|\nabla e^{\tau\Delta}\varphi\|_{L^p(\Omega)} \leq k_1\|\nabla\varphi\|_{L^p(\Omega)},$$

and for all $\tau > 0$ and each $\varphi \in L^\infty(\Omega)$

$$\|\nabla e^{\tau\Delta}\varphi\|_{L^p(\Omega)} \leq k_2(1 + \tau^{-\frac{1}{2}})\|\varphi\|_{L^\infty(\Omega)},$$

as well as for all $\tau > 0$ and all $\varphi \in C^1(\bar{\Omega}; \mathbb{R}^3)$ fulfilling $\varphi \cdot \nu = 0$ on $\partial\Omega$

$$\|e^{\tau\Delta}\nabla \cdot \varphi\|_{L^\infty(\Omega)} \leq k_3(1 + \tau^{-\frac{1}{2}-\frac{3}{2p}})\|\varphi\|_{L^p(\Omega)}.$$

2.3　Blow-Up Prevention by Nonlinear Diffusion to a Two-Dimensional Keller–Segel–Navier–Stokes System

2.3.1　Some Basic a Priori Estimates

In order to establish the global solvability of system (2.2.1), this section is to derive some necessary estimates for the approximate system (2.2.1). Let us first state two basic estimates on n_ε and c_ε.

Lemma 2.5 (Ke and Zheng 2019) *The solution of (2.2.1) satisfies*

$$\int_\Omega n_\varepsilon = \int_\Omega n_0 \quad for \ all \ \ t \in (0, T_{max,\varepsilon}) \tag{2.3.1}$$

as well as

$$\int_\Omega c_\varepsilon \leq \max\{\int_\Omega n_0, \int_\Omega c_0\} \ \text{for all} \ t \in (0, T_{max,\varepsilon}).$$

According to Lemma 2.5, we will derive some information on $(n_\varepsilon + \varepsilon)^{m-1}$, $\nabla(n_\varepsilon + \varepsilon)^{m-1}$, c_ε^2 and $|\nabla c_\varepsilon|^2$. This approach has been undertaken previously in, e.g., Wang et al. (2018), Ke and Zheng (2019) and Liu and Wang (2016).

Lemma 2.6 *Let $m > 1$. Then there exists $C > 0$ independent of ε such that the solution of (2.2.1) satisfies*

$$\int_\Omega (n_\varepsilon + \varepsilon)^{m-1} + \int_\Omega c_\varepsilon^2 + \int_\Omega |u_\varepsilon|^2 \leq C \ \text{for all} \ t \in (0, T_{max,\varepsilon}) \tag{2.3.2}$$

as well as

$$\int_t^{t+\tau} \int_\Omega \left[(n_\varepsilon + \varepsilon)^{2m-4} |\nabla n_\varepsilon|^2 + |\nabla c_\varepsilon|^2 + |\nabla u_\varepsilon|^2 \right] \leq C \tag{2.3.3}$$

for all $t \in (0, T_{max,\varepsilon} - \tau)$ with $\tau = \min\{1, \frac{1}{6} T_{max,\varepsilon}\}$.

Utilizing the latter spatio-temporal bound for $\nabla(n_\varepsilon + \varepsilon)^{m-1}$, we can establish the regularity of c_ε beyond Lemma 2.6.

Lemma 2.7 *Let $(n_\varepsilon, c_\varepsilon, u_\varepsilon, P_\varepsilon)$ be the solution of (2.2.1). Then for any $q > 2$, there exists $C := C(q)$ independent of ε such that*

$$\|c_\varepsilon(\cdot, t)\|_{L^q(\Omega)} \leq C \ \text{for all} \ t \in (0, T_{max,\varepsilon}). \tag{2.3.4}$$

Proof Multiplying the second equation in (2.2.1) by c_ε^{p-1} with $p > 3 + 4(m-1)$, using the fact $\nabla \cdot u_\varepsilon = 0$, we have

$$
\begin{aligned}
&\frac{1}{p}\frac{d}{dt}\int_\Omega c_\varepsilon^p + (p-1)\int_\Omega c_\varepsilon^{p-2}|\nabla c_\varepsilon|^2 + \int_\Omega c_\varepsilon^p \\
&\leq \int_\Omega c_\varepsilon^{p-1}(n_\varepsilon + \varepsilon) \\
&\leq \|n_\varepsilon + \varepsilon\|_{L^{\frac{p-2(m-1)}{p-4(m-1)}}(\Omega)} \left(\int_\Omega c_\varepsilon^{\frac{(p-1)[p-2(m-1)]}{m-1}} \right)^{\frac{m-1}{p-2(m-1)}} \\
&\leq C_1 \|n_\varepsilon + \varepsilon\|_{L^{\frac{p-2(m-1)}{p-4(m-1)}}(\Omega)} (\|\nabla c_\varepsilon^{\frac{p}{2}}\|_{L^2(\Omega)}^{\frac{p[p-2(m-1)-1]}{[p-1][p-2(m-1)]}} \|c_\varepsilon^{\frac{p}{2}}\|_{L^{\frac{2}{p}}(\Omega)}^{\frac{2(m-1)}{(p-1)[p-2(m-1)]}} + \|c_\varepsilon^{\frac{p}{2}}\|_{L^{\frac{2}{p}}(\Omega)}^{\frac{2(p-1)}{p}} \\
&\leq C_2 \|n_\varepsilon + \varepsilon\|_{L^{\frac{p-2(m-1)}{p-4(m-1)}}(\Omega)} (\|\nabla c_\varepsilon^{\frac{p}{2}}\|_{L^2(\Omega)}^{\frac{2[p-2(m-1)-1]}{p-2(m-1)}} + 1) \\
&\leq \frac{(p-1)}{2} \int_\Omega c_\varepsilon^{p-2}|\nabla c_\varepsilon|^2 + C_3 \|n_\varepsilon + \varepsilon\|_{L^{\frac{p-2(m-1)}{p-4(m-1)}}(\Omega)}^{p-2(m-1)} + C_3
\end{aligned}
\tag{2.3.5}
$$

for some positive constants C_i, $(i = 1, 2, 3)$, due to the Gagliardo–Nirenberg inequality.

By appropriate reformulation, if follows from (2.3.5) that for all $t \in (0, T_{max,\varepsilon})$,

$$\frac{1}{p}\frac{d}{dt}\int_{\Omega} c_{\varepsilon}^p + \frac{(p-1)}{2}\int_{\Omega} c_{\varepsilon}^{p-2}|\nabla c_{\varepsilon}|^2 + \int_{\Omega} c_{\varepsilon}^p$$
$$\leq C_3 \|n_{\varepsilon} + \varepsilon\|^{p-2(m-1)}_{L^{\frac{p-2(m-1)}{p-4(m-1)}}(\Omega)} + C_3 \text{ for all } t \in (0, T_{max,\varepsilon}). \tag{2.3.6}$$

In the following, we will estimate the integrals on the right-hand side of (2.3.6). In view of the Gagliardo–Nirenberg inequality, for some $C_4 > 0$ and $C_5 > 0$, we may derive from (2.3.3) that

$$\|n_{\varepsilon} + \varepsilon\|^{p-2(m-1)}_{L^{\frac{p-2(m-1)}{p-4(m-1)}}(\Omega)}$$

$$=\|(n_{\varepsilon} + \varepsilon)^{m-1}\|^{\frac{p-2(m-1)}{m-1}}_{L^{\frac{p-2(m-1)}{[p-4(m-1)](m-1)}}(\Omega)}$$

$$\leq\|\nabla(n_{\varepsilon} + \varepsilon)^{m-1}\|^2_{L^2(\Omega)}\|(n_{\varepsilon} + \varepsilon)^{m-1}\|^{\frac{p}{m-1}}_{L^{\frac{1}{m-1}}(\Omega)} + \|(n_{\varepsilon} + \varepsilon)^{m-1}\|^{\frac{p-2(m-1)}{m-1}}_{L^{\frac{1}{m-1}}(\Omega)}$$

$$\leq\|\nabla(n_{\varepsilon} + \varepsilon)^{m-1}\|^2_{L^2(\Omega)}$$

which along with (2.3.6) and Lemma 2.2 leads to (2.3.4).

Based on the information from Lemma 2.7, we can derive the more regularity property of solutions than that in Lemma 2.6 asserted in the following lemma.

Lemma 2.8 *Let $m > 1$. Then the solution of (2.2.1) satisfies*

$$\int_{\Omega}(n_{\varepsilon} + \varepsilon)^m + \int_{\Omega}|\nabla c_{\varepsilon}|^2 \leq C \quad \text{for all } t \in (0, T_{max,\varepsilon}) \tag{2.3.7}$$

as well as

$$\int_t^{t+\tau}\int_{\Omega}|\nabla(n_{\varepsilon} + \varepsilon)^{\frac{2m-1}{2}}|^2 \leq C \quad \text{for all } t \in (0, T_{max,\varepsilon} - \tau), \tag{2.3.8}$$

where $\tau = \min\{1, \frac{1}{6}T_{max,\varepsilon}\}$.

Proof Multiplying the first equation of (2.2.1) by $(n_{\varepsilon} + \varepsilon)^{m-1}$ and noticing $\nabla \cdot u_{\varepsilon} = 0$, one obtains

$$\frac{1}{m}\frac{d}{dt}\|n_{\varepsilon} + \varepsilon\|^m_{L^m(\Omega)} + (m-1)\int_{\Omega}(n_{\varepsilon} + \varepsilon)^{2m-3}|\nabla n_{\varepsilon}|^2$$

$$=-\int_{\Omega}(n_{\varepsilon} + \varepsilon)^{m-1}\nabla \cdot (n_{\varepsilon} S_{\varepsilon}(x, n_{\varepsilon}, c_{\varepsilon})\nabla c_{\varepsilon})$$

$$\leq C_S(m-1)\int_{\Omega}(n_{\varepsilon} + \varepsilon)^{m-1}|\nabla n_{\varepsilon}||\nabla c_{\varepsilon}| \quad \text{for all } t \in (0, T_{max,\varepsilon})$$

by using (2.1.10). From the Young inequality, it follows that for any $\eta > 0$, there exists $C_1(\eta) > 0$ such that

$$
\frac{1}{m}\frac{d}{dt}\|n_\varepsilon + \varepsilon\|^m_{L^m(\Omega)} + (m-1)\int_\Omega (n_\varepsilon + \varepsilon)^{2m-3}|\nabla n_\varepsilon|^2
$$

$$
\leq \frac{m-1}{2}\int_\Omega (n_\varepsilon + \varepsilon)^{2m-3}|\nabla n_\varepsilon|^2 + \frac{(m-1)C_S^2}{2}\int_\Omega (n_\varepsilon + \varepsilon)|\nabla c_\varepsilon|^2 \qquad (2.3.9)
$$

$$
\leq \frac{m-1}{2}\int_\Omega (n_\varepsilon + \varepsilon)^{2m-3}|\nabla n_\varepsilon|^2 + \eta\int_\Omega (n_\varepsilon + \varepsilon)^{2m} + C_1(\eta)\int_\Omega |\nabla c_\varepsilon|^{\frac{4m}{2m-1}}.
$$

On the other hand, in view of Lemma 2.5 and the Gagliardo–Nirenberg inequality, we infer that for some $C_2 > 0$,

$$
\int_\Omega (n_\varepsilon + \varepsilon)^{2m} = \|(n_\varepsilon + \varepsilon)^{\frac{2m-1}{2}}\|^{\frac{4m}{2m-1}}_{L^{\frac{4m}{2m-1}}(\Omega)} \leq C_2\|\nabla(n_\varepsilon + \varepsilon)^{\frac{2m-1}{2}}\|^2_{L^2(\Omega)} + C_2.
$$
$$(2.3.10)$$

Inserting (2.3.10) into (2.3.9) and choosing η appropriately small, we then get

$$
\frac{1}{m}\frac{d}{dt}\|n_\varepsilon + \varepsilon\|^m_{L^m(\Omega)} + \frac{m-1}{4}\int_\Omega (n_\varepsilon + \varepsilon)^{2m-3}|\nabla n_\varepsilon|^2
$$

$$
\leq C_3\int_\Omega |\nabla c_\varepsilon|^{\frac{4m}{2m-1}}.
$$

In light of (2.3.4), there exist positive constants $l_0 > \frac{1}{m-1}$ and C_2, such that

$$
\|c_\varepsilon(\cdot, t)\|_{L^{l_0}(\Omega)} \leq C_2 \quad \text{for all } t \in (0, T_{max,\varepsilon}). \qquad (2.3.11)
$$

Next, with the help of the Gagliardo–Nirenberg inequality and (2.3.11), we derive that

$$
\int_\Omega |\nabla c_\varepsilon|^{\frac{4m}{2m-1}} \leq C_4\|\Delta c_\varepsilon\|^{a\frac{4m}{2m-1}}_{L^2(\Omega)}\|c_\varepsilon\|^{(1-a)\frac{4m}{2m-1}}_{L^{l_0}(\Omega)} + C_4\|c_\varepsilon\|^{\frac{4m}{2m-1}}_{L^{l_0}(\Omega)}
$$

$$
\leq C_5\|\Delta c_\varepsilon\|^{a\frac{4m}{2m-1}}_{L^2(\Omega)} + C_5
$$

with some positive constants C_3 and C_4, and

$$
a = \frac{\frac{1}{2} + \frac{1}{l_0} - \frac{2m-1}{4m}}{\frac{1}{2} + \frac{1}{l_0}} \in (0, 1),
$$

which together with the fact that $\frac{4am}{2m-1} < 2$ (due to $l_0 > \frac{1}{m-1}$), yields

$$
C_3\int_\Omega |\nabla c_\varepsilon|^{\frac{4m}{2m-1}} \leq \frac{1}{8}\|\Delta c_\varepsilon\|^2_{L^2(\Omega)} + C_6. \qquad (2.3.12)
$$

On the other hand, taking $-\Delta c_\varepsilon$ as the test function for the second equation of (2.2.1), and using the Young inequality, it yields that for all $t \in (0, T_{max,\varepsilon})$

$$
\begin{aligned}
&\frac{1}{2}\frac{d}{dt}\|\nabla c_\varepsilon\|^2_{L^2(\Omega)} + \int_\Omega |\Delta c_\varepsilon|^2 + \int_\Omega |\nabla c_\varepsilon|^2 \\
&= -\int_\Omega n_\varepsilon \Delta c_\varepsilon - \int_\Omega \nabla c_\varepsilon \nabla(u_\varepsilon \cdot \nabla c_\varepsilon) \\
&= -\int_\Omega n_\varepsilon \Delta c_\varepsilon - \int_\Omega \nabla c_\varepsilon (\nabla u_\varepsilon \cdot \nabla c_\varepsilon) \\
&\leq \frac{1}{2}\int_\Omega |\Delta c_\varepsilon|^2 + \frac{1}{2}\int_\Omega n_\varepsilon^2 + \|\nabla u_\varepsilon\|_{L^2(\Omega)}\|\nabla c_\varepsilon\|^2_{L^4(\Omega)} \\
&\leq \frac{1}{2}\int_\Omega |\Delta c_\varepsilon|^2 + \frac{1}{2}\int_\Omega n_\varepsilon^2 + C_7\|\nabla u_\varepsilon\|_{L^2(\Omega)}\|\Delta c_\varepsilon\|_{L^2(\Omega)}\|\nabla c_\varepsilon\|_{L^2(\Omega)} \\
&\leq \frac{3}{4}\int_\Omega |\Delta c_\varepsilon|^2 + \frac{1}{2}\int_\Omega n_\varepsilon^2 + C_8\|\nabla u_\varepsilon\|^2_{L^2(\Omega)}\|\nabla c_\varepsilon\|^2_{L^2(\Omega)}
\end{aligned}
\tag{2.3.13}
$$

where we have used the fact that

$$
\int_\Omega \nabla c_\varepsilon \cdot (D^2 c_\varepsilon \cdot u_\varepsilon) = \frac{1}{2}\int_\Omega u_\varepsilon \cdot \nabla|\nabla c_\varepsilon|^2 = 0 \ \text{ for all } t \in (0, T_{max,\varepsilon})
$$

as well as

$$
\|\nabla c_\varepsilon\|^2_{L^4(\Omega)} \leq C_7\|\Delta c_\varepsilon\|_{L^2(\Omega)}\|\nabla c_\varepsilon\|_{L^2(\Omega)} \ \text{ for all } t \in (0, T_{max,\varepsilon})
$$

for some $C_7 > 0$ by the elliptic regularity (Gilbarg and Trudinger 2001).

Hence by appropriate reformulation, (2.3.9), (2.3.12) and (2.3.13), we derive that for all $t \in (0, T_{max,\varepsilon})$,

$$
\begin{aligned}
&\frac{d}{dt}(\|n_\varepsilon + \varepsilon\|^m_{L^m(\Omega)} + \|\nabla c_\varepsilon\|^2_{L^2(\Omega)}) + C_8\int_\Omega |\nabla(n_\varepsilon + \varepsilon)^{\frac{2m-1}{2}}|^2 \\
&\quad + C_8\int_\Omega (|\Delta c_\varepsilon|^2 + |\nabla c_\varepsilon|^2) \\
&\leq C_9\int_\Omega (n_\varepsilon + \varepsilon)^2 + C_9\|\nabla u_\varepsilon\|^2_{L^2(\Omega)}\|\nabla c_\varepsilon\|^2_{L^2(\Omega)} + C_9 \ \text{ for all } t \in (0, T_{max,\varepsilon})
\end{aligned}
\tag{2.3.14}
$$

with some $C_8 > 0$ and $C_9 > 0$. So by (2.3.10) and $m > 1$, we can see that

$$
\begin{aligned}
&\frac{d}{dt}(\|n_\varepsilon + \varepsilon\|^m_{L^m(\Omega)} + \|\nabla c_\varepsilon\|^2_{L^2(\Omega)}) + \frac{C_8}{2}\int_\Omega |\nabla(n_\varepsilon + \varepsilon)^{\frac{2m-1}{2}}|^2 \\
&\quad + C_8\int_\Omega (|\Delta c_\varepsilon|^2 + |\nabla c_\varepsilon|^2) \\
&\leq C_9\|\nabla u_\varepsilon\|^2_{L^2(\Omega)}\|\nabla c_\varepsilon\|^2_{L^2(\Omega)} + C_{10} \ \text{ for all } t \in (0, T_{max,\varepsilon})
\end{aligned}
\tag{2.3.15}
$$

for some constant $C_{10} > 0$.

Note that from the Gagliardo–Nirenberg inequality and (2.3.3), it follows that there exist constants $C_i > 0$, ($i = 11, 12, 13$), such that

$$
\begin{aligned}
&\int_t^{t+\tau} \int_\Omega (n_\varepsilon + \varepsilon)^m \\
&= \int_t^{t+\tau} \|(n_\varepsilon + \varepsilon)^{m-1}\|_{L^{\frac{m}{m-1}}(\Omega)}^{\frac{m}{m-1}} \\
&\leq C_{11} \left(\int_t^{t+\tau} \|\nabla(n_\varepsilon + \varepsilon)^{m-1}\|_{L^2(\Omega)}^{\frac{m-1}{m}} \|(n_\varepsilon + \varepsilon)^{m-1}\|_{L^{\frac{1}{m-1}}(\Omega)}^{\frac{1}{m}} \right. \\
&\quad \left. + \int_t^{t+\tau} \|(n_\varepsilon + \varepsilon)^{m-1}\|_{L^{\frac{1}{m-1}}(\Omega)} \right)^{\frac{m}{m-1}} \\
&\leq C_{12} \int_t^{t+\tau} \|\nabla(n_\varepsilon + \varepsilon)^{m-1}\|_{L^2(\Omega)}^2 + C_{12} \\
&\leq C_{13}.
\end{aligned}
$$

(2.3.16)

Therefore, if we write $y(t) := \|n_\varepsilon(\cdot, t) + \varepsilon\|_{L^m(\Omega)}^m + \|\nabla c_\varepsilon(\cdot, t)\|_{L^2(\Omega)}^2 + 1$ and $\rho(t) = C_9 \int_\Omega |\nabla u_\varepsilon(\cdot, t)|^2 + C_{10}$, (2.3.15) implies that

$$
y'(t) + h(t) \leq \rho(t) y(t) \quad \text{for all } t \in (0, T_{max,\varepsilon}),
$$

(2.3.17)

with

$$
h(t) = \frac{C_8}{2} \int_\Omega |\nabla(n_\varepsilon + \varepsilon)^{\frac{2m-1}{2}}|^2 + C_8 \int_\Omega |\Delta c_\varepsilon|^2.
$$

Next, by estimates (2.3.3) and (2.3.16), one obtains

$$
\int_t^{t+\tau} \rho(s) ds \leq C_{14}
$$

(2.3.18)

as well as

$$
\int_t^{t+\tau} y(s) ds \leq C_{14},
$$

(2.3.19)

for all $t \in (0, T_{max,\varepsilon} - \tau)$ and some $C_{14} > 0$. For given $t \in (0, T_{max,\varepsilon})$, by estimate (2.3.19), one can see that there exists $t_0 \geq 0$ such that $t_0 \in [t - \tau, t]$, $y(\cdot, t_0) \leq \frac{C_{14}}{\tau}$ and hence

$$
y(t) \leq y(t_0) e^{\int_{t_0}^t \rho(s) ds} \leq \frac{C_{14}}{\tau} e^{C_{14}}
$$

(2.3.20)

due to (2.3.18) and the Gronwall lemma. Moreover, combining (2.3.17), (2.3.18) and (2.3.20), we immediately get the desired inequality (2.3.8).

In order to obtain the global boundedness of n_ε, a further regularity of ∇n_ε beyond (2.3.7) seems to be required. Indeed, drawing on (2.3.3) and (2.3.8), we first have the following.

Lemma 2.9 *Let $m > 1$. There exists a positive constant C independent of ε, such that*

$$\int_\Omega |\nabla u_\varepsilon(\cdot, t)|^2 \leq C \text{ for all } t \in (0, T_{max,\varepsilon}). \tag{2.3.21}$$

Proof Applying the Helmholtz projection to the third equation in (2.2.1), and also multiplying the result identified by $A u_\varepsilon$, we then find that

$$
\begin{aligned}
&\frac{1}{2}\frac{d}{dt}\|A^{\frac{1}{2}} u_\varepsilon\|^2_{L^2(\Omega)} + \int_\Omega |A u_\varepsilon|^2 \\
&= \int_\Omega A u_\varepsilon \mathscr{P}(-\kappa(Y_\varepsilon u_\varepsilon \cdot \nabla)u_\varepsilon) + \int_\Omega \mathscr{P}(n_\varepsilon \nabla \phi) A u_\varepsilon \\
&\leq \frac{1}{2} \int_\Omega |A u_\varepsilon|^2 + \kappa^2 \int_\Omega |(Y_\varepsilon u_\varepsilon \cdot \nabla)u_\varepsilon|^2 \\
&\quad + \|\nabla\phi\|^2_{L^\infty(\Omega)} \int_\Omega n_\varepsilon^2 \text{ for all } t \in (0, T_{max,\varepsilon}).
\end{aligned}
\tag{2.3.22}
$$

Noticing that since $\|Y_\varepsilon u_\varepsilon\|_{L^2(\Omega)} \leq \|u_\varepsilon\|_{L^2(\Omega)}$, it follows from the Gagliardo–Nirenberg inequality and the Cauchy–Schwarz inequality that with some $C_1 > 0$ and $C_2 > 0$

$$
\begin{aligned}
&\kappa^2 \int_\Omega |(Y_\varepsilon u_\varepsilon \cdot \nabla)u_\varepsilon|^2 \\
&\leq \kappa^2 \|Y_\varepsilon u_\varepsilon\|^2_{L^4(\Omega)} \|\nabla u_\varepsilon\|^2_{L^4(\Omega)} \\
&\leq \kappa^2 C_1 [\|\nabla Y_\varepsilon u_\varepsilon\|_{L^2(\Omega)} \|Y_\varepsilon u_\varepsilon\|_{L^2(\Omega)}][\|A u_\varepsilon\|_{L^2(\Omega)} \|\nabla u_\varepsilon\|_{L^2(\Omega)}] \\
&\leq \kappa^2 C_2 \|\nabla Y_\varepsilon u_\varepsilon\|_{L^2(\Omega)} \|A u_\varepsilon\|_{L^2(\Omega)} \|\nabla u_\varepsilon\|_{L^2(\Omega)} \text{ for all } t \in (0, T_{max,\varepsilon}).
\end{aligned}
\tag{2.3.23}
$$

Now, from the fact that $D(A^{\frac{1}{2}}) := W_0^{1,2}(\Omega; \mathbb{R}^2) \cap L^2_\sigma(\Omega)$ and (2.3.2), it follows that

$$\|\nabla Y_\varepsilon u_\varepsilon\|_{L^2(\Omega)} = \|A^{\frac{1}{2}} Y_\varepsilon u_\varepsilon\|_{L^2(\Omega)} = \|Y_\varepsilon A^{\frac{1}{2}} u_\varepsilon\|_{L^2(\Omega)} \leq \|A^{\frac{1}{2}} u_\varepsilon\|_{L^2(\Omega)} \leq \|\nabla u_\varepsilon\|_{L^2(\Omega)}.$$

This, together with (2.3.23), yields

$$
\begin{aligned}
&\kappa^2 \int_\Omega |(Y_\varepsilon u_\varepsilon \cdot \nabla)u_\varepsilon|^2 \\
&\leq C_3 \|A u_\varepsilon\|_{L^2(\Omega)} \|\nabla u_\varepsilon\|^2_{L^2(\Omega)} \\
&\leq \frac{1}{4} \|A u_\varepsilon\|^2_{L^2(\Omega)} + C_3^2 \|\nabla u_\varepsilon\|^4_{L^2(\Omega)} \text{ for all } t \in (0, T_{max,\varepsilon}),
\end{aligned}
$$

which combining with (2.3.22) implies that

$$\frac{1}{2}\frac{d}{dt}\|A^{\frac{1}{2}}u_\varepsilon\|_{L^2(\Omega)}^2 \le C_3^2\|\nabla u_\varepsilon\|_{L^2(\Omega)}^4 + \|\nabla\phi\|_{L^\infty(\Omega)}^2\int_\Omega n_\varepsilon^2 \text{ for all } t \in (0, T_{max,\varepsilon}).$$

By the fact that $\|A^{\frac{1}{2}}u_\varepsilon\|_{L^2(\Omega)}^2 = \|\nabla u_\varepsilon\|_{L^2(\Omega)}^2$, we conclude that

$$z'(t) \le \rho(t)z(t) + h(t) \text{ for all } t \in (0, T_{max,\varepsilon}), \tag{2.3.24}$$

where $z(t) := \int_\Omega |\nabla u_\varepsilon(\cdot, t)|^2$, as well as $\rho(t) = 2C_3^2\int_\Omega |\nabla u_\varepsilon(\cdot, t)|^2$ and

$$h(t) = 2\|\nabla\phi\|_{L^\infty(\Omega)}^2\int_\Omega n_\varepsilon^2(\cdot, t).$$

Note that (2.3.3), (2.3.8) and (2.3.10) warrant that for some positive constant C_4,

$$\int_t^{t+\tau} (\rho(s) + h(s))ds \le C_4.$$

Therefore, by an argument similar to the proof of (2.3.8), we can arrive at (2.3.21) and thus complete the proof of this lemma.

Lemma 2.10 *Let $m > 1$. Then there exists a positive constant C independent of ε such that the solution of (2.2.1) satisfies*

$$\|\nabla c_\varepsilon(\cdot, t)\|_{L^{2m}(\Omega)} \le C \text{ for all } t \in (0, T_{max,\varepsilon}). \tag{2.3.25}$$

Proof Considering the fact that $\nabla c_\varepsilon \cdot \nabla\Delta c_\varepsilon = \frac{1}{2}\Delta|\nabla c_\varepsilon|^2 - |D^2 c_\varepsilon|^2$, the straightforward computation implies that

$$\frac{1}{2m}\frac{d}{dt}\|\nabla c_\varepsilon\|_{L^{2m}(\Omega)}^{2m}$$

$$= \int_\Omega |\nabla c_\varepsilon|^{2m-2}\nabla c_\varepsilon \cdot \nabla(\Delta c_\varepsilon - c_\varepsilon + n_\varepsilon - u_\varepsilon \cdot \nabla c_\varepsilon)$$

$$= \frac{1}{2}\int_\Omega |\nabla c_\varepsilon|^{2m-2}\Delta|\nabla c_\varepsilon|^2 - \int_\Omega |\nabla c_\varepsilon|^{2m-2}|D^2 c_\varepsilon|^2 - \int_\Omega |\nabla c_\varepsilon|^{2m}$$

$$\quad - \int_\Omega n_\varepsilon\nabla \cdot (|\nabla c_\varepsilon|^{2m-2}\nabla c_\varepsilon) + \int_\Omega (u_\varepsilon \cdot \nabla c_\varepsilon)\nabla \cdot (|\nabla c_\varepsilon|^{2m-2}\nabla c_\varepsilon) \tag{2.3.26}$$

$$= -\frac{m-1}{2}\int_\Omega |\nabla c_\varepsilon|^{2m-4}\left|\nabla|\nabla c_\varepsilon|^2\right|^2 + \frac{1}{2}\int_{\partial\Omega} |\nabla c_\varepsilon|^{2m-2}\frac{\partial|\nabla c_\varepsilon|^2}{\partial\nu} - \int_\Omega |\nabla c_\varepsilon|^{2m}$$

$$\quad - \int_\Omega |\nabla c_\varepsilon|^{2m-2}|D^2 c_\varepsilon|^2 - \int_\Omega n_\varepsilon|\nabla c_\varepsilon|^{2m-2}\Delta c_\varepsilon - \int_\Omega n_\varepsilon\nabla c_\varepsilon \cdot \nabla(|\nabla c_\varepsilon|^{2m-2})$$

$$\quad + \int_\Omega (u_\varepsilon \cdot \nabla c_\varepsilon)|\nabla c_\varepsilon|^{2m-2}\Delta c_\varepsilon + \int_\Omega (u_\varepsilon \cdot \nabla c_\varepsilon)\nabla c_\varepsilon \cdot \nabla(|\nabla c_\varepsilon|^{2m-2})$$

$$= -\frac{2(m-1)}{m^2}\int_\Omega \left|\nabla|\nabla c_\varepsilon|^m\right|^2 + \frac{1}{2}\int_{\partial\Omega} |\nabla c_\varepsilon|^{2m-2}\frac{\partial|\nabla c_\varepsilon|^2}{\partial\nu} - \int_\Omega |\nabla c_\varepsilon|^{2m-2}|D^2 c_\varepsilon|^2$$

$$- \int_\Omega n_\varepsilon |\nabla c_\varepsilon|^{2m-2} \Delta c_\varepsilon - \int_\Omega n_\varepsilon \nabla c_\varepsilon \cdot \nabla(|\nabla c_\varepsilon|^{2m-2}) - \int_\Omega |\nabla c_\varepsilon|^{2m}$$

$$+ \int_\Omega (u_\varepsilon \cdot \nabla c_\varepsilon)|\nabla c_\varepsilon|^{2m-2} \Delta c_\varepsilon + \int_\Omega (u_\varepsilon \cdot \nabla c_\varepsilon)\nabla c_\varepsilon \cdot \nabla(|\nabla c_\varepsilon|^{2m-2})$$

for all $t \in (0, T_{max,\varepsilon})$. Since $|\Delta c_\varepsilon| \le \sqrt{2}|D^2 c_\varepsilon|$, we can estimate

$$\int_\Omega n_\varepsilon |\nabla c_\varepsilon|^{2m-2} \Delta c_\varepsilon \le \sqrt{2} \int_\Omega n_\varepsilon |\nabla c_\varepsilon|^{2m-2}|D^2 c_\varepsilon|$$

$$\le \frac{1}{4} \int_\Omega |\nabla c_\varepsilon|^{2m-2}|D^2 c_\varepsilon|^2 + 2 \int_\Omega n_\varepsilon^2 |\nabla c_\varepsilon|^{2m-2}$$

$$\le \frac{1}{4} \int_\Omega |\nabla c_\varepsilon|^{2m-2}|D^2 c_\varepsilon|^2 + 2 \int_\Omega (n_\varepsilon + \varepsilon)^2 |\nabla c_\varepsilon|^{2m-2} \tag{2.3.27}$$

and

$$\int_\Omega (u_\varepsilon \cdot \nabla c_\varepsilon)|\nabla c_\varepsilon|^{2m-2} \Delta c_\varepsilon \le \sqrt{2} \int_\Omega |u_\varepsilon \cdot \nabla c_\varepsilon||\nabla c_\varepsilon|^{2m-2}|D^2 c_\varepsilon|$$

$$\le \frac{1}{4} \int_\Omega |\nabla c_\varepsilon|^{2m-2}|D^2 c_\varepsilon|^2 + 2 \int_\Omega |u_\varepsilon \cdot \nabla c_\varepsilon|^2 |\nabla c_\varepsilon|^{2m-2}$$

$$\le \frac{1}{4} \int_\Omega |\nabla c_\varepsilon|^{2m-2}|D^2 c_\varepsilon|^2 + 2 \int_\Omega |u_\varepsilon|^2 |\nabla c_\varepsilon|^{2m}$$

$$\le \frac{1}{4} \int_\Omega |\nabla c_\varepsilon|^{2m-2}|D^2 c_\varepsilon|^2 + 2 \int_\Omega |u_\varepsilon|^2 |\nabla c_\varepsilon|^{2m} \tag{2.3.28}$$

for all $t \in (0, T_{max,\varepsilon})$. Again, from the Young inequality, we have

$$- \int_\Omega n_\varepsilon \nabla c_\varepsilon \cdot \nabla(|\nabla c_\varepsilon|^{2m-2})$$

$$= -(m-1) \int_\Omega n_\varepsilon |\nabla c_\varepsilon|^{2(m-2)} \nabla c_\varepsilon \cdot \nabla|\nabla c_\varepsilon|^2$$

$$\le \frac{m-1}{8} \int_\Omega |\nabla c_\varepsilon|^{2m-4} \left||\nabla|\nabla c_\varepsilon|^2\right|^2 + 2(m-1) \int_\Omega |n_\varepsilon|^2 |\nabla c_\varepsilon|^{2m-2} \tag{2.3.29}$$

$$\le \frac{(m-1)}{2m^2} \int_\Omega \left|\nabla|\nabla c_\varepsilon|^m\right|^2 + 2(m-1) \int_\Omega |n_\varepsilon|^2 |\nabla c_\varepsilon|^{2m-2}$$

and

$$\int_\Omega (u_\varepsilon \cdot \nabla c_\varepsilon)\nabla c_\varepsilon \cdot \nabla(|\nabla c_\varepsilon|^{2m-2})$$

$$= (m-1) \int_\Omega (u_\varepsilon \cdot \nabla c_\varepsilon)|\nabla c_\varepsilon|^{2(m-2)} \nabla c_\varepsilon \cdot \nabla|\nabla c_\varepsilon|^2$$

$$\leq \frac{m-1}{8} \int_{\Omega} |\nabla c_\varepsilon|^{2m-4} \left| \nabla |\nabla c_\varepsilon|^2 \right|^2 \tag{2.3.30}$$

$$+ 2(m-1) \int_{\Omega} |u_\varepsilon \cdot \nabla c_\varepsilon|^2 |\nabla c_\varepsilon|^{2m-2}$$

$$\leq \frac{(m-1)}{2m^2} \int_{\Omega} \left| \nabla |\nabla c_\varepsilon|^m \right|^2 + 2(m-1) \int_{\Omega} |u_\varepsilon|^2 |\nabla c_\varepsilon|^{2m}.$$

Observe that

$$\int_{\partial\Omega} \frac{\partial |\nabla c_\varepsilon|^2}{\partial \nu} |\nabla c_\varepsilon|^{2m-2} \leq C_\Omega \int_{\partial\Omega} |\nabla c_\varepsilon|^{2m} \tag{2.3.31}$$

$$= C_\Omega \| |\nabla c_\varepsilon|^m \|^2_{L^2(\partial\Omega)}.$$

Due to Proposition 4.22 (ii) in Haroske and Triebel (2008), $W^{r+\frac{1}{2},2}(\Omega) \hookrightarrow L^2(\partial\Omega)$ with $r \in (0, \frac{1}{2})$ is compact and thus

$$\| |\nabla c_\varepsilon|^m \|^2_{L^2(\partial\Omega)} \leq C_1 \| |\nabla c_\varepsilon|^m \|^2_{W^{r+\frac{1}{2},2}(\Omega)}. \tag{2.3.32}$$

Therefore, from the fractional Gagliardo–Nirenberg inequality and Lemma 2.8, it follows that for some positive constants C_2 and C_3,

$$\| |\nabla c_\varepsilon|^m \|^2_{W^{r+\frac{1}{2},2}(\Omega)}$$

$$\leq C_2 \| \nabla |\nabla c_\varepsilon|^m \|^{2a}_{L^2(\Omega)} \| |\nabla c_\varepsilon|^m \|^{2-2a}_{L^{\frac{2}{m}}(\Omega)} + \delta_1 \| |\nabla c_2|^m \|^2_{L^{\frac{2}{m}}(\Omega)} \tag{2.3.33}$$

$$\leq C_3 \| \nabla |\nabla c_\varepsilon|^m \|^a_{L^2(\Omega)} + C_3$$

with $a = \frac{2m+2r-1}{2m}$. Note that $r \in (0, \frac{1}{2})$ and $m > 1$, $0 < a < 1$. Hence, combining (2.3.31)–(2.3.33) and using the fact that $a \in (0, 1)$, we can see that

$$\int_{\partial\Omega} \frac{\partial |\nabla c_\varepsilon|^2}{\partial \nu} |\nabla c_\varepsilon|^{2m-2} \leq \frac{(m-1)}{2m^2} \int_{\Omega} \left| \nabla |\nabla c_\varepsilon|^m \right|^2 + C_4. \tag{2.3.34}$$

Now from (2.3.26)–(2.3.30) and (2.3.34), we obtain that for some positive constant C_5,

$$\frac{1}{2m} \frac{d}{dt} \|\nabla c_\varepsilon\|^{2m}_{L^{2m}(\Omega)} + \frac{m-1}{2m^2} \int_{\Omega} \left| \nabla |\nabla c_\varepsilon|^m \right|^2 + \frac{1}{2} \int_{\Omega} |\nabla c_\varepsilon|^{2m-2} |D^2 c_\varepsilon|^2 + \int_{\Omega} |\nabla c_\varepsilon|^{2m}$$

$$\leq 2m \int_{\Omega} n_\varepsilon^2 |\nabla c_\varepsilon|^{2m-2} + 2m \int_{\Omega} |u_\varepsilon|^2 |\nabla c_\varepsilon|^{2m} + C_5 \quad \text{for all } t \in (0, T_{max,\varepsilon}).$$

$$\tag{2.3.35}$$

Next we turn to estimate terms on the right of (2.3.35). By the Young inequality, we have

$$2m \int_\Omega n_\varepsilon^2 |\nabla c_\varepsilon|^{2m-2} \leq 2m \int_\Omega (n_\varepsilon + \varepsilon)^2 |\nabla c_\varepsilon|^{2m-2}$$

$$\leq \frac{1}{2} \int_\Omega |\nabla c_\varepsilon|^{2m} + C_5 \int_\Omega (n_\varepsilon + \varepsilon)^{2m} \quad \text{for all } t \in (0, T_{max,\varepsilon})$$

$$(2.3.36)$$

and

$$2m \int_\Omega |u_\varepsilon|^2 |\nabla c_\varepsilon|^{2m} \leq \int_\Omega |\nabla c_\varepsilon|^{2m+1} + C_6 \int_\Omega u_\varepsilon^{4m+2} \quad \text{for all } t \in (0, T_{max,\varepsilon}),$$

$$(2.3.37)$$

where $C_5 = \frac{m}{m-1} \left(\frac{1}{2}m\right)^{-\frac{1}{m-1}} (2m)^m$ and $C_6 = (2m)^{2m+1}$. On the other hand due to (2.3.7), we derive from the Gagliardo–Nirenberg inequality that for some positive constants C_i, $(i = 7, 8, 9)$,

$$\int_\Omega |\nabla c_\varepsilon|^{2m+1} = \||\nabla c_\varepsilon|^m\|_{L^{\frac{2m+1}{m}}(\Omega)}^{\frac{2m+1}{m}}$$

$$\leq C_7 (\||\nabla|\nabla c_\varepsilon|^m\|_{L^2(\Omega)}^{\frac{2m-1}{2m+1}} \||\nabla c_\varepsilon|^m\|_{L^{\frac{2}{m}}(\Omega)}^{\frac{2}{2m+1}} + \||\nabla c_\varepsilon|^m\|_{L^{\frac{2}{m}}(\Omega)})^{\frac{2m+1}{m}}$$

$$\leq C_8 (\||\nabla|\nabla c_\varepsilon|^m\|_{L^2(\Omega)}^{\frac{2m-1}{m}} + 1),$$

$$\leq \frac{m-1}{2m^2} \int_\Omega |\nabla|\nabla c_\varepsilon|^m|^2 + C_9$$

which along with (2.3.37) implies that

$$2m \int_\Omega |u_\varepsilon|^2 |\nabla c_\varepsilon|^{2m}$$

$$\leq \frac{m-1}{2m^2} \int_\Omega |\nabla|\nabla c_\varepsilon|^m|^2 + C_6 \int_\Omega u_\varepsilon^{4m+2} + C_9 \quad \text{for all } t \in (0, T_{max,\varepsilon}).$$

$$(2.3.38)$$

Substituting (2.3.36) and (2.3.38) into (2.3.35), we have

$$\frac{1}{2m} \frac{d}{dt} \|\nabla c_\varepsilon\|_{L^{2m}(\Omega)}^{2m} + \frac{1}{2} \int_\Omega |\nabla c_\varepsilon|^{2m}$$

$$\leq C_5 \int_\Omega (n_\varepsilon + \varepsilon)^{2m} + C_6 \int_\Omega u_\varepsilon^{4m+2} + C_{10} \quad \text{for all } t \in (0, T_{max,\varepsilon}).$$

Hence, due to $W^{1,2}(\Omega) \hookrightarrow L^p(\Omega)$ for any $p > 1$, the boundedness of $\|\nabla u_\varepsilon(\cdot, t)\|_{L^2(\Omega)}$ (see Lemma 2.9) implies that there exists a positive constant C_{11} such that $\|u_\varepsilon(\cdot, t)\|_{L^{4m+2}(\Omega)} \leq C_{11}$ for all $t \in (0, T_{max,\varepsilon})$, which together with (2.3.8), (2.3.10) leads to (2.3.25) by Lemma 2.2. This completes the proof of Lemma 2.10.

Lemma 2.11 *Let $m > 1$. Then for all $p > 1$, there exists a positive constant C independent of ε, such that the solution of (2.2.1) from Lemma 2.2 satisfies*

$$\|n_\varepsilon(\cdot, t)\|_{L^p(\Omega)} \le C \text{ for all } t \in (0, T_{max,\varepsilon}). \tag{2.3.39}$$

Proof Taking $(n_\varepsilon + \varepsilon)^{p-1}$ with $p > \max\{1, m-1\}$ as the test function for the first equation of (2.2.1), combining with the second equation, and using $\nabla \cdot u_\varepsilon = 0$, we obtain that for all $t \in (0, T_{max,\varepsilon})$,

$$\frac{1}{p}\frac{d}{dt}\|n_\varepsilon + \varepsilon\|_{L^p(\Omega)}^p + m(p-1)\int_\Omega (n_\varepsilon + \varepsilon)^{m+p-3}|\nabla n_\varepsilon|^2$$

$$\le (p-1)\int_\Omega (n_\varepsilon + \varepsilon)^{p-2}n_\varepsilon|\nabla n_\varepsilon||S_\varepsilon(x, n_\varepsilon, c_\varepsilon)||\nabla c_\varepsilon|$$

$$\le (p-1)C_S \int_\Omega (n_\varepsilon + \varepsilon)^{p-1}|\nabla n_\varepsilon||\nabla c_\varepsilon|$$

$$\le \frac{m(p-1)}{2}\int_\Omega (n_\varepsilon + \varepsilon)^{m+p-3}|\nabla n_\varepsilon|^2 + \frac{(p-1)C_S^2}{2m}\int_\Omega (n_\varepsilon + \varepsilon)^{p+1-m}|\nabla c_\varepsilon|^2,$$

and hence

$$\frac{1}{p}\frac{d}{dt}\|n_\varepsilon + \varepsilon\|_{L^p(\Omega)}^p + \frac{m(p-1)}{2}\int_\Omega (n_\varepsilon + \varepsilon)^{m+p-3}|\nabla n_\varepsilon|^2$$
$$\le \frac{(p-1)C_S^2}{2m}\int_\Omega (n_\varepsilon + \varepsilon)^{p+1-m}|\nabla c_\varepsilon|^2 \tag{2.3.40}$$

for all $t \in (0, T_{max,\varepsilon})$. In the following, we will estimate the right-hand side of (2.3.40). In fact, due to $m > 1$, we conclude from (2.3.25) that

$$\int_\Omega (n_\varepsilon + \varepsilon)^{p+1-m}|\nabla c_\varepsilon|^2$$

$$\le \left(\int_\Omega (n_\varepsilon + \varepsilon)^{\frac{m(p+1-m)}{m-1}}\right)^{\frac{m-1}{m}} \left(\int_\Omega |\nabla c_\varepsilon|^{2m}\right)^{\frac{1}{m}} \tag{2.3.41}$$

$$\le C_1 \left(\int_\Omega (n_\varepsilon + \varepsilon)^{\frac{m(p+1-m)}{m-1}}\right)^{\frac{m-1}{m}} \text{ for all } t \in (0, T_{max,\varepsilon}).$$

Further, noticing that $\frac{2(mp-m^2+1)}{m(p+m-1)} < 2$ due to $m > 1$, an application of the Gagliardo–Nirenberg inequality then leads to

$$C_1 \left(\int_\Omega (n_\varepsilon + \varepsilon)^{\frac{m(p+1-m)}{m-1}} \right)^{\frac{m-1}{m}}$$

$$= C_1 \| (n_\varepsilon + \varepsilon)^{\frac{m(p+1-m)}{m-1}} \|_{L^{\frac{2m(p+1-m)}{(m-1)(m+p-1)}}(\Omega)}^{\frac{2(p+1-m)}{m+p-1}}$$

$$\leq C_2 (\| \nabla(n_\varepsilon + \varepsilon)^{\frac{p+m-1}{2}} \|_{L^2(\Omega)}^{\frac{mp-m^2+1}{m(p+1-m)}} \| (n_\varepsilon + \varepsilon)^{\frac{p+m-1}{2}} \|_{L^{\frac{2}{p+m-1}}(\Omega)}^{\frac{m-1}{m(p+1-m)}} \tag{2.3.42}$$

$$+ \| (n_\varepsilon + \varepsilon)^{\frac{p+m-1}{2}} \|_{L^{\frac{2}{p+m-1}}(\Omega)})^{\frac{2(p+1-m)}{m+p-1}}$$

$$\leq C_3 (\| \nabla(n_\varepsilon + \varepsilon)^{\frac{p+m-1}{2}} \|_{L^2(\Omega)}^{\frac{2(mp-m^2+1)}{m(p+m-1)}} + 1)$$

$$\leq \frac{m(p-1)}{4} \int_\Omega (n_\varepsilon + \varepsilon)^{m+p-3} |\nabla n_\varepsilon|^2 + C_4$$

for some $C_3 > 0$ and $C_4 > 0$. Therefore, combining (2.3.40), (2.3.41) with (2.3.42), we arrive at

$$\frac{1}{p} \frac{d}{dt} \| n_\varepsilon + \varepsilon \|_{L^p(\Omega)}^p + \frac{m(p-1)}{4} \int_\Omega (n_\varepsilon + \varepsilon)^{m+p-3} |\nabla n_\varepsilon|^2 \leq C_5,$$

which along with the fact that for some $C_6 > 0$,

$$\| n_\varepsilon + \varepsilon \|_{L^p(\Omega)}^p \leq \frac{m(p-1)}{8} \int_\Omega (n_\varepsilon + \varepsilon)^{m+p-3} |\nabla n_\varepsilon|^2 + C_6,$$

and Lemma 2.2 implies that (2.3.39) holds.

Now we can rely on standard reasoning to obtain the following.

Lemma 2.12 *Let $m > 1$ and $\gamma \in (\frac{1}{2}, 1)$. Then one can find a positive constant C independent of ε, such that*

$$\| n_\varepsilon(\cdot, t) \|_{L^\infty(\Omega)} \leq C \text{ for all } t \in (0, T_{max,\varepsilon}),$$

$$\| c_\varepsilon(\cdot, t) \|_{W^{1,\infty}(\Omega)} \leq C \text{ for all } t \in (0, T_{max,\varepsilon})$$

as well as

$$\| A^\gamma u_\varepsilon(\cdot, t) \|_{L^2(\Omega)} \leq C \text{ for all } t \in (0, T_{max,\varepsilon}).$$

Proof Firstly, applying the variation-of-constants formula to the projected version of the third equation in (2.2.1), we derive that

$$u_\varepsilon(\cdot, t) = e^{-tA} u_0 + \int_0^t e^{-(t-\tau)A} \mathcal{P}[n_\varepsilon(\cdot, t) \nabla \phi - \kappa(Y_\varepsilon u_\varepsilon \cdot \nabla) u_\varepsilon] d\tau \text{ for all } t \in (0, T_{max,\varepsilon}).$$

Let $h_\varepsilon = \mathscr{P}[n_\varepsilon(\cdot, t)\nabla\phi - \kappa(Y_\varepsilon u_\varepsilon \cdot \nabla)u_\varepsilon]$. Then in view of the standard smoothing properties of the Stokes semigroup, we can conclude that for $\gamma \in (\frac{1}{2}, 1)$ and $p_0 \in (\frac{2}{3-2\gamma}, 2)$, there exists $C_1 > 0$ such that

$$\|A^\gamma u_\varepsilon(\cdot, t)\|_{L^2(\Omega)}$$
$$\leq \|A^\gamma e^{-tA} u_0\|_{L^2(\Omega)} + \int_0^t \|A^\gamma e^{-(t-\tau)A} h_\varepsilon(\cdot, \tau) d\tau\|_{L^2(\Omega)} d\tau \tag{2.3.43}$$
$$\leq \|A^\gamma u_0\|_{L^2(\Omega)} + C_1 \int_0^t (t-\tau)^{-\gamma-\frac{1}{p_0}+\frac{1}{2}} e^{-\lambda(t-\tau)} \|h_\varepsilon(\cdot, \tau)\|_{L^{p_0}(\Omega)} d\tau$$

for all $t \in (0, T_{max,\varepsilon})$.

In light of (2.3.39), for some positive constant C_2, it has

$$\|n_\varepsilon(\cdot, t)\|_{L^{p_0}(\Omega)} \leq C_2 \text{ for all } t \in (0, T_{max,\varepsilon}).$$

So employing the Hölder inequality and the continuity of $\mathscr{P}$ in $L^p(\Omega; \mathbb{R}^2)$ (see Fujiwara and Morimoto 1977), there exist positive constants $C_i, (i = 3, 4, 5, 6)$, such that

$$\|h_\varepsilon(\cdot, t)\|_{L^{p_0}(\Omega)}$$
$$\leq C_3 \|(Y_\varepsilon u_\varepsilon \cdot \nabla)u_\varepsilon(\cdot, t)\|_{L^{p_0}(\Omega)} + C_3 \|n_\varepsilon(\cdot, t)\|_{L^{p_0}(\Omega)}$$
$$\leq C_4 \|Y_\varepsilon u_\varepsilon\|_{L^{\frac{2p_0}{2-p_0}}(\Omega)} \|\nabla u_\varepsilon(\cdot, t)\|_{L^2(\Omega)} + C_4 \tag{2.3.44}$$
$$\leq C_5 \|\nabla Y_\varepsilon u_\varepsilon\|_{L^2(\Omega)} \|\nabla u_\varepsilon(\cdot, t)\|_{L^2(\Omega)} + C_4$$
$$\leq C_6 \|\nabla u_\varepsilon(\cdot, t)\|_{L^2(\Omega)}^2 + C_4 \text{ for all } t \in (0, T_{max,\varepsilon}),$$

where we have used the fact that $W^{1,2}(\Omega) \hookrightarrow L^{\frac{2p_0}{2-p_0}}(\Omega)$. Collecting (2.3.43), (2.3.21) and (2.3.44), we conclude that

$$\|A^\gamma u_\varepsilon(\cdot, t)\|_{L^2(\Omega)} \leq C_7 \text{ for all } t \in (0, T_{max,\varepsilon})$$

which together with the fact that $D(A^\gamma)$ is continuously embedded into $L^\infty(\Omega)$ by $\gamma > \frac{1}{2}$ yields

$$\|u_\varepsilon(\cdot, t)\|_{L^\infty(\Omega)} \leq C_8 \text{ for all } t \in (0, T_{max,\varepsilon}). \tag{2.3.45}$$

Further, in view of (2.3.25), one may use (2.1.11), $m > 1$ and the smoothing properties of the Neumann heat semigroup $(e^{t\Delta})_{t\geq 0}$ to obtain that there exists $C_9 > 0$ such that

$$\|\nabla c_\varepsilon(\cdot, t)\|_{L^\infty(\Omega)} \leq C_9 \text{ for all } t \in (0, T_{max,\varepsilon}). \tag{2.3.46}$$

Moreover, the boundedness of n_ε can be archived by the well-known Moser–Alikakos iteration procedure (see, e.g., Lemma A.1 in Tao and Winkler 2012a). Indeed, by (2.3.45) and (2.3.46), we see that the hypotheses of Lemma A.1 in Tao and Winkler

(2012a) are valid provided that the parameter p in Lemma 2.11 is appropriately large. The proof of Lemma 2.12 is complete.

With all the above regularization properties of n_ε, c_ε and u_ε at hand, we can show the existence of global bounded solutions to the regularized system (2.2.1).

Lemma 2.13 *Let $m > 1$ and $\gamma \in (\frac{1}{2}, 1)$ and $(n_\varepsilon, c_\varepsilon, u_\varepsilon, P_\varepsilon)_{\varepsilon\in(0,1)}$ be classical solutions of (2.2.1) constructed in Lemma 2.2 on $[0, T_{max,\varepsilon})$. Then the solution is global on $[0, \infty)$. Moreover, one can find $C > 0$ independent of $\varepsilon \in (0, 1)$ such that*

$$\|n_\varepsilon(\cdot, t)\|_{L^\infty(\Omega)} \leq C \ \text{for all} \ t \in (0, \infty)$$

and

$$\|c_\varepsilon(\cdot, t)\|_{W^{1,\infty}(\Omega)} \leq C \ \text{for all} \ t \in (0, \infty)$$

as well as

$$\|A^\gamma u_\varepsilon(\cdot, t)\|_{L^2(\Omega)} \leq C \ \text{for all} \ t \in (0, \infty).$$

Then, with the help of Lemma 2.13, we can straightforwardly deduce the uniform Hölder properties of c_ε, ∇c_ε and u_ε by the standard parabolic regularity theory as the proof of Lemmas 3.18–3.19 in Winkler (2015b) (see also Zheng 2016).

Lemma 2.14 *Let $m > 1$. Then one can find $\mu \in (0, 1)$ such that for some $C > 0$*

$$\|c_\varepsilon(\cdot, t)\|_{C^{\mu, \frac{\mu}{2}}(\Omega \times [t, t+1])} \leq C \ \text{for all} \ t \in (0, \infty)$$

as well as

$$\|u_\varepsilon(\cdot, t)\|_{C^{\mu, \frac{\mu}{2}}(\Omega \times [t, t+1])} \leq C \ \text{for all} \ t \in (0, \infty),$$

and for any $\tau > 0$, there exists $C(\tau) > 0$ fulfilling

$$\|\nabla c_\varepsilon(\cdot, t)\|_{C^{\mu, \frac{\mu}{2}}(\Omega \times [t, t+1])} \leq C(\tau) \ \text{for all} \ t \in (\tau, \infty).$$

2.3.2 Global Boundedness of Weak Solutions

Based on the above lemmas, the weak solution of (2.1.6)–(2.1.8) can be obtained as the limitation of classical solutions to the systems (2.2.1). Applying the idea of Zheng (2016) (see also Liu and Wang 2016 and Winkler 2015b), we first state the definition of the solution as follows.

Definition 2.2 Let $T > 0$ and (n_0, c_0, u_0) fulfill (2.1.11). Then a triple of functions (n, c, u) is called a weak solution of (2.1.6)–(2.1.8) in $\Omega \times (0, T)$ if the following conditions are satisfied:

$$\begin{cases} n \in L^1_{loc}(\bar{\Omega} \times [0, T)), \\ c \in L^1_{loc}([0, T); W^{1,1}(\Omega)), \\ u \in L^1_{loc}([0, T); W^{1,1}(\Omega)), \end{cases}$$

where $n \geq 0$ and $c \geq 0$ in $\Omega \times (0, T)$ as well as $\nabla \cdot u = 0$ in the distributional sense in $\Omega \times (0, T)$, moreover, $n^m \in L^1_{loc}(\bar{\Omega} \times [0, \infty))$, cu, nu and $n\nabla c$ belong to $L^1_{loc}(\bar{\Omega} \times [0, \infty); \mathbb{R}^2)$ and $u \otimes u \in L^1_{loc}(\bar{\Omega} \times [0, \infty); \mathbb{R}^{2 \times 2})$; and

$$-\int_0^T \int_\Omega n\varphi_t - \int_\Omega n_0\varphi(\cdot, 0) = \int_0^T \int_\Omega n^m \Delta\varphi + \int_0^T \int_\Omega n\nabla c \cdot \nabla\varphi + \int_0^T \int_\Omega nu \cdot \nabla\varphi$$

for any $\varphi \in C_0^\infty(\bar{\Omega} \times [0, T))$ satisfying $\frac{\partial \varphi}{\partial \nu} = 0$ on $\partial\Omega \times (0, T)$, as well as

$$-\int_0^T \int_\Omega c\varphi_t - \int_\Omega c_0\varphi(\cdot, 0)$$
$$= -\int_0^T \int_\Omega \nabla c \cdot \nabla\varphi - \int_0^T \int_\Omega c\varphi + \int_0^T \int_\Omega n\varphi + \int_0^T \int_\Omega cu \cdot \nabla\varphi$$

for any $\varphi \in C_0^\infty(\bar{\Omega} \times [0, T))$ and

$$-\int_0^T \int_\Omega u\varphi_t - \int_\Omega u_0\varphi(\cdot, 0)$$
$$= \kappa \int_0^T \int_\Omega u \otimes u \cdot \nabla\varphi - \int_0^T \int_\Omega \nabla u \cdot \nabla\varphi - \int_0^T \int_\Omega n\nabla\phi \cdot \varphi$$

for any $\varphi \in C_0^\infty(\bar{\Omega} \times [0, T); \mathbb{R}^2)$ fulfilling $\nabla\varphi \equiv 0$ in $\Omega \times (0, T)$.

If for each $T > 0$, $(n, c, u) : \Omega \times (0, \infty) \longrightarrow \mathbb{R}^4$ is a weak solution of (2.1.6)–(2.1.8) in $\Omega \times (0, T)$, then we call (n, c, u) a global weak solution of (2.1.6)–(2.1.8).

In order to apply the Aubin–Lions Lemma (Simon 1986), we will need the regularity of the time derivative of bounded solutions. Employing almost exactly the same arguments as that in the proof of Lemmas 3.22–3.23 in Winkler (2015b) (the minor necessary changes are left as an exercise to the reader), and taking advantage of Lemma 2.13, we conclude the following lemma.

Lemma 2.15 Let $m > 1$ and let $\varsigma > \max\{m, 2(m-1)\}$. Then for every $T > 0$ and $\varepsilon \in (0, 1)$, one can find $C(T) > 0$ independent of ε such that $\int_0^T \|\partial_t(n_\varepsilon + \varepsilon)^\varsigma(\cdot, t)\|_{(W_0^{2,2}(\Omega))^*} dt \leq C(T)$ as well as $\int_0^T \int_\Omega |\nabla(n_\varepsilon + \varepsilon)^\varsigma|^2 \leq C(T)$.

At this position, the main result can be proved as follows.

Proof of Theorem 2.1. In conjunction with Lemmas 2.13, 2.11 and the Aubin–Lions compactness lemma (see Simon 1986), one can infer the existence of a sequence of numbers $\varepsilon = \varepsilon_j \searrow 0$ along which

$$n_\varepsilon \longrightarrow n \quad \text{a.e. in } \Omega \times (0, \infty), \tag{2.3.47}$$

$$\nabla n_\varepsilon^m \rightharpoonup \nabla n^m \ \text{ in } \ L_{loc}^2(\Omega \times [0, \infty)), \tag{2.3.48}$$

$$c_\varepsilon \to c \ \text{ in } \ C_{loc}^0(\bar{\Omega} \times [0, \infty)), \tag{2.3.49}$$

$$\nabla c_\varepsilon \to \nabla c \ \text{ in } \ C_{loc}^0(\bar{\Omega} \times (0, \infty)), \tag{2.3.50}$$

$$\nabla c_\varepsilon \overset{*}{\rightharpoonup} \nabla c \ \text{ weakly star in } \ L^\infty(\Omega \times (0, \infty)) \tag{2.3.51}$$

as well as

$$u_\varepsilon \to u \ \text{ in } \ C_{loc}^0(\bar{\Omega} \times [0, \infty)) \tag{2.3.52}$$

and

$$Du_\varepsilon \overset{*}{\rightharpoonup} Du \ \text{ weakly star in } \ L^\infty(\Omega \times (0, \infty)) \tag{2.3.53}$$

holds for some limit $(n, c, u) \in (L^\infty(\Omega \times (0, \infty)))^4$ with nonnegative n and c. Indeed, Lemma 2.15 implies that for each $T > 0$, $(n_\varepsilon^\varsigma)_{\varepsilon \in (0,1)}$ is bounded in $L^2((0, T);$ $W^{1,2}(\Omega))$. By using Aubin–Lions lemma, one then obtains $n_\varepsilon^\varsigma \to z^\varsigma$ for some non-negative measurable $z : \Omega \times (0, \Omega) \to \mathbb{R}$. Further by Lemma 2.11 and the Egorov theorem, one has (2.3.47) and (2.3.48).

Due to these convergence properties (see (2.3.47)–(2.3.53)), by applying the standard arguments, we may take $\varepsilon = \varepsilon_j \searrow 0$ in each term of the natural weak formulation of (2.2.1) separately. Then we can verify that (n, c, u) can be complemented by some pressure function P in such a way that (n, c, u, P) is a weak solution of (2.1.6)–(2.1.8). Finally, we can infer from the boundedness of $(n_\varepsilon, c_\varepsilon, u_\varepsilon)$ and the Banach–Alaoglu theorem that (n, c, u) is bounded.

2.4 Global Existence of Solutions to a Three-Dimensional Keller–Segel–Navier–Stokes System

2.4.1 A Priori Estimates for Approximate Solutions

In this subsection, we are going to establish an iteration step to develop the main ingredients of our result. The iteration depends on a series of a priori estimates. We first recall some properties of F_ε and F_ε', which play an important role in the proof of Theorem 2.2.

Lemma 2.16 *Assume F_ε is given by (2.2.9). Then*

$$0 \leq F'_\varepsilon(s) = \frac{1}{1 + \varepsilon s} \leq 1 \text{ for all } s \geq 0 \text{ and } \varepsilon > 0 \tag{2.4.1}$$

as well as

$$\lim_{\varepsilon \to 0^+} F_\varepsilon(s) = s, \quad \lim_{\varepsilon \to 0^+} F'_\varepsilon(s) = 1 \text{ for all } s \geq 0 \tag{2.4.2}$$

and

$$0 \leq F_\varepsilon(s) \leq s \text{ for all } s \geq 0. \tag{2.4.3}$$

Proof Recalling (2.2.9), by tedious and simple calculations, we can derive (2.4.1)–(2.4.3).

The proof of this lemma is very similar to that of Lemmas 2.2 and 2.6 of Tao and Winkler (2015b) (see also Lemma 3.2 of Wang 2017), so we omit it here.

Lemma 2.17 *There exists $\lambda > 0$ independent of ε such that the solution of (2.2.8) satisfies*

$$\int_\Omega n_\varepsilon + \int_\Omega c_\varepsilon \leq \lambda \text{ for all } t \in (0, T_{max,\varepsilon}). \tag{2.4.4}$$

Lemma 2.18 *Let $\alpha > \frac{1}{3}$. Then there exists $C > 0$ independent of ε such that the solution of (2.2.8) satisfies*

$$\int_\Omega n_\varepsilon^{2\alpha} + \int_\Omega c_\varepsilon^2 + \int_\Omega |u_\varepsilon|^2 \leq C \text{ for all } t \in (0, T_{max,\varepsilon}). \tag{2.4.5}$$

Moreover, for $T \in (0, T_{max,\varepsilon})$, one can find a constant $C > 0$ independent of ε such that

$$\int_0^T \int_\Omega \left[n_\varepsilon^{2\alpha-2} |\nabla n_\varepsilon|^2 + |\nabla c_\varepsilon|^2 + |\nabla u_\varepsilon|^2 \right] \leq C. \tag{2.4.6}$$

Proof The proof consists of two cases.

Case (I) $2\alpha \neq 1$: We first obtain from $\nabla \cdot u_\varepsilon = 0$ in $\Omega \times (0, T_{max,\varepsilon})$ and straightforward calculations that

$$
\begin{aligned}
&\text{sign}(2\alpha - 1)\frac{1}{2\alpha}\frac{d}{dt}\|n_\varepsilon\|_{L^{2\alpha}(\Omega)}^{2\alpha} \\
&+\text{sign}(2\alpha - 1)(2\alpha - 1)\int_\Omega n_\varepsilon^{2\alpha-2}|\nabla n_\varepsilon|^2 \\
&=-\int_\Omega \text{sign}(2\alpha - 1)n_\varepsilon^{2\alpha-1}\nabla \cdot (n_\varepsilon F'_\varepsilon(n_\varepsilon)S_\varepsilon(x, n_\varepsilon, c_\varepsilon) \cdot \nabla c_\varepsilon \\
&\leq\text{sign}(2\alpha - 1)(2\alpha - 1)\int_\Omega n_\varepsilon^{2\alpha-2}n_\varepsilon F'_\varepsilon(n_\varepsilon)|S_\varepsilon(x, n_\varepsilon, c_\varepsilon)||\nabla n_\varepsilon||\nabla c_\varepsilon|
\end{aligned}
\tag{2.4.7}
$$

for all $t \in (0, T_{max,\varepsilon})$. Therefore, from (2.4.1), in light of (2.1.5) and (2.2.9), we can estimate the right-hand side of (2.4.7) as follows:

$$
\begin{aligned}
&\text{sign}(2\alpha - 1)(2\alpha - 1) \int_\Omega n_\varepsilon^{2\alpha-2} n_\varepsilon F_\varepsilon'(n_\varepsilon) |S_\varepsilon(x, n_\varepsilon, c_\varepsilon)| |\nabla n_\varepsilon| |\nabla c_\varepsilon| \\
&\leq \text{sign}(2\alpha - 1)(2\alpha - 1) \int_\Omega n_\varepsilon^{2\alpha-2} n_\varepsilon C_S (1 + n_\varepsilon)^{-\alpha} |\nabla n_\varepsilon| |\nabla c_\varepsilon| \\
&\leq \text{sign}(2\alpha - 1) \frac{2\alpha - 1}{2} \int_\Omega n_\varepsilon^{2\alpha-2} |\nabla n_\varepsilon|^2 \\
&\quad + \frac{|2\alpha - 1|}{2} C_S^2 \int_\Omega n_\varepsilon^{2\alpha-2} n_\varepsilon^2 (1 + n_\varepsilon)^{-2\alpha} |\nabla c_\varepsilon|^2 \\
&\leq \text{sign}(2\alpha - 1) \frac{2\alpha - 1}{2} \int_\Omega n_\varepsilon^{2\alpha-2} |\nabla n_\varepsilon|^2 \\
&\quad + \frac{|2\alpha - 1|}{2} C_S^2 \int_\Omega |\nabla c_\varepsilon|^2 \quad \text{for all } t \in (0, T_{max,\varepsilon})
\end{aligned}
\tag{2.4.8}
$$

by using Young's inequality, where in the last inequality we have used the fact that $n_\varepsilon^{2\alpha-2} n_\varepsilon^2 (1 + n_\varepsilon)^{-2\alpha} \leq 1$ for all $\varepsilon \geq 0$ and $(x, t) \in \Omega \times (0, T_{max,\varepsilon})$. Inserting (2.4.8) into (2.4.7), we conclude that

$$
\begin{aligned}
&\text{sign}(2\alpha - 1) \frac{1}{2\alpha} \frac{d}{dt} \|n_\varepsilon\|_{L^{2\alpha}(\Omega)}^{2\alpha} + \text{sign}(2\alpha - 1) \frac{2\alpha - 1}{2} \int_\Omega n_\varepsilon^{2\alpha-2} |\nabla n_\varepsilon|^2 \\
&\leq \frac{|2\alpha - 1|}{2} C_S^2 \int_\Omega |\nabla c_\varepsilon|^2 \quad \text{for all } t \in (0, T_{max,\varepsilon}).
\end{aligned}
\tag{2.4.9}
$$

To track the time evolution of c_ε, taking c_ε as the test function for the second equation of (2.2.8) and using $\nabla \cdot u_\varepsilon = 0$ and (2.4.3) together with Hölder's inequality yields

$$
\begin{aligned}
&\frac{1}{2} \frac{d}{dt} \|c_\varepsilon\|_{L^2(\Omega)}^2 + \int_\Omega |\nabla c_\varepsilon|^2 + \int_\Omega |c_\varepsilon|^2 \\
&= \int_\Omega F_\varepsilon(n_\varepsilon) c_\varepsilon \\
&\leq \int_\Omega n_\varepsilon c_\varepsilon \\
&\leq \|n_\varepsilon\|_{L^{\frac{6}{5}}(\Omega)} \|c_\varepsilon\|_{L^6(\Omega)} \quad \text{for all } t \in (0, T_{max,\varepsilon}).
\end{aligned}
\tag{2.4.10}
$$

By applying Sobolev embedding $W^{1,2}(\Omega) \hookrightarrow L^6(\Omega)$ in the three-dimensional setting, in view of (2.4.4), there exist positive constants C_1 and C_2 such that

$$
\begin{aligned}
\|c_\varepsilon\|_{L^6(\Omega)}^2 &\leq C_1 \|\nabla c_\varepsilon\|_{L^2(\Omega)}^2 + C_1 \|c_\varepsilon\|_{L^1(\Omega)}^2 \\
&\leq C_1 \|\nabla c_\varepsilon\|_{L^2(\Omega)}^2 + C_2 \quad \text{for all } t \in (0, T_{max,\varepsilon}).
\end{aligned}
\tag{2.4.11}
$$

Thus, by means of Young's inequality and (2.4.11), we proceed to estimate

$$\frac{1}{2}\frac{d}{dt}\|c_\varepsilon\|^2_{L^2(\Omega)} + \int_\Omega |\nabla c_\varepsilon|^2 + \int_\Omega |c_\varepsilon|^2$$
$$\leq \frac{1}{2C_1}\|c_\varepsilon\|^2_{L^6(\Omega)} + \frac{C_1}{2}\|n_\varepsilon\|^2_{L^{\frac{6}{5}}(\Omega)} \tag{2.4.12}$$
$$\leq \frac{1}{2}\|\nabla c_\varepsilon\|^2_{L^2(\Omega)} + \frac{C_1}{2}\|n_\varepsilon\|^2_{L^{\frac{6}{5}}(\Omega)} + C_3 \text{ for all } t \in (0, T_{max,\varepsilon})$$

and some positive constant C_3 independent of ε. Therefore,

$$\frac{1}{2}\frac{d}{dt}\|c_\varepsilon\|^2_{L^2(\Omega)} + \frac{1}{2}\int_\Omega |\nabla c_\varepsilon|^2 + \int_\Omega |c_\varepsilon|^2 \leq \frac{C_1}{2}\|n_\varepsilon\|^2_{L^{\frac{6}{5}}(\Omega)} + C_3 \text{ for all } t \in (0, T_{max,\varepsilon}).$$
$$\tag{2.4.13}$$

To estimate $\|n_\varepsilon\|_{L^{\frac{6}{5}}(\Omega)}$ for all $t \in (0, T_{max,\varepsilon})$, we should notice that $\alpha > \frac{1}{3}$ ensures that $\frac{2}{6\alpha-1} < 2$, so that in light of (2.4.4), the Gagliardo–Nirenberg inequality and Young's inequality allow us to estimate that

$$\|n_\varepsilon\|^2_{L^{\frac{6}{5}}(\Omega)}$$
$$= \|n_\varepsilon^\alpha\|^{\frac{2}{\alpha}}_{L^{\frac{6}{5\alpha}}(\Omega)}$$
$$\leq C_4(\|\nabla n_\varepsilon^\alpha\|^{\frac{2}{6\alpha-1}}_{L^2(\Omega)}\|n_\varepsilon^\alpha\|^{\frac{2}{\alpha}-\frac{2}{6\alpha-1}}_{L^{\frac{1}{\alpha}}(\Omega)} + \|n_\varepsilon^\alpha\|^{\frac{2}{\alpha}}_{L^{\frac{1}{\alpha}}(\Omega)}) \tag{2.4.14}$$
$$\leq \frac{1}{4}\frac{1}{C_1\alpha^2 C_S^2}\|\nabla n_\varepsilon^\alpha\|^2_{L^2(\Omega)} + C_5 \text{ for all } t \in (0, T_{max,\varepsilon})$$

with some positive constants C_4 and C_5 independent of ε. This together with (2.4.13) contributes to

$$\frac{1}{2}\frac{d}{dt}\|c_\varepsilon\|^2_{L^2(\Omega)} + \frac{1}{2}\int_\Omega |\nabla c_\varepsilon|^2 + \int_\Omega |c_\varepsilon|^2$$
$$\leq \frac{1}{8}\frac{1}{\alpha^2 C_S^2}\|\nabla n_\varepsilon^\alpha\|^2_{L^2(\Omega)} + C_6 \text{ for all } t \in (0, T_{max,\varepsilon}) \tag{2.4.15}$$

and some positive constant C_6. Taking an evident linear combination of the inequalities provided by (2.4.9) and (2.4.15), one can obtain

$$\text{sign}(2\alpha - 1)\frac{1}{2\alpha}\frac{d}{dt}\|n_\varepsilon\|^{2\alpha}_{L^{2\alpha}(\Omega)} + |2\alpha - 1|C_S^2\frac{d}{dt}\|c_\varepsilon\|^2_{L^2(\Omega)}$$
$$+ \frac{|2\alpha - 1|}{2}C_S^2\int_\Omega |\nabla c_\varepsilon|^2 + 2|2\alpha - 1|C_S^2\int_\Omega |c_\varepsilon|^2$$
$$+ \left(\text{sign}(2\alpha - 1)\frac{2\alpha - 1}{2} - \frac{1}{4}|2\alpha - 1|\right)\int_\Omega n_\varepsilon^{2\alpha-2}|\nabla n_\varepsilon|^2 \tag{2.4.16}$$
$$\leq C_7 \text{ for all } t \in (0, T_{max,\varepsilon})$$

and some positive constant C_7. Since $\mathrm{sign}(2\alpha - 1)\dfrac{2\alpha - 1}{2} = \dfrac{|2\alpha - 1|}{2}$, (2.4.16)
implies that

$$
\begin{aligned}
&\mathrm{sign}(2\alpha - 1)\frac{1}{2\alpha}\frac{d}{dt}\|n_\varepsilon\|_{L^{2\alpha}(\Omega)}^{2\alpha} + |2\alpha - 1|C_S^2\frac{d}{dt}\|c_\varepsilon\|_{L^2(\Omega)}^2 \\
&\quad + \frac{|2\alpha - 1|}{2}C_S^2\int_\Omega|\nabla c_\varepsilon|^2 + 2|2\alpha - 1|C_S^2\int_\Omega|c_\varepsilon|^2 \\
&\quad + \frac{|2\alpha - 1|}{4}\int_\Omega n_\varepsilon^{2\alpha-2}|\nabla n_\varepsilon|^2 \\
&\leq C_7 \text{ for all } t \in (0, T_{max,\varepsilon}).
\end{aligned}
\tag{2.4.17}
$$

If $2\alpha > 1$, then $\mathrm{sign}(2\alpha - 1) = 1 > 0$, thus, integrating (2.4.17) over time, we can
obtain

$$
\int_\Omega n_\varepsilon^{2\alpha} + \int_\Omega c_\varepsilon^2 \leq C_8 \text{ for all } t \in (0, T_{max,\varepsilon})
\tag{2.4.18}
$$

and

$$
\int_0^T\int_\Omega\left[n_\varepsilon^{2\alpha-2}|\nabla n_\varepsilon|^2 + |\nabla c_\varepsilon|^2\right] \leq C_8(T + 1) \text{ for all } T \in (0, T_{max,\varepsilon})
\tag{2.4.19}
$$

and some positive constant C_8. If $2\alpha < 1$, then $\mathrm{sign}(2\alpha - 1) = -1 < 0$; hence, in
view of (2.4.4), integrating (2.4.17) over time and employing Hölder's inequality,
we also conclude that there exists a positive constant C_9 such that

$$
\int_\Omega n_\varepsilon^{2\alpha} + \int_\Omega c_\varepsilon^2 \leq C_9 \text{ for all } t \in (0, T_{max,\varepsilon})
\tag{2.4.20}
$$

and

$$
\int_0^T\int_\Omega\left[n_\varepsilon^{2\alpha-2}|\nabla n_\varepsilon|^2 + |\nabla c_\varepsilon|^2\right] \leq C_9(T + 1) \text{ for all } T \in (0, T_{max,\varepsilon}).
\tag{2.4.21}
$$

Case (II) $2\alpha = 1$: Using the first equation of (2.2.8) and (2.2.9), integrating by
parts, and applying (2.1.5) and (2.4.1), we obtain

$$
\begin{aligned}
&\frac{d}{dt}\int_\Omega n_\varepsilon \ln n_\varepsilon \\
&= \int_\Omega n_{\varepsilon t}\ln n_\varepsilon + \int_\Omega n_{\varepsilon t} \\
&= \int_\Omega \Delta n_\varepsilon \ln n_\varepsilon - \int_\Omega \ln n_\varepsilon \nabla\cdot(n_\varepsilon F_\varepsilon'(n_\varepsilon)S_\varepsilon(x, n_\varepsilon, c_\varepsilon)\cdot\nabla c_\varepsilon) \\
&\leq -\int_\Omega\frac{|\nabla n_\varepsilon|^2}{n_\varepsilon} + \int_\Omega C_S(1 + n_\varepsilon)^{-\alpha}\frac{n_\varepsilon}{n_\varepsilon}|\nabla n_\varepsilon||\nabla c_\varepsilon| \text{ for all } t \in (0, T_{max,\varepsilon}),
\end{aligned}
$$

which combined with Young's inequality and $2\alpha = 1$ implies that

$$\frac{d}{dt} \int_\Omega n_\varepsilon \ln n_\varepsilon + \frac{1}{2} \int_\Omega \frac{|\nabla n_\varepsilon|^2}{n_\varepsilon} \leq \frac{1}{2} C_S^2 \int_\Omega |\nabla c_\varepsilon|^2 \text{ for all } t \in (0, T_{max,\varepsilon}).$$

However, since $2\alpha = 1$ yields $\alpha > \frac{1}{3}$, by employing almost exactly the same arguments as in the proof of (2.4.10)–(2.4.16) (with the minor necessary changes being left as an easy exercise to the reader), we conclude an estimate of

$$\int_\Omega n_\varepsilon \ln n_\varepsilon + \int_\Omega c_\varepsilon^2 \leq C_{10} \text{ for all } t \in (0, T_{max,\varepsilon}) \tag{2.4.22}$$

and

$$\int_0^T \int_\Omega \left[\frac{|\nabla n_\varepsilon|^2}{n_\varepsilon} + |\nabla c_\varepsilon|^2 \right] \leq C_{10}(T + 1) \text{ for all } T \in (0, T_{max,\varepsilon}). \tag{2.4.23}$$

Now, multiplying the third equation of (2.2.8) by u_ε, integrating by parts, and using $\nabla \cdot u_\varepsilon = 0$ give

$$\frac{1}{2} \frac{d}{dt} \int_\Omega |u_\varepsilon|^2 + \int_\Omega |\nabla u_\varepsilon|^2 = \int_\Omega n_\varepsilon u_\varepsilon \cdot \nabla \phi \text{ for all } t \in (0, T_{max,\varepsilon}). \tag{2.4.24}$$

Here, we use Hölder's inequality, Young's inequality and the continuity of the embedding $W^{1,2}(\Omega) \hookrightarrow L^6(\Omega)$ to find C_{11} and $C_{12} > 0$ such that

$$\begin{aligned}
\int_\Omega n_\varepsilon u_\varepsilon \cdot \nabla \phi &\leq \|\nabla \phi\|_{L^\infty(\Omega)} \|n_\varepsilon\|_{L^{\frac{6}{5}}(\Omega)} \|u_\varepsilon\|_{L^6(\Omega)} \\
&\leq C_{11} \|\nabla \phi\|_{L^\infty(\Omega)} \|n_\varepsilon\|_{L^{\frac{6}{5}}(\Omega)} \|\nabla u_\varepsilon\|_{L^2(\Omega)} \\
&\leq \frac{1}{2} \|\nabla u_\varepsilon\|_{L^2(\Omega)}^2 + C_{12} \|n_\varepsilon\|_{L^{\frac{6}{5}}(\Omega)}^2 \text{ for all } t \in (0, T_{max,\varepsilon}).
\end{aligned} \tag{2.4.25}$$

Next, in view of (2.4.4) and $\alpha > \frac{1}{3}$, (2.4.14) and Young's inequality along with the Gagliardo–Nirenberg inequality yield

$$\begin{aligned}
\int_\Omega n_\varepsilon u_\varepsilon \cdot \nabla \phi &\leq \frac{1}{2} \|\nabla u_\varepsilon\|_{L^2(\Omega)}^2 + C_8 \|\nabla n_\varepsilon^\alpha\|_{L^2(\Omega)}^{\frac{2}{6\alpha-1}} \|n_\varepsilon^\alpha\|_{L^{\frac{1}{\alpha}}(\Omega)}^{\frac{2}{\alpha} - \frac{2}{6\alpha-1}} \\
&\leq \frac{1}{2} \|\nabla u_\varepsilon\|_{L^2(\Omega)}^2 + \|\nabla n_\varepsilon^\alpha\|_{L^2(\Omega)}^2 + C_{13} \text{ for all } t \in (0, T_{max,\varepsilon})
\end{aligned} \tag{2.4.26}$$

and some positive constant C_{13}. Now, inserting (2.4.25) and (2.4.26) into (2.4.24) and using (2.4.19) and (2.4.23), one has

$$\int_\Omega |u_\varepsilon|^2 \leq C_{14} \text{ for all } t \in (0, T_{max,\varepsilon}) \tag{2.4.27}$$

and

$$\int_0^T \int_\Omega |\nabla u_\varepsilon|^2 \leq C_{14}(T+1) \quad \text{for all } T \in (0, T_{max,\varepsilon}) \tag{2.4.28}$$

and some positive constant C_{14}. Finally, collecting (2.4.18)–(2.4.21), (2.4.22)–(2.4.23) and (2.4.27)–(2.4.28), we can get (2.4.5) and (2.4.6).

With the help of Lemma 2.18, based on the Gagliardo–Nirenberg inequality and an application of well-known arguments from parabolic regularity theory, we can derive the following lemmas.

Lemma 2.19 *Let $\alpha > \frac{1}{3}$. Then there exists $C > 0$ independent of ε such that, for each $T \in (0, T_{max,\varepsilon})$, the solution of (2.2.8) satisfies*

$$\int_0^T \int_\Omega \left[|\nabla n_\varepsilon|^{\frac{3\alpha+1}{2}} + n_\varepsilon^{\frac{6\alpha+2}{3}} \right] \leq C(T+1) \text{ if } \frac{1}{3} < \alpha \leq \frac{1}{2}, \tag{2.4.29}$$

$$\int_0^T \int_\Omega \left[|\nabla n_\varepsilon|^{\frac{10\alpha}{3+2\alpha}} + n_\varepsilon^{\frac{10\alpha}{3}} \right] \leq C(T+1) \text{ if } \frac{1}{2} < \alpha < 1, \tag{2.4.30}$$

as well as

$$\int_0^T \int_\Omega \left[|\nabla n_\varepsilon|^2 + n_\varepsilon^{\frac{10}{3}} \right] \leq C(T+1) \text{ if } \alpha \geq 1 \tag{2.4.31}$$

and

$$\int_0^T \left\{ \int_\Omega [c_\varepsilon^{\frac{10}{3}} + |u_\varepsilon|^{\frac{10}{3}}] + \|u_\varepsilon\|_{L^6(\Omega)}^2 \right\} \leq C(T+1). \tag{2.4.32}$$

Proof Case $\frac{1}{3} < \alpha \leq \frac{1}{2}$: From (2.4.4), (2.4.5) and (2.4.6), in light of the Gagliardo–Nirenberg inequality, for some C_1 and $C_2 > 0$ that are independent of ε, one may verify that

$$\int_0^T \int_\Omega n_\varepsilon^{\frac{6\alpha+2}{3}}$$
$$= \int_0^T \|n_\varepsilon^\alpha\|_{L^{\frac{6\alpha+2}{3\alpha}}(\Omega)}^{\frac{6\alpha+2}{3\alpha}} \tag{2.4.33}$$
$$\leq C_1 \int_0^T \left(\|\nabla n_\varepsilon^\alpha\|_{L^2(\Omega)}^2 \|n_\varepsilon^\alpha\|_{L^{\frac{1}{\alpha}}(\Omega)}^{\frac{2}{3\alpha}} + \|n_\varepsilon^\alpha\|_{L^{\frac{1}{\alpha}}(\Omega)}^{\frac{6\alpha+2}{3\alpha}} \right)$$
$$\leq C_2(T+1) \text{ for all } T > 0.$$

Therefore, employing Hölder's inequality (with two exponents $\frac{4}{3\alpha+1}$ and $\frac{4}{3-3\alpha}$), we conclude that there exists a positive constant C_3 such that

$$\int_0^T \int_\Omega |\nabla n_\varepsilon|^{\frac{3\alpha+1}{2}} \leq \left[\int_0^T \int_\Omega n_\varepsilon^{2\alpha-2} |\nabla n_\varepsilon|^2 \right]^{\frac{3\alpha+1}{4}} \left[\int_0^T \int_\Omega n_\varepsilon^{\frac{6\alpha+2}{3}} \right]^{\frac{3-3\alpha}{4}}$$

$$\leq C_3 (T+1) \text{ for all } T > 0. \tag{2.4.34}$$

Case $\frac{1}{2} < \alpha < 1$: Again by (2.4.4), (2.4.5) and (2.4.6) and the Gagliardo–Nirenberg inequality and Hölder's inequality (with two exponents $\frac{3+2\alpha}{5\alpha}$ and $\frac{3+2\alpha}{3-3\alpha}$), we derive that there exist positive constants C_4, C_5 and C_6 such that

$$\int_0^T \int_\Omega n_\varepsilon^{\frac{10\alpha}{3}}$$

$$= \int_0^T \| n_\varepsilon^\alpha \|_{L^{\frac{10}{3}}(\Omega)}^{\frac{10}{3}} \tag{2.4.35}$$

$$\leq C_4 \int_0^T \left(\| \nabla n_\varepsilon^\alpha \|_{L^2(\Omega)}^2 \| n_\varepsilon^\alpha \|_{L^2(\Omega)}^{\frac{4}{3}} + \| n_\varepsilon^\alpha \|_{L^2(\Omega)}^{\frac{10\alpha}{3}} \right)$$

$$\leq C_5 (T+1) \text{ for all } T > 0$$

and

$$\int_0^T \int_\Omega |\nabla n_\varepsilon|^{\frac{10\alpha}{3+2\alpha}} \leq \left[\int_0^T \int_\Omega n_\varepsilon^{2\alpha-2} |\nabla n_\varepsilon|^2 \right]^{\frac{5\alpha}{3+2\alpha}} \left[\int_0^T \int_\Omega n_\varepsilon^{\frac{10\alpha}{3}} \right]^{\frac{3-3\alpha}{3+2\alpha}}$$

$$\leq C_6 (T+1) \text{ for all } T > 0. \tag{2.4.36}$$

Case $\alpha \geq 1$: Multiplying the first equation in (2.2.8) by n_ε, in view of (2.2.9) and using $\nabla \cdot u_\varepsilon = 0$, we derive

$$\frac{1}{2} \frac{d}{dt} \| n_\varepsilon \|_{L^2(\Omega)}^2 + \int_\Omega |\nabla n_\varepsilon|^2$$

$$= - \int_\Omega n_\varepsilon \nabla \cdot (n_\varepsilon F_\varepsilon'(n_\varepsilon) S_\varepsilon(x, n_\varepsilon, c_\varepsilon) \cdot \nabla c_\varepsilon) \tag{2.4.37}$$

$$\leq \int_\Omega n_\varepsilon F_\varepsilon'(n_\varepsilon) |S_\varepsilon(x, n_\varepsilon, c_\varepsilon)| |\nabla n_\varepsilon| |\nabla c_\varepsilon| \text{ for all } t \in (0, T_{max,\varepsilon}).$$

Recalling (2.1.5) and (2.2.9) and using $\alpha \geq 1$, via Young's inequality, we derive

$$\int_\Omega n_\varepsilon F_\varepsilon'(n_\varepsilon) |S_\varepsilon(x, n_\varepsilon, c_\varepsilon)| |\nabla n_\varepsilon| |\nabla c_\varepsilon|$$

$$\leq C_S \int_\Omega |\nabla n_\varepsilon| |\nabla c_\varepsilon| \tag{2.4.38}$$

$$\leq \frac{1}{2} \int_\Omega |\nabla n_\varepsilon|^2 + \frac{C_S^2}{2} \int_\Omega |\nabla c_\varepsilon|^2 \text{ for all } t \in (0, T_{max,\varepsilon}).$$

Here, we have used the fact that

$$n_\varepsilon F'_\varepsilon(n_\varepsilon)|S_\varepsilon(x, n_\varepsilon, c_\varepsilon)| \le C_S n_\varepsilon (1 + n_\varepsilon)^{-1} \le C_S$$

by using (2.1.5). Therefore, collecting (2.4.37), (2.4.38) and using (2.4.6), we conclude that

$$\int_\Omega n_\varepsilon^2 \le C_7 \text{ for all } t \in (0, T_{max,\varepsilon}) \tag{2.4.39}$$

and

$$\int_0^T \int_\Omega |\nabla n_\varepsilon|^2 \le C_7(T + 1). \tag{2.4.40}$$

Hence, from (2.4.39)–(2.4.40) and (2.4.5)–(2.4.6), in light of the Gagliardo–Nirenberg inequality, we derive that there exist positive constants C_i, $(i = 8, \cdots, 17)$ such that

$$\int_0^T \int_\Omega n_\varepsilon^{\frac{10}{3}} \le C_8 \int_0^T \left(\|\nabla n_\varepsilon\|_{L^2(\Omega)}^2 \|n_\varepsilon\|_{L^2(\Omega)}^{\frac{4}{3}} + \|n_\varepsilon\|_{L^2(\Omega)}^{\frac{10}{3}} \right)$$
$$\le C_9(T + 1) \text{ for all } T > 0, \tag{2.4.41}$$

$$\int_0^T \int_\Omega c_\varepsilon^{\frac{10}{3}} \le C_{10} \int_0^T \left(\|\nabla c_\varepsilon\|_{L^2(\Omega)}^2 \|c_\varepsilon\|_{L^2(\Omega)}^{\frac{4}{3}} + \|c_\varepsilon\|_{L^2(\Omega)}^{\frac{10}{3}} \right)$$
$$\le C_{11}(T + 1) \text{ for all } T > 0 \tag{2.4.42}$$

as well as

$$\int_0^T \int_\Omega |u_\varepsilon|^{\frac{10}{3}} \le C_{14} \int_0^T \left(\|\nabla u_\varepsilon\|_{L^2(\Omega)}^2 \|u_\varepsilon\|_{L^2(\Omega)}^{\frac{4}{3}} + \|u_\varepsilon\|_{L^2(\Omega)}^{\frac{10}{3}} \right)$$
$$\le C_{15}(T + 1) \text{ for all } T > 0 \tag{2.4.43}$$

and

$$\int_0^T \|u_\varepsilon\|_{L^6(\Omega)}^2 \le C_{16} \int_0^T \|\nabla u_\varepsilon\|_{L^2(\Omega)}^2$$
$$\le C_{17}(T + 1) \text{ for all } T > 0, \tag{2.4.44}$$

where in the last inequality we have used the embedding $W_{0,\sigma}^{1,2}(\Omega) \hookrightarrow L^6(\Omega)$ and the Poincaré inequality. Finally, combining (2.4.33)–(2.4.36) with (2.4.40)–(2.4.44), we can obtain the results.

Lemma 2.20 Let $\frac{1}{3} < \alpha \le \frac{8}{21}$. Then there exist $\gamma = \frac{2\alpha + \frac{2}{3}}{\alpha + 1} \in (1, 2)$ and $C > 0$ independent of ε such that, for each $T \in (0, T_{max,\varepsilon})$, the solution of (2.2.8) satisfies

$$\int_0^T \|n_\varepsilon\|_{L^{\frac{6\gamma}{6-\gamma}}(\Omega)}^{\frac{2\gamma}{2-\gamma}} \le C(T + 1). \tag{2.4.45}$$

Proof To this end, we first prove that for all $p \in (1, 6\alpha)$, there exists a positive constant C_1 independent of ε such that, for each $T \in (0, T_{max,\varepsilon})$, the solution of

(2.2.8) satisfies

$$\int_0^T \|n_\varepsilon\|_{L^p(\Omega)}^{\frac{2p(\alpha-\frac{1}{6})}{p-1}} \le C_1(T+1). \tag{2.4.46}$$

In fact, by (2.4.4) and (2.4.6), we derive that for some positive constants C_2 and C_3 independent of ε such that

$$\int_0^T \|n_\varepsilon\|_{L^p(\Omega)}^{\frac{2p(\alpha-\frac{1}{6})}{p-1}}$$

$$= \int_0^T \|n_\varepsilon^\alpha\|_{L^{\frac{p}{\alpha}}(\Omega)}^{\frac{2p}{p-1}\cdot\frac{6\alpha-1}{6\alpha}}$$

$$\le C_2 \int_0^T \left(\|\nabla n_\varepsilon^\alpha\|_{L^2(\Omega)}^2 \|n_\varepsilon^\alpha\|_{L^{\frac{1}{\alpha}}(\Omega)}^{\frac{2p}{p-1}\cdot\frac{6\alpha-1}{6\alpha}-2} + \|n_\varepsilon^\alpha\|_{L^{\frac{1}{\alpha}}(\Omega)}^{\frac{2p}{p-1}\cdot\frac{6\alpha-1}{6\alpha}} \right)$$

$$\le C_3(T+1) \text{ for all } T > 0,$$

which implies that (2.4.46) holds. Next, by $\alpha \in (\frac{1}{3}, \frac{8}{21}]$, we may choose $\gamma = \frac{2\alpha+\frac{2}{3}}{\alpha+1}$ such that

$$1 < \gamma < \min\{\frac{6\alpha}{\alpha+1}, 2\} \tag{2.4.47}$$

as well as

$$p := \frac{6\gamma}{6-\gamma} \in (1, 6\alpha) \tag{2.4.48}$$

and

$$\frac{2p(\alpha-\frac{1}{6})}{p-1} = \frac{12\gamma(\alpha-\frac{1}{6})}{7\gamma-6} > \frac{2\gamma}{2-\gamma}. \tag{2.4.49}$$

Collecting (2.4.46)–(2.4.49), one can derive (2.4.45) by the Young inequality.

2.4.2 Global Solvability of the Approximate System

The main task of this subsection is to prove the global solvability of the regularized problem (2.2.8). To this end, first, we need to establish some ε-dependent estimates for n_ε, c_ε and u_ε.

Lemma 2.21 Let $\alpha > \frac{1}{3}$. Then there exists $C = C(\varepsilon) > 0$ depending on ε such that the solution of (2.2.8) satisfies

$$\int_\Omega n_\varepsilon^{2\alpha+2} + \int_\Omega |\nabla u_\varepsilon|^2 \le C \text{ for all } t \in (0, T_{max,\varepsilon}). \tag{2.4.50}$$

In addition, for each $T \in (0, T_{max,\varepsilon}]$ with $T < \infty$, one can find a constant $C > 0$ depending on ε such that

$$\int_0^T \int_\Omega \left[n_\varepsilon^{2\alpha} |\nabla n_\varepsilon|^2 + |\Delta u_\varepsilon|^2 \right] \leq C. \tag{2.4.51}$$

Proof In view of (2.2.9), we derive

$$F_\varepsilon'(n_\varepsilon) \leq \frac{1}{\varepsilon n_\varepsilon},$$

so that, by multiplying the first equation in (2.2.8) by $n_\varepsilon^{1+2\alpha}$, using $\nabla \cdot u_\varepsilon = 0$, and applying the same argument as in the proof of (2.4.7)–(2.4.21), one can obtain that there exist positive constants C_1 and C_2 depending on ε such that

$$\int_\Omega n_\varepsilon^{2\alpha+2} \leq C_1 \text{ for all } t \in (0, T_{max,\varepsilon}) \tag{2.4.52}$$

and

$$\int_0^T \int_\Omega n_\varepsilon^{2\alpha} |\nabla n_\varepsilon|^2 \leq C_2 \text{ for all } T \in (0, T_{max,\varepsilon}] \text{ with } T < \infty.$$

Now, from $D(1 + \varepsilon A) := W^{2,2}(\Omega) \cap W_{0,\sigma}^{1,2}(\Omega) \hookrightarrow L^\infty(\Omega)$ and (2.4.5), it follows that, for some $C_3 > 0$ and $C_4 > 0$,

$$\|Y_\varepsilon u_\varepsilon\|_{L^\infty(\Omega)} = \|(I + \varepsilon A)^{-1} u_\varepsilon\|_{L^\infty(\Omega)} \leq C_3 \|u_\varepsilon(\cdot, t)\|_{L^2(\Omega)} \leq C_4 \text{ for all } t \in (0, T_{max,\varepsilon}). \tag{2.4.53}$$

Next, testing the projected Stokes equation $u_{\varepsilon t} + Au_\varepsilon = \mathscr{P}[-\kappa(Y_\varepsilon u_\varepsilon \cdot \nabla)u_\varepsilon + n_\varepsilon \nabla\phi]$ by Au_ε, we derive

$$\frac{1}{2}\frac{d}{dt}\|A^{\frac{1}{2}}u_\varepsilon\|_{L^2(\Omega)}^2 + \int_\Omega |Au_\varepsilon|^2$$

$$= \int_\Omega Au_\varepsilon \mathscr{P}(-\kappa(Y_\varepsilon u_\varepsilon \cdot \nabla)u_\varepsilon) + \int_\Omega \mathscr{P}(n_\varepsilon \nabla\phi) Au_\varepsilon$$

$$\leq \frac{1}{2}\int_\Omega |Au_\varepsilon|^2 + \kappa^2 \int_\Omega |(Y_\varepsilon u_\varepsilon \cdot \nabla)u_\varepsilon|^2 + \|\nabla\phi\|_{L^\infty(\Omega)}^2 \int_\Omega n_\varepsilon^2 \text{ for all } t \in (0, T_{max,\varepsilon}). \tag{2.4.54}$$

However, in light of the Gagliardo–Nirenberg inequality, Young's inequality and (2.4.53), there exists a positive constant C_5 such that

$$\kappa^2 \int_\Omega |(Y_\varepsilon u_\varepsilon \cdot \nabla)u_\varepsilon|^2 \le \kappa^2 \|Y_\varepsilon u_\varepsilon\|_{L^\infty(\Omega)}^2 \int_\Omega |\nabla u_\varepsilon|^2$$

$$\le \kappa^2 \|Y_\varepsilon u_\varepsilon\|_{L^\infty(\Omega)}^2 \int_\Omega |\nabla u_\varepsilon|^2 \tag{2.4.55}$$

$$\le C_5 \int_\Omega |\nabla u_\varepsilon|^2 \text{ for all } t \in (0, T_{max,\varepsilon}).$$

Here, we have used the well-known fact that $\|A(\cdot)\|_{L^2(\Omega)}$ defines a norm equivalent to $\|\cdot\|_{W^{2,2}(\Omega)}$ on $D(A)$ (see Theorem 2.c2.2-1.3 of Sohr 2001). Now, recall that $\|A^{\frac{1}{2}} u_\varepsilon\|_{L^2(\Omega)}^2 = \|\nabla u_\varepsilon\|_{L^2(\Omega)}^2$. Substituting (2.4.55) into (2.4.54) yields

$$\frac{1}{2}\frac{d}{dt}\|\nabla u_\varepsilon\|_{L^2(\Omega)}^2 + \int_\Omega |\Delta u_\varepsilon|^2$$
$$\le C_6 \int_\Omega |\nabla u_\varepsilon|^2 + \|\nabla\phi\|_{L^\infty(\Omega)}^2 \int_\Omega n_\varepsilon^2 \text{ for all } t \in (0, T_{max,\varepsilon}). \tag{2.4.56}$$

Since $\alpha > \frac{1}{3}$ yields $2\alpha + 2 > \frac{8}{3} > 2$, by collecting (2.4.52) and (2.4.56) and performing some basic calculations, we can get the results.

Lemma 2.22 *Under the assumptions of Theorem 2.2, one can find that there exists* $C = C(\varepsilon) > 0$ *depending on* ε *such that*

$$\int_\Omega |\nabla c_\varepsilon(\cdot, t)|^2 \le C \text{ for all } t \in (0, T_{max,\varepsilon}) \tag{2.4.57}$$

and

$$\int_0^T \int_\Omega |\Delta c_\varepsilon|^2 \le C \text{ for all } T \in (0, T_{max,\varepsilon}] \text{ with } T < \infty. \tag{2.4.58}$$

Proof First, testing the second equation in (2.2.8) against $-\Delta c_\varepsilon$, employing Young's inequality, and using (2.4.3) yields

$$\frac{1}{2}\frac{d}{dt}\|\nabla c_\varepsilon\|_{L^2(\Omega)}^2 = \int_\Omega -\Delta c_\varepsilon(\Delta c_\varepsilon - c_\varepsilon + F_\varepsilon(n_\varepsilon) - u_\varepsilon \cdot \nabla c_\varepsilon)$$

$$= -\int_\Omega |\Delta c_\varepsilon|^2 - \int_\Omega |\nabla c_\varepsilon|^2 - \int_\Omega F_\varepsilon(n_\varepsilon)\Delta c_\varepsilon - \int_\Omega (u_\varepsilon \cdot \nabla c_\varepsilon)\Delta c_\varepsilon$$

$$\le -\frac{1}{4}\int_\Omega |\Delta c_\varepsilon|^2 - \int_\Omega |\nabla c_\varepsilon|^2 + \int_\Omega n_\varepsilon^2 + \int_\Omega |u_\varepsilon \cdot \nabla c_\varepsilon||\Delta c_\varepsilon|$$
$$\tag{2.4.59}$$

for all $t \in (0, T_{max,\varepsilon})$. Next, one needs to estimate the last term on the right-hand side of (2.4.59). Indeed, in view of Sobolev's embedding theorem ($W^{1,2}(\Omega) \hookrightarrow L^6(\Omega)$) and applying (2.4.50) and (2.4.5), we derive from Hölder's inequality, the Gagliardo–Nirenberg inequality, and Young's inequality that there exist positive constants C_1, C_2, C_3 and C_4 such that

$$\int_{\Omega} |u_{\varepsilon} \cdot \nabla c_{\varepsilon}| |\Delta c_{\varepsilon}| \leq \|u_{\varepsilon}\|_{L^6(\Omega)} \|\nabla c_{\varepsilon}\|_{L^3(\Omega)} \|\Delta c_{\varepsilon}\|_{L^2(\Omega)}$$

$$\leq C_1 \|\nabla c_{\varepsilon}\|_{L^3(\Omega)} \|\Delta c_{\varepsilon}\|_{L^2(\Omega)}$$

$$\leq C_2 (\|\Delta c_{\varepsilon}\|_{L^2(\Omega)}^{\frac{3}{4}} \|c_{\varepsilon}\|_{L^2(\Omega)}^{\frac{1}{4}} + \|c_{\varepsilon}\|_{L^2(\Omega)}^2) \|\Delta c_{\varepsilon}\|_{L^2(\Omega)} \quad (2.4.60)$$

$$\leq C_3 (\|\Delta c_{\varepsilon}\|_{L^2(\Omega)}^{\frac{7}{4}} + \|\Delta c_{\varepsilon}\|_{L^2(\Omega)})$$

$$\leq \frac{1}{4} \|\Delta c_{\varepsilon}\|_{L^2(\Omega)}^2 + C_4 \text{ for all } t \in (0, T_{max,\varepsilon}).$$

Inserting (2.4.60) into (2.4.59) and using (2.4.50), one obtains (2.4.57) and (2.4.58). This completes the proof of Lemma 2.22.

Lemma 2.23 *Let* $\alpha > \frac{1}{3}$. *Assume that the hypothesis of Theorem 2.2 holds. Then there exists a positive constant* $C = C(\varepsilon)$ *depending on* ε *such that, for any* $3 < q < 6$, *the solution of (2.2.8) from Lemma 2.3 satisfies*

$$\|A^{\gamma} u_{\varepsilon}(\cdot, t)\|_{L^2(\Omega)} \leq C \text{ for all } t \in (0, T_{max,\varepsilon}) \quad (2.4.61)$$

as well as

$$\|u_{\varepsilon}(\cdot, t)\|_{L^{\infty}(\Omega)} \leq C \text{ for all } t \in (0, T_{max,\varepsilon}) \quad (2.4.62)$$

and

$$\|\nabla c_{\varepsilon}(\cdot, t)\|_{L^q(\Omega)} \leq C \text{ for all } t \in (0, T_{max,\varepsilon}), \quad (2.4.63)$$

where γ *is the same as in (2.1.13).*

Proof Let $h_{\varepsilon}(x, t) = \mathscr{P}[n_{\varepsilon} \nabla \phi - \kappa (Y_{\varepsilon} u_{\varepsilon} \cdot \nabla) u_{\varepsilon}]$. Because $\alpha > \frac{1}{3}$, then along with (2.4.50), and (2.4.53), there exist positive constants $q_0 > \frac{3}{2}$ and C_1 such that

$$\|n_{\varepsilon}(\cdot, t)\|_{L^{q_0}(\Omega)} \leq C_1 \text{ for all } t \in (0, T_{max,\varepsilon}) \quad (2.4.64)$$

and

$$\|h_{\varepsilon}(\cdot, t)\|_{L^{q_0}(\Omega)} \leq C_1 \text{ for all } t \in (0, T_{max,\varepsilon}). \quad (2.4.65)$$

Hence, because $q_0 > \frac{3}{2}$, we pick an arbitrary $\gamma \in (\frac{3}{4}, 1)$ and, then, $-\gamma - \frac{3}{2}(\frac{1}{q_0} - \frac{1}{2}) > -1$. Therefore, in view of the smoothing properties of the Stokes semigroup Giga (1986), we derive that, for some $\lambda, C_2 > 0$, and $C_3 > 0$,

$$\|A^{\gamma} u_{\varepsilon}(\cdot, t)\|_{L^2(\Omega)}$$

$$\leq \|A^{\gamma} e^{-tA} u_0\|_{L^2(\Omega)} + \int_0^t \|A^{\gamma} e^{-(t-\tau)A} h_{\varepsilon}(\cdot, \tau) d\tau\|_{L^2(\Omega)} d\tau$$

$$\leq \|A^{\gamma} u_0\|_{L^2(\Omega)} + C_2 \int_0^t (t - \tau)^{-\gamma - \frac{3}{2}(\frac{1}{q_0} - \frac{1}{2})} e^{-\lambda(t-\tau)} \|h_{\varepsilon}(\cdot, \tau)\|_{L^{q_0}(\Omega)} d\tau \quad (2.4.66)$$

$$\leq C_3 \text{ for all } t \in (0, T_{max,\varepsilon}).$$

Observe that $\gamma > \frac{3}{4}$, and $D(A^\gamma)$ is continuously embedded into $L^\infty(\Omega)$. Therefore, we derive that there exists a positive constant C_4 such that

$$\|u_\varepsilon(\cdot, t)\|_{L^\infty(\Omega)} \le C_4 \text{ for all } t \in (0, T_{max,\varepsilon}) \tag{2.4.67}$$

from (2.4.66). However, from (2.4.57), with the help of Sobolev's embedding theorem, it follows that, for any fixed $\tilde{q} \in (3, 6)$,

$$\|c_\varepsilon(\cdot, t)\|_{L^{\tilde{q}}(\Omega)} \le C_5 \text{ for all } t \in (0, T_{max,\varepsilon}). \tag{2.4.68}$$

Now, involving the variation-of-constants formula for c_ε and applying $\nabla \cdot u_\varepsilon = 0$ in $x \in \Omega, t > 0$, we have

$$c_\varepsilon(t) = e^{t(\Delta-1)}c_0 + \int_0^t e^{(t-s)(\Delta-1)}(F_\varepsilon(n_\varepsilon(s)) + \nabla \cdot (u_\varepsilon(s)c_\varepsilon(s))ds, \ t \in (0, T_{max,\varepsilon}), \tag{2.4.69}$$

so that, for any $3 < q < \min\{\frac{3q_0}{(3-q_0)_+}, \tilde{q}\}$, we have

$$\|\nabla c_\varepsilon(\cdot, t)\|_{L^q(\Omega)}$$

$$\le \|\nabla e^{t(\Delta-1)}c_0\|_{L^q(\Omega)} + \int_0^t \|\nabla e^{(t-s)(\Delta-1)}F_\varepsilon(n_\varepsilon(s))\|_{L^q(\Omega)}ds \tag{2.4.70}$$

$$+ \int_0^t \|\nabla e^{(t-s)(\Delta-1)}\nabla \cdot (u_\varepsilon(s)c_\varepsilon(s))\|_{L^q(\Omega)}ds.$$

To address the right-hand side of (2.4.70), in view of (2.1.13), we first use Lemma 2.4 to get

$$\|\nabla e^{t(\Delta-1)}c_0\|_{L^q(\Omega)} \le C_6 \text{ for all } t \in (0, T_{max,\varepsilon}). \tag{2.4.71}$$

Since (2.4.64) and (2.4.68) yields

$$-\frac{1}{2} - \frac{3}{2}\left(\frac{1}{q_0} - \frac{1}{q}\right) > -1,$$

together with this and (2.4.3), by using Lemma 2.4 again, the second term of the right-hand side is estimated as

$$\int_0^t \|\nabla e^{(t-s)(\Delta-1)}F_\varepsilon(n_\varepsilon(s))\|_{L^q(\Omega)}ds$$

$$\le C_7 \int_0^t [1 + (t - s)^{-\frac{1}{2} - \frac{3}{2}(\frac{1}{q_0} - \frac{1}{q})}]e^{-(t-s)}\|n_\varepsilon(s)\|_{L^{q_0}(\Omega)}ds$$

$$\le C_8 \text{ for all } t \in (0, T_{max,\varepsilon}).$$

Finally, we will address the third term on the right-hand side of (2.4.70). To this end, we choose $0 < \iota < \frac{1}{2}$ satisfying $\frac{1}{2} + \frac{3}{2}(\frac{1}{\bar{q}} - \frac{1}{q}) < \iota$ and $\tilde{\kappa} \in (0, \frac{1}{2} - \iota)$. In view of Hölder's inequality, we derive from Lemma 2.4, (2.4.68) and (2.4.67) that there exist constants C_9, C_{10}, C_{11} and C_{12} such that

$$
\begin{aligned}
&\int_0^t \|\nabla e^{(t-s)(\Delta-1)} \nabla \cdot (u_\varepsilon(s)c_\varepsilon(s))\|_{L^{\bar{q}}(\Omega)} ds \\
&\leq C_9 \int_0^t \|(-\Delta + 1)^\iota e^{(t-s)(\Delta-1)} \nabla \cdot (u_\varepsilon(s)c_\varepsilon(s))\|_{L^q(\Omega)} ds \\
&\leq C_{10} \int_0^t (t-s)^{-\iota-\frac{1}{2}-\tilde{\kappa}} e^{-\lambda(t-s)} \|u_\varepsilon(s)c_\varepsilon(s)\|_{L^{\bar{q}}(\Omega)} ds \\
&\leq C_{11} \int_0^t (t-s)^{-\iota-\frac{1}{2}-\tilde{\kappa}} e^{-\lambda(t-s)} \|u_\varepsilon(s)\|_{L^\infty(\Omega)} \|c_\varepsilon(s)\|_{L^{\bar{q}}(\Omega)} ds \\
&\leq C_{12} \text{ for all } t \in (0, T_{max,\varepsilon}).
\end{aligned}
\tag{2.4.72}
$$

Here, we have used the fact that

$$
\int_0^t (t-s)^{-\iota-\frac{1}{2}-\tilde{\kappa}} e^{-\lambda(t-s)} ds \leq \int_0^\infty \sigma^{-\iota-\frac{1}{2}-\tilde{\kappa}} e^{-\lambda\sigma} d\sigma < +\infty.
$$

Finally, collecting (2.4.70)–(2.4.72), we can obtain that there exists a positive constant C_{13} such that

$$
\int_\Omega |\nabla c_\varepsilon(t)|^q \leq C_{13} \text{ for all } t \in (0, T_{max,\varepsilon}) \text{ and some } q \in \left(3, \min\left\{\frac{3q_0}{(3-q_0)_+}, \tilde{q}\right\}\right).
\tag{2.4.73}
$$

The proof of Lemma 2.23 is complete.

Then we can establish global existence in the approximate problem (2.2.8) by using Lemmas 2.21 and 2.22.

Lemma 2.24 *Let $\alpha > \frac{1}{3}$. Then, for all $\varepsilon \in (0, 1)$, the solution of (2.2.8) is global in time.*

Proof Assume that $T_{max,\varepsilon}$ is finite for some $\varepsilon \in (0, 1)$. Fix $T \in (0, T_{max,\varepsilon})$, and let $M(T) := \sup_{t \in (0,T)} \|n_\varepsilon(\cdot, t)\|_{L^\infty(\Omega)}$ and $\tilde{h}_\varepsilon := F'_\varepsilon(n_\varepsilon) S_\varepsilon(x, n_\varepsilon, c_\varepsilon) \nabla c_\varepsilon + u_\varepsilon$. Then, by Lemma 2.23, (2.1.5) and (2.4.1), there exists $C_1 > 0$ such that

$$
\|\tilde{h}_\varepsilon(\cdot, t)\|_{L^q(\Omega)} \leq C_1 \text{ for all } t \in (0, T_{max,\varepsilon}) \text{ and some } 3 < q < 6.
\tag{2.4.74}
$$

Hence, because $\nabla \cdot u_\varepsilon = 0$, we can derive

$$
n_\varepsilon(t) = e^{(t-t_0)\Delta} n_\varepsilon(\cdot, t_0) - \int_{t_0}^t e^{(t-s)\Delta} \nabla \cdot (n_\varepsilon(\cdot, s)\tilde{h}_\varepsilon(\cdot, s)) ds, \quad t \in (t_0, T)
\tag{2.4.75}
$$

by means of an associate variation-of-constants formula for n, where $t_0 := (t-1)_+$. If $t \in (0, 1]$, by virtue of the maximum principle, we can derive

$$\|e^{(t-t_0)\Delta}n_\varepsilon(\cdot, t_0)\|_{L^\infty(\Omega)} \leq \|n_0\|_{L^\infty(\Omega)}, \tag{2.4.76}$$

while if $t > 1$ then, with the help of the L^p–L^q estimates for the Neumann heat semigroup and Lemma 2.17, we conclude that

$$\|e^{(t-t_0)\Delta}n_\varepsilon(\cdot, t_0)\|_{L^\infty(\Omega)} \leq C_2(t-t_0)^{-\frac{3}{2}}\|n_\varepsilon(\cdot, t_0)\|_{L^1(\Omega)} \leq C_3. \tag{2.4.77}$$

Finally, we fix an arbitrary $p \in (3, q)$ and then once more invoke known smoothing properties of the Stokes semigroup (see P. 201 of Giga 1986) and Hölder's inequality to find $C_4 > 0$ such that

$$\begin{aligned}
&\int_{t_0}^t \|e^{(t-s)\Delta}\nabla \cdot (n_\varepsilon(\cdot, s)\tilde{h}_\varepsilon(\cdot, s)\|_{L^\infty(\Omega)}ds \\
&\leq C_4 \int_{t_0}^t (t-s)^{-\frac{1}{2}-\frac{3}{2p}}\|n_\varepsilon(\cdot, s)\tilde{h}_\varepsilon(\cdot, s)\|_{L^p(\Omega)}ds \\
&\leq C_4 \int_{t_0}^t (t-s)^{-\frac{1}{2}-\frac{3}{2p}}\|n_\varepsilon(\cdot, s)\|_{L^{\frac{pq}{q-p}}(\Omega)}\|\tilde{h}_\varepsilon(\cdot, s)\|_{L^q(\Omega)}ds \\
&\leq C_4 \int_{t_0}^t (t-s)^{-\frac{1}{2}-\frac{3}{2p}}\|u_\varepsilon(\cdot, s)\|_{L^\infty(\Omega)}^b\|u_\varepsilon(\cdot, s)\|_{L^1(\Omega)}^{1-b}\|\tilde{h}_\varepsilon(\cdot, s)\|_{L^q(\Omega)}ds \\
&\leq C_5 M^b(T) \text{ for all } t \in (0, T),
\end{aligned} \tag{2.4.78}$$

where $b := \frac{pq-q+p}{pq} \in (0, 1)$ and

$$C_5 := C_4 C_1^{2-b} \int_0^1 \sigma^{-\frac{1}{2}-\frac{3}{2p}}d\sigma.$$

Since $p > 3$, we conclude that $-\frac{1}{2} - \frac{3}{2p} > -1$. In combination with (2.4.75)–(2.4.78) and using the definition of $M(T)$, we obtain $C_6 > 0$ such that

$$M(T) \leq C_6 + C_6 M^b(T) \text{ for all } T \in (0, T_{max,\varepsilon}). \tag{2.4.79}$$

Hence, in view of $b < 1$, with some basic calculation, since $T \in (0, T_{max,\varepsilon})$ was arbitrary, we can obtain there exists a positive constant C_7 such that

$$\|n_\varepsilon(\cdot, t)\|_{L^\infty(\Omega)} \leq C_7 \text{ for all } t \in (0, T_{max,\varepsilon}). \tag{2.4.80}$$

To prove the boundedness of $\|\nabla c_\varepsilon(\cdot, t)\|_{L^\infty(\Omega)}$, we rewrite the variation-of-constants formula for c_ε in the form

$$c_\varepsilon(\cdot, t) = e^{t(\Delta-1)}c_0 + \int_0^t e^{(t-s)(\Delta-1)}[F_\varepsilon(n_\varepsilon)(s) - u_\varepsilon(s) \cdot \nabla c_\varepsilon(s)]ds \text{ for all } t \in (0, T_{max,\varepsilon}).$$

Now, we choose $\theta \in (\frac{1}{2} + \frac{3}{2q}, 1)$, where $3 < q < 6$ (see (2.4.73)), then the domain of the fractional power $D((-\Delta + 1)^\theta) \hookrightarrow W^{1,\infty}(\Omega)$ (see Horstmann and Winkler 2005). Hence, in view of L^p–L^q estimates associated with the heat semigroup, (2.4.62), (2.4.63) and (2.4.3), we derive that there exist positive constants λ, C_8, C_9, C_{10} and C_{11} such that

$$\|c_\varepsilon(\cdot, t)\|_{W^{1,\infty}(\Omega)}$$
$$\leq C_8\|(-\Delta + 1)^\theta c_\varepsilon(\cdot, t)\|_{L^q(\Omega)}$$
$$\leq C_9 t^{-\theta}e^{-\lambda t}\|c_0\|_{L^q(\Omega)} + C_9 \int_0^t (t-s)^{-\theta}e^{-\lambda(t-s)}\|(F_\varepsilon(n_\varepsilon) - u_\varepsilon \cdot \nabla c_\varepsilon)(s)\|_{L^q(\Omega)}ds$$
$$\leq C_{10} + C_{10} \int_0^t (t-s)^{-\theta}e^{-\lambda(t-s)}[\|n_\varepsilon(s)\|_{L^q(\Omega)} + \|u_\varepsilon(s)\|_{L^\infty(\Omega)}\|\nabla c_\varepsilon(s)\|_{L^q(\Omega)}]ds$$
$$\leq C_{11} \text{ for all } t \in (0, T_{max,\varepsilon}).$$

$$(2.4.81)$$

Here, we have used Hölder's inequality as well as

$$\int_0^t (t-s)^{-\theta}e^{-\lambda(t-s)} \leq \int_0^\infty \sigma^{-\theta}e^{-\lambda\sigma}d\sigma < +\infty.$$

In view of (2.4.61), (2.4.80) and (2.4.81), we apply Lemma 2.3 to reach a contradiction.

2.4.3 Regularity Property of Time Derivatives

In preparation of an Aubin–Lions-type compactness argument, we will rely on an additional regularity estimate for $n_\varepsilon F_\varepsilon'(n_\varepsilon)S_\varepsilon(x, n_\varepsilon, c_\varepsilon)\nabla c_\varepsilon$, $u_\varepsilon \cdot \nabla c_\varepsilon$, $n_\varepsilon u_\varepsilon$ and $c_\varepsilon u_\varepsilon$.

Lemma 2.25 *Let $\alpha > \frac{1}{3}$, and assume that (2.1.13) holds. Then one can find $C > 0$ independent of ε such that, for all $T \in (0, \infty)$,*

$$\int_0^T \int_\Omega \left[|n_\varepsilon F_\varepsilon'(n_\varepsilon)S_\varepsilon(x, n_\varepsilon, c_\varepsilon)\nabla c_\varepsilon|^{\frac{3\alpha+1}{2}} + |n_\varepsilon u_\varepsilon|^{\frac{2\alpha+\frac{2}{3}}{\alpha+1}} \right]$$
$$\leq C(T + 1), \quad if \frac{1}{3} < \alpha \leq \frac{8}{21},$$

$$(2.4.82)$$

$$\int_0^T \int_\Omega \left[|n_\varepsilon F_\varepsilon'(n_\varepsilon)S_\varepsilon(x, n_\varepsilon, c_\varepsilon)\nabla c_\varepsilon|^{\frac{3\alpha+1}{2}} + |n_\varepsilon u_\varepsilon|^{\frac{10(3\alpha+1)}{9(\alpha+2)}} \right]$$
$$\leq C(T + 1), \quad if \frac{8}{21} < \alpha \leq \frac{1}{2},$$

$$(2.4.83)$$

$$\int_0^T \int_\Omega \left[|n_\varepsilon F_\varepsilon'(n_\varepsilon) S_\varepsilon(x, n_\varepsilon, c_\varepsilon) \nabla c_\varepsilon|^{\frac{10\alpha}{3+2\alpha}} + |n_\varepsilon u_\varepsilon|^{\frac{10\alpha}{3(\alpha+1)}} \right],$$
$$\leq C(T+1) \quad if \frac{1}{2} < \alpha < 1 \tag{2.4.84}$$

as well as

$$\int_0^T \int_\Omega \left[|n_\varepsilon F_\varepsilon'(n_\varepsilon) S_\varepsilon(x, n_\varepsilon, c_\varepsilon) \nabla c_\varepsilon|^2 + |n_\varepsilon u_\varepsilon|^{\frac{5}{3}} \right] \leq C(T+1), \ if \ \alpha \geq 1 \tag{2.4.85}$$

and

$$\int_0^T \int_\Omega \left[|u_\varepsilon \cdot \nabla c_\varepsilon|^{\frac{5}{4}} + |c_\varepsilon u_\varepsilon|^{\frac{5}{3}} \right] \leq C(T+1). \tag{2.4.86}$$

Proof First, by (2.1.5), (2.4.1) and (2.2.10), we derive

$$n_\varepsilon F_\varepsilon'(n_\varepsilon) S_\varepsilon(x, n_\varepsilon, c_\varepsilon) \leq C_S n_\varepsilon^{(1-\alpha)_+}$$

with $(1-\alpha)_+ = \max\{0, 1-\alpha\}$. Case $\frac{8}{21} < \alpha \leq \frac{1}{2}$: It is not difficult to verify that

$$\frac{2}{3\alpha+1} = \frac{1}{2} + \frac{3}{6\alpha+2}(1-\alpha)$$

and

$$\frac{9(\alpha+2)}{10(3\alpha+1)} = \frac{3}{10} + \frac{3}{6\alpha+2},$$

so that, recalling (2.4.29), (2.4.44) and Hölder's inequality, we can obtain (2.4.83). While if $\frac{1}{3} < \alpha \leq \frac{8}{21}$, in light of (2.4.6), (2.4.29), (2.4.32), (2.4.45), an employment of the Hölder and Young inequalities to shows that

$$\int_0^T \int_\Omega \left[|n_\varepsilon F_\varepsilon'(n_\varepsilon) S_\varepsilon(x, n_\varepsilon, c_\varepsilon) \nabla c_\varepsilon|^{\frac{3\alpha+1}{2}} + |n_\varepsilon u_\varepsilon|^\gamma \right]$$
$$\leq C_1 \left[\int_0^T \int_\Omega n_\varepsilon^{\frac{6\alpha+2}{3}} \right]^{\frac{3-3\alpha}{4}} \left[\int_0^T \int_\Omega |\nabla c_\varepsilon|^2 \right]^{\frac{3\alpha+1}{4}}$$
$$+ C_1 \int_0^T \|n_\varepsilon\|_{L^{\frac{6\gamma}{6-\gamma}}(\Omega)}^\gamma \|u_\varepsilon\|_{L^6(\Omega)}^\gamma$$
$$\leq C_2(T+1),$$

where $\gamma = \frac{2\alpha+\frac{2}{3}}{\alpha+1}$ is given by Lemma 2.20.

Other cases can be proved very similarly. Therefore, we omit their proofs.

To prepare our subsequent compactness properties of $(n_\varepsilon, c_\varepsilon, u_\varepsilon)$ by means of the Aubin–Lions lemma (see Simon 1986), we use Lemmas 2.17–2.19 to obtain the following regularity property with respect to the time variable.

Lemma 2.26 *Let $\alpha > \frac{1}{3}$, and assume that (2.1.13) holds. Then there exists $C > 0$ independent of ε such that*

$$\int_0^T \|\partial_t n_\varepsilon(\cdot, t)\|_{(W^{2,4}(\Omega))^*} dt \leq C(T+1) \text{ for all } T \in (0, \infty) \tag{2.4.87}$$

as well as

$$\int_0^T \|\partial_t c_\varepsilon(\cdot, t)\|_{(W^{1,5}(\Omega))^*}^{\frac{5}{4}} dt \leq C(T+1) \text{ for all } T \in (0, \infty) \tag{2.4.88}$$

and

$$\int_0^T \|\partial_t u_\varepsilon(\cdot, t)\|_{(W^{1,5}_{0,\sigma}(\Omega))^*}^{\frac{5}{4}} dt \leq C(T+1) \text{ for all } T \in (0, \infty). \tag{2.4.89}$$

Proof Firstly, testing the first equation of (2.2.8) by certain $\varphi \in C^\infty(\bar{\Omega})$, we have

$$\left| \int_\Omega (n_{\varepsilon,t}) \varphi \right|$$

$$= \left| \int_\Omega \left[\Delta n_\varepsilon - \nabla \cdot (n_\varepsilon F_\varepsilon'(n_\varepsilon) S_\varepsilon(x, n_\varepsilon, c_\varepsilon) \nabla c_\varepsilon) - u_\varepsilon \cdot \nabla n_\varepsilon \right] \varphi \right|$$

$$= \left| \int_\Omega \left[-\nabla n_\varepsilon \cdot \nabla \varphi + n_\varepsilon F_\varepsilon'(n_\varepsilon) S_\varepsilon(x, n_\varepsilon, c_\varepsilon) \nabla c_\varepsilon \cdot \nabla \varphi + n_\varepsilon u_\varepsilon \cdot \nabla \varphi \right] \right|$$

$$\leq \left| \int_\Omega \left[|\nabla n_\varepsilon| + |n_\varepsilon F_\varepsilon'(n_\varepsilon) S_\varepsilon(x, n_\varepsilon, c_\varepsilon) \nabla c_\varepsilon| + |n_\varepsilon u_\varepsilon| \right] \right| \|\varphi\|_{W^{1,\infty}(\Omega)}$$

for all $t > 0$.

Observe that the embedding $W^{2,4}(\Omega) \hookrightarrow W^{1,\infty}(\Omega)$, so that, in view of $\alpha > \frac{1}{3}$, Lemmas 2.19 and 2.25, we deduce from the Young inequality that for some C_1 and C_2 such that

$$\int_0^T \|\partial_t n_\varepsilon(\cdot, t)\|_{(W^{2,4}(\Omega))^*} dt$$

$$\leq C_1 \left\{ \int_0^T \int_\Omega |\nabla n_\varepsilon|^{r_1} + \int_0^T \int_\Omega |n_\varepsilon F_\varepsilon'(n_\varepsilon) S_\varepsilon(x, n_\varepsilon, c_\varepsilon) \nabla c_\varepsilon|^{r_1} + \int_0^T \int_\Omega |n_\varepsilon u_\varepsilon|^{r_2} + T \right\}$$

$$\leq C_2(T+1) \text{ for all } T > 0, \tag{2.4.90}$$

where

$$
r_1 = \begin{cases} \dfrac{3\alpha + 1}{2} & \text{if } \dfrac{1}{3} < \alpha \le \dfrac{1}{2}, \\[3mm] \dfrac{10\alpha}{3 + 2\alpha} & \text{if } \dfrac{1}{2} < \alpha < 1, \\[3mm] 2 & \text{if } \alpha \ge 1 \end{cases}
$$

and

$$
r_2 = \begin{cases} \dfrac{2\alpha + \frac{2}{3}}{\alpha + 1} & \text{if } \dfrac{1}{3} < \alpha \le \dfrac{8}{21}, \\[3mm] \dfrac{10(3\alpha + 1)}{9(\alpha + 2)} & \text{if } \dfrac{8}{21} < \alpha \le \dfrac{1}{2}, \\[3mm] \dfrac{10\alpha}{3(\alpha + 1)} & \text{if } \dfrac{1}{2} < \alpha < 1, \\[3mm] \dfrac{5}{3} & \text{if } \alpha \ge 1, \end{cases}
$$

Likewise, given any $\varphi \in C^\infty(\bar{\Omega})$, we may test the second equation in (2.2.8) against φ to conclude that

$$
\left| \int_\Omega \partial_t c_\varepsilon(\cdot, t)\varphi \right|
$$
$$
= \left| \int_\Omega [\Delta c_\varepsilon - c_\varepsilon + n_\varepsilon - u_\varepsilon \cdot \nabla c_\varepsilon] \cdot \varphi \right|
$$
$$
= \left| -\int_\Omega \nabla c_\varepsilon \cdot \nabla \varphi - \int_\Omega c_\varepsilon \varphi + \int_\Omega n_\varepsilon \varphi + \int_\Omega c_\varepsilon u_\varepsilon \cdot \nabla \varphi \right|
$$
$$
\le \left\{ \|\nabla c_\varepsilon\|_{L^{\frac{5}{4}}(\Omega)} + \|c_\varepsilon\|_{L^{\frac{5}{4}}(\Omega)} + \|n_\varepsilon\|_{L^{\frac{5}{4}}(\Omega)} + \|c_\varepsilon u_\varepsilon\|_{L^{\frac{5}{4}}(\Omega)} \right\} \|\varphi\|_{W^{1.5}(\Omega)}
$$

for all $t > 0$. Thus, from Lemmas 2.19 and 2.25 again, in light of $\alpha > \frac{1}{3}$, we invoke the Young inequality again and obtain that there exist positive constant C_3 and C_4 such that

$$
\int_0^T \|\partial_t c_\varepsilon(\cdot, t)\|_{(W^{1.5}(\Omega))^*}^{\frac{5}{4}} dt
$$
$$
\le C_3 \left(\int_0^T \int_\Omega |\nabla c_\varepsilon|^2 + \int_0^T \int_\Omega n_\varepsilon^{r_3} + \int_0^T \int_\Omega c_\varepsilon^{\frac{10}{3}} + \int_0^T \int_\Omega |u_\varepsilon|^{\frac{10}{3}} + T \right)
$$
$$
\le C_4(T + 1) \quad \text{for all } T > 0
$$

with

$$
r_3 = \begin{cases} \dfrac{6\alpha + 2}{3} & \text{if } \dfrac{1}{3} < \alpha \le \dfrac{1}{2}, \\[3mm] \dfrac{10\alpha}{3} & \text{if } \dfrac{1}{2} < \alpha < 1, \\[3mm] \dfrac{10}{3} & \text{if } \alpha \ge 1. \end{cases} \tag{2.4.91}
$$

Finally, for any given $\varphi \in C_{0,\sigma}^{\infty}(\Omega; \mathbb{R}^3)$, we infer from the third equation in (2.2.8) that for all $t > 0$

$$
\left| \int_{\Omega} \partial_t u_{\varepsilon}(\cdot, t) \varphi \right| = \left| -\int_{\Omega} \nabla u_{\varepsilon} \cdot \nabla \varphi - \kappa \int_{\Omega} (Y_{\varepsilon} u_{\varepsilon} \otimes u_{\varepsilon}) \cdot \nabla \varphi + \int_{\Omega} n_{\varepsilon} \nabla \phi \cdot \varphi \right|.
$$

Now, by virtue of (2.4.6), Lemmas 2.19 and 2.25, we also get that there exist positive constants C_5, C_6 and C_7 such that

$$
\int_0^T \| \partial_t u_{\varepsilon}(\cdot, t) \|_{(W_{0,\sigma}^{1.5}(\Omega))^*}^{\frac{5}{4}} dt
$$
$$
\le C_5 \left(\int_0^T \int_{\Omega} |\nabla u_{\varepsilon}|^{\frac{5}{4}} + \int_0^T \int_{\Omega} |Y_{\varepsilon} u_{\varepsilon} \otimes u_{\varepsilon}|^{\frac{5}{4}} + \int_0^T \int_{\Omega} n_{\varepsilon}^{\frac{5}{4}} \right)
$$
$$
F \le C_6 \left(\int_0^T \int_{\Omega} |\nabla u_{\varepsilon}|^2 + \int_0^T \int_{\Omega} |Y_{\varepsilon} u_{\varepsilon}|^2 + \int_0^T \int_{\Omega} |u_{\varepsilon}|^{\frac{10}{3}} + \int_0^T \int_{\Omega} n_{\varepsilon}^{r_3} + T \right)
$$
$$
\le C_7 (T + 1) \quad \text{for all } T > 0,
$$

which implies (2.4.89). Here r_3 is the same as (2.4.91).

2.4.4 Global Existence of Weak Solutions

Based on the above lemmas and by extracting suitable subsequences in a standard way, we can prove Theorem 2.2.

Lemma 2.27 *Let (2.1.4), (2.1.5) and (2.1.13) hold, and suppose that $\alpha > \frac{1}{3}$. There exists $(\varepsilon_j)_{j \in \mathbb{N}} \subset (0, 1)$ such that $\varepsilon_j \searrow 0$ as $j \to \infty$ and such that as $\varepsilon = \varepsilon_j \searrow 0$ we have*

$$
n_{\varepsilon} \to n \text{ a.e. in } \Omega \times (0, \infty) \text{ and in } L_{loc}^r(\bar{\Omega} \times [0, \infty)) \text{ with } r = \begin{cases} \dfrac{3\alpha + 1}{2} & \text{if } \dfrac{1}{3} < \alpha \le \dfrac{1}{2}, \\[3mm] \dfrac{10\alpha}{3 + 2\alpha} & \text{if } \dfrac{1}{2} < \alpha < 1, \\[3mm] 2 & \text{if } \alpha \ge 1, \end{cases} \tag{2.4.92}
$$

$$\nabla n_\varepsilon \rightharpoonup \nabla n \text{ in } L^r_{loc}(\bar{\Omega} \times [0, \infty)) \text{ with } r = \begin{cases} \dfrac{3\alpha + 1}{2} & \text{if } \dfrac{1}{3} < \alpha \leq \dfrac{1}{2}, \\[2mm] \dfrac{10\alpha}{3 + 2\alpha} & \text{if } \dfrac{1}{2} < \alpha < 1, \\[2mm] 2 & \text{if } \alpha \geq 1, \end{cases} \tag{2.4.93}$$

$$c_\varepsilon \to c \text{ in } L^2_{loc}(\bar{\Omega} \times [0, \infty)) \text{ and a.e. in } \Omega \times (0, \infty), \tag{2.4.94}$$

$$\nabla c_\varepsilon \to \nabla c \text{ a.e. in } \Omega \times (0, \infty), \tag{2.4.95}$$

$$u_\varepsilon \to u \text{ in } L^2_{loc}(\bar{\Omega} \times [0, \infty)) \text{ and a.e. in } \Omega \times (0, \infty) \tag{2.4.96}$$

as well as

$$\nabla c_\varepsilon \rightharpoonup \nabla c \text{ in } L^2_{loc}(\bar{\Omega} \times [0, \infty)) \tag{2.4.97}$$

and

$$\nabla u_\varepsilon \rightharpoonup \nabla u \text{ in } L^2_{loc}(\bar{\Omega} \times [0, \infty)) \tag{2.4.98}$$

and

$$u_\varepsilon \rightharpoonup u \text{ in } L^{\frac{10}{3}}_{loc}(\bar{\Omega} \times [0, \infty)) \tag{2.4.99}$$

with some triple (n, c, u) that is a global weak solution of (2.1.3) in the sense of Definition 2.1.

Proof First, from Lemma 2.19 and (2.4.87), we derive that there exists a positive constant C_0 such that

$$\|n_\varepsilon\|_{L^r_{loc}([0,\infty);W^{1,r}(\Omega))} \leq C_0(T + 1) \text{ and } \|\partial_t n_\varepsilon\|_{L^1_{loc}([0,\infty);(W^{2,4}(\Omega))^*)} \leq C_0(T + 1), \tag{2.4.100}$$

where r is given by (2.4.92). Hence, from (2.4.100) and the Aubin–Lions lemma (see, e.g., Simon 1986), we conclude that

$$(n_\varepsilon)_{\varepsilon\in(0,1)} \text{ is strongly precompact in } L^r_{loc}(\bar{\Omega} \times [0, \infty)), \tag{2.4.101}$$

so that, there exists a sequence $(\varepsilon_j)_{j\in\mathbb{N}} \subset (0, 1)$ such that $\varepsilon = \varepsilon_j \searrow 0$ as $j \to \infty$ and

$$n_\varepsilon \to n \text{ a.e. in } \Omega \times (0, \infty) \text{ and in } L^r_{loc}(\bar{\Omega} \times [0, \infty)) \text{ as } \varepsilon = \varepsilon_j \searrow 0, \tag{2.4.102}$$

where r is the same as (2.4.92). Now, in view of Lemmas 2.18, 2.19, 2.25 and 2.26, employing the same arguments as in the proof of (2.4.100)–(2.4.102), we can derive (2.4.92)–(2.4.94) and (2.4.96)–(2.4.99) hold. Next, let $g_\varepsilon(x, t) := -c_\varepsilon + F_\varepsilon(n_\varepsilon) - u_\varepsilon \cdot \nabla c_\varepsilon$. With this notation, the second equation of (2.2.8) can be rewritten in component form as

$$c_{\varepsilon t} - \Delta c_\varepsilon = g_\varepsilon. \tag{2.4.103}$$

Case $\frac{1}{3} < \alpha \le \frac{1}{2}$: Observe that

$$\frac{5}{4} < \frac{4}{3} < \min \left\{ \frac{6\alpha + 2}{3}, \frac{10}{3} \right\} \text{ for } \frac{1}{3} < \alpha \le \frac{1}{2}.$$

Thus, recalling (2.4.29), (2.4.32) and (2.4.86) and applying Hölder's inequality, we conclude that, for any $\varepsilon \in (0, 1)$, g_ε is bounded in $L^{\frac{5}{4}}(\Omega \times (0, T))$, and we may invoke the standard parabolic regularity theory to (2.4.103) and infer that $(c_\varepsilon)_{\varepsilon \in (0,1)}$ is bounded in $L^{\frac{5}{4}}((0, T); W^{2, \frac{5}{4}}(\Omega))$. Hence, by virtue of (2.4.88) and the Aubin–Lions lemma, we derive the relative compactness of $(c_\varepsilon)_{\varepsilon \in (0,1)}$ in $L^{\frac{5}{4}}((0, T); W^{1, \frac{5}{4}}(\Omega))$. We can pick an appropriate subsequence that is still written as $(\varepsilon_j)_{j \in \mathbb{N}}$ such that $\nabla c_{\varepsilon_j} \to z_1$ in $L^{\frac{5}{4}}(\Omega \times (0, T))$ for all $T \in (0, \infty)$ and some $z_1 \in L^{\frac{5}{4}}(\Omega \times (0, T))$ as $j \to \infty$. Therefore, by (2.4.88), we can also derive that $\nabla c_{\varepsilon_j} \to z_1$ a.e. in $\Omega \times (0, \infty)$ as $j \to \infty$. In view of (2.4.97) and Egorov's theorem, we conclude that $z_1 = \nabla c$ and hence (2.4.95) holds. Next, we pay attention to the case $\frac{1}{2} < \alpha < 1$: By straightforward calculations, and using relation $\frac{1}{2} < \alpha < 1$, one has

$$\frac{5}{4} < \frac{5}{3} < \min \left\{ \frac{10\alpha}{3}, \frac{10}{3} \right\}.$$

Consequently, based on (2.4.30), (2.4.32) and (2.4.86), it follows from Hölder's inequality that

$$c_{\varepsilon t} - \Delta c_\varepsilon = g_\varepsilon \text{ is bounded in } L^{\frac{5}{4}}(\Omega \times (0, T)) \text{ for any } \varepsilon \in (0, 1). \tag{2.4.104}$$

Employing almost exactly the same arguments as in the proof of the case $\frac{1}{3} < \alpha \le \frac{1}{2}$, and taking advantage of (2.4.104), we conclude the estimate (2.4.97). The proof of case $\alpha \ge 1$ is similar to that of case $\frac{1}{3} < \alpha \le \frac{1}{2}$, so we omit it.

In the following proof, we shall prove that (n, c, u) is a weak solution of problem (2.1.3) in Definition 2.1. In fact, by $\alpha > \frac{1}{3}$, we conclude that

$$r > 1,$$

where r is given by (2.4.92). Therefore, with the help of (2.4.92)–(2.4.94) and (2.4.96)–(2.4.98), we can derive (2.2.3). Now, by the nonnegativity of n_ε and c_ε, we obtain $n \ge 0$ and $c \ge 0$. Next, from (2.4.98) and $\nabla \cdot u_\varepsilon = 0$, we conclude that $\nabla \cdot u = 0$ a.e. in $\Omega \times (0, \infty)$. However, in view of (2.4.83), (2.4.84) and (2.4.85), we conclude that

$$n_\varepsilon F'_\varepsilon(n_\varepsilon) S_\varepsilon(x, n_\varepsilon, c_\varepsilon) \nabla c_\varepsilon \rightharpoonup z_2 \text{ in } L^r(\Omega \times (0, T)) \tag{2.4.105}$$

as $\varepsilon = \varepsilon_j \searrow 0$ for each $T \in (0, \infty)$, where r is given by (2.4.92). However, it follows from (2.1.4), (2.2.10), (2.4.2), (2.4.92), (2.4.94) and (2.4.95) that

$$n_\varepsilon F'_\varepsilon(n_\varepsilon) S_\varepsilon(x, n_\varepsilon, c_\varepsilon) \nabla c_\varepsilon \to n S(x, n, c) \nabla c \text{ a.e. in } \Omega \times (0, \infty) \text{ as } \varepsilon = \varepsilon_j \searrow 0.$$
(2.4.106)

Again by Egorov's theorem, we gain $z_2 = n S(x, n, c) \nabla c$, and therefore (2.4.105) can be rewritten as

$$n_\varepsilon F'_\varepsilon(n_\varepsilon) S_\varepsilon(x, n_\varepsilon, c_\varepsilon) \nabla c_\varepsilon \rightharpoonup n S(x, n, c) \nabla c \text{ in } L^r(\Omega \times (0, T)) \qquad (2.4.107)$$

as $\varepsilon = \varepsilon_j \searrow 0$ for each $T \in (0, \infty)$, which together with $r > 1$ implies the integrability of $n S(x, n, c) \nabla c$ in (2.2.4) as well. It is not difficult to check that

$$\frac{2\alpha + \frac{2}{3}}{\alpha + 1} > 1 \text{ if } \frac{1}{3} < \alpha \le \frac{8}{21},$$

$$\frac{10(3\alpha + 1)}{9(\alpha + 2)} > 1 \text{ if } \frac{8}{21} < \alpha \le \frac{1}{2},$$

$$\frac{10\alpha}{3(\alpha + 1)} > 1 \text{ if } \frac{1}{2} < \alpha < 1.$$

Thereupon, recalling (2.4.83), (2.4.84) and (2.4.85), we infer that, for each $T \in (0, \infty)$, when $\varepsilon = \varepsilon_j \searrow 0$,

$$n_\varepsilon u_\varepsilon \rightharpoonup z_3 \text{ in } L^{\tilde{r}}(\Omega \times (0, T)) \text{ with } \tilde{r} = \begin{cases} \dfrac{2\alpha + \frac{2}{3}}{\alpha + 1} & \text{if } \dfrac{1}{3} < \alpha \le \dfrac{8}{21}, \\[2mm] \dfrac{10(3\alpha + 1)}{9(\alpha + 2)} & \text{if } \dfrac{8}{21} < \alpha \le \dfrac{1}{2}, \\[2mm] \dfrac{10\alpha}{3(\alpha + 1)} & \text{if } \dfrac{1}{2} < \alpha < 1, \\[2mm] \dfrac{5}{3} & \text{if } \alpha \ge 1. \end{cases} \qquad (2.4.108)$$

(2.4.108) together with (2.4.92) and (2.4.96) implies

$$n_\varepsilon u_\varepsilon \to n u \text{ a.e. in } \Omega \times (0, \infty) \text{ as } \varepsilon = \varepsilon_j \searrow 0. \qquad (2.4.109)$$

(2.4.108) along with (2.4.109) and Egorov's theorem guarantees that $z_3 = n u$, whereupon we derive from (2.4.108) that

$$n_\varepsilon u_\varepsilon \rightharpoonup nu \text{ in } L^{\tilde{r}}(\Omega \times (0, T)) \text{ with } \tilde{r} = \begin{cases} \dfrac{2\alpha + \frac{2}{3}}{\alpha + 1} & \text{if } \dfrac{1}{3} < \alpha \le \dfrac{8}{21}, \\ \dfrac{10(3\alpha + 1)}{9(\alpha + 2)} & \text{if } \dfrac{8}{21} < \alpha \le \dfrac{1}{2}, \\ \dfrac{10\alpha}{3(\alpha + 1)} & \text{if } \dfrac{1}{2} < \alpha < 1, \\ \dfrac{5}{3} & \text{if } \alpha \ge 1 \end{cases} \tag{2.4.110}$$

as $\varepsilon = \varepsilon_j \searrow 0$, for each $T \in (0, \infty)$.

As a straightforward consequence of (2.4.94) and (2.4.96), it holds that

$$c_\varepsilon u_\varepsilon \to cu \text{ in } L^1_{loc}(\bar{\Omega} \times (0, \infty)) \text{ as } \varepsilon = \varepsilon_j \searrow 0. \tag{2.4.111}$$

Thus, the integrability of nu and cu in (2.2.4) is verified by (2.4.94) and (2.4.96).

Next, by (2.4.96) and the fact that $\|Y_\varepsilon \varphi\|_{L^2(\Omega)} \le \|\varphi\|_{L^2(\Omega)} (\varphi \in L^2_\sigma(\Omega))$ and $Y_\varepsilon \varphi \to \varphi$ in $L^2(\Omega)$ as $\varepsilon \searrow 0$, we can get that there exists a positive constant C_1 such that, for any $\varepsilon \in (0, 1)$,

$$\begin{aligned} \|Y_\varepsilon u_\varepsilon(\cdot, t) - u(\cdot, t)\|_{L^2(\Omega)} &\le \|Y_\varepsilon[u_\varepsilon(\cdot, t) - u(\cdot, t)]\|_{L^2(\Omega)} + \|Y_\varepsilon u(\cdot, t) - u(\cdot, t)\|_{L^2(\Omega)} \\ &\le \|u_\varepsilon(\cdot, t) - u(\cdot, t)\|_{L^2(\Omega)} + \|Y_\varepsilon u(\cdot, t) - u(\cdot, t)\|_{L^2(\Omega)} \\ &\to 0 \text{ as } \varepsilon = \varepsilon_j \searrow 0 \end{aligned}$$

and

$$\begin{aligned} \|Y_\varepsilon u_\varepsilon(\cdot, t) - u(\cdot, t)\|^2_{L^2(\Omega)} &\le \left(\|Y_\varepsilon u_\varepsilon(\cdot, t)\|_{L^2(\Omega)} + \|u(\cdot, t)\|_{L^2(\Omega)} \right)^2 \\ &\le \left(\|u_\varepsilon(\cdot, t)\|_{L^2(\Omega)} + \|u(\cdot, t)\|_{L^2(\Omega)} \right)^2 \\ &\le C_1 \end{aligned}$$

for all $t \in (0, \infty)/N$ with some null set $N \subset (0, \infty)$, and thus by the dominated convergence theorem, we can find that

$$\int_0^T \|Y_\varepsilon u_\varepsilon(\cdot, t) - u(\cdot, t)\|^2_{L^2(\Omega)} dt \to 0 \text{ as } \varepsilon = \varepsilon_j \searrow 0 \text{ for all } T > 0.$$

Therefore,

$$Y_\varepsilon u_\varepsilon \to u \text{ in } L^2_{loc}([0, \infty); L^2(\Omega)). \tag{2.4.112}$$

Now, combining (2.4.96) with (2.4.112), we derive

$$Y_\varepsilon u_\varepsilon \otimes u_\varepsilon \to u \otimes u \text{ in } L^1_{loc}(\bar{\Omega} \times [0, \infty)) \text{ as } \varepsilon = \varepsilon_j \searrow 0. \tag{2.4.113}$$

Therefore, the integrability of $nS(x, n, c)\nabla c$, nu, cu and $u \otimes u$ in (2.2.4) is verified by (2.4.107), (2.4.110), (2.4.111) and (2.4.113). Finally, for any fixed $T \in (0, \infty)$, applying (2.4.92), one can get

$$\int_0^T \|F_\varepsilon(n_\varepsilon(\cdot, t)) - n(\cdot, t)\|_{L^r(\Omega)}^r \, dt$$

$$\leq \int_0^T \|F_\varepsilon(n_\varepsilon(\cdot, t)) - F_\varepsilon(n(\cdot, t))\|_{L^r(\Omega)}^r \, dt$$

$$+ \int_0^T \|F_\varepsilon(n(\cdot, t)) - n(\cdot, t)\|_{L^r(\Omega)}^r \, dt \qquad (2.4.114)$$

$$\leq \|F_\varepsilon'\|_{L^\infty(\Omega \times (0,\infty))} \int_0^T \|n_\varepsilon(\cdot, t) - n(\cdot, t)\|_{L^r(\Omega)}^r \, dt$$

$$+ \int_0^T \|F_\varepsilon(n(\cdot, t)) - n(\cdot, t)\|_{L^r(\Omega)}^r \, dt,$$

where r is the same as in (2.4.92). Besides that, we also deduce from (2.4.3) and $r > 1$ that

$$\|F_\varepsilon(n(\cdot, t)) - n(\cdot, t)\|_{L^r(\Omega \times (0,T))}^r \leq 2^r \|n(\cdot, t)\|$$

for each $t \in (0, T)$, which together with (2.4.92) shows the integrability of

$$\|F_\varepsilon(n(\cdot, t)) - n(\cdot, t)\|_{L^r(\Omega)}^r$$

on $(0, T)$. Thereupon, by virtue of (2.4.2), we infer from the dominated convergence theorem that

$$\int_0^T \|F_\varepsilon(n) - n\|_{L^r(\Omega)}^r \, dt \to 0 \text{ as } \varepsilon = \varepsilon_j \searrow 0 \qquad (2.4.115)$$

for each $T \in (0, \infty)$. Inserting (2.4.115) into (2.4.114) and using (2.4.92) and (2.4.1), we can see clearly that

$$F_\varepsilon(n) \to n \text{ in } L_{loc}^r(\bar{\Omega} \times [0, \infty)) \text{ as } \varepsilon = \varepsilon_j \searrow 0. \qquad (2.4.116)$$

Finally, according to (2.4.92)–(2.4.94), (2.4.96)–(2.4.98), (2.4.107), (2.4.110)–(2.4.113) and (2.4.116), we may pass to the limit in the respective weak formulations associated with the regularized system (2.2.8) and obtain the integral identities (2.2.5)–(2.2.7).

Chapter 3
Chemotaxis–Haptotaxis System

3.1 Introduction

Cancer invasion and metastasis are influenced by a plethora of biochemical processes and involve many biochemical mechanisms, among which chemotaxis and haptotaxis are two of the main mechanisms directing the migration of cancer cells Chaplain and Lolas (2005). Evidence has been found that cancer cells release complex enzymes such as the urokinase-type plasminogen activator (uPA), which degrade the surrounding extracellular matrix (ECM), and thereby allow the migration of cells following the concentration gradient of such diffusive enzymes. This process is referred to as chemotaxis Chaplain and Lolas (2006). On the other hand, in addition to random diffusion, the movement of cancer cells is biased toward the gradient of an immovable stimulus (density of tissue fiber) by finding matrix molecules such as vitronectin adhered therein. This process is called haptotaxis Perumpanani and Byrne (1999).

Recently, a variety of mathematical models have been proposed for various aspects of cancer invasion and metastasis Aznavoorian et al. (1990); Chaplain and Lolas (2005, 2006); Friedman and Lolas (2005); Gatenby and Gawlinski (1996); Meral et al. (2015); Szymańska et al. (2009). Gatenby and Gawlinski (1996) used reaction–diffusion equations to describe the interaction between the density of normal cells, tumor cells and the concentration of H^+-ions produced by the latter. They suggested that cancer cells up-regulate certain mechanisms, which allow for the extrusion of excessive protons and hence acidify the environment. This triggers apoptosis of normal cells and thus allows the neoplastic tissue to extend into the space made available. Later on, Meral et al. (2015) proposed a population-based micro–macro model for acid-mediated tumor invasion, which involves the the microscopic dynamics of intracellular protons and their exchange with extracellular counterparts. The continuum micro–macro models explicitly accounting for subcellular events are rather new, especially in the context of cancer cell migration Stinner et al. (2014, 2016).

The analytical results on various models of cancer invasion are mathematically interesting Bellomo et al. (2015); Engwer et al. (2017); Jin (2018); Li and Lankeit (2016); Liţcanu and Morales-Rodrigo (2010b); Morales-Rodrigo and Tello (2014);

© The Author(s) 2022
Y. Ke et al., *Analysis of Reaction-Diffusion Models with the Taxis Mechanism*,
Financial Mathematics and Fintech, https://doi.org/10.1007/978-981-19-3763-7_3

Stinner et al. (2014, 2016); Szymańska et al. (2009); Tao and Wang (2008); Walker and Webb (2007); Zhigun et al. (2016). From a mathematical point of view, the system under consideration comprises a strong coupling of reaction–diffusion equations and an ordinary differential equation (ODE) in two or three space dimensions. Since ODE corresponds to an everywhere degenerate reaction–diffusion equation and has no regularizing effect, this amounts to considerable difficulty for the analysis. Indeed, analytical results on the cancer invasion model are yet quite fragmentary, so far mainly concentrating on the global existence and boundedness of solutions. For example, Stinner et al. (2014) proved the global existence of weak solutions to a PDE-ODE system modeling the multiscale invasion of tumor cells through the surrounding tissue matrix. Very recently, Engwer et al. (2017) studied the global existence of weak solutions to a multiscale model for tumor cell migration in a tissue network. The more detailed answers have been given only in some special cases Hillen et al. (2013); Liţcanu and Morales-Rodrigo (2010b); Tao and Wang (2009); Wang and Ke (2016).

The first part of this chapter is concerned with the Chaplain–Lolas model of cancer invasion Chaplain and Lolas (2005, 2006)

$$
\begin{cases}
u_t = \Delta u - \chi \nabla \cdot (u \nabla v) - \xi \nabla \cdot (u \nabla w) + \mu u(r - u - w), & x \in \Omega, t > 0, \\
\sigma v_t = \Delta v - v + u, & x \in \Omega, t > 0, \\
w_t = -vw + \eta w(1 - w - u), & x \in \Omega, t > 0, \\
\dfrac{\partial u}{\partial \nu} - \chi u \dfrac{\partial v}{\partial \nu} - \xi u \dfrac{\partial w}{\partial \nu} = \dfrac{\partial v}{\partial \nu} = 0, & x \in \partial \Omega, t > 0, \\
u(x, 0) = u_0(x), \sigma v(x, 0) = \sigma v_0(x), w(x, 0) = w_0(x), & x \in \Omega
\end{cases}
$$

$$(3.1.1)$$

in a bounded domain $\Omega \subset \mathbb{R}^n$ ($n = 2, 3$) with smooth boundary $\partial \Omega$, where $\partial / \partial \nu$ denotes the outward normal derivative on $\partial \Omega$, u denotes the density of cancer cells, v represents the concentration of the matrix degrading enzyme (MDE) and w describes the concentration of the extracellular matrix (ECM), respectively; and χ and ξ measure the chemotactic and haptotactic sensitivities, respectively. The term $\mu u(r - u - w)$ assumes that in the absence of the ECM, cancer cell proliferation satisfies a logistic law, and $\eta > 0$ embodies the ability of the ECM to remodel back to a normal level. The parameter σ may take on the value of 0 or 1. When $\sigma = 0$, we are making the simplifying assumption that the diffusion rate of the MDE is much greater than that of cancer cells, which is supported by evidence Chaplain and Lolas (2006). Indeed, similar quasi-steady approximations for corresponding chemoattrac-tant equations are frequently used to study classical chemotaxis systems (see Jäger and Luckhaus (1992)). As for the initial data (u_0, v_0, w_0), we suppose throughout this section that, for some $\vartheta \in (0, 1)$,

$$\begin{cases} u_0 \in C^1(\bar{\Omega}) \text{ with } u_0 \geq 0 \text{ in } \Omega, \ u_0 \not\equiv 0, \\ v_0 \in W^{1,\infty}(\Omega) \text{ with } v_0 \geq 0 \text{ in } \Omega, \\ w_0 \in C^{2+\vartheta}(\bar{\Omega}) \text{ with } w_0 \geq 0 \text{ in } \bar{\Omega} \text{ and } \dfrac{\partial w_0}{\partial \nu} = 0 \text{ on } \partial\Omega. \end{cases} \qquad (3.1.2)$$

It is observed that when letting $w \equiv 0$, (3.1.1) is reduced to the Keller–Segel system with logistic source. This chemotaxis-only system has been extensively studied by many authors during the last decades. In this context, the particular attention focuses on the question of whether the solutions of the models are bounded or blow-up (see, e.g., Cieślak and Stinner (2012), Cieślak and Winkler (2008), Ishida et al. (2014), Painter and Hillen (2002) and Winkler (2008, 2010a, 2011b)). In particular, solutions may blow up in finite time when $n \geq 2$, $\mu = 0$ Herrero and Velázquez (1997); Nagai (2001). It is known that arbitrarily small $\mu > 0$ guarantees the boundedness of solutions when $n = 2$ Osaki et al. (2002), while when $n \geq 3$, appropriately large μ (compared with the chemotactic coefficient χ) is required to exclude unbounded solution Winkler (2010a). It is still unknown whether finite-time blow-up may occur if $\mu > 0$ is small, though global weak solutions are known to exist and will become smooth after some time Lankeit (2015). On the other hand, the nonlinear self-diffusion of cells may prevent blow-up of solutions Cieślak and Winkler (2008); Ishida et al. (2014); Wang et al. (2014).

When $\chi = 0$, (3.1.1) becomes the haptotaxis-only system. For $\chi = \mu = \eta = 0$, $\sigma = 1$, the local existence and uniqueness of classical solutions have been shown in Morales-Rodrigo (2008). The global existence and asymptotic behavior of weak solutions have been proven in Liţcanu and Morales-Rodrigo (2010b); Marciniak-Czochra and Ptashnyk (2010); Walker and Webb (2007) when $\eta = 0$, and global existence and uniqueness of classical solutions have been shown in Tao (2011) when $\eta > 0$, respectively.

Note that in contrast of the chemotaxis-only system, haptotaxis-only system and the chemotaxis–haptotaxis system, the chemotaxis–haptotaxis system **with remodeling of non-diffusible attractant** ($\eta > 0$ in (3.1.1)) is much less understood (Chaplain and Lolas (2006), Pang and Wang (2017) and Tao and Winkler (2014b)). The main technical difficulty in their proof lies in the effects of the strong coupling in (3.1.1) on the spatial regularity of u, v and w when $\eta > 0$. When $\eta = 0$, one can build a one-sided pointwise estimate which connects Δw to v (see Lemma 2.2 of Cao (2016) or (3.10) of Wang (2016)). Relying on such a pointwise estimate, one can derive two useful energy-type inequalities which can help us to bypass the term $\int_\Omega u^{p-1} \nabla \cdot (u\nabla w)$ (see Lemma 3.2 of Zheng (2017b)). Using such information along with coupled estimate techniques and the boundedness of the $\|\nabla v(\cdot, t)\|_{L^2(\Omega)}$, one can establish the estimates on $\int_\Omega u^p + |\nabla v|^{2q}$ for any p and $q > 1$ (see Lemmas 3.3 and 3.4 of Zheng (2017b)), which combined with the standard regularity theory of parabolic equation and the Moser iteration procedure implies the boundedness of u in $L^\infty(\Omega)$ (see Lemma 3.5 of Zheng (2017b)). However, for the model (3.1.1) with $\eta > 0$, one needs to estimate the chemotaxis-related integral term $\int_\Omega a^p |\nabla v|^2$ (see

(3.28) in Tao and Winkler (2014b)) or $\int_\Omega e^{-(p+1)(t-s)} a^p |\nabla v|^2$ (see (3.8) of Pang and Wang (2017)) with $a := u e^{-\xi w}$, which requires much more technical demanding. In Pang and Wang (2017), assuming that $\mu > \xi \eta \max\{\|u_0\|_{L^\infty(\Omega)}, 1\} + \mu^*(\chi^2, \xi)$ (the hypothesis cannot be dropped (see the proof of Lemma 3.2 of Pang and Wang (2017))), Pang and Wang (2017) proved that the problem 3.1.1 admits a unique global solution $(u, v, w) \in (C^{2,1}(\bar\Omega \times (0, \infty)))^3$. Moreover, u is bounded in $\Omega \times (0, \infty)$.

This chapter consists of three parts. The first part shows the global boundedness of classical solutions to the chemotaxis–haptotaxis model with any $\eta > 0$ (Ke and Zheng (2018)).

Theorem 3.1 *Let* $\sigma > 0, \chi > 0, \xi > 0, r = 1$ *and* $\eta > 0$. *Assume that* $\Omega \subseteq \mathbb{R}^2$ *is a bounded domain with smooth boundary and the initial data* (u_0, v_0, w_0) *satisfy*

$$
\begin{cases}
u_0 \in C^{2+\vartheta}(\bar\Omega) \text{ with } u_0 \geq 0 \text{ in } \Omega \text{ and } \dfrac{\partial u_0}{\partial v} = 0 \text{ on } \partial\Omega, \\[2mm]
v_0 \in C^{2+\vartheta}(\bar\Omega) \text{ with } v_0 \geq 0 \text{ in } \Omega \text{ and } \dfrac{\partial v_0}{\partial v} = 0 \text{ on } \partial\Omega, \\[2mm]
w_0 \in C^{2+\vartheta}(\bar\Omega) \text{ with } w_0 \geq 0 \text{ in } \bar\Omega \text{ and } \dfrac{\partial w_0}{\partial v} = 0 \text{ on } \partial\Omega
\end{cases}
$$

with some $\vartheta \in (0, 1)$. *If* $\mu > 0$, *then there exists a triple* $(u, v, w) \in (C^0(\bar\Omega \times [0, \infty)) \times C^{2,1}(\bar\Omega \times (0, \infty)))^3$ *which solves (3.1.1) in the classical sense. Moreover,* u *and* v *are bounded in* $\Omega \times (0, \infty)$.

Remark 3.1 (i) If $w \equiv 0$, it is not difficult to obtain that the solutions under the conditions of Theorem 3.1 are uniformly bounded when $n = 2$, which coincides with the results of Osaki et al. (2002).

(ii) From Theorem 3.1, it follows that solutions of model (3.1.1) are global and bounded for any $\eta = 0, \mu > 0$ and $n \leq 2$, which coincides with the result of Tao (2014).

The second part of this chapter is devoted to the integrative interactions of chemotaxis, haptotaxis, logistic growth and remodeling mechanisms, and establishes the global existence of classical solutions to the chemotaxis–haptotaxis model (3.1.1) with the remodeling of the ECM. It is noticed that the authors of Tao and Winkler (2014b) made appropriate use of the dampening effect of $-\eta u w$ in the third equation of (3.1.1) to derive an energy-like inequality, which yields an a priori estimate of $\int_\Omega u \ln u$ in bounded time intervals. The latter is the starting point for a bootstrap argument used to derive higher regularity estimates. In this part, thanks to the variable transformation $a = u e^{-\xi w}$ (Tao and Wang (2009, 2008); Tao and Winkler (2014b)), making use of the damping effect in the first equation of cancer cells, one derives a priori estimate of $\int_\Omega u \ln u$ for all time $t > 0$ and thus proves the global boundedness of solutions thereof rather comprehensively. The result in this respect is the following (Pang and Wang (2018)).

Theorem 3.2 *Let $\Omega \subset \mathbb{R}^2$ be a bounded smooth domain, and suppose that $\chi > 0, \xi > 0, \eta > 0$ and $\mu > 0$. Then for any $r > 0$, the problem (3.1.1) admits a unique global classical solution (u, v, w), where $\|u(\cdot, t)\|_{L^\infty(\Omega)}$ is uniformly bounded for $t \in (0, \infty)$.*

The key step of our analysis of (3.1.1) consists of identifying a certain dissipative property of the functional $\int_\Omega e^{\xi w} a^2$ with $a = e^{-\xi w} u$. Indeed, we shall see in Lemma 3.17 below that a certain variant thereof satisfies an inequality of the form

$$\frac{d}{dt} \int_\Omega e^{\xi w} a^2 + \frac{1}{\varepsilon} \int_\Omega e^{\xi w} a^2 \leq c(\|\Delta v\|_{L^2(\Omega)}^2 + \|a\|_{L^2(\Omega)}^2) \int_\Omega e^{\xi w} a^2 + c(\varepsilon)$$

with some $c > 0, c(\varepsilon) > 0$ for any $\varepsilon > 0$ (see (3.4.17)), whereupon Lemma 3.5 will provide the bound of $\int_\Omega u^2$, which provides a starting point for the higher regularity estimates of solutions. On the other hand, in the case of $\sigma = 0$, the key step in our proof of theorem is to identify

$$\frac{d}{dt} \int_\Omega e^{\xi w} a \ln a + \frac{\mu}{2} \int_\Omega e^{\xi w} a \ln a$$
$$\leq \varepsilon c(\|\Delta v\|_{L^2(\Omega)}^2 + \|\nabla v\|_{L^2(\Omega)}^2) \int_\Omega e^{\xi w} a \ln a \qquad (3.1.3)$$
$$+ c(\|\Delta v\|_{L^2(\Omega)}^2 + \|\nabla v\|_{L^2(\Omega)}^2) + c(\varepsilon)$$

with some $c > 0, c(\varepsilon) > 0$ for all $\varepsilon > 0$, which along with Lemma 3.5 enables us to obtain an a priori estimate for u in the space $L \log L(\Omega)$ for all time. Notice that in the two-dimensional space, the global boundedness of solutions (p, c, w) to a tumor angiogenesis model was established in Morales-Rodrigo and Tello (2014) if the initial data w_0 of the fibronectin concentration satisfies either $w_0 > 1$ or $\|w_0 - 1\|_{L^\infty(\Omega)} < \delta$ with some $\delta > 0$. It should be mentioned that by the estimate technique above, one can remove the extra assumption on w_0.

In the three-dimensional setting, the problem of the global existence of solutions to (3.1.1) seems to be more delicate. Indeed, the only result that we are aware of is presented in the recent paper Bellomo et al. (2015), where a certain global weak solution was constructed for (3.1.1) with $\sigma = 1$. To the best of our knowledge, the existence of global classical solutions to (3.1.1) is still open. As mentioned previously, some weak solutions to the three-dimensional chemotaxis system including logistic growth eventually become classical solutions after some waiting time when smallness conditions on the growth rate of the cells are imposed Lankeit (2015); Winkler (2008). A natural question is whether the chemotaxis–haptotaxis system (3.1.1) possesses global classical solutions under some smallness conditions. Our result in this direction is as follows (Pang and Wang (2018)).

Theorem 3.3 *Let $\Omega \subset \mathbb{R}^3$ be a bounded convex domain with smooth boundary and $\chi > 0, \xi > 0, \eta > 0$ and $\mu > 0$. For any given w_0, there exist a constant $r_0 = r_0(\mu, |\Omega|, \|w_0\|_{L^\infty(\Omega)}) > 0$ and appropriate small $\|u_0\|_{L^2(\Omega)}$ and $\|v_0\|_{W^{1,4}(\Omega)}$*

such that the problem (3.1.1) possesses a unique global classical solution (u, v, w)
provided that $r < r_0$.

Relying on the mass evolution of solutions to (3.1.1), the quantity $\int_\Omega a^2(t) +$ $\int_\Omega |\nabla v(t)|^4$ is shown to satisfy an autonomous ordinary differential inequality, and thereby is bounded whenever $r > 0$ and initial data are suitably small by the comparison argument of the corresponding ordinary differential equation. This serves as a starting point for the bootstrap procedure to yield a bound for a in $L^\infty(\Omega)$.

As a physiological process, angiogenesis involves the formation of new capillary networks sprouting from a pre-existing vascular network and plays an important role in embryo development, wound healing and tumor growth. For example, it has been recognized that capillary growth through angiogenesis leads to the vascularization of a tumor, providing it with its own dedicated blood supply and consequently allowing for rapid growth and metastasis.

The process of tumor angiogenesis can be divided into three main stages (which may be overlapping): (i) changes within existing blood vessels; (ii) formation of new vessels; and (iii) maturation of new vessels. Over the past decade, a lot of work has been done on the mathematical modeling of tumor growth; see, for example, Anderson and Chaplain (1998b); Bellomo et al. (2015); Chaplain and Lolas (2005, 2006); Li et al. (2015); Stinner et al. (2015, 2016) and the references cited therein. In particular, the role of angiogenesis in tumor growth has also attracted a great deal of attention; see, for example, Anderson and Chaplain (1998a); Chaplain and Stuart (1993); Levine et al. (2001); Paweletz and Knierim (1989); Sleeman (1997) and the references cited therein. For example, in Levine et al. (2001), a system of PDEs using reinforced random walks was deployed to model the first stage of angiogenesis, in which chemotactic substances from the tumor combine with the receptors on the endothelial cell wall to release proteolytic enzymes that can degrade the basal membrane of the blood vessels eventually.

The third part of this chapter considers a variation of the model proposed in Anderson and Chaplain (1998b), namely

$$\begin{cases} p_t = \Delta p - \nabla \cdot p(\dfrac{\alpha}{1+c}\nabla c + \rho \nabla w) + \lambda p(1 - p), & x \in \Omega, t > 0, \\ c_t = \Delta c - c - \mu p c, & x \in \Omega, t > 0, \\ w_t = \gamma p(1 - w), & x \in \Omega, t > 0, \\ \dfrac{\partial p}{\partial v} - p(\dfrac{\alpha}{1+c}\dfrac{\partial c}{\partial v} + \rho \dfrac{\partial w}{\partial v}) = \dfrac{\partial c}{\partial v} = 0, & x \in \partial\Omega, t > 0, \\ p(x, 0) = p_0(x), \ c(x, 0) = c_0(x), \ w(x, 0) = w_0(x), & x \in \Omega, \end{cases}$$

$$(3.1.4)$$

in a bounded smooth domain $\Omega \subset \mathbb{R}^N (N = 1, 2)$, where, in addition to random motion, the existing blood vessels' endothelial cells p migrate in response to the concentration gradient of a chemical signal c (called Tumor Angiogenic Factor, or TAF) secreted by tumor cells as well as the concentration gradient of non-diffusible glycoprotein fibronectin w produced by the endothelial cells Morales-Rodrigo and Tello (2014). The formerly directed migration is a chemotactic process, whereas the latter is a haptotactic process. In this model, it is assumed that the endothelial cells proliferate according to a logistic law, that the spatio-temporal evolution of TAF occurs through diffusion, natural decay and degradation upon binding to the endothelial cells, and that the fibronectin is produced by the endothelial cells and degrades upon binding to the endothelial cells.

For the remainder of this chapter, the initial data are assumed to satisfy

$$\begin{cases} (p_0, c_0, w_0) \in (C^{2+\beta}(\overline{\Omega}))^3 \text{ is nonnegative for some } \beta \in (0, 1) \text{ with } p_0 \not\equiv 0, \\ \dfrac{\partial p_0}{\partial v} - p_0\left(\dfrac{\alpha}{1+c_0}\dfrac{\partial c_0}{\partial v} + \rho\dfrac{\partial w_0}{\partial v}\right) = \dfrac{\partial c_0}{\partial v} = 0. \end{cases}$$

(3.1.5)

The third part focuses on the global existence and asymptotic behavior of classical solutions to (3.1.4). Let us look at two subsystems contained in (3.1.4). The first is a Keller–Segel-type chemotaxis system with signal absorption:

$$\begin{cases} p_t = \Delta p - \nabla \cdot (p\nabla c) + \lambda p(1-p), & x \in \Omega, t > 0, \\ c_t = \Delta c - pc, & x \in \Omega, t > 0. \end{cases}$$

(3.1.6)

It is known that, unlike the standard Keller–Segel model, (3.1.6) with $\lambda = 0$ possesses global, bounded classical solutions in two-dimensional bounded convex domains for arbitrarily large initial data; while in three spatial dimensions, it admits at least global weak solutions which eventually become smooth and bounded after some waiting time Tao and Winkler (2012c). In the high-dimensional setting, it has been proved that global bounded classical solutions exist for suitably large $\lambda > 0$, while only certain weak solutions are known to exist for arbitrary $\lambda > 0$ Lankeit and Wang (2017).

Another delicate subsystem of (3.1.4) is the haptotaxis-only system obtained by letting $\alpha = 0$ in (3.1.4):

$$\begin{cases} p_t = \Delta p - \rho\nabla\cdot(p\nabla w) + \lambda p(1-p), & x \in \Omega, t > 0, \\ w_t = \gamma p(1-w), & x \in \Omega, t > 0. \end{cases}$$

Here, since the quantity w satisfies an ODE without any diffusion, the smoothing effect on the spatial regularity of w during evolution cannot be expected. To the best of our knowledge, unlike the study of chemotaxis systems, the mathematical literature on haptotaxis systems is comparatively thin. Indeed, the literature provides only some results on global solvability in various special models, and the detailed description of qualitative properties such as long-time behaviors of solutions is available only in very particular cases (see, for example, Corrias et al. (2004); Liţcanu and Morales-Rodrigo (2010b, a); Marciniak-Czochra and Ptashnyk (2010); Tao (2011); Tao and Winkler (2019a); Walker and Webb (2007); Winkler (2018b)).

More recently, some results on global existence and asymptotic behavior for certain chemotaxis–haptotaxis models of cancer invasion have been obtained (see, for example, Li and Lankeit (2016); Pang and Wang (2017, 2018); Stinner et al. (2014); Tao and Winkler (2014b, 2015a); Wang and Ke (2016)). Particularly, Hillen et al. (2013) have shown the convergence of a cancer invasion model in one-dimensional domains, and the result has been subsequently extended to higher dimensions Li and Lankeit (2016); Tao and Winkler (2015a); Wang and Ke (2016).

In Morales-Rodrigo and Tello (2014), in two spatial dimensions, the authors showed the global existence and long-time behavior of classical solutions to (3.1.4) when the initial data (p_0, c_0, w_0) satisfy either $w_0 > 1$ or $\|w_0 - 1\|_{L^\infty(\Omega)} < \delta$ for some $\delta > 0$ (see Lemma 5.8 of Morales-Rodrigo and Tello (2014)). Generalizing this result, our first main result establishes that, for any choice of reasonably regular initial data (p_0, c_0, w_0), the L^∞-norm of p is globally bounded. This is done via an iterative method (Pang and Wang (2019)).

Theorem 3.4 *Let $\alpha, \rho, \lambda, \mu$ and γ be positive parameters. Then for any initial data (p_0, c_0, w_0) satisfying (3.1.5), the problem (3.1.4) possesses a unique classical solution (p, c, w) comprising nonnegative functions in $C(\bar{\Omega} \times [0, \infty)) \cap C^{2,1}(\bar{\Omega} \times (0, \infty))$ such that $\|p(\cdot, t)\|_{L^\infty(\Omega)} \leq C$ for all $t > 0$.*

Next, we investigate the asymptotic behavior of solutions to (3.1.4). Under an additional mild condition on the initial data w_0, we will show that the solution (p, c, w) converges to the spatially homogeneous equilibrium $(1, 0, 1)$ as time tends to infinity (Pang and Wang (2019)).

Theorem 3.5 *Let $\alpha, \rho, \lambda, \mu$ and γ be positive parameters, and suppose that (3.1.5) is satisfied and $w_0 > 1 - \dfrac{1}{\rho}$. Then the solution $(p, c, w) \in C(\bar{\Omega} \times [0, \infty)) \cap C^{2,1}$ $(\bar{\Omega} \times (0, \infty))$ of (3.1.4) satisfies*

$$\lim_{t \to \infty} \|p(\cdot, t) - 1\|_{L^r(\Omega)} + \|c(\cdot, t)\|_{W^{1,2}(\Omega)} + \|w(\cdot, t) - 1\|_{L^r(\Omega)} = 0 \qquad (3.1.7)$$

for any $r \geq 2$. In particular, if $N = 1$, then for any $\epsilon \in (0, \min\{\lambda_1, 1, \gamma, \lambda\})$ there exists $C(\epsilon) > 0$ such that

$$\|p(\cdot, t) - 1\|_{L^\infty(\Omega)} \leq C(\epsilon) e^{-(\min\{\lambda_1, 1, \gamma, \lambda\} - \epsilon)t}, \qquad (3.1.8)$$

$$\|c(\cdot, t)\|_{W^{1,2}(\Omega)} \leq C(\epsilon) e^{-(1-\epsilon)t}, \qquad (3.1.9)$$

$$\|w(\cdot, t) - 1\|_{W^{1,2}(\Omega)} \leq C(\epsilon) e^{-(\gamma-\epsilon)t}, \qquad (3.1.10)$$

where $\lambda_1 > 0$ is the first nonzero eigenvalue of $-\Delta$ in Ω with the homogeneous Neumann boundary condition.

The main mathematical challenge of the full chemotaxis–haptotaxis system is the strong coupling between the migratory cells p and the haptotactic agent w. This strong coupling has an important effect on the spatial regularity of p and w. In fact, the lack of regularization effect in the spatial variable in the w-equation and the presence of p therein demand tedious estimates on the solution. The key ideas behind this result are as follows.

As pointed out in Tao and Winkler (2015a), the variable transformation $z := pe^{-\rho w}$ plays an important role in the examination of global solvability for the full chemotaxis–haptotaxis model in the two- and higher dimensional setting. However, due to the presence of the additional chemotaxis term in our model, this approach is not directly applicable to our problem. Instead, in the derivation of Theorem 3.4, we introduce the variable transformation $q := p(c + 1)^{-\alpha} e^{-\rho w}$ as in Morales-Rodrigo and Tello (2014), and thereby ensure that $q(\cdot, t)$ is bounded in $L^n(\Omega)$ for any finite n (see Lemma 3.27). It is essential to our approach to derive a bound for $\int_\Omega q^{2^{m+1}} + \int_t^{t+\tau} \int_\Omega |\nabla q^{2^m}|^2$ from the bound of $\int_t^{t+\tau} \int_\Omega q^{2^m}$ ($m = 1, 2, \ldots$) by making appropriate use of (3.5.3)–(3.5.4) in Lemma 3.26 (see (3.5.7) below).

3.2 Preliminaries

Before formulating our main results, we recall some preliminary lemmas used throughout this chapter. Some basic properties of solution can be found in Horstmann and Winkler (2005) (see also Winkler (2010), Zhang and Li (2015b)).

Lemma 3.1 (Horstmann and Winkler (2005)) *For $p \in (1, \infty)$, let $A := A_p$ denote the sectorial operator defined by*

$$A_p u := -\Delta u \text{ for all } u \in D(A_p) := \{\varphi \in W^{2,p}(\Omega) | \frac{\partial \varphi}{\partial \nu}|_{\partial \Omega} = 0\}.$$

The operator $A + 1$ possesses fractional powers $(A + 1)^\alpha (\alpha \geq 0)$, the domains of which have the embedding properties

$$D((A + 1)^\alpha) \hookrightarrow W^{1,p}(\Omega) \text{ if } \alpha > \frac{1}{2}.$$

If $m \in \{0, 1\}$, $p \in [1, \infty]$ and $q \in (1, \infty)$ with $m - \frac{n}{p} < 2\alpha - \frac{n}{q}$, then we have

$$\|u\|_{W^{m,p}(\Omega)} \leq C\|(A + 1)^\alpha u\|_{L^q(\Omega)} \text{ for all } u \in D((A + 1)^\alpha),$$

where C is a positive constant. The fact that the spectrum of A is a p-independent countable set of positive real numbers $0 = \kappa_0 < \kappa_1 < \kappa_2 < \cdots$ entails the following consequences: for all $1 \leq p < q < \infty$ and $u \in L^p(\Omega)$, it has

$$\|(A + 1)^\alpha e^{-tA} u\|_{L^q(\Omega)} \leq ct^{-\alpha - \frac{n}{2}(\frac{1}{p} - \frac{1}{q})} e^{(1-\kappa)t} \|u\|_{L^p(\Omega)}$$

for any $t > 0$ and $\alpha \geq 0$ with some $\kappa > 0$.

In deriving some preliminary estimates for v, we shall make use of the following property referred to as a variation of Maximal Sobolev Regularity.

Lemma 3.2 (Hieber and Prüss (1997, Theorem 3.1) Li and Wang (2018, Lemma 2.2)) *Let $r \in (1, \infty)$ and $\kappa > 0$; consider the following evolution equation:*

$$\begin{cases} h_t = \Delta h - \kappa h + f, & (x, t) \in \Omega \times (0, T), \\ \nabla h \cdot v = 0, & (x, t) \in \partial\Omega \times (0, T), \\ h(x, 0) = h_0(x), & x \in \Omega. \end{cases} \quad (3.2.1)$$

Then for each $h_0 \in W^{2,r}(\Omega)$ with $\nabla h_0 \cdot v = 0$ on $\partial\Omega$ and any $f \in L^r((0, T),$ $L^r(\Omega))$, (3.2.1) admits a unique mild solution $h \in W^{1,r}((0, T); L^r(\Omega)) \cap L^r((0, T); W^{2,r}(\Omega))$. Moreover, for any $\varepsilon \in (0, \kappa]$, there exists $C_r > 0$ such that

$$\int_0^T e^{\varepsilon rs} \|h(\cdot, s)\|_{W^{2,r}(\Omega)}^r ds \leq C_r \left(\int_0^T e^{\varepsilon rs} \|f(\cdot, s)\|_{L^r(\Omega)}^r ds + (\|h_0(\cdot)\|_{W^{2,r}(\Omega)}^r) \right). \quad (3.2.2)$$

Let us also recall the well-known Gagliardo–Nirenberg inequality Friedman (1969); Tao and Wang (2009).

Lemma 3.3 *Let $\Omega \subset \mathbb{R}^n$ be a bounded domain with smooth boundary. Let l, k be any integers satisfying $0 \leq l < k$, $1 \leq q, r \leq \infty$ and $p \in \mathbb{R}^+$, $\frac{l}{k} \leq \theta \leq 1$ such that*

$$\frac{1}{p} - \frac{l}{n} = \theta\left(\frac{1}{q} - \frac{k}{n}\right) + \frac{1 - \theta}{r}. \quad (3.2.3)$$

Then there are positive constants C_{GN} and C_1 depending only on Ω, q, k, r and n such that for any function $\varphi \in W^{k,q}(\Omega) \cap L^r(\Omega)$,

$$\|\nabla^l \varphi\|_{L^p(\Omega)} \leq C_{GN} \|\nabla^k \varphi\|_{L^q(\Omega)}^\theta \|\varphi\|_{L^r(\Omega)}^{1-\theta} + C_1 \|\varphi\|_{L^r(\Omega)} \quad (3.2.4)$$

with the following exception: If $1 < q < \infty$ and $k - l - \frac{n}{q}$ is a nonnegative integer, then we assume that (3.2.3) holds for θ satisfying $\frac{l}{k} \leq \theta < 1, r > 1$.

To estimate $\int_\Omega a^2 + \int_\Omega |\nabla v|^4$ with $a = e^{-\xi w} u$ in the proof Theorem 3.3, we will have to get a handle on $\int_\Omega |\nabla v|^6$ and $\int_\Omega a^3$. The above Gagliardo–Nirenberg inequality enables us to replace them by more convenient terms.

Lemma 3.4 *Let $\Omega \subset \mathbb{R}^3$ be a bounded domain with smooth boundary. For any $\varepsilon > 0$, there are $C(\varepsilon) > 0$ and $C_2 > 0$ such that for any $v \in C^2(\Omega)$*

$$\int_\Omega |\nabla v|^6 \leq \varepsilon \int_\Omega |\nabla|\nabla v|^2|^2 + C(\varepsilon) \left(\left(\int_\Omega |\nabla v|^4 \right)^3 + \left(\int_\Omega |\nabla v|^4 \right)^{\frac{3}{2}} \right), \quad (3.2.5)$$

and for any $a \in W^{1,2}(\Omega)$

$$\int_{\Omega} a^3 \leq \varepsilon \int_{\Omega} |\nabla a|^2 + C(\varepsilon) \left(\int_{\Omega} a^2 \right)^3 + C_2 \left(\int_{\Omega} a \right)^3. \qquad (3.2.6)$$

Proof We would like refer the reader to Lemma 4.3 of Lankeit (2015) for (3.2.5) and (2.7) with $\gamma = 2$ of Winkler (2008) for (3.2.6), respectively.

The following statement generalizing that of Lemma 3.4 in Stinner et al. (2014) plays an important role in the proofs of Lemmas 3.17 and 3.22 below.

Lemma 3.5 *Let $T \in (0, \infty], 0 < \tau < T$ and suppose that y is a nonnegative absolutely continuous function satisfying*

$$y'(t) + a(t)y(t) \leq b(t)y(t) + c(t) \qquad \text{for a.e. } t \in (0, T) \qquad (3.2.7)$$

with some functions $a(t) > 0, b(t) \geq 0, c(t) \geq 0$ and $a, b, c \in L^1_{loc}(0, T)$ for which there exist $b_1, c_1 > 0$ and $\rho > 0$ such that

$$\sup_{0 \leq t \leq T-\tau} \int_t^{t+\tau} b(s)ds \leq b_1, \qquad \sup_{0 \leq t \leq T-\tau} \int_t^{t+\tau} c(s)ds \leq c_1$$

and

$$\int_t^{t+\tau} a(s)ds - \int_t^{t+\tau} b(s)ds \geq \rho \quad \text{for any } t \in (0, T-\tau).$$

Then

$$y(t) \leq y(0)e^{b_1} + \frac{c_1 e^{2b_1}}{1 - e^{-\rho}} + c_1 e^{b_1} \quad \text{for all } t \in (0, T).$$

Proof From (3.2.7) and a comparison argument, we obtain that for any $\tau \leq t < T$,

$$y(t) \leq y(t-\tau)e^{\int_{t-\tau}^t (b(s)-a(s))ds} + \int_{t-\tau}^t c(s)e^{\int_s^t (b(\sigma)-a(\sigma))d\sigma} ds$$

$$\leq y(t-\tau)e^{-\rho} + \int_{t-\tau}^t c(s)e^{\int_s^t b(\sigma)d\sigma} ds$$

$$\leq y(t-\tau)e^{-\rho} + \int_{t-\tau}^t c(s)e^{\int_{t-\tau}^t b(\sigma)d\sigma} ds$$

$$\leq y(t-\tau)e^{-\rho} + c_1 e^{b_1}.$$

Hence, taking $t = k\tau$ $(k = 1, 2, \ldots, [\frac{T}{\tau}])$, we have

$$y(k\tau) \leq y((k-1)\tau)e^{-\rho} + c_1 e^{b_1}$$
$$\leq e^{-\rho}(y((k-2)\tau)e^{-\rho} + c_1 e^{b_1}) + c_1 e^{b_1}$$
$$= e^{-2\rho} y((k-2)\tau) + e^{-\rho} c_1 e^{b_1} + c_1 e^{b_1}$$
$$= e^{-k\rho} y(0) + c_1 e^{b_1} \sum_{j=0}^{k-1} e^{-j\rho}$$
$$\leq e^{-k\rho} y(0) + \frac{c_1 e^{b_1}}{1 - e^{-\rho}}.$$

Now for any given $t \in (0, T) > 0$, we can fix $k \in \mathbb{N}$ such that $k\tau < t \leq (k+1)\tau$, i.e., $k = [\frac{t}{\tau}]$, and thus get

$$y(t) \leq y(k\tau)e^{\int_{k\tau}^{t}(b(s)-a(s))ds} + \int_{k\tau}^{t} c(s)e^{\int_{s}^{t}(b(\sigma)-a(\sigma))d\sigma} ds$$
$$\leq y(k\tau)e^{\int_{k\tau}^{t} b(s)ds} + \int_{k\tau}^{t} c(s)e^{\int_{s}^{t} b(\sigma)d\sigma} ds$$
$$\leq y(k\tau)e^{b_1} + c_1 e^{b_1}$$
$$\leq y(0)e^{b_1}e^{-[\frac{t}{\tau}]\rho} + \frac{c_1 e^{2b_1}}{1 - e^{-\rho}} + c_1 e^{b_1}.$$

Apart from the asserted results in Lemma 3.2, we also need some fundamental estimates for the inhomogeneous linear heat equation

$$\begin{cases} v_t = \Delta v - v + u, & x \in \Omega, t > 0, \\ \dfrac{\partial v}{\partial \nu} = 0, & x \in \partial\Omega, t > 0, \\ v(x, 0) = v_0(x), & x \in \Omega, \end{cases} \tag{3.2.8}$$

which can be derived from a standard regularity argument involving the variation-of-constants formula for v and $L^q - L^p$ estimates for the heat semigroup (see Horstmann and Winkler (2005) for instance).

Lemma 3.6 *Ishida et al. (2014, Lemma 2.1) Yang et al. (2015, Lemma 2.2) Let $T > 0$, $1 \leq p \leq \infty$, $v_0 \in L^p(\Omega)$ and $u \in L^1(0, T; L^p(\Omega))$. Then (3.2.8) has a unique mild solution $v \in C([0, T]; L^p(\Omega))$ given by*

$$v(t) = e^{-t}e^{t\Delta} v_0 + \int_0^t e^{-(t-s)}e^{(t-s)\Delta} u(s)ds \quad \text{for all } t \in [0, T],$$

where $e^{t\Delta}$ is the semigroup generated by the Neumann Laplacian. In addition, let $1 \leq q \leq p < \frac{nq}{n-q}$, $v_0 \in W^{1,p}(\Omega)$ and $u \in L^\infty(0, T; L^q(\Omega))$. Then for every $t \in (0, T)$,

$$\|v(t)\|_{L^p(\Omega)} \leq \|v_0\|_{L^p(\Omega)} + c_2 \|u\|_{L^\infty((0,T);L^q(\Omega))}, \tag{3.2.9}$$

$$\|\nabla v(t)\|_{L^p(\Omega)} \leq \|\nabla v_0\|_{L^p(\Omega)} + c_2 \|u\|_{L^\infty((0,T);L^q(\Omega))}, \qquad (3.2.10)$$

where c_2 is a positive constant depending on p, q and n.

The following statement can be found in Appendix A of Tao and Winkler (2014b).

Lemma 3.7 *Tao and Winkler (2014b, Lemma A.3 and Lemma A.4) Let $\Omega \subset \mathbb{R}^2$ be a bounded domain with smooth boundary. Then for all $M > 0$, there exist constants $\alpha > 0$, $\beta > 0$ depending only upon M such that for any nonnegative function $u \in L^2(\Omega)$ and $\int_\Omega u \leq M$, the solution v of*

$$\begin{cases} -\Delta v + v = u, & x \in \Omega, \\ \dfrac{\partial v}{\partial \nu} = 0, & x \in \partial\Omega \end{cases} \qquad (3.2.11)$$

satisfies

$$\int_\Omega |\nabla v|^2 + \int_\Omega |v|^2 \leq \alpha \int_\Omega u \ln u + \beta. \qquad (3.2.12)$$

3.3 Global Boundedness of Solutions to a Chemotaxis–Haptotaxis Model

In some parts of our subsequent analysis, we introduce the variable transformation (see Tao and Wang (2009); Tao and Winkler (2011, 2014b), Pang and Wang (2017))

$$a = ue^{-\xi w}, \qquad (3.3.1)$$

upon which (3.1.1) takes the form

$$\begin{cases} a_t = e^{-\xi w}\nabla \cdot (e^{\xi w}\nabla a) - \chi e^{-\xi w}\nabla \cdot (e^{\xi w} a\nabla v) + \xi avw \\ \qquad + a(\mu - \xi\eta w)(1 - e^{\xi w}a - w), & x \in \Omega, t > 0, \\ v_t = \Delta v + ae^{\xi w} - v, & x \in \Omega, t > 0, \\ w_t = -vw + \eta w(1 - ae^{\xi w} - w), & x \in \Omega, t > 0, \\ \dfrac{\partial a}{\partial \nu} = \dfrac{\partial v}{\partial \nu} = 0, & x \in \partial\Omega, t > 0, \\ a(x,0) := a_0(x) = u_0(x)e^{-\xi w_0(x)}, v(x,0) = v_0(x), w(x,0) = w_0(x), & x \in \Omega. \end{cases}$$
$$(3.3.2)$$

The following lemma deals with local-in-time existence and the uniqueness of a classical solution for the problem (3.1.1).

Lemma 3.8 (Pang and Wang (2017)) *Assume that the nonnegative functions u_0, v_0, and w_0 satisfy (3.3.2) for some $\vartheta \in (0, 1)$. Then there exists a maximal existence* *time $T_{max} \in (0, \infty]$ and a triple of nonnegative functions*

$$\begin{cases} a \in C^0(\bar{\Omega} \times [0, T_{max})) \cap C^{2,1}(\bar{\Omega} \times (0, T_{max})), \\ v \in C^0(\bar{\Omega} \times [0, T_{max})) \cap C^{2,1}(\bar{\Omega} \times (0, T_{max})), \\ w \in C^{2,1}(\bar{\Omega} \times [0, T_{max})), \end{cases}$$

which solves (3.3.2) classically and satisfies

$$0 \le w \le \rho := \max\{1, \|w_0\|_{L^\infty(\Omega)}\} \ in \ \Omega \times (0, T_{max}). \tag{3.3.3}$$

Moreover, if $T_{max} < +\infty$, then

$$\|a(\cdot, t)\|_{L^\infty(\Omega)} + \|\nabla w(\cdot, t)\|_{L^5(\Omega)} \to \infty \ as \ t \nearrow T_{max}. \tag{3.3.4}$$

In this subsection, we are going to establish an iteration step to develop the main ingredient of our result. Firstly, based on the ideas of Lemma 3.1 in Pang and Wang (2017) (see also Lemma 2.1 of Winkler (2010a)), we can derive the following properties of solutions of (3.1.1).

Lemma 3.9 *Under the assumptions in Theorem 3.1, we derive that there exists a positive constant C such that the solution of (3.1.1) satisfies*

$$\int_\Omega u(x, t) + \int_\Omega v^2(x, t) + \int_\Omega |\nabla v(x, t)|^2 \le C \ for \ all \ t \in (0, T_{max}).$$

Lemma 3.10 *Let*

$$A_1 = \frac{1}{\delta + 1}\left(\frac{\delta + 1}{\delta}\right)^{-\delta}\left[\frac{\delta(\delta - 1)}{2}\chi^2\right]^{\delta+1}C_7 C_{\delta+1}$$

and $H(y) = y + A_1 y^{-\delta}$ for $y > 0$. For any fixed $\delta \ge 1, C_7, \chi, C_{\delta+1} > 0$,

$$\min_{y>0} H(y) = \frac{\delta(\delta - 1)\chi^2}{2}(C_7 C_{\delta+1})^{\frac{1}{\delta+1}}.$$

Proof It is easy to verify that $H'(y) = 1 - A_1\delta y^{-\delta-1}$ and $H'((A_1\delta)^{\frac{1}{\delta+1}}) = 0$. On the other hand, $\lim_{y\to 0^+} H(y) = +\infty$ and $\lim_{y\to+\infty} H(y) = +\infty$. Hence, we have

$$\min_{y>0} H(y) = H[(A_1\delta)^{\frac{1}{\delta+1}}] = \frac{\delta(\delta - 1)\chi^2}{2}(C_7 C_{\delta+1})^{\frac{1}{\delta+1}},$$

whereby the proof is completed.

Lemma 3.11 *Let $h(p) := \dfrac{p\mu}{2} - \dfrac{p(p - 1)\chi^2}{2}(C_7 C_{p+1})^{\frac{1}{p+1}} - (p - 1)\xi\eta\rho$, where $p \ge 1, \xi, \chi, \eta, \rho, \mu, C_7$ and $\tilde{C}_{p+1}$ are positive constants. Then there exists a positive* *constant $p_0 > 1$ such that*

$$h(p_0) > 0. \tag{3.3.5}$$

Proof Since $h(1) = \frac{\mu}{2} > 0$, from the continuity of h it follows that for each $\mu > 0$, there is some $p_0 > 1$ such that (3.3.5) holds.

According to the local existence results in Lemma 3.8, for any fixed $s \in (0, T_{max})$, it yields $(u(\cdot, s), v(\cdot, s), w(\cdot, s)) \in (C^2(\bar{\Omega}))^3$. Therefore, without loss of generality, we can assume that there exists a constant $\beta > 0$ such that

$$\|u_0\|_{C^2(\bar{\Omega})} \le \beta, \quad \|v_0\|_{C^2(\bar{\Omega})} \le \beta \text{ and } \|w_0\|_{C^2(\bar{\Omega})} \le \beta. \tag{3.3.6}$$

Lemma 3.12 *Let μ, χ, η and ξ be the positive constants. Assume that (a, v, w) is a solution of (3.3.2) on $(0, T_{max})$. Then there exists a positive constant $C = C(p_0, |\Omega|, \mu, \chi, \xi, \eta, \beta)$ such that*

$$\int_\Omega a^{p_0}(x, t)dx \le C \text{ for all } t \in (0, T_{max}), \tag{3.3.7}$$

where $p_0 > 1$ is the same as in Lemma 3.11.

Proof By using (3.3.2) and integration by parts, we get

$$\frac{d}{dt} \int_\Omega e^{\xi w} a^{p_0} + (p_0 + 1) \int_\Omega e^{\xi w} a^{p_0}$$

$$= \xi \int_\Omega e^{\xi w} a^{p_0} \cdot \{-vw + \eta w(1 - ae^{\xi w} - w)\}$$

$$+ p_0 \int_\Omega e^{\xi w} a^{p_0 - 1} \cdot \{e^{-\xi w} \nabla \cdot (e^{\xi w} \nabla a) - \chi e^{-\xi w} \nabla \cdot (e^{\xi w} a \nabla v)\}$$

$$+ a\xi vw + a(\mu - \xi \eta w)(1 - ae^{\xi w} - w)\} + (p_0 + 1) \int_\Omega e^{\xi w} a^{p_0}$$

$$= -p_0(p_0 - 1) \int_\Omega e^{\xi w} a^{p_0 - 2} |\nabla a|^2 + p_0(p_0 - 1)\chi \int_\Omega e^{\xi w} a^{p_0 - 1} \nabla a \cdot \nabla v \tag{3.3.8}$$

$$+ (p_0 - 1)\xi \int_\Omega e^{\xi w} a^{p_0} vw$$

$$+ \int_\Omega e^{\xi w} a^{p_0} \{(p_0 + 1) + (p_0 - 1)\xi \eta w(w - 1) + p_0 \mu(1 - w)\}$$

$$+ \int_\Omega e^{2\xi w} a^{p_0 + 1} [(p_0 - 1)\xi \eta w - p_0 \mu]$$

$$:= J_1 + J_2 + J_3 + J_4 + J_5 \text{ for all } t \in (0, T_{max}).$$

Now, in light of (3.3.3), (3.3.4) and the Young inequality, we derive that

$$J_3 \leq \varepsilon_1 \int_\Omega e^{2\xi w} a^{p_0+1} + \frac{1}{p_0+1} \left(\varepsilon_1 \cdot \frac{p_0+1}{p_0} \right)^{-p_0} [(p_0-1)\xi]^{p_0+1} \int_\Omega e^{\xi w(1-p_0)} v^{p_0+1}$$

$$\leq \varepsilon_1 \int_\Omega e^{2\xi w} a^{p_0+1} + \frac{1}{p_0+1} \left(\frac{\varepsilon_1(p_0+1)}{p_0} \right)^{-p_0} [(p_0-1)\xi]^{p_0+1} \int_\Omega v^{p_0+1}, \qquad (3.3.9)$$

$$J_4 \leq [(p_0+1) + (p_0-1)\xi \eta \rho^2 + p_0 \mu] \int_\Omega e^{\xi w} a^{p_0}$$

$$\leq (p_0+1)[1 + \xi \eta \rho^2 + \mu] \int_\Omega e^{\xi w} a^{p_0} \qquad (3.3.10)$$

$$\leq \varepsilon_2 \int_\Omega e^{2\xi w} a^{p_0+1} + \frac{1}{p_0+1} \left(\frac{\varepsilon_2(p_0+1)}{p_0} \right)^{-p_0} (p_0+1)^{p_0+1}[1 + \xi \eta \rho^2 + \mu]^{p_0+1} |\Omega|$$

as well as

$$J_5 \leq \int_\Omega e^{2\xi w} a^{p_0+1}[(p_0-1)\xi \eta \rho - p_0 \mu] \quad \text{for all } t \in (0, T_{max})$$

and

$$J_2 \leq \frac{p_0(p_0-1)}{2} \int_\Omega e^{\xi w} a^{p_0-2} |\nabla a|^2 + \frac{p_0(p_0-1)}{2} \chi^2 \int_\Omega e^{\xi w} a^{p_0} |\nabla v|^2$$

$$\leq \frac{p_0(p_0-1)}{2} \int_\Omega e^{\xi w} a^{p_0-2} |\nabla a|^2 + \lambda_0 \int_\Omega e^{2\xi w} a^{p_0+1}$$

$$+ \frac{1}{p_0+1} \left(\frac{\lambda_0(p_0+1)}{p_0} \right)^{-p_0} \left[\frac{p_0(p_0-1)}{2} \chi^2 \right]^{p_0+1} \int_\Omega e^{(1-p_0)\xi w} |\nabla v|^{2(p_0+1)}$$

$$\qquad (3.3.11)$$

$$\leq \frac{p_0(p_0-1)}{2} \int_\Omega e^{\xi w} a^{p_0-2} |\nabla a|^2 + \lambda_0 \int_\Omega e^{2\xi w} a^{p_0+1}$$

$$+ \frac{1}{p_0+1} \left(\frac{\lambda_0(p_0+1)}{p_0} \right)^{-p_0} \left[\frac{p_0(p_0-1)}{2} \chi^2 \right]^{p_0+1} \int_\Omega |\nabla v|^{2(p_0+1)}$$

with any small positive constants ε_1, ε_2 and λ_0.

Inserting (3.3.9)–(3.3.11) into (3.3.8), we derive that

$$\frac{d}{dt} \int_\Omega e^{\xi w} a^{p_0}$$

$$+ (p_0+1) \int_\Omega e^{\xi w} a^{p_0} + \int_\Omega e^{2\xi w} a^{p_0+1}[p_0 \mu - \varepsilon_1 - \varepsilon_2 - \lambda_0 - (p_0-1)\xi \eta \rho]$$

$$\leq \frac{1}{p_0+1} \left(\frac{\lambda_0(p_0+1)}{p_0} \right)^{-p_0} [\frac{p_0(p_0-1)}{2} \chi^2]^{p_0+1} \int_\Omega |\nabla v|^{2(p_0+1)}$$

$$+ C_1(\varepsilon_1, \varepsilon_2) \quad \text{for all } t \in (0, T_{max}),$$

$$\qquad (3.3.12)$$

where

$$C_1(\varepsilon_1, \varepsilon_2) := \frac{1}{p_0 + 1}\left(\frac{\varepsilon_2(p_0 + 1)}{p_0}\right)^{-p_0}(p_0 + 1)^{p_0+1}[1 + \xi\eta\rho^2 + \mu]^{p_0+1}|\Omega|$$
$$+ \frac{1}{p_0 + 1}\left(\frac{\varepsilon_1(p_0 + 1)}{p_0}\right)^{-p_0}[(p_0 - 1)\xi]^{p_0+1}\int_\Omega v^{p_0+1}.$$

$$(3.3.13)$$

Next, from Lemma 3.9, $n = 2$ and the Gagliardo–Nirenberg inequality, it follows that

$$\|v(\cdot, t)\|_{L^{p_0}(\Omega)} \le C_2 \quad \text{for all} \quad t \in (0, T_{max}). \tag{3.3.14}$$

This along with (3.3.13) follows

$$C_1(\varepsilon_1, \varepsilon_2) \le C_3(\varepsilon_1, \varepsilon_2)$$
$$:= \frac{1}{p_0 + 1}\left(\varepsilon_2 \times \frac{p_0 + 1}{p_0}\right)^{-p_0}(p_0 + 1)^{p_0+1}[1 + \xi\eta\rho^2 + \mu]^{p_0+1}|\Omega|$$
$$+ C_2\frac{1}{p_0 + 1}\left(\varepsilon_1 \times \frac{p_0 + 1}{p_0}\right)^{-p_0}[(p_0 - 1)\xi]^{p_0+1}.$$

From this and (3.3.12), we also obtain

$$\frac{d}{dt}\int_\Omega e^{\xi w}a^{p_0} + (p_0 + 1)\int_\Omega e^{\xi w}a^{p_0}$$
$$+ \int_\Omega e^{2\xi w}a^{p_0+1}[p_0\mu - \varepsilon_1 - \varepsilon_2 - \lambda_0 - (p_0 - 1)\xi\eta\rho]$$
$$\le \frac{1}{p_0 + 1}\left(\frac{\lambda_0(p_0 + 1)}{p_0}\right)^{-p_0}\left[\frac{p_0(p_0 - 1)}{2}\chi^2\right]^{p_0+1}\int_\Omega |\nabla v|^{2(p_0+1)}\|$$
$$+ C_3(\varepsilon_1, \varepsilon_2) \quad \text{for all} \quad t \in (0, T_{max}).$$

Then for any $t \in (0, T_{max})$, by means of the variation-of-constants representation for the above inequality, we can estimate

$$\int_\Omega e^{\xi w}a^{p_0}(\cdot, t) + [p_0\mu - \varepsilon_1 - \varepsilon_2 - \lambda_0 - (p_0 - 1)\xi\eta\rho]$$
$$\cdot \int_0^t\int_\Omega e^{-(p_0-1)(t-s)}e^{2\xi w}a^{p_0+1}$$
$$\le \int_\Omega u_0^p + \frac{1}{p_0 + 1}\left(\frac{\lambda_0(p_0 + 1)}{p_0}\right)^{-p_0}\left[\frac{p_0(p_0 - 1)}{2}\chi^2\right]^{p_0+1} \tag{3.3.15}$$
$$\cdot \int_0^t\int_\Omega e^{-(p_0-1)(t-s)}|\nabla v|^{2(p_0+1)}$$
$$+ C_3(\varepsilon_1, \varepsilon_2) \quad \text{for all} \quad t \in (0, T_{max}).$$

Next, according to the Gagliardo–Nirenberg inequality, (3.3.14) and Lemma 3.9, we can choose C_4 and C_5 such that

$$\|\nabla v(\cdot,s)\|_{L^{2(p_0+1)}(\Omega)}^{2(p_0+1)} \leq C_4 \|v(\cdot,s)\|_{W^{2,p_0+1}(\Omega)}^{p_0+1} \|\nabla v(\cdot,s)\|_{L^2(\Omega)}^{p_0+1}$$

$$\leq C_5 \|v(\cdot,s)\|_{W^{2,p_0+1}(\Omega)}^{p_0+1} \quad \text{for all } t \in (0, T_{max}). \tag{3.3.16}$$

Therefore, with the help of (3.3.16), applying (3.2.2) of Lemma 3.2 with $\gamma = p_0 + 1$, we obtain

$$\frac{1}{p_0+1}\left(\frac{\lambda_0(p_0+1)}{p_0}\right)^{-p_0}\left[\frac{p_0(p_0-1)}{2}\chi^2\right]^{p_0+1} \int_0^t \int_\Omega e^{-(p_0-1)(t-s)} |\nabla v|^{2(p_0+1)}$$

$$\leq \frac{1}{p_0+1}\left(\frac{\lambda_0(p_0+1)}{p_0}\right)^{-p_0}\left[\frac{p_0(p_0-1)}{2}\chi^2\right]^{p_0+1} C_5 \int_0^t e^{-(p_0-1)(t-s)} \|v(\cdot,s)\|_{W^{2,p_0+1}(\Omega)}^{p_0+1}$$

$$\leq \frac{1}{p_0+1}\left(\frac{\lambda_0(p_0+1)}{p_0}\right)^{-p_0}\left[\frac{p_0(p_0-1)}{2}\chi^2\right]^{p_0+1} C_5 C_{p_0+1}$$

$$\cdot \int_0^t \int_\Omega e^{-(p_0-1)(t-s)} u^{p_0+1} + C_6$$

$$\leq \frac{1}{p_0+1}\left(\frac{\lambda_0(p_0+1)}{p_0}\right)^{-p_0}\left[\frac{p_0(p_0-1)}{2}\chi^2\right]^{p_0+1} C_5 C_{p_0+1} e^{\xi(p_0-1)} \tag{3.3.17}$$

$$\cdot \int_0^t \int_\Omega e^{-(p_0-1)(t-s)} e^{2\xi w} a^{p_0+1} + C_6$$

$$\leq \frac{1}{p_0+1}\left(\frac{\lambda_0(p_0+1)}{p_0}\right)^{-p_0}\left[\frac{p_0(p_0-1)}{2}\chi^2\right]^{p_0+1} C_7 C_{p_0+1}$$

$$\cdot \int_0^t \int_\Omega e^{-(p_0-1)(t-s)} e^{2\xi w} a^{p_0+1} + C_6$$

for all $t \in (0, T_{max})$, where

$$C_6 := \frac{1}{p_0+1}\left(\frac{\lambda_0(p_0+1)}{p_0}\right)^{-p_0}\left[\frac{p_0(p_0-1)}{2}\chi^2\right]^{p_0+1} C_5 C_{p_0+1} \|v_0\|_{W^{2,\gamma}(\Omega)}^{\gamma}$$

and

$$C_7 := C_5 e^{\xi(p_0-1)}.$$

Substituting (3.3.17) into (3.3.15), we derive

$$\int_\Omega e^{\xi w} a^{p_0}(\cdot,t) + [p_0\mu - \varepsilon_1 - \varepsilon_2 - \lambda_0 - (p_0-1)\xi\eta\rho]$$

$$\cdot \int_0^t \int_\Omega e^{-(p_0-1)(t-s)} e^{2\xi w} a^{p_0+1}$$

$$\leq \frac{1}{p_0+1}\left(\frac{\lambda_0(p_0+1)}{p_0}\right)^{-p_0}\left[\frac{p_0(p_0-1)}{2}\chi^2\right]^{p_0+1} C_7 C_{p_0+1} \tag{3.3.18}$$

$$\cdot \int_0^t \int_\Omega e^{-(p_0-1)(t-s)} e^{2\xi w} a^{p_0+1} + C_8(\varepsilon_1, \varepsilon_2) \quad \text{for all } t \in (0, T_{max}),$$

where $C_8(\varepsilon_1, \varepsilon_2) := C_3(\varepsilon_1, \varepsilon_2) + C_6$. Choosing $\lambda_0 = (A_1 p_0)^{\frac{1}{p_0+1}}$ in (3.3.18) and using Lemma 3.10, we derive

$$\int_\Omega e^{\xi w} a^{p_0}(\cdot, t) + [p_0\mu - \varepsilon_1 - \varepsilon_2 - \frac{p_0(p_0-1)\chi^2}{2}(C_7 C_{p_0+1})^{\frac{1}{p_0+1}} - (p_0-1)\xi\eta\rho]$$

$$\cdot \int_0^t \int_\Omega e^{-(p_0-1)(t-s)} e^{2\xi w} a^{p_0+1} \qquad\qquad (3.3.19)$$

$$\leq C_8(\varepsilon_1, \varepsilon_2) \text{ for all } t \in (0, T_{max}).$$

Now, for the above positive constants μ, χ, ξ and η, due to Lemma 3.11, it has

$$p_0\mu - \frac{p_0(p_0-1)\chi^2}{2}(C_7 C_{p_0+1})^{\frac{1}{p_0+1}} - (p_0-1)\xi\eta\rho > \frac{p_0\mu}{2} > 0,$$

thus one can choose ε_1 and ε_2 appropriately small (e.g., $\varepsilon_1 = \varepsilon_2 = \frac{p_0\mu}{8}$) such that

$$0 < \varepsilon_1 + \varepsilon_2 < p_0\mu - \frac{p_0(p_0-1)\chi^2}{2}(C_7 C_{p_0+1})^{\frac{1}{p_0+1}} - (p_0-1)\xi\eta\rho. \qquad (3.3.20)$$

Collecting (3.3.19) and (3.3.20), we derive that there exists a positive constant C_9 such that

$$\int_\Omega u^{p_0}(x, t)dx \leq C_9 \text{ for all } t \in (0, T_{max}).$$

The proof of Lemma 3.12 is completed.

Lemma 3.13 *Assume the hypothesis of Lemma 3.12 holds. Then for all $p > 1$, there exists a positive constant $C = C(p, |\Omega|, \mu, \chi, \xi, \eta, \beta)$ such that $\int_\Omega a^p(x, t)dx \leq C$ for all $t \in (0, T_{max})$.*

Proof Firstly, from Lemma 3.12 (see (3.3.7)) and (3.3.1), there exists a positive constant C_1 such that

$$\int_\Omega u^{p_0}(x, t)dx \leq C_1 \text{ for all } t \in (0, T_{max}), \qquad (3.3.21)$$

where $p_0 > 1$ is the same as that in Lemma 3.11. Next, we fix $q < \frac{2p_0}{(2-p_0)^+}$ and choose some $\alpha > \frac{1}{2}$ such that

$$q < \frac{1}{\frac{1}{p_0} - \frac{1}{2} + \frac{2}{3}(\alpha - \frac{1}{2})} \leq \frac{2p_0}{(2-p_0)^+}. \qquad (3.3.22)$$

Now, involving the variation-of-constants formula for v, we have

$$v(t) = e^{-(A+1)}v_0 + \int_0^t e^{-(t-s)(A+1)}u(s)ds, \quad t \in (0, T_{max}). \qquad (3.3.23)$$

Hence, it follows from (3.3.6), (3.3.21)–(3.3.23) that

$$\|(A+1)^\alpha v(t)\|_{L^q(\Omega)}$$
$$\leq c \int_0^t (t-s)^{-\alpha-\frac{2}{2}(\frac{1}{p_0}-\frac{1}{q})} e^{-\kappa(t-s)} \|u(s)\|_{L^{p_0}(\Omega)} ds + c e^{-\kappa t} t^{-\alpha+\frac{1}{q}} \|v_0\|_{L^\infty(\Omega)}$$
$$\leq C_2 \int_0^{+\infty} \sigma^{-\alpha-\frac{2}{2}(\frac{1}{p_0}-\frac{1}{q})} e^{-\kappa\sigma} d\sigma + C_3 t^{-\alpha+\frac{1}{q}} \quad \text{for all } t \in (0, T_{max}),$$

(3.3.24)

where $c > 0$ is given by Lemma 3.1. Hence, in light of Lemmas 3.1 and 3.8, due to (3.3.22) and (3.3.24), we have

$$\int_\Omega |\nabla v(t)|^q \leq C_4 \text{ for all } t \in (0, T_{max}) \text{ and } q \in [1, \frac{2p_0}{(2-p_0)^+}) \quad (3.3.25)$$

with some positive constant C_4. Now, due to the Sobolev embedding theorems and $N = 2$, we conclude that

$$\|v(\cdot, t)\|_{L^\infty(\Omega)} \leq C_5 \text{ for all } t \in (0, T_{max}). \quad (3.3.26)$$

Applying the Young inequality, one obtains from (3.3.3), (3.3.2) and (3.3.26) that for any $p > \max\{2, p_0 - 1\}$

$$\frac{d}{dt} \int_\Omega e^{\xi w} a^p + p(p-1) \int_\Omega e^{\xi w} a^{p-2}|\nabla a|^2 + p\mu \int_\Omega e^{2\xi w} a^{p+1}$$
$$= p(p-1)\chi \int_\Omega e^{\xi w} a^{p-1}\nabla a \cdot \nabla v + (p-1)\xi \int_\Omega e^{\xi w} a^p vw$$
$$+ \int_\Omega e^{\xi w} a^p \{(p+1) + (p-1)\xi\eta w(w-1) + p\mu(1-w)\}$$
$$+ \int_\Omega e^{2\xi w} a^{p+1}(p-1)\xi\eta w \quad (3.3.27)$$
$$\leq \frac{p(p-1)}{2} \int_\Omega e^{\xi w} a^{p-2}|\nabla a|^2 + \frac{p(p-1)}{2}\chi^2 \int_\Omega e^{\xi w} a^p |\nabla v|^2 + (p-1)\xi \int_\Omega e^{\xi w} a^p vw$$
$$+ \int_\Omega e^{\xi w} a^p \{(p+1) + (p-1)\xi\eta w(w-1) + p\mu(1-w)\}$$
$$+ \int_\Omega e^{2\xi w} a^{p+1}(p-1)\xi\eta w$$
$$\leq \frac{p(p-1)}{2} \int_\Omega e^{\xi w} a^{p-2}|\nabla a|^2 + \frac{p(p-1)}{2}\chi^2 \int_\Omega e^{\xi w} a^p |\nabla v|^2 + C_6 \int_\Omega a^{p+1}$$
$$\leq \frac{p(p-1)}{2} \int_\Omega e^{\xi w} a^{p-2}|\nabla a|^2 + \frac{p(p-1)}{2}\chi^2 e^{\xi\rho} \int_\Omega a^p |\nabla v|^2$$
$$+ C_6 \int_\Omega a^{p+1} \quad \text{for all } t \in (0, T_{max}).$$

Next, with the help of the Gagliardo–Nirenberg inequality (see, e.g., Zheng (2015)), it yields that

$$C_6 \int_\Omega a^{p+1} = C_6 \|a^{\frac{p}{2}}\|_{L^{2\frac{(p+1)}{p}}(\Omega)}^{2\frac{(p+1)}{p}}$$

$$\leq C_7 (\|\nabla a^{\frac{p}{2}}\|_{L^2(\Omega)}^{\mu_1} \|a^{\frac{p}{2}}\|_{L^{\frac{2p_0}{p}}(\Omega)}^{1-\mu_1} + \|a^{\frac{p}{2}}\|_{L^{\frac{2p_0}{p}}(\Omega)})^{2\frac{(p+1)}{p}}$$

$$\leq C_8 (\|\nabla a^{\frac{p}{2}}\|_{L^2(\Omega)}^{2\mu_1} + 1)$$

$$= C_8 (\|\nabla a^{\frac{p}{2}}\|_{L^2(\Omega)}^{\frac{2(p-p_0+1)}{p+1}} + 1)$$

with some positive constants C_7, C_8 and

$$\mu_1 = \frac{\frac{p}{p_0} - \frac{p}{p+1}}{\frac{p}{p_0}} = \frac{p+1-p_0}{p+1} \in (0, 1).$$

Since, $p_0 > 1$ yields $p_0 < \frac{2p_0}{2(2-p_0)^+}$, in light of the Hölder inequality and (3.3.25), we derive

$$\frac{\chi^2 p(p-1)}{2} e^{\xi \rho} \int_\Omega a^p |\nabla v|^2 \leq \frac{\chi^2 p(p-1)}{2} e^{\xi \rho} \left(\int_\Omega a^{\frac{p_0}{p_0-1} p} \right)^{\frac{p_0-1}{p_0}} \left(\int_\Omega |\nabla v|^{2p_0} \right)^{\frac{1}{p_0}}$$

$$\leq C_9 \|a^{\frac{p}{2}}\|_{L^{2\frac{p_0}{p_0-1}}(\Omega)}^2,$$

where C_9 is a positive constant. Since $p_0 > 1$ and $p > p_0 - 1$, we have

$$\frac{p_0}{p} \leq \frac{p_0}{p_0 - 1} < +\infty,$$

which together with the Gagliardo–Nirenberg inequality (see, e.g., Zheng (2015)) implies that

$$C_9 \|a^{\frac{p}{2}}\|_{L^{2\frac{p_0}{p_0-1}}(\Omega)}^2 \leq C_{10} (\|\nabla a^{\frac{p}{2}}\|_{L^2(\Omega)}^{\mu_2} \|a^{\frac{p}{2}}\|_{L^{\frac{2p_0}{p}}(\Omega)}^{1-\mu_2} + \|a^{\frac{p}{2}}\|_{L^{\frac{2p_0}{p}}(\Omega)})^2$$

$$\leq C_{11} (\|\nabla a^{\frac{p}{2}}\|_{L^2(\Omega)}^{2\mu_2} + 1)$$

$$= C_{11} (\|\nabla a^{\frac{p}{2}}\|_{L^2(\Omega)}^{\frac{2(p-p_0+1)}{p}} + 1)$$

with some positive constants C_{10}, C_{11} and

$$\mu_2 = \frac{\frac{p}{p_0} - \frac{\frac{p}{p_0-1} p}{\frac{p}{p_0-1} p}}{\frac{p}{p_0}} \in (0, 1).$$

Moreover, an application of the Young inequality shows that

$$C_6 \int_\Omega a^{p+1} + \frac{\chi^2 p(p-1)}{2} e^{\xi\rho} \int_\Omega a^p |\nabla v|^2$$

$$\leq \frac{p(p-1)}{4} \int_\Omega a^{p-2} |\nabla a|^2 + C_{12} \qquad (3.3.28)$$

$$\leq \frac{p(p-1)}{4} \int_\Omega e^{\xi w} a^{p-2} |\nabla a|^2 + C_{12}.$$

Inserting (3.3.28) into (3.3.27), we conclude that

$$\frac{d}{dt} \int_\Omega e^{\xi w} a^p + \frac{p(p-1)}{4} \int_\Omega e^{\xi w} a^{p-2} |\nabla a|^2 + p\mu \int_\Omega e^{2\xi w} a^{p+1} \leq C_{13}.$$

Therefore, integrating the above inequality with respect to t yields

$$\|a(\cdot, t)\|_{L^p(\Omega)} \leq C_{14} \text{ for all } p \geq 1 \text{ and } t \in (0, T_{max})$$

for some positive constant C_{14}.

Remark 3.2 It only assumes that $\mu > 0$ which is different from that in Pang and Wang (2017). Indeed, by the technical lemma (see Lemma 3.10), one could conclude the boundedness of $\int_\Omega a^{q_0}$ (for some $q_0 > 1$),, and further in light of the variation-of-constants formula and L^q-L^p estimates for the heat semigroup, one may derive the boundedness of $\int_\Omega a^p$ (for any $p > 1$).

Our main result on global existence and boundedness thereby becomes a straight-forward consequence of Lemma 3.8 and Lemma 3.13.

The proof of Theorem 3.1: The proof of Theorem 3.1 consists of the following steps.

Step 1. $\|a(\cdot, t)\|_{L^\infty(\Omega)}$: Firstly, in light of (3.3.3), due to Lemma 3.13, we derive that there exist positive constants $p_0 > 2$ and C_1 such that

$$\|u(\cdot, t)\|_{L^{p_0}(\Omega)} \leq C_1 \text{ for all } t \in (0, T_{max}).$$

Next, since $p_0 > 2$ and $n = 2$ yield to $+\infty = \frac{np_0}{(n-p_0)_+}$, therefore, by using Lemma 3.1 (see also Lemma 2.1 of Ishida et al. (2014)), we conclude that

$$\|\nabla v(t)\|_{L^\infty(\Omega)} \leq C_2 \text{ for all } t \in (0, T_{max}). \qquad (3.3.29)$$

Applying the Young inequality, in light of (3.3.3) and the first equation of (3.3.2), one obtains from (3.3.29) that for any $p \geq 4$

$$
\frac{d}{dt}\int_\Omega e^{\xi w}a^p + p(p-1)\int_\Omega e^{\xi w}a^{p-2}|\nabla a|^2 + \int_\Omega e^{\xi w}a^p
$$

$$
=\xi\int_\Omega e^{\xi w}a^p\cdot\{-vw+\eta w(1-ae^{\xi w}-w)\}
$$

$$
+p\int_\Omega e^{\xi w}a^{p-1}\cdot\{e^{-\xi w}\nabla\cdot(e^{\xi w}\nabla a)-\chi e^{-\xi w}\nabla\cdot(e^{\xi w}a\nabla v)\}
$$

$$
+a\xi vw+a(\mu-\xi\eta w)(1-ae^{\xi w}-w)\}+p\int_\Omega e^{\xi w}a^p
$$

$$
\leq\frac{p(p-1)}{4}\int_\Omega e^{\xi w}a^{p-2}|\nabla a|^2 + p(p-1)\chi^2 C_3\int_\Omega e^{\xi w}a^p
$$

$$
+(p-1)\xi\int_\Omega e^{\xi w}a^p vw + \int_\Omega e^{\xi w}a^p\{(p+1)+(p-1)\xi\eta w(w-1)+p\mu(1-w)\}
$$

$$
+\int_\Omega e^{2\xi w}a^{p+1}[(p-1)\xi\eta w-p\mu]
$$

$$
\leq\frac{p(p-1)}{4}\int_\Omega e^{\xi w}a^{p-2}|\nabla a|^2 + C_4p^2(\int_\Omega a^{p+1}+1)\ \text{for all } t\in(0,T_{max}),
$$

(3.3.30)

where $C_3>0$ and $C_4>0$ are independent of p. Here and throughout the proof of Theorem 3.1, we shall denote by $C_i\,(i\in\mathbb{N})$ the several positive constants independent of p. Therefore, (3.3.30) implies that

$$
\frac{d}{dt}\int_\Omega e^{\xi w}a^p + C_5\int_\Omega|\nabla a^{\frac{p}{2}}|^2 + \int_\Omega e^{\xi w}a^p \leq C_4p^2(\int_\Omega a^{p+1}+1)\ \text{for all } t\in(0,T_{max}).
$$

(3.3.31)

Next, once more by means of the Gagliardo–Nirenberg inequality, we can estimate

$$
C_4p^2\int_\Omega a^{p+1} = C_4p^2\|a^{\frac{p}{2}}\|_{L^{\frac{2(p+1)}{p}}(\Omega)}^{\frac{2(p+1)}{p}}
$$

$$
\leq C_6p^2(\|\nabla a^{\frac{p}{2}}\|_{L^2(\Omega)}^{\frac{2(p+1)}{p}\varsigma_1}\|a^{\frac{p}{2}}\|_{L^1(\Omega)}^{\frac{2(p+1)}{p}(1-\varsigma_1)} + \|a^{\frac{p}{2}}\|_{L^1(\Omega)}^{\frac{2(p+1)}{p}})
$$

$$
= C_6p^2(\|\nabla a^{\frac{p}{2}}\|_{L^2(\Omega)}^{\frac{p+2}{p}}\|a^{\frac{p}{2}}\|_{L^1(\Omega)} + \|a^{\frac{p}{2}}\|_{L^1(\Omega)}^{\frac{2(p+1)}{p}})
$$

(3.3.32)

$$
\leq C_5\|\nabla a^{\frac{p}{2}}\|_{L^2(\Omega)}^2 + C_7p^{\frac{4p}{p-2}}\|a^{\frac{p}{2}}\|_{L^1(\Omega)}^{\frac{2p}{p-2}} + C_6p^2\|a^{\frac{p}{2}}\|_{L^1(\Omega)}^{\frac{2(p+1)}{p}}
$$

$$
\leq C_5\|\nabla a^{\frac{p}{2}}\|_{L^2(\Omega)}^2 + C_8p^{\frac{4p}{p-2}}\|a^{\frac{p}{2}}\|_{L^1(\Omega)}^{\frac{2p}{p-2}},
$$

where

$$
0<\varsigma_1 = \frac{2-\frac{2p}{2(p+1)}}{1-\frac{2}{2}+2} = \frac{p+2}{2(p+1)} < 1.
$$

Here, we have used the fact that $\frac{4p}{p-2}\geq 2$. Therefore, inserting (3.3.32) into (3.3.31), we derive that

$$
\frac{d}{dt}\int_\Omega e^{\xi w}a^p + \int_\Omega e^{\xi w}a^p \leq C_8p^{\frac{4p}{p-2}}\|a^{\frac{p}{2}}\|_{L^1(\Omega)}^{\frac{2p}{p-2}} + C_4p^2
$$

(3.3.33)

$$
\leq C_9p^{\frac{4p}{p-2}}\left(\max\{1,\|u^{\frac{p}{2}}\|_{L^1(\Omega)}\}\right)^{\frac{2p}{p-2}}.
$$

Now, choosing $p_i = 2^{i+2}$ and letting $M_i = \max\{1, \sup_{t\in(0,T)} \int_\Omega a^{\frac{p_i}{2}}\}$ for $T \in (0, T_{max})$ and $i = 1, 2, 3, \ldots$, we then obtain from (3.3.33) that

$$\frac{d}{dt}\int_\Omega e^{\xi w} a^{p_i} + \int_\Omega e^{\xi w} a^{p_i} \le C_{10} p_i^{\frac{2p_i}{p_i-2}} M_{i-1}^{\frac{2p_i}{p_i-2}}(T),$$

which together with the comparison argument entails that there exists $\lambda > 1$ independent of i such that

$$M_i(T) \le \max\{\lambda^i M_{i-1}^{\frac{2p_i}{p_i-2}}(T), e^\xi |\Omega| \|a_0\|_{L^\infty(\Omega)}^{p_i}\}. \tag{3.3.34}$$

Here, we use the fact that $\kappa_i := \frac{2p_i}{p_i-2} \le 4$. Now, if $\lambda^i M_{i-1}^{\kappa_i}(T) \le e^{\xi\rho}|\Omega|\|a_0\|_{L^\infty(\Omega)}^{p_i}$ for infinitely many $i \ge 1$, we get

$$\left(\sup_{t\in(0,T)}\int_\Omega a^{p_{i-1}}(\cdot,t)\right)^{\frac{1}{p_{i-1}}} \le \left(\frac{e^{\xi\rho}|\Omega|\|a_0\|_{L^\infty(\Omega)}^{p_i}}{\lambda^i}\right)^{\frac{1}{p_{i-1}\kappa_i}}$$

for such i, which entails that

$$\sup_{t\in(0,T)} \|a(\cdot,t)\|_{L^\infty(\Omega)} \le \|a_0\|_{L^\infty(\Omega)}. \tag{3.3.35}$$

Otherwise, if $\lambda^i M_{i-1}^{\kappa_i}(T) > e^\xi |\Omega| \|a_0\|_{L^\infty(\Omega)}^{p_i}$ for all sufficiently large i, then by (3.3.34), we derive that

$$M_i(T) \le \lambda^i M_{i-1}^{\kappa_i}(T) \quad \text{for all sufficiently large } i, \tag{3.3.36}$$

and thus (3.3.36) is still valid for all $i \ge 1$ upon enlarging λ if necessary. That is,

$$M_i(T) \le \lambda^i M_{i-1}^{\kappa_i}(T) \quad \text{for all } i \ge 1.$$

Therefore, based on a straightforward induction (see, e.g., Lemma 3.12 of Tao and Winkler (2014b)), we have

$$M_i(T) \le \lambda^{i+\sum_{j=2}^i (j-1)\cdot\Pi_{k=j}^i \kappa_k} M_0^{\Pi_{k=1}^i \kappa_k} \quad \text{for all } i \ge 1, \tag{3.3.37}$$

where $\kappa_k = 2(1 + \varepsilon_k)$ satisfies $\varepsilon_k = \frac{4}{p_k-2} \le \frac{C_{11}}{2^k}$ for all $k \ge 1$ with some $C_{11} > 0$. Therefore, due to the fact that $\ln(1 + x) \le x (x \ge 0)$, we derive

$$\Pi_{k=j}^i \kappa_k = 2^{i+1-j} e^{\sum_{k=j}^i \ln(1+\varepsilon_j)}$$

$$\le 2^{i+1-j} e^{\sum_{k=j}^i \varepsilon_j}$$

$$\le 2^{i+1-j} e^{C_{11}} \quad \text{for all } i \ge 1 \text{ and } j \in \{1, \ldots, i\},$$

which implies that

$$
\frac{\sum_{j=2}^{i}(j-1)\cdot \Pi_{k=j}^{i}\kappa_k}{2^{i+2}} \leq \frac{\sum_{j=2}^{i}(j-1)2^{i+1-j}e^{C_{11}}}{2^{i+2}}
$$

$$
\leq \frac{e^{C_{11}}}{2}\sum_{j=2}^{i}\frac{(j-1)}{2^j}
$$

$$
\leq \frac{3e^{C_{11}}}{8}.
$$

By the definition of p_i, we easily deduce from (3.3.37) that

$$
M_i^{\frac{1}{p_i}}(T) \leq \lambda^{\frac{i}{2^{i+2}}+\frac{\sum_{j=2}^{i}(j-1)\cdot \Pi_{k=j}^{i}\kappa_k}{2^{i+2}}}M_0^{\frac{\Pi_{k=1}^{i}\kappa_k}{2^{i+2}}} \leq \lambda^{\frac{i}{2^{i+2}}}\lambda^{\frac{3e^{C_{11}}}{8}}M_0^{\frac{e^{C_{11}}}{4}},
$$

which after taking $i \to \infty$ and $T \nearrow T_{max}$ readily implies that

$$
\|a(\cdot,t)\|_{L^{\infty}(\Omega)} \leq \lambda^{\frac{3e^{C_{11}}}{8}}M_0^{\frac{e^{C_{11}}}{4}} \quad \text{for all } t \in (0, T_{max}). \tag{3.3.38}
$$

Step 2:$\|\nabla w(\cdot,t)\|_{L^5(\Omega)}$ Employing almost exactly the same arguments as that in the proof of Lemmas 3.5–3.6 in Pang and Wang (2017) (the minor necessary changes are left as an easy exercise to the reader), and taking advantage of (3.3.29) and (3.3.38), we conclude the estimate for any $T < T_{max}$,

$$
\|\nabla w(\cdot,t)\|_{L^5(\Omega)} \leq C \quad \text{for all } t \in (0, T).
$$

Now, with the above estimate in hand, using (3.3.35) and (3.3.38), employing the extendibility criterion provided by Lemma 3.8, we may prove Theorem 3.1.

Remark 3.3 If $\mu > \xi\eta\max\{\|u_0\|_{L^{\infty}(\Omega)}, 1\} + \mu^*(\chi^2, \xi)$ (see the proof of Lemma 3.4 to Pang and Wang (2017)), one only needs to estimate $Cp^2\int_{\Omega}a^p$ other than $Cp^2(\int_{\Omega}a^{p+1}+1)$.

3.4 Global Boundedness of Solutions to a Chemotaxis–Haptotaxis Model with Tissue Remodeling

3.4.1 A Convenient Extensibility Criterion

For the convenience in some parts of our subsequent analysis, we introduce the variable transformation Tao and Wang (2009, 2008); Tao and Winkler (2014b)

$$a = ue^{-\xi w},$$

upon which (3.1.1) takes the following form:

$$
\begin{cases}
a_t = e^{-\xi w}\nabla \cdot (e^{\xi w}\nabla a) - e^{-\xi w}\chi\nabla \cdot (e^{\xi w}a\nabla v) + \xi avw \\
\quad + a\mu(r - e^{\xi w}a - w) - a\xi\eta w(1 - e^{\xi w}a - w), & x \in \Omega, t > 0, \\
\sigma v_t = \Delta v - v + e^{\xi w}a, & x \in \Omega, t > 0, \\
w_t = -vw + \eta w(1 - w - e^{\xi w}a), & x \in \Omega, t > 0, \\
\dfrac{\partial a}{\partial \nu} = \dfrac{\partial v}{\partial \nu} = \dfrac{\partial w}{\partial \nu} = 0, & x \in \partial\Omega, t > 0, \\
a(x,0) = a_0(x) = u_0(x)e^{-\xi w_0(x)}, \; \sigma v_0(x,0) = \sigma v_0(x), \quad w(x,0) = w_0(x), \quad x \in \Omega.
\end{cases}
\tag{3.4.1}
$$

We note that (3.1.1) and (3.4.1) are equivalent within the concept of classical solutions.

The following result is concerned with the local existence and uniqueness of classical solutions to the problem (3.4.1), along with a convenient extensibility criterion for such solutions.

Lemma 3.14 *Let $\chi > 0, \xi > 0, \mu > 0$ and $r > 0$, and suppose that u_0, v_0 and w_0 satisfy (3.1.2) with some $\vartheta \in (0, 1)$. Then the problem (3.4.1) admits a unique classical solution*

$$
\begin{cases}
a \in C^0(\bar{\Omega} \times [0, T_{max})) \cap C^{2,1}(\bar{\Omega} \times (0, T_{max})) \\
v \in C^0(\bar{\Omega} \times [0, T_{max})) \cap C^{2,1}(\bar{\Omega} \times (0, T_{max})) \\
\quad w \in C^{2,1}(\bar{\Omega} \times [0, T_{max}))
\end{cases}
\tag{3.4.2}
$$

with $a \geq 0, v \geq 0$ and $0 \leq w \leq A := \max\{1, \|w_0\|_{L^\infty(\Omega)}\}$, where T_{max} denotes the maximal existence time. In addition, if $T_{max} < +\infty$, then

$$\|a(\cdot, t)\|_{L^\infty(\Omega)} + \|\nabla w(\cdot, t)\|_{L^5(\Omega)} \to \infty \quad as \quad t \nearrow T_{max}. \tag{3.4.3}$$

Proof Invoking well-established fixed point arguments and applying the standard parabolic regularity theory, one can readily verify the local existence and uniqueness of classical solutions, as well as the extensibility criterion (3.4.3) (see Pang and Wang (2017); Tao and Winkler (2014a, b) for instance). With the help of the maximum principle, we can also verify the asserted nonnegativity of the solutions.

It should be pointed out that the extensibility criterion in (3.4.3) involves the L^5-norm of $|\nabla w|$. Although the L^5-norm of $|\nabla w|$ is time-dependent, it is sufficient to enable us to apply standard parabolic regularity theory to the first equation of (3.4.1) in the two-dimensional setting (see Lemma 2.2 of Pang and Wang (2017) and Tao and Winkler (2014b) for instance).

For the classical solution of (3.4.1), the following observation will be used frequently below.

Lemma 3.15 *Let (a, v, w) be the classical solution of (3.4.1) in $\Omega \times [0, T_{max})$. Then for any $p > 1$, we have*

$$\frac{d}{dt} \int_{\Omega} e^{\xi w} a^p + \frac{2(p-1)}{p} \int_{\Omega} e^{\xi w} |\nabla a^{\frac{p}{2}}|^2 + (p\mu - (p-1)\xi \eta A) \int_{\Omega} a^{p+1} e^{2\xi w}$$

$$\leq \frac{\chi^2 p(p-1)}{2} \int_{\Omega} e^{\xi w} a^p |\nabla v|^2 + \xi A(p-1) \int_{\Omega} e^{\xi w} a^p v$$

$$+ (\mu pr + \xi \eta A^2 (p-1)) \int_{\Omega} e^{\xi w} a^p$$

$$(3.4.4)$$

with $A = \max\{1, \|w_0\|_{L^\infty(\Omega)}\}$.

Proof Testing the first equation in (3.4.1) by a^{p-1} with $p > 1$ and integrating by parts yields

$$\frac{d}{dt} \int_{\Omega} e^{\xi w} a^p + \frac{4(p-1)}{p} \int_{\Omega} e^{\xi w} |\nabla a^{\frac{p}{2}}|^2 + (p\mu - (p-1)\xi \eta A) \int_{\Omega} a^{p+1} e^{2\xi w}$$

$$\leq \chi p(p-1) \int_{\Omega} e^{\xi w} a^{p-1} \nabla a \cdot \nabla v + \xi A(p-1) \int_{\Omega} e^{\xi w} a^p v$$

$$+ (\mu pr + \xi \eta A^2 (p-1)) \int_{\Omega} e^{\xi w} a^p.$$

$$(3.4.5)$$

Here, we note that $0 \leq w \leq A$ in $\Omega \times [0, T_{max})$. By the Young inequality, we estimate

$$\chi p(p-1) \int_{\Omega} e^{\xi w} a^{p-1} \nabla a \cdot \nabla v$$

$$\leq \frac{p(p-1)}{2} \int_{\Omega} e^{\xi w} a^{p-2} |\nabla a|^2 + \frac{\chi^2 p(p-1)}{2} \int_{\Omega} e^{\xi w} a^p |\nabla v|^2.$$

This together with (3.4.5) proves (3.4.4).

3.4.2 Global Existence in Two-Dimensional Domains

According to Lemma 2.6, the key step in the proof of Theorem 3.2 is to establish a priori estimates of $\|a(\cdot, t)\|_{L^\infty(\Omega)}$ and $\|\nabla w(\cdot, t)\|_{L^5(\Omega)}$. As pointed out in Tao and Winkler (2014b), one essential analytic difficulty stems from the fact that the chemotaxis and haptotaxis terms in the first equation in (3.1.1) require different L^p-estimate techniques, since ECM density satisfies an ordinary differential equation (ODE) whereas MDE concentration satisfies a parabolic equation (PDE). This part establishes the crucial a priori estimates of solutions via identifying a certain dissipative property of the functionals $\int_{\Omega} e^{\xi w} a^2$ and $\int_{\Omega} e^{\xi w} a \ln a$ with $a = e^{-\xi w} u$.

1. The Case of $\sigma = 1$

According to the above local existence result, $(u(\cdot, s), v(\cdot, s), w(\cdot, s)) \in (C^2(\bar{\Omega}))^3$ for any $s \in (0, T_{max})$. Hence without loss of generality, we may assume that there exists a constant $C > 0$ such that

$$\|u_0\|_{C^2(\bar{\Omega})} + \|v_0\|_{C^2(\bar{\Omega})} + \|w_0\|_{C^2(\bar{\Omega})} \le C. \tag{3.4.6}$$

From now on, (u, v, w) is the unique maximal solution provided by Lemma 3.14. In order to avoid regularity problems, we assume in the rest of this section that the initial data satisfies (3.1.2). Some basic but important properties of solutions of (3.1.1) are summarized in the next lemmas.

Lemma 3.16 *Let (u, v, w) be the classical solution of (3.1.1) with $\sigma = 1$. Then we have*

(i) $\|u(\cdot, t)\|_{L^1(\Omega)} \le m_0 := \max\{r|\Omega|, \|u_0\|_{L^1(\Omega)}\}$ *for all* $t \in (0, T_{max})$;

(ii) $\int_t^{t+\tau} \|u(\cdot, s)\|^2_{L^2(\Omega)} ds \le m_1 := r^2|\Omega| + \dfrac{2m_0}{\mu}$ *for any* $0 < \tau \le \min\{1, \frac{T_{max}}{3}\}$ *and all* $t \in (0, T_{max} - \tau)$;

(iii) $\|v(\cdot, t)\|_{L^1(\Omega)} \le \tilde{m}_0 := \max\{m_0, \|v_0\|_{L^1(\Omega)}\}$ *for all* $t \in (0, T_{max})$;

(iv) $\|\nabla v(\cdot, t)\|^2_{L^2(\Omega)} \le m_2 := \dfrac{r\mu + 2}{\mu} m_0 + \|\nabla v_0\|^2_{L^2(\Omega)}$ *for all* $t \in (0, T_{max})$;

(v) $\int_t^{t+\tau} \|\Delta v(\cdot, s)\|^2_{L^2(\Omega)} ds \le m_3 := m_2 + m_1$ *for any* $0 < \tau \le \min\{1, \frac{T_{max}}{3}\}$ *and all* $t \in (0, T_{max} - \tau)$.

Proof (i) Integrating the first equation in (3.1.1) with respect to $x \in \Omega$ yields

$$\frac{d}{dt} \int_\Omega u(x, t) \le r\mu \int_\Omega u(x, t) - \mu \int_\Omega u^2(x, t), \tag{3.4.7}$$

since $w \ge 0$ by Lemma 3.14. Moreover, by $2\mu r u \le \mu u^2 + \mu r^2$, we get

$$\frac{d}{dt} \int_\Omega u(x, t) + r\mu \int_\Omega u(x, t) \le \mu r^2 |\Omega|,$$

which implies that $\|u(\cdot, t)\|_{L^1(\Omega)} \le \max\{r|\Omega|, \|u_0\|_{L^1(\Omega)}\}$.

(ii) By (3.4.7) and the Cauchy–Schwartz inequality, we also have

$$\frac{d}{dt} \int_\Omega u(x, t) + \frac{\mu}{2} \int_\Omega u^2(x, t) \le \frac{\mu r^2}{2} |\Omega|. \tag{3.4.8}$$

Then we integrate (3.4.8) over $(t, t + \tau)$ to get

$$\frac{\mu}{2} \int_t^{t+\tau} \int_\Omega u^2 \le \frac{r^2\mu}{2} |\Omega| \tau + \int_\Omega u(x, t),$$

which along with (i) yields (ii) of the lemma.

(iii) Integrating the second equation in (3.1.1) with respect to $x \in \Omega$ yields

$$\frac{d}{dt} \int_\Omega v(x, t) + \int_\Omega v(x, t) \le \int_\Omega u(x, t) \le \sup_{t \ge 0} \int_\Omega u(x, t).$$

So (iii) follows from the nonnegativity of v and (i).

(iv) Multiplying the second equation in (3.1.1) by $-\Delta v$ and integrating over Ω, we find

$$\frac{1}{2} \frac{d}{dt} \int_\Omega |\nabla v(x, t)|^2 + \int_\Omega |\Delta v(x, t)|^2 + \int_\Omega |\nabla v(x, t)|^2$$
$$= -\int_\Omega u \Delta v$$
$$\le \frac{1}{2} \int_\Omega |\Delta v(x, t)|^2 + \frac{1}{2} \int_\Omega u^2(x, t)$$

and thus

$$\frac{d}{dt} \int_\Omega |\nabla v(x, t)|^2 + \int_\Omega |\Delta v(x, t)|^2 + \int_\Omega |\nabla v(x, t)|^2 \le \int_\Omega u^2(x, t). \qquad (3.4.9)$$

Combining (3.4.9) with (3.4.7), we can obtain

$$\frac{d}{dt} \int_\Omega (u(x, t) + \mu |\nabla v(x, t)|^2) + \int_\Omega (u(x, t) + \mu |\nabla v(x, t)|^2)$$
$$\le (r\mu + 1) \int_\Omega u(x, t) \qquad (3.4.10)$$
$$\le (r\mu + 1)m_0,$$

which, together with the Gronwall lemma, yields

$$\mu \int_\Omega |\nabla v(x, t)|^2 \le (r\mu + 2)m_0 + \mu \|\nabla v_0\|_{L^2(\Omega)}^2$$

and hence (iv) holds.

(v) In view of (3.4.9), we have

$$\int_t^{t+\tau} \|\Delta v(\cdot, s)\|_{L^2(\Omega)}^2 ds \le \int_\Omega |\nabla v(x, t)|^2 + \int_t^{t+\tau} \|u(\cdot, s)\|_{L^2(\Omega)}^2 ds.$$

So (v) follows from (ii) and (iv).

Lemma 3.17 *Let (a, v, w) be a classical solution of (3.4.1) with $\sigma = 1$ in $(0, T_{max})$. Then there exists some $C > 0$ such that*

$$\|a(\cdot, t)\|_{L^3(\Omega)} \le C \qquad (3.4.11)$$

is valid for all $t \in (0, T_{max})$.

Proof Applying Lemma 3.15 with $p = 2$, one can find $k_1(A) > 0$ such that

$$\frac{d}{dt} \int_\Omega e^{\xi w} a^2 + \int_\Omega e^{\xi w} |\nabla a|^2 + (2\mu - \xi \eta A) \int_\Omega e^{2\xi w} a^3$$

$$\leq \chi^2 \int_\Omega e^{\xi w} a^2 |\nabla v|^2 + k_1(A) \int_\Omega e^{\xi w} a^2 + k_1(A) \int_\Omega e^{\xi w} a^2 v.$$

Therefore, by means of the Young inequality, we can get

$$\frac{d}{dt} \int_\Omega e^{\xi w} a^2 + \int_\Omega e^{\xi w} |\nabla a|^2 + \int_\Omega e^{\xi w} a^2$$

$$\leq k_2 \|a\|_{L^4(\Omega)}^2 \|\nabla v\|_{L^4(\Omega)}^2 + k_2 \|a\|_{L^3(\Omega)}^3 + k_2 \|v\|_{L^3(\Omega)}^3 + k_2,$$

(3.4.12)

where $k_2 > 0$ is the constant only depending upon A, ξ, η, χ.

On applying Lemma 3.3 with $n = 2$, we have

$$\|a\|_{L^4(\Omega)}^2 \leq k_3 \|\nabla a\|_{L^2(\Omega)} \|a\|_{L^2(\Omega)} + k_3 \|a\|_{L^2(\Omega)}^2$$

and

$$\|\nabla v\|_{L^4(\Omega)}^2 \leq k_3 \|\Delta v\|_{L^2(\Omega)} \|\nabla v\|_{L^2(\Omega)} + k_3 \|\nabla v\|_{L^2(\Omega)}^2,$$

which along with Lemma 3.16 (iv), implies that $\|\nabla v\|_{L^4(\Omega)}^2 \leq k_4 \|\Delta v\|_{L^2(\Omega)} + k_4$. Hence, combining above inequalities and by the Young inequality, we have

$$k_2 \|a\|_{L^4(\Omega)}^2 \|\nabla v\|_{L^4(\Omega)}^2 \leq \frac{1}{2} \|\nabla a\|_{L^2(\Omega)}^2 + k_5 \|a\|_{L^2(\Omega)}^2 (1 + \|\Delta v\|_{L^2(\Omega)}^2). \quad (3.4.13)$$

Therefore, inserting (3.4.13) into (3.4.12), and noting the fact $\|v\|_{L^3(\Omega)} \leq \|v\|_{W^{1,2}(\Omega)} \leq k_6$ by Lemma 3.16 (iv) (iii), we can conclude that

$$\frac{d}{dt} \int_\Omega e^{\xi w} a^2 + \frac{1}{2} \int_\Omega e^{\xi w} |\nabla a|^2 + \int_\Omega e^{\xi w} a^2$$

$$\leq k_7 \|\Delta v\|_{L^2(\Omega)}^2 \int_\Omega e^{\xi w} a^2 + k_7 \int_\Omega a^3 + k_7.$$

(3.4.14)

Now applying Lemma 3.3 and the Young inequality, we have

$$\|a\|_{W^{1,2}(\Omega)}^2 \geq \frac{1}{\varepsilon} \|a\|_{L^3(\Omega)}^3 - \frac{C_{GN}^6}{\varepsilon^2} \|a\|_{L^2(\Omega)}^4$$

for any $\varepsilon > 0$, which after inserting into (3.4.14) and taking $\varepsilon = \frac{1}{4k_7}$ says that

$$\frac{d}{dt}\int_\Omega e^{\xi w}a^2 + k_7\int_\Omega a^3 \le k_7\|\Delta v\|^2_{L^2(\Omega)}\int_\Omega e^{\xi w}a^2 + k_7 + k_8\|a\|^4_{L^2(\Omega)}$$

with $k_8 = 16k_7^2 C^6_{GN} + k_7 e^{\xi A}$.

In view of $a^3 \ge \frac{1}{\varepsilon}a^2 - \frac{1}{\varepsilon^3}$ for any $\varepsilon > 0$, we can obtain

$$\frac{d}{dt}\int_\Omega e^{\xi w}a^2 + \frac{1}{\varepsilon}\int_\Omega e^{\xi w}a^2$$

$$\le (k_7\|\Delta v\|^2_{L^2(\Omega)} + k_8\|a\|^2_{L^2(\Omega)})\int_\Omega e^{\xi w}a^2 + k_7 + \frac{e^{3\xi A}}{\varepsilon^3 k_7^2}|\Omega|. \tag{3.4.15}$$

Now, let $\tau = \min\{1, \frac{T_{max}}{6}\}$ and $\varepsilon = \frac{\tau}{1+k_7m_3+k_8m_1}$. Then (3.4.15) implies that writing $a(t) := \frac{1}{\varepsilon}$, $b(t) := k_7\|\Delta v\|^2_{L^2(\Omega)} + k_8\|a\|^2_{L^2(\Omega)}$ and $c(t) := k_7 + \frac{e^{3\xi A}}{\varepsilon^3 k_7^2}|\Omega|$, function $y(t) := \int_\Omega e^{\xi w}a^2$ satisfies

$$y'(t) + a(t)y(t) \le b(t)y(t) + c(t). \tag{3.4.16}$$

Hence, the application of Lemma 3.5 to (3.4.16) with $b_1 = k_7m_3 + k_8m_1$, $k_1 = k_7 + \frac{e^{3\xi A}}{\varepsilon^3 k_7^2}|\Omega|$ and $\rho = 1$ yields

$$\int_\Omega e^{\xi w}a^2 \le C := e^{b_1}e^{\xi A}\|a_0\|^2_{L^2(\Omega)} + \frac{k_1 e^{2b_1}}{1-e^{-1}} + k_1 e^{b_1}.$$

Now we turn to estimate $\|a(\cdot, t)\|_{L^3(\Omega)}$. Applying the Young inequality, one obtains from (3.4.4) that

$$\frac{d}{dt}\int_\Omega e^{\xi w}a^3 + \frac{3}{2}\int_\Omega e^{\xi w}a|\nabla a|^2 + \int_\Omega e^{\xi w}a^3$$

$$\le 6\chi^2\int_\Omega e^{\xi w}a^3|\nabla v|^2 + 3(\xi\eta A - \mu)\int_\Omega e^{2\xi w}a^4 + 2\xi A\int_\Omega e^{\xi w}a^3 v + 2\xi\eta A^2\int_\Omega e^{\xi w}a^3 \tag{3.4.17}$$

$$\le 6\chi^2\int_\Omega e^{\xi w}a^3|\nabla v|^2 + k_9(\sup_{0\le t<T_{max}}\|v(\cdot, t)\|_{L^\infty(\Omega)} + 1)\int_\Omega e^{\xi w}a^3 + k_9\int_\Omega e^{\xi w}a^4.$$

On the other hand, by $\int_\Omega a^2 \le C$ and Lemma 3.6, one can find some constant $k_{10} > 0$ such that $\|v(\cdot, t)\|_{L^\infty(\Omega)} \le k_{10}$ and $\|\nabla v(\cdot, t)\|_{L^8(\Omega)} \le k_{10}$ for all $t < T_{max}$. Hence, (3.4.17) shows that there exists $k_{11} > 0$ such that

$$\frac{d}{dt}\int_\Omega e^{\xi w}a^3 + \frac{2}{3}\int_\Omega e^{\xi w}|\nabla a^{\frac{3}{2}}|^2 + \int_\Omega e^{\xi w}a^3 \le k_{11}\int_\Omega a^4 + k_{11}. \tag{3.4.18}$$

By means of the Gagliardo–Nirenberg inequality, we have

$$\|a\|^4_{L^4(\Omega)} = \|a^{\frac{3}{2}}\|^{\frac{8}{3}}_{L^{\frac{8}{3}}(\Omega)} \le 2C^{\frac{8}{3}}_{c3.2-2.1}\|\nabla a^{\frac{3}{2}}\|^{\frac{4}{3}}_{L^2(\Omega)}\|a^{\frac{3}{2}}\|^{\frac{4}{3}}_{L^{\frac{4}{3}}(\Omega)} + 2C^{\frac{8}{3}}_1\|a^{\frac{3}{2}}\|^{\frac{8}{3}}_{L^{\frac{4}{3}}(\Omega)}.$$

Therefore by (3.4.18), $\|a(\cdot, t)\|_{L^2(\Omega)} \leq C$ and the Young inequality, one can arrive at

$$\frac{d}{dt} \int_\Omega e^{\xi w} a^3 + \int_\Omega e^{\xi w} a^3 \leq k_{12}.$$

Finally, (3.4.11) follows from the Gronwall inequality.

On applying Lemma 3.6, the following result is an immediate consequence of $0 \leq w(x, t) \leq A$ and Lemma 3.17.

Lemma 3.18 *Under the same assumptions as in Theorem 3.2, there exists $C > 0$ such that the classical solution (u, v, w) of (3.1.1) satisfies*

$$\|v(\cdot, t)\|_{W^{1,\infty}(\Omega)} \leq C \quad \text{for all} \quad t \in (0, T_{max}). \tag{3.4.19}$$

Note that $\|\nabla v(\cdot, t)\|_{L^\infty(\Omega)}$ is bounded by (3.4.19). However, $\|\nabla w(\cdot, t)\|_{L^\infty(\Omega)}$ might become unbounded. Therefore, Lemma A.1 of Tao and Winkler (2012a) as the result of the well-known Moser–Alikakos iteration Alikakos (1979) cannot be directly applied to the first equation in (3.1.1) to get the boundedness of $\|u(\cdot, t)\|_{L^\infty(\Omega)}$. At this position, arguing as in Lemma 3.6 of Pang and Wang (2017), Lemma 4.2 of Tao (2011) or Lemma 3.5 of Tao and Winkler (2014b), we can establish the following estimates.

Lemma 3.19 *Under the assumptions of Theorem 3.2, there exists $C > 0$ such that the classical solution (u, v, w) of (3.1.1) satisfies*

$$\|u(\cdot, t)\|_{L^\infty(\Omega)} \leq C \quad \text{for all} \quad t \in (0, T_{max}). \tag{3.4.20}$$

Lemma 3.20 *Under the assumptions of Theorem 3.2, for all $T > 0$ there exists $C(T) > 0$ such that the classical solution (u, v, w) of (3.1.1) satisfies*

$$\|\nabla w(\cdot, t)\|_{L^5(\Omega)} \leq C(T) \quad \text{for all} \quad t \in (0, \min\{T, T_{max}\}). \tag{3.4.21}$$

We are now in the position to prove Theorem 3.2 in the case $\sigma = 1$.

Proof of Theorem 3.2 *in the case of $\sigma = 1$.* By a rather standard argument, we can show the global existence of classical solutions to (3.1.1) with $\sigma = 1$, i.e., $T_{max} = +\infty$. In view of Lemma 3.18, $\|a(\cdot, t)\|_{L^\infty(\Omega)}$ is bounded uniformly with respect to $t \in (0, T_{max})$. Combining this with Lemma 3.20, we can obtain $\|\nabla w(\cdot, t)\|_{L^5(\Omega)} \leq C(T_{max})$ for all $t \in (0, T_{max})$. Hence, the statement of global existence and boundedness of classical solutions to (3.1.1) is a straightforward consequence of Lemma 3.14. Now by retracing the proof of Lemma 3.17, one can find that $\tau = 1$, and thereby there exists a constant $C > 0$ which is time-independent such that $\|a(\cdot, t)\|_{L^3(\Omega)} \leq C$ for all $t > 0$. Therefore, $\|a(\cdot, t)\|_{L^\infty(\Omega)} \leq C$ for some $C > 0$ and all $t > 0$.

2. The Case of $\sigma = 0$

Now we turn to proving Theorem 3.2 in the case $\sigma = 0$.

Lemma 3.21 *Let (u, v, w) be the classical solution of (3.1.1) with $\sigma = 0$. Then we have*

(i) $\|u(\cdot, t)\|_{L^1(\Omega)} \leq m_0$ *for all* $t \in (0, T_{max})$;

(ii) $\displaystyle\int_t^{t+\tau} \|u(\cdot, s)\|_{L^2(\Omega)}^2 ds \leq m_1$ *for any* $0 < \tau \leq \min\{1, \frac{T_{max}}{3}\}$ *and all* $t \in (0, T_{max} - \tau)$;

(iii) $\|v(\cdot, t)\|_{L^1(\Omega)} \leq m_0$ *for all* $t \in (0, T_{max})$;

(iv) $\displaystyle\int_t^{t+\tau} \|\nabla v(\cdot, s)\|_{L^2(\Omega)}^2 ds \leq m_1$ *for any* $0 < \tau \leq \min\{1, \frac{T_{max}}{3}\}$ *and all* $t \in (0, T_{max} - \tau)$;

(v) $\displaystyle\int_t^{t+\tau} \|\Delta v(\cdot, s)\|_{L^2(\Omega)}^2 ds \leq m_1$ *for any* $0 < \tau \leq \min\{1, \frac{T_{max}}{3}\}$ *and all* $t \in (0, T_{max} - \tau)$.

Proof We note that we only need to show (iii), (iv) and (v) here. Integrating the elliptic equation in (3.1.1) with respect to $x \in \Omega$ yields

$$\int_\Omega v(x, t) = \int_\Omega u(x, t), \tag{3.4.22}$$

so (iii) is the consequence of (i).

Testing the equation for v in (3.1.1) by $-\Delta v$ and integrating over Ω, we can see that

$$\int_\Omega |\Delta v(x, t)|^2 + \int_\Omega |\nabla v(x, t)|^2 = -\int_\Omega u \Delta v \leq \frac{1}{2}\int_\Omega |\Delta v(x, t)|^2 + \frac{1}{2}\int_\Omega u^2(x, t)$$

and thus $\int_\Omega |\Delta v(x, t)|^2 + \int_\Omega |\nabla v(x, t)|^2 \leq \int_\Omega u^2(x, t)$. Hence (iv) and (v) follow from (ii). $\blacksquare$

As pointed out in Tao and Winkler (2014b), with the help of a well-known regularity result on semilinear second-order elliptic equations, one can only infer that $\|\nabla v(\cdot, t)\|_{L^q(\Omega)} \leq C$ for any $1 < q < 2$ and all $t \in (0, T_{max})$ from Lemma 3.21 (i) and the second equation in (3.1.1) with $\sigma = 0$, hence in order to allow for the choice $q = 2$, some additional efforts are needed. It should be remarked that for (3.1.1) with $\sigma = 1$, $\|\nabla v(\cdot, t)\|_{L^2(\Omega)} \leq C$ can be obtained directly (see Lemma 3.16 (iv)). It is observed in Tao and Winkler (2014b) that a key step toward this is to estimate $\int_\Omega u(\cdot, t) \ln u(\cdot, t)$ (see (3.16) of Tao and Winkler (2014b)). However, the estimate of the latter involves the compensation of the term $\int_\Omega \nabla u \cdot \nabla w$, which makes the bound of $\int_\Omega u(\cdot, t) \ln u(\cdot, t)$ to be time-dependent. Here, we make use of Lemmas 3.5 and 3.21 to derive the global boundedness of $\int_\Omega u(\cdot, t) \ln u(\cdot, t)$.

Lemma 3.22 *There exists some $C > 0$ such that for any (u_0, w_0) fulfilling (3.1.2), the corresponding classical solution (u, v, w) of (3.1.1) with $\sigma = 0$ satisfies*

$$\int_\Omega u(\cdot, t) \ln u(\cdot, t) \le C \text{ for all } t \in (0, T_{max}).$$

Proof From the first equation in (3.4.1), it follows

$$(ae^{\xi w})_t = \nabla \cdot (e^{\xi w} \nabla a) - \chi \nabla \cdot (e^{\xi w} a \nabla v) + \mu a e^{\xi w}(r - w - ae^{\xi w})$$

and thus

$$
\begin{aligned}
\frac{d}{dt} & \int_\Omega e^{\xi w} a \ln a + \int_\Omega e^{\xi w} \frac{|\nabla a|^2}{a} + \mu \int_\Omega a^2 e^{2\xi w} \ln a \\
& = \int_\Omega (e^{\xi w} a)_t \ln a + \int_\Omega e^{\xi w} a_t + \int_\Omega e^{\xi w} \frac{|\nabla a|^2}{a} + \mu \int_\Omega a^2 e^{2\xi w} \ln a \\
& = \chi \int_\Omega e^{\xi w} \nabla a \cdot \nabla v + \int_\Omega a e^{\xi w}[\mu(r - w - ae^{\xi w}) - \xi \eta w(1 - w - ae^{\xi w})]
\end{aligned}
\tag{3.4.23}
$$

$$
\begin{aligned}
& + \int_\Omega a e^{\xi w}[\mu \ln a(r - w) + \xi v w] \\
& \le \frac{1}{2} \int_\Omega e^{\xi w} \frac{|\nabla a|^2}{a} + \frac{\chi^2}{2} \int_\Omega e^{\xi w} a |\nabla v|^2 + \frac{\mu}{4} \int_\Omega a^2 e^{2\xi w} \ln a + k_1
\end{aligned}
$$

for some $k_1 > 0$ and $t \in (0, T_{max})$. Here, we may use the facts that $0 \le w \le A := \max\{\|w_0\|, 1\}$, $a^2 \le \varepsilon a^2 \ln a + e^{\frac{2}{\varepsilon}}$, $a \ln a \le \varepsilon a^2 \ln a - \varepsilon^{-1} \ln \varepsilon$ and $a \le \varepsilon a^2 \ln a + 2e^{\frac{2}{\varepsilon}}$ for any $\varepsilon \in (0, 1)$.

By Young's inequality and applying $a^2 \le \varepsilon a^2 \ln a + e^{\frac{2}{\varepsilon}}$ again, we have

$$\frac{\chi^2}{2} \int_\Omega e^{\xi w} a |\nabla v|^2 \le \varepsilon \int_\Omega |\nabla v|^4 + \frac{\mu}{4} \int_\Omega a^2 e^{2\xi w} \ln a + k_2(\varepsilon). \tag{3.4.24}$$

Along with $ae^{\xi w} \ln a \le a^2 e^{2\xi w} \ln a + \xi A e^{2\xi A}$, combining (3.4.24) with (3.4.23) gives

$$
\begin{aligned}
\frac{d}{dt} & \int_\Omega e^{\xi w} a \ln a + \frac{1}{2} \int_\Omega e^{\xi w} \frac{|\nabla a|^2}{a} + \frac{\mu}{2} \int_\Omega a e^{\xi w} \ln a \\
& \le \varepsilon \int_\Omega |\nabla v|^4 + k_1 + k_2(\varepsilon) + \frac{\mu \xi A}{2} e^{2\xi A} |\Omega|,
\end{aligned}
\tag{3.4.25}
$$

which along with the Gagliardo–Nirenberg interpolation inequality

$$\|\nabla v\|_{L^4(\Omega)}^4 \le C_{c3.2-2.1}^4 (\|\Delta v\|_{L^2(\Omega)}^2 + \|\nabla v\|_{L^2(\Omega)}^2) \|\nabla v\|_{L^2(\Omega)}^2$$

and Lemma 3.7 entails

$$\frac{d}{dt}\int_{\Omega} e^{\xi w} a \ln a + \frac{1}{2}\int_{\Omega} e^{\xi w}\frac{|\nabla a|^2}{a} + \frac{\mu}{2}\int_{\Omega} e^{\xi w} a \ln a$$

$$\leq \varepsilon C_{c3.2-2.1}^4 (\|\Delta v\|_{L^2(\Omega)}^2 + \|\nabla v\|_{L^2(\Omega)}^2)\|\nabla v\|_{L^2(\Omega)}^2 + k_3(\varepsilon)$$

$$\leq \varepsilon C_{c3.2-2.1}^4 (\|\Delta v\|_{L^2(\Omega)}^2 + \|\nabla v\|_{L^2(\Omega)}^2)(\alpha\int_{\Omega} u \ln u + \beta) + k_3(\varepsilon).$$

Therefore, by the fact $u \ln u = e^{\xi w} a \ln a + a\xi w e^{\xi w} \leq 2e^{\xi w} a \ln a + k_4$ for some $k_4 > 0$, we have

$$\frac{d}{dt}(\int_{\Omega} e^{\xi w} a \ln a + e^{\xi A}|\Omega|) + \frac{1}{2}\int_{\Omega} e^{\xi w}\frac{|\nabla a|^2}{a} + \frac{\mu}{2}(\int_{\Omega} e^{\xi w} a \ln a + e^{\xi A}|\Omega|)$$

$$\leq 2\varepsilon C_{c3.2-2.1}^4 (\|\Delta v\|_{L^2(\Omega)}^2 + \|\nabla v\|_{L^2(\Omega)}^2)(\alpha\int_{\Omega} e^{\xi w} a \ln a + \alpha|\Omega|k_4 + \beta) + k_5(\varepsilon)$$

$$\leq \varepsilon k_6(\|\Delta v\|_{L^2(\Omega)}^2 + \|\nabla v\|_{L^2(\Omega)}^2)(\int_{\Omega} e^{\xi w} a \ln a + e^{\xi A}|\Omega|)$$

$$+ k_7(\|\Delta v\|_{L^2(\Omega)}^2 + \|\nabla v\|_{L^2(\Omega)}^2) + k_5(\varepsilon)$$

for $t \in (0, T_{max})$.

Now, we let the nonnegative functions $a(t)$, $b(t)$ and $c(t)$ be defined by $a(t) := \frac{\mu}{2}$, $b(t) := \varepsilon k_6(\|\Delta v\|_{L^2(\Omega)}^2 + \|\nabla v\|_{L^2(\Omega)}^2)$, $c(t):=k_7(\|\Delta v\|_{L^2(\Omega)}^2 + \|\nabla v\|_{L^2(\Omega)}^2) + k_5(\varepsilon)$. Then, we see that the nonnegative function

$$y(t) := \int_{\Omega} e^{\xi w} a \ln a + e^{\xi A}|\Omega|$$

satisfies $y'(t) + a(t)y(t) \leq b(t)y(t) + c(t)$. With the help of Lemma 3.21 (iv) and (v), we can conclude that when fixing $\tau := \min\{1, \frac{T_{max}}{6}\}$ and taking $\varepsilon = \frac{\mu\tau}{8k_6m_1}$, applying Lemma 3.5 with $\rho = \frac{\mu}{4}\tau$ yields $\int_{\Omega} e^{\xi w} a \ln a \leq C(\tau)$ with some constant $C(\tau) > 0$, which completes the proof of this lemma as $0 \leq w \leq A$.

Based on the above $L\log L(\Omega)$ estimate of u, we have the following.

Corollary 3.1 *There exists some $C > 0$ such that for any (u_0, w_0) fulfilling (3.1.2), the corresponding classical solution (u, v, w) of (3.1.1) with $\sigma = 0$ satisfies*

$$\|v(\cdot, t)\|_{W^{1,2}(\Omega)} \leq C \quad \text{for all } t \in (0, T_{max}).$$

Proof This follows from Lemma 3.21 (i), Lemmas 3.22 and 3.7 immediately.

At this position, we can proceed as in the proof of Lemmas 3.2–3.5 or Lemmas 3.10–3.12 of Tao and Winkler (2014b) to derive the a priori estimates below. It should be pointed out that the bounds of $\int_{\Omega} a \ln a$ play an essential role in the proof of Lemma 3.11 in Tao and Winkler (2014b), while it is not necessary for our argument in the proof of Lemma 3.17 whenever $\|v(\cdot, t)\|_{W^{1,2}(\Omega)}$ is bounded.

Lemma 3.23 *Let* (u, v, w) *be the classical solution of (3.1.1) with* $\sigma = 0$. *Then there exists some* $C > 0$ *such that*

$$\|u(\cdot, t)\|_{L^\infty(\Omega)} \le C \tag{3.4.26}$$

is valid for all $t \in (0, T_{max})$.

Proof of Theorem 3.2 *in the case of* $\sigma = 0$. Since the proof is very similar to that of Theorem 3.2 in the case $\sigma = 1$, we omit it here.

3.4.3   Global Existence in Three-Dimensional Domains

In this subsection, inspired by Lankeit (2015); Winkler (2011b), we prove Theorem 3.3. The main idea of the proof is to verify that the quantity $\int_\Omega a^2(t) + \int_\Omega |\nabla v(t)|^4$ satisfies an autonomous ordinary differential inequality, and then the comparison argument can be applied to the corresponding ordinary differential equation when $r > 0$ and the initial data are suitably small.

Lemma 3.24 *Let* (a, v, w) *be the classical solution of problem (3.4.1) in* $\Omega \times [0, T_{max})$. *Then*

$$\sigma \frac{d}{dt} \int_\Omega |\nabla v|^4 + \int_\Omega |\nabla |\nabla v|^2|^2 + 4 \int_\Omega |\nabla v|^4 \le 7 \int_\Omega e^{2\xi w} a^2 |\nabla v|^2. \tag{3.4.27}$$

Proof We refer the interested reader to Lemma 3.2 of Tao and Winkler (2015b), (24)–(26) in Viglialoro (2017), and Lemma 4.6 in Lankeit (2015) for the proof.

Proof of Theorem 3.3 *in the case of* $\sigma = 1$. By Lemma 3.15, we have

$$\frac{d}{dt} \int_\Omega e^{\xi w} a^2 + \int_\Omega e^{\xi w} |\nabla a|^2 + (2\mu - \xi \eta A) \int_\Omega a^3 e^{2\xi w}$$
$$\le \chi^2 \int_\Omega e^{\xi w} a^2 |\nabla v|^2 + \xi A \int_\Omega e^{\xi w} a^2 v + (2\mu r + \xi \eta A^2) \int_\Omega e^{\xi w} a^2.$$

Furthermore, the Young inequality entails that there is a constant $k_1 > 0$ depending upon ξ, χ, η and A only such that

$$\frac{d}{dt} \int_\Omega e^{\xi w} a^2 + \int_\Omega e^{\xi w} |\nabla a|^2 + \int_\Omega e^{\xi w} a^2$$
$$\le k_1 \int_\Omega e^{2\xi w} a^3 + \int_\Omega |\nabla v|^6 + \int_\Omega v^3 + r^2 \mu \int_\Omega a,$$

which along with Lemmas 3.4 and 3.16 implies that

$$\frac{d}{dt}\int_\Omega e^{\xi w}a^2 + \frac{1}{2}\int_\Omega e^{\xi w}|\nabla a|^2 + \int_\Omega e^{\xi w}a^2$$

$$\leq k_2\left(\int_\Omega e^{\xi w}a^2\right)^3 + \int_\Omega |\nabla v|^6 + \int_\Omega v^3 + r^2\mu\int_\Omega a + k_3(A)\left(\int_\Omega a\right)^3$$

$$\leq k_2\left(\int_\Omega e^{\xi w}a^2\right)^3 + \int_\Omega |\nabla v|^6 + k_4(\|\nabla v\|_{L^2(\Omega)}^{\frac{12}{5}} + \|v\|_{L^1(\Omega)}^{\frac{12}{5}})\|v\|_{L^1(\Omega)}^{\frac{3}{5}}$$

$$+ r^2\mu\int_\Omega a + k_3(A)\left(\int_\Omega a\right)^3$$

$$\leq k_2\left(\int_\Omega e^{\xi w}a^2\right)^3 + \int_\Omega |\nabla v|^6 + k_4\tilde{m}_0^{\frac{3}{5}}(m_2^{\frac{6}{5}} + \tilde{m}_0^{\frac{12}{5}}) + r^2\mu\int_\Omega a + k_3(A)\left(\int_\Omega a\right)^3.$$

$$(3.4.28)$$

On the other hand, from Lemmas 3.24 and 3.4, it follows that

$$\frac{d}{dt}\int_\Omega |\nabla v|^4 + \int_\Omega |\nabla|\nabla v|^2|^2 + 4\int_\Omega |\nabla v|^4$$

$$\leq \int_\Omega a^3 + k_5(A)\int_\Omega |\nabla v|^6$$

$$\leq \int_\Omega a^3 + \frac{1}{2}\int_\Omega |\nabla|\nabla v|^2|^2 + k_6(A)\left((\int_\Omega |\nabla v|^4)^3 + (\int_\Omega |\nabla v|^4)^{\frac{3}{2}}\right),$$

and thus

$$\frac{d}{dt}\int_\Omega |\nabla v|^4 + \frac{1}{2}\int_\Omega |\nabla|\nabla v|^2|^2 + 4\int_\Omega |\nabla v|^4$$

$$\leq \int_\Omega a^3 + k_6(A)\left((\int_\Omega |\nabla v|^4)^3 + (\int_\Omega |\nabla v|^4)^{\frac{3}{2}}\right).$$

$$(3.4.29)$$

Combining (3.4.29) with (3.4.28) and using Lemma 3.4 again, we can see

$$\frac{d}{dt}(\int_\Omega e^{\xi w}a^2 + \int_\Omega |\nabla v|^4) + \frac{1}{2}\int_\Omega e^{\xi w}|\nabla a|^2 + \frac{1}{2}\int_\Omega |\nabla|\nabla v|^2|^2 + \int_\Omega e^{\xi w}a^2$$

$$+ 4\int_\Omega |\nabla v|^4$$

$$\leq \int_\Omega a^3 + \int_\Omega |\nabla v|^6 + k_2\left(\int_\Omega e^{\xi w}a^2\right)^3 + k_6(A)\left((\int_\Omega |\nabla v|^4)^3 + (\int_\Omega |\nabla v|^4)^{\frac{3}{2}}\right)$$

$$+ k_4\tilde{m}_0^{\frac{3}{5}}(m_2^{\frac{6}{5}} + \tilde{m}_0^{\frac{12}{5}}) + r^2\mu\int_\Omega a + k_3(A)(\int_\Omega a)^3$$

$$\leq \frac{1}{4}\int_\Omega e^{\xi w}|\nabla a|^2 + \frac{1}{4}\int_\Omega |\nabla|\nabla v|^2|^2 + k_7\left(\int_\Omega e^{\xi w}a^2\right)^3$$

$$+ k_8(A)\left((\int_\Omega |\nabla v|^4)^3 + (\int_\Omega |\nabla v|^4)^{\frac{3}{2}}\right)$$

$$+ k_4\tilde{m}_0^{\frac{3}{5}}(m_2^{\frac{6}{5}} + \tilde{m}_0^{\frac{12}{5}}) + r^2\mu\int_\Omega a + k_9(A)(\int_\Omega a)^3.$$

$$(3.4.30)$$

Finally, by the Young inequality, we obtain that

$$k_8(A)\left(\int_\Omega |\nabla v|^4\right)^{\frac{3}{2}} \le 2\int_\Omega |\nabla v|^4 + k_{10}(A)\left(\int_\Omega |\nabla v|^4\right)^3.$$

Hence, applying Lemma 3.16, (3.4.30) shows that $y(t) := \int_\Omega e^{\xi w}a^2 + \int_\Omega |\nabla v|^4$, $t > 0$, satisfies

$$y'(t) + y(t) \le k_{11}(A)y^3(t) + k_4\tilde{m}_0^{\frac{3}{5}}(m_2^{\frac{6}{5}} + \tilde{m}_0^{\frac{12}{5}}) + r^2\mu\int_\Omega a(\cdot, t) + k_9(A)\left(\int_\Omega a(\cdot, t)\right)^3$$

$$\le k_{11}(A)y^3(t) + k_4\tilde{m}_0^{\frac{3}{5}}(m_2^{\frac{6}{5}} + \tilde{m}_0^{\frac{12}{5}}) + r^2\mu m_0 + k_9(A)m_0^3$$

for some $k_{11}(A) > 0$.

Now we can conclude that there is a positive constant r_0 such that function $\Theta(\varsigma) := -\varsigma + k_{11}(A)\varsigma^3 + k_4\tilde{m}_0^{\frac{3}{5}}(m_2^{\frac{6}{5}} + \tilde{m}_0^{\frac{12}{5}}) + r^2\mu m_0 + k_9(A)m_0^3$, $\varsigma \ge 0$, attains its minimum at $\varsigma_0 = (\frac{1}{3k_{11}(A)})^{\frac{1}{2}}$, and $\Theta(\varsigma_0) < 0$ when $r < r_0$, and $\|u_0\|_{L^1(\Omega)}$ and $\|v_0\|_{W^{1,2}(\Omega)}$ are suitably small. In fact, it is observed that $\Theta(\varsigma_0) < 0$ provided that $k_4\tilde{m}_0^{\frac{3}{5}}(m_2^{\frac{6}{5}} + \tilde{m}_0^{\frac{12}{5}}) + r^2\mu m_0 + k_9(A)m_0^3 < \frac{2\varsigma_0}{3}$. To this end, taking

$$r_0 = \min\{1, \frac{1}{|\Omega|}, \frac{\varsigma_0}{2|\Omega|}(\mu + c_9(A) + 2c_4(1 + \frac{1}{\mu}))^{-1}\}$$

and by continuity of the expressions m_0, m_2 and $\tilde{m}_0$, one can verify that $k_4\tilde{m}_0^{\frac{3}{5}}(m_2^{\frac{6}{5}} + \tilde{m}_0^{\frac{12}{5}}) + r^2\mu m_0 + k_9(A)m_0^3 < \frac{2\varsigma_0}{3}$ is indeed valid if $r < r_0$, and $\|u_0\|_{L^1(\Omega)}$ and $\|v_0\|_{W^{1,2}(\Omega)}$ are suitably small. The comparison principle for ordinary differential equations $y'(t) \le \Theta(y(t))$ therefore shows by means of comparison with $y \equiv \varsigma_0$ that $y(t) \le \varsigma_0$ for all $t \ge 0$ when $y(0) \le \varsigma_0$, which can be satisfied whenever $\|u_0\|_{L^2(\Omega)}$ and $\|v_0\|_{W^{1,4}(\Omega)}$ are sufficiently small.

The next step is to obtain a bound for a with respect to the norm in $L^\infty(\Omega)$ by a bootstrap procedure, on the basis of the bounds on $\|a\|_{L^2(\Omega)}$ and $\|v\|_{W^{1,4}(\Omega)}$.

By Lemma 3.15 with $p = 3$, we get

$$\frac{d}{dt}\int_\Omega e^{\xi w}a^3 + \frac{4}{3}\int_\Omega e^{\xi w}|\nabla a^{\frac{3}{2}}|^2 + 3(\mu - \xi\eta A)\int_\Omega a^4 e^{2\xi w}$$

$$\le 3\chi^2\int_\Omega e^{\xi w}a^3|\nabla v|^2 + 2\xi A\int_\Omega e^{\xi w}a^3 v + (3\mu r + 2\xi\eta A^2)\int_\Omega e^{\xi w}a^3.$$

Moreover, due to $W^{1,4}(\Omega) \hookrightarrow L^\infty(\Omega)$ and applying Lemma 3.3, we have

$$\frac{d}{dt} \int_\Omega e^{\xi w} a^3 + \int_\Omega |\nabla a^{\frac{3}{2}}|^2 + \int_\Omega e^{\xi w} a^3$$

$$\leq 3\chi^2 e^{\xi A} \int_\Omega a^3 |\nabla v|^2 + \int_\Omega a^4 + k_{12}(A)$$

$$\leq 3\chi^2 e^{\xi A} (\int_\Omega a^6)^{\frac{1}{2}} (\int_\Omega |\nabla v|^4)^{\frac{1}{2}} + \int_\Omega a^4 + k_{12}(A)$$

$$\leq k_{13}(A) \left((\int_\Omega |\nabla a^{\frac{3}{2}}|^2)^{\frac{6}{7}} (\int_\Omega a^2)^{\frac{3}{14}} + (\int_\Omega a^2)^{\frac{3}{2}} \right) (\int_\Omega |\nabla v|^4)^{\frac{1}{2}}$$

$$+ (\int_\Omega |\nabla a^{\frac{3}{2}}|^2)^{\frac{6}{7}} (\int_\Omega a^2)^{\frac{5}{7}} + (\int_\Omega a^2)^2 + k_{12}(A)$$

$$\leq \frac{1}{2} \int_\Omega |\nabla a^{\frac{3}{2}}|^2 + k_{14}(A),$$

which implies that $\int_\Omega e^{\xi w} a^3 \leq C$ for some $C > 0$. At this position, similarly as in the proof of Theorem 3.2, one can derive $\|a(\cdot, t)\|_{L^\infty(\Omega)} \leq C$ and $\|v(\cdot, t)\|_{W^{1,\infty}(\Omega)} \leq C$ for some time-independent constant $C > 0$ and all $t > 0$, and then $\|\nabla w(\cdot, t)\|_{L^5(\Omega)} \leq C(T)$ for all $t \in (0, T)$, which along with Lemma 3.14 completes the proof of this theorem.

Proof of Theorem 3.3 *in the case of* $\sigma = 0$. Since the proof is similar to that of the case $\sigma = 1$, we may confine ourselves to an outline, giving only details in places which are characteristic for the present setting.

By Lemma 3.15 and Young's inequality, we have

$$\frac{d}{dt} \int_\Omega e^{\xi w} a^2 + \int_\Omega e^{\xi w} |\nabla a|^2 + 2 \int_\Omega e^{\xi w} a^2$$
$$\leq k_1(A) \int_\Omega a^3 + \int_\Omega |\nabla v|^6 + \int_\Omega v^3 + r^2 \mu \int_\Omega a. \tag{3.4.31}$$

On the other hand, from Lemma 3.24, it follows that

$$\int_\Omega |\nabla |\nabla v|^2|^2 + 4 \int_\Omega |\nabla v|^4 \leq k_2(A) \int_\Omega a^2 |\nabla v|^2 \leq k_3(A) \int_\Omega a^3 + \int_\Omega |\nabla v|^6. \tag{3.4.32}$$

Combining (3.4.32) with (3.4.31) and applying Lemma 3.4, we can obtain

$$\frac{d}{dt}\int_\Omega e^{\xi w}a^2 + \int_\Omega e^{\xi w}|\nabla a|^2 + 2\int_\Omega e^{\xi w}a^2 + \int_\Omega |\nabla|\nabla v|^2|^2 + 4\int_\Omega |\nabla v|^4$$

$$\le k_4(A)\int_\Omega a^3 + 2\int_\Omega |\nabla v|^6 + \int_\Omega v^3 + r^2\mu\int_\Omega a$$

$$\le \frac{1}{2}\int_\Omega e^{\xi w}|\nabla a|^2 + \frac{1}{2}\int_\Omega |\nabla|\nabla v|^2|^2 + k_5(A)\left(\int_\Omega e^{\xi w}a^2\right)^3 + k_6(A)(\int_\Omega a)^3$$

$$+ r^2\mu\int_\Omega a + k_7(\int_\Omega |\nabla v|^4)^3 + \int_\Omega |\nabla v|^4 + \int_\Omega v^3$$

$$\le \frac{1}{2}\int_\Omega e^{\xi w}|\nabla a|^2 + \frac{1}{2}\int_\Omega |\nabla|\nabla v|^2|^2 + k_5(A)\left(\int_\Omega e^{\xi w}a^2\right)^3 + k_6(A)(\int_\Omega a)^3$$

$$+ r^2\mu\int_\Omega a + k_7(\int_\Omega |\nabla v|^4)^3 + 2\int_\Omega |\nabla v|^4 + k_8(\int_\Omega v)^3.$$

$$(3.4.33)$$

By the Gagliardo–Nirenberg inequality,

$$\|v\|^3_{L^3(\Omega)} \le C^3_{c3.2-2.1}\|\nabla v\|^{\frac{24}{13}}_{L^4(\Omega)}\|v\|^{\frac{15}{13}}_{L^1(\Omega)} + k_9\|v\|^3_{L^1(\Omega)}$$

$$\le \|\nabla v\|^4_{L^4(\Omega)} + k_{10}\|v\|^3_{L^1(\Omega)}.$$

Hence from (3.4.33), it follows that

$$\frac{d}{dt}\int_\Omega e^{\xi w}a^2 + 2\int_\Omega e^{\xi w}a^2 + 2\int_\Omega |\nabla v|^4$$

$$\le k_5(A)\left(\int_\Omega e^{\xi w}a^2\right)^3 + k_6(A)(\int_\Omega a)^3 + r^2\mu\int_\Omega a + k_7(\int_\Omega |\nabla v|^4)^3 + k_8(\int_\Omega v)^3.$$

$$(3.4.34)$$

On the other hand, applying the standard elliptic regularity theory in the three-dimensional setting to the second equation in (3.4.1), we have

$$\int_\Omega |\nabla v|^4 \le \|v\|^4_{W^{2,2}(\Omega)} \le k_{11}\|e^{\xi w}a\|^4_{L^2(\Omega)}$$

for some $k_{11} > 0$, which combined with (3.4.34) and the Young inequality says that

$$\frac{d}{dt}\int_\Omega e^{\xi w}a^2 + 2\int_\Omega e^{\xi w}a^2$$

$$\le k_5(A)\left(\int_\Omega e^{\xi w}a^2\right)^3 + k_6(A)(\int_\Omega a)^3 + r^2\mu\int_\Omega a + k_{12}(A)\left(\int_\Omega e^{\xi w}a^2\right)^6 + k_8(\int_\Omega v)^3$$

$$\le \int_\Omega e^{\xi w}a^2 + k_{13}(A)\left(\int_\Omega e^{\xi w}a^2\right)^6 + k_{14}(A)(\int_\Omega a)^3 + r^2\mu\int_\Omega a,$$

$$(3.4.35)$$

where we have used the fact $\int_\Omega v = \int_\Omega u$. Therefore along with Lemma 3.21 (i), (3.4.35) shows that $y(t) := \int_\Omega e^{\xi w}a^2$, $t > 0$, satisfies

$$y'(t) + y(t) \le k_{13}(A)y^6(t) + k_{14}(A)\left(\int_{\Omega} a\right)^3 + r^2\mu \int_{\Omega} a$$

$$\le k_{13}(A)y^6(t) + k_{14}(A)m_0^3 + r^2\mu m_0.$$

At this point, the proof can be completed by arguments similar to those for the case $\sigma = 1$.

3.5 Asymptotic Behavior of Solutions to a Chemotaxis–Haptotaxis Model

3.5.1 Global Boundedness

In this part, we first recall the result on local existence and uniqueness of classical solutions to (3.1.4) as well as a convenient extensibility criterion, which follows from Theorem 3.1, Lemma 5.9 and Theorem 5.1 of Morales-Rodrigo and Tello (2014).

Lemma 3.25 (Morales-Rodrigo and Tello (2014)) *Let $\Omega \subset \mathbb{R}^n$ be a smooth bounded domain. There exists $T_{max} \in (0, \infty]$ such that the problem (3.1.4) possesses a unique classical solution satisfying $(p, c, w) \in (C(\bar{\Omega} \times [0, T_{max})) \cap C^{2,1}(\bar{\Omega} \times (0, T_{max}))^3$. Moreover, for any $s > n + 2$,*

$$\limsup_{t \nearrow T_{max}} \|p(\cdot, t)\|_{W^{1,s}(\Omega)} \to \infty \tag{3.5.1}$$

if $T_{max} < +\infty$.

From now on, let (p, c, w) be the local classical solution of (3.1.4) on $(0, T_{max})$ provided by Lemma 3.25, and $\tau := \min\{1, \frac{T_{max}}{6}\}$.

The following basic but important properties of the solution to (3.1.4) can be directly obtained via standard arguments.

Lemma 3.26 (Morales-Rodrigo and Tello (2014)) *There exists a positive constant C independent of time such that*

$$\int_{\Omega} p(\cdot, t) \le C_1 := \max\{\int_{\Omega} p_0, |\Omega|\}, \quad \int_0^t e^{-2s} \int_{\Omega} p^2 ds \le C \text{ for all } t \in (0, T_{max}), \tag{3.5.2}$$

$$\int_t^{t+\tau} \int_{\Omega} p^2 \le C_1(1 + \frac{1}{\lambda}) \text{ for all } t \in (0, T_{max} - \tau), \tag{3.5.3}$$

$$c(t) \le \|c_0\|_{L^{\infty}(\Omega)} e^{-t}, \quad \int_{\Omega} |\nabla c(t)|^2 + \int_0^t \int_{\Omega} (|\nabla c|^2 + |\Delta c|^2) \le C \text{ for all } t \in (0, T_{max}), \tag{3.5.4}$$

$$0 \le w(t) \le \max\{\|w_0\|_{L^{\infty}(\Omega)}, 1\} \text{ for all } t \in (0, T_{max}). \tag{3.5.5}$$

As the proof of Theorem 3.4 in the one-dimensional case is similar to that for two dimensions, henceforth in this section, we shall focus on the case $n = 2$.

First, we shall show that p remains bounded in $L^q(\Omega)$ for any finite q. We note that the $L^q(\Omega)$-bound in Lemma 3.10 of Morales-Rodrigo and Tello (2014) depends on the time variable.

Lemma 3.27 *For any $r \in (1, \infty)$, there exists a positive constant $C(r, \tau)$ indepen-dent of t, such that $\|p(\cdot, t)\|_{L^r(\Omega)} \leq C(r, \tau)$ for all $t \in (0, T_{max})$.*

Proof Let $q := p(c + 1)^{-\alpha} e^{-\rho w}$. As in the proof of Lemma 3.10 in Morales-Rodrigo and Tello (2014), we infer that for any $m = 1, 2, \ldots$ there exist constants $c(m) > 0$ depending upon m and $C_1 > 0$ such that

$$\frac{d}{dt} \int_\Omega q^{2^m} (c + 1)^\alpha e^{\rho w} + \int_\Omega |\nabla q^{2^{m-1}}|^2 \qquad (3.5.6)$$
$$\leq c(m) \left(\int_\Omega |\Delta c|^2 + 1 \right) \int_\Omega q^{2^m} + c(m) \left(\int_\Omega q^{2^m} \right)^2 + C_1.$$

Next, we use induction to show

$$\int_\Omega q^{2^m} + \int_t^{t+\tau} \int_\Omega |\nabla q^{2^{m-1}}|^2 \leq C(m). \qquad (3.5.7)$$

Taking $m = 1$ in (3.5.6), we get

$$\frac{d}{dt} \int_\Omega q^2 (c + 1)^\alpha e^{\rho w} + \int_\Omega |\nabla q|^2 \qquad (3.5.8)$$
$$\leq c(1) \left(\int_\Omega |\Delta c|^2 + 1 + \int_\Omega q^2 \right) \int_\Omega q^2 (c + 1)^\alpha e^{\rho w} + C_1,$$

which implies that for the functions

$$y(t) = \int_\Omega q^2 (c + 1)^\alpha e^{\rho w} \quad \text{and} \quad a(t) = c(1) \left(\int_\Omega |\Delta c|^2 + 1 + \int_\Omega q^2 \right),$$

we have $\frac{dy}{dt} \leq a(t) y + C_1$. On the other hand, for any given $t > \tau$, it follows from (3.5.2) that there exists some $t_0 \in [t - \tau, t]$ such that $y(t_0) \leq \frac{C_1}{\tau} (1 + \frac{1}{\lambda})$. Hence, by ODE comparison argument we get

$$y(t) \leq y(t_0) e^{\int_{t_0}^t a(s)ds} + C_1 \int_{t_0}^t e^{\int_s^t a(\tau)d\tau} ds \leq C_2. \qquad (3.5.9)$$

In this inequality, we have taken $t_0 = 0$ if $t \leq \tau$ and noticed that $\int_{t-\tau}^t a(s)ds \leq C_3$ for all $t < T_{max}$ by Lemma 3.25. Combining (3.5.8) with (3.5.9), one can see that (3.5.7) is indeed valid for $m = 1$.

Now, suppose that (3.5.7) is valid for an integer $m + 1 = k \geq 2$, i.e.,

$$\int_\Omega q^{2^{k-1}} + \int_t^{t+\tau} \int_\Omega |\nabla q^{2^{k-2}}|^2 \leq C(k). \tag{3.5.10}$$

By the Gagliardo–Nirenberg inequality in two dimensions

$$\|z\|^4_{L^4(\Omega)} \leq C_3 \|\nabla z\|^2_{L^2(\Omega)} \|z\|^2_{L^2(\Omega)} + C_4 \|z\|^4_{L^2(\Omega)},$$

and hence

$$\int_\Omega q^{2^k} \leq C_3 \int_\Omega |\nabla q^{2^{k-2}}|^2 \int_\Omega q^{2^{k-1}} + C_4 (\int_\Omega q^{2^{k-1}})^2. \tag{3.5.11}$$

Integrating (3.5.10) between t and $t + \tau$, we have

$$\int_t^{t+\tau} \int_\Omega q^{2^k} \leq C_5(k),$$

which implies that for any $t \geq \tau$, there exists some $t_0 \in [t - \tau, t]$ such that $\int_\Omega q^{2^k}(t_0) \leq C_6$. At this point, let

$$y(t) := \int_\Omega q^{2^k}(c+1)^\alpha e^{\rho w} \quad \text{and} \quad b(t) = c(k)(\int_\Omega |\Delta c|^2 + 1 + \int_\Omega q^{2^k}).$$

Then (3.5.6) can be rewritten as

$$\frac{dy}{dt} + \int_\Omega |\nabla q^{2^{k-1}}|^2 \leq b(t)y + C_1.$$

By the argument above, one can obtain

$$\int_\Omega q^{2^k} + \int_t^{t+\tau} \int_\Omega |\nabla q^{2^{k-1}}|^2 \leq C,$$

and thereby conclude that (3.5.7) is valid for all integers $m \geq 1$. The proof of Lemma 3.27 is now complete in view of the boundedness of the weight $(c+1)^\alpha e^{\rho w}$.

To establish a priori estimates of $\|p(\cdot, t)\|_{L^\infty(\Omega)}$, we need some fundamental estimates for the solution of the following problem:

$$\begin{cases} c_t = \Delta c - c + f, & x \in \Omega, t > 0, \\ \dfrac{\partial c}{\partial v} = 0, & x \in \partial\Omega, t > 0, \\ c(x, 0) = c_0(x), & x \in \Omega. \end{cases} \tag{3.5.12}$$

Lemma 3.28 (Li and Wang (2018, Lemma 2.2)) *Let $T > 0$, $r \in (1, \infty)$. Then for each $c_0 \in W^{2,r}(\Omega)$ with $\dfrac{\partial c_0}{\partial \nu} = 0$ on $\partial \Omega$ and $f \in L^r(0, T; L^r(\Omega))$, (3.5.12) has a unique solution $c \in W^{1,r}(0, T; L^r(\Omega)) \cap L^r(0, T; W^{2,r}(\Omega))$ given by*

$$c(t) = e^{-t}e^{t\Delta}c_0 + \int_0^t e^{-(t-s)}e^{(t-s)\Delta}f(s)ds, \quad t \in [0, T],$$

where $e^{t\Delta}$ is the semigroup generated by the Neumann Laplacian, and there is $C_r > 0$ such that

$$\int_0^t \int_\Omega e^{rs}|\Delta c(x,s)|^r \, dx \, ds \leq C_r \int_0^t \int_\Omega e^{rs}|f(x,s)|^r \, dx \, ds + C_r \|v_0\|_{W^{2,r}(\Omega)}.$$

Now applying these estimates to control the cross-diffusive flux appropriately, we can derive the boundedness of p in $\Omega \times (0, T_{max})$.

Lemma 3.29 *There exists a constant $C > 0$ independent of t such that $\|p(\cdot, t)\|_{L^\infty(\Omega)} \leq C$ for all $t \in (0, T_{max})$.*

Proof We will only give a sketch of the proof, which is similar to that of Lemma c3.3–3.13 of Morales-Rodrigo and Tello (2014). For $k \geq \max\{2, \|p_0\|_{L^\infty(\Omega)}\}$, let $q_k = \max\{q - k, 0\}$ and $\Omega_k(t) = \{x \in \Omega : q(x, t) > k\}$. Multiplying the equation of q by q_k, we obtain

$$\begin{aligned}
&\frac{d}{dt}\int_\Omega q_k^2(c+1)^\alpha e^{\rho w} + 2\int_\Omega |\nabla q_k|^2 + 2\int_\Omega q_k^2 + 8\int_\Omega q_k^2(c+1)^\alpha e^{\rho w}\\
&\leq C_1 \int_\Omega q_k^3 + C_1 k \int_\Omega q_k^2 + C_1 k^2 \int_\Omega q_k + C_1 \int_\Omega (q_k^2 + kq_k)|\Delta c|
\end{aligned} \tag{3.5.13}$$

for some $C_1 > 0$ independent of k. By the boundedness of q in $L^r(\Omega)$ for any $r > 1$, the Gagliardo–Nirenberg inequality and Young inequality, we obtain

$$C_1 \|q_k\|_{L^3(\Omega)}^3 \leq \frac{1}{4}\|q_k\|_{H^1(\Omega)}^2 + C_2\|q_k\|_{L^1(\Omega)},$$

$$C_1 k \|q_k\|_{L^2(\Omega)}^2 \leq \frac{1}{4}\|q_k\|_{H^1(\Omega)}^2 + C_2 k^2 \|q_k\|_{L^1(\Omega)},$$

$$C_1 \int_\Omega q_k^2 |\Delta c| \leq \frac{1}{4}\|q_k\|_{H^1(\Omega)}^2 + C_2\|q_k\|_{L^2(\Omega)}^2 \|\Delta c\|_{L^2(\Omega)}^2,$$

$$C_1 k \int_\Omega q_k |\Delta c| \leq \frac{1}{4}\|q_k\|_{H^1(\Omega)}^2 + C_2 k^2 (1 + \|\Delta c\|_{L^8(\Omega)}^8)|\Omega_k|^{\frac{3}{2}}.$$

Inserting the above estimates into (3.5.10), we have

$$
\frac{d}{dt} \int_\Omega q_k^2 (c+1)^\alpha e^{\rho w} + \int_\Omega |\nabla q_k|^2 + \int_\Omega q_k^2 + 8 \int_\Omega q_k^2 (c+1)^\alpha e^{\rho w}
$$

$$
\leq C_2 \|q_k\|_{L^2(\Omega)}^2 \|\Delta c\|_{L^2(\Omega)}^2 + C_2 k^2 (1 + \|\Delta c\|_{L^8(\Omega)}^8) |\Omega_k|^{\frac{3}{2}} + (C_1 + 2C_2) k^2 \|q_k\|_{L^1(\Omega)}
$$

$$
\leq C_2 \|q_k\|_{L^2(\Omega)}^2 \|\Delta c\|_{L^2(\Omega)}^2 + \frac{1}{2} \|q_k\|_{H^1(\Omega)}^2 + C_3 k^4 (1 + \|\Delta c\|_{L^8(\Omega)}^8) |\Omega_k|^{\frac{3}{2}}.
$$

On the other hand, according to the relation between distribution functions and L^p integrals (see, e.g., (2.6) of Reyes and Vázquez (2006)), we can see that

$$
(r+1) \int_0^\infty s^r |\Omega_s(t)| ds = \|q(t)\|_{L^{r+1}(\Omega)}^{r+1}.
$$

Hence, taking into account Lemma 3.26, we get

$$
(k-1)^{16} |\Omega_k(t)| < \int_{k-1}^k s^{16} |\Omega_s(t)| ds < \int_0^\infty s^{16} |\Omega_s(t)| ds \leq \frac{1}{17} \|q(\cdot, t)\|_{L^{17}(\Omega)}^{17}
$$

and thus

$$
\frac{d}{dt} \int_\Omega q_k^2 (c+1)^\alpha e^{\rho w} + 8 \int_\Omega q_k^2 (c+1)^\alpha e^{\rho w}
$$

$$
\leq C_2 \|\Delta c\|_{L^2(\Omega)}^2 \int_\Omega q_k^2 (c+1)^\alpha e^{\rho w} + C_4 (1 + \|\Delta c\|_{L^8(\Omega)}^8) |\Omega_k|^{\frac{5}{4}}.
$$

Therefore, if $h(t) = 8 - C_2 \|\Delta c\|_{L^2(\Omega)}^2$, then

$$
\int_\Omega q_k^2 (c+1)^\alpha e^{\rho w} \leq C_4 e^{-\int_0^t h(s) ds} \int_0^t (1 + \|\Delta c\|_{L^8(\Omega)}^8) e^{\int_0^s h(\sigma) d\sigma} |\Omega_k(s)|^{\frac{5}{4}} ds.
$$

Furthermore, since $e^{-\int_0^t h(s) ds} = e^{-8t} e^{C_2 \int_0^t \|\Delta c\|_{L^2(\Omega)}^2 ds} \leq C_5 e^{-8t}$ by Lemma 3.25 and $e^{\int_0^s h(\sigma) d\sigma} \leq e^{8s}$, we get

$$
\int_\Omega q_k^2 \leq C_6 \int_0^t e^{-8(t-s)} (1 + \|\Delta c\|_{L^8(\Omega)}^8) |\Omega_k(s)|^{\frac{5}{4}} ds
$$

$$
\leq C_6 \int_0^t e^{-8(t-s)} (1 + \|\Delta c\|_{L^8(\Omega)}^8) ds \cdot \sup_{t \geq 0} |\Omega_k(t)|^{\frac{5}{4}}.
$$

To estimate the integral term in the right-hand side of the above inequality, we apply Lemma 3.28 with $r = 8$ and Lemma 3.27 to get

$$
\int_0^t e^{-8(t-s)} \|\Delta c\|_{L^8(\Omega)}^8 ds \leq C_7
$$

and thus $\int_{\Omega} q_k^2 \leq C_8 (\sup_{t \geq 0} |\Omega_k(t)|)^{\frac{5}{4}}$.

On the other hand, $\int_{\Omega} q_k^2(t) \geq \int_{\Omega_j(t)} q_k^2(t) \geq (j - k)^2 |\Omega_j(t)|$ for $j > k$. Consequently

$$(j - k)^2 \sup_{t \geq 0} |\Omega_j(t)| \leq C_8 (\sup_{t \geq 0} |\Omega_k(t)|)^{\frac{5}{4}} |\Omega_k(t)|.$$

According to Lemma B.1 of Kinderlehrer and Stampacchia (1980), there exists $k_0 < \infty$ such that $|\Omega_{k_0}(t)| = 0$ for all $t \in (0, T_{max})$. Therefore, $\|q(\cdot, t)\|_{L^{\infty}(\Omega)} \leq k_0$ for any $t \in (0, T_{max})$ and thereby the proof is complete.

Proof of Theorem 3.4. By the boundedness of p in $L^{\infty}((0, T_{max}), L^{\infty}(\Omega))$ from Lemma 3.29 and a bootstrap argument as in Morales-Rodrigo and Tello (2014), we can see that the global existence of classical solutions to (3.1.4) is an immediate consequence of Lemma 3.25, i.e., $T_{max} = \infty$. Indeed, suppose that $T_{max} < \infty$, then by Lemmas 3.15 and 3.19 of Morales-Rodrigo and Tello (2014), we can see that for any $s > n + 2$ and $t \leq T_{max}$

$$\|c(\cdot, t)\|_{W^{1,s}(\Omega)} + \|w(\cdot, t)\|_{W^{1,s}(\Omega)} \leq C.$$

Further by Lemma 3.20 of Morales-Rodrigo and Tello (2014), we have $\|p(\cdot, t)\|_{W^{1,s}(\Omega)} \leq C$ which contradicts (3.5.1) and thus implies that $T_{max} = \infty$. Moreover, since $\tau := \min\{1, \frac{T_{max}}{6}\} = 1$, there exists a constant $C > 0$ independent of time t such that $\|p(\cdot, t)\|_{L^{\infty}(\Omega)} \leq C$ for all $t \geq 0$ by retracing the proofs of Lemmas 3.27 and 3.28. This completes the proof of Theorem 3.4.

3.5.2 Asymptotic Behavior

In this part, on the basis of the L^{∞}-bound of p provided by Theorem 3.4, we shall look at the asymptotic behavior of the solution (p, c, w) of the problem (3.1.4).

1. L^r-convergence of solutions in two dimensions
When either $w_0 > 1$ or $\|w_0 - 1\|_{L^{\infty}(\Omega)} < \delta$ for some $\delta > 0$, the authors of Morales-Rodrigo and Tello (2014) removed the time dependence of the L^{∞}-bound of p (see Lemma 5.8 of Morales-Rodrigo and Tello (2014)) and thereby investigated the asymptotic behavior of solutions to (3.1.4). In this subsection, on the basis of the L^{∞}-bound of p being independent of time as provided by Theorem 3.4, we shall derive the same estimates as in Lemmas 5.6 and 5.7 of Morales-Rodrigo and Tello (2014) under the weaker assumption that $w_0 > 1 - \frac{1}{\rho}$. We shall show that the solution (p, c, w) to (3.1.4) converges to the homogeneous steady state $(1, 0, 1)$ as $t \to \infty$.

Before going into the details, let us first collect some useful related estimates. It should be noted that no other assumptions on the initial data (p_0, c_0, w_0) are made except for reasonable regularity, i.e., (3.1.5).

Lemma 3.30 (Morales-Rodrigo and Tello (2014, Lemmas 3.4, 3.8, 5.1, 5.2)) *Let* (p, c, w) *be the global, classical solution of (3.1.4). Then*

$$\left| \int_0^\infty \int_\Omega p(1-p) \right| \leq \max\{ \int_\Omega p_0, |\Omega|\}/\lambda; \tag{3.5.14}$$

$$\int_0^\infty \int_\Omega p|w - 1| \leq \|w_0 - 1\|_{L^1(\Omega)}; \tag{3.5.15}$$

$$\int_0^\infty \int_\Omega p|\nabla c|^2 < \infty; \tag{3.5.16}$$

$$\int_\Omega |\nabla c(t)|^2 \leq e^{-2t} \left(\int_\Omega |\nabla c_0|^2 + \mu^2 \|c_0\|_{L^\infty(\Omega)}^2 \max\{ \int_\Omega p_0, |\Omega|\}(t + \frac{1}{\lambda}) \right). \tag{3.5.17}$$

Lemma 3.31 *Under the assumptions of Theorem 3.4, we have*

$$\sup_{t \geq 0} \|c(t)\|_{W^{1,\infty}(\Omega)} \leq C. \tag{3.5.18}$$

Proof We know that c solves the linear equation

$$c_t = \Delta c - c + f$$

under the Neumann boundary condition with $f := -\mu pc$. Since $p \geq 0$, we know that $0 \leq c(x, t) \leq \|c_0\|_{L^\infty(\Omega)} e^{-t}$ by the standard sub-super solutions method. On the other hand, by Theorem 3.4, $\sup_{t \geq 0} \|p(t)\|_{L^\infty(\Omega)} \leq C_1$, which readily implies that $\|f\|_{L^\infty((0,\infty);L^\infty(\Omega))} \leq C_1$. Now upon a standard regularity argument, we can deduce the desired result. For the reader's convenience, we only give a brief sketch of the main ideas, and would like to refer to the proof of Lemma 1 in Kowalczyk and Szymańska (2008) or Lemma 4.1 in Horstmann and Winkler (2005) for more details. Indeed, according to the variation-of-constants formula of c, we have for $t > 2$

$$c(\cdot, t) = e^{(t-1)(\Delta-1)} c(\cdot, 1) + \int_1^t e^{(t-s)(\Delta-1)} f(\cdot, s) ds.$$

So by Lemma 1.1 (ii), we infer that

$$\|\nabla c(\cdot, t)\|_{L^\infty(\Omega)}$$

$$\leq 2c_2 \|c(\cdot, 1)\|_{L^1(\Omega)} + c_2 \int_1^t (1 + (t-s)^{-\frac{1}{2}}) e^{-(t-s)(\lambda_1+1)} \|f(\cdot, s)\|_{L^\infty(\Omega)} ds$$

$$\leq 2c_2 \|c(\cdot, 1)\|_{L^1(\Omega)} + c_2 C_1 \int_0^\infty (1 + \sigma^{-\frac{1}{2}}) e^{-\sigma} d\sigma.$$

Lemma 3.32 (Morales-Rodrigo and Tello (2014, Lemma 5.4)) *Let (p, c, w) be the global, classical solution of* (3.1.4). *Then for every $t \geq 0$ and $\kappa > 0$,*

$$\frac{d}{dt} F(p(t), w(t)) = G(p(t), w(t), c(t)),$$

where

$$F(p, w) = \kappa \int_{\Omega} |\nabla w|^2 + \int_{\Omega} p(\ln p - 1) + \int_{\Omega} p(w - 1) - \gamma \kappa \int_{\Omega} p(w - 1)^2,$$

and

$$\begin{aligned}
G(p, w, c) = &-\int_{\Omega} \frac{|\nabla p|^2}{p} + \int_{\Omega} \frac{\alpha}{1+c} \nabla p \cdot \nabla c + \int_{\Omega} (2\alpha\gamma\kappa(1 - w) + \alpha\rho) \frac{p}{1+c} \nabla c \cdot \nabla w \\
&+ \int_{\Omega} (\rho^2 - 2\gamma\kappa + 2\rho\gamma\kappa(1 - w)) p |\nabla w|^2 + \lambda \int_{\Omega} p(1 - p) \ln p \\
&+ \lambda\rho \int_{\Omega} p(1 - p)(w - 1) + \gamma\rho \int_{\Omega} p^2(1 - w) \\
&+ 2\gamma^2\kappa \int_{\Omega} p^2(w - 1)^2 - \lambda\gamma\kappa \int_{\Omega} p(1 - p)(w - 1)^2.
\end{aligned}$$

Lemma 3.33 *If $w_0 > 1 - \dfrac{1}{\rho}$, then there exists $\kappa > 0$ such that*

$$G(p, w, c) \leq -\frac{1}{2} \int_{\Omega} \frac{|\nabla p|^2}{p} - \frac{1}{2} \int_{\Omega} p |\nabla w|^2 + C \int_{\Omega} p|w - 1| + C \int_{\Omega} p|\nabla c|^2 \tag{3.5.19}$$

for some $C > 0$.

Proof By the Hölder and Young inequalities, we have

$$\int_{\Omega} \frac{\alpha}{1+c} \nabla p \cdot \nabla c \leq \frac{1}{2} \int_{\Omega} \frac{|\nabla p|^2}{p} + \frac{\alpha^2}{2} \int_{\Omega} p|\nabla c|^2,$$

and

$$\int_{\Omega} (2\alpha\gamma\kappa(1 - w) + \alpha\rho) \frac{p}{1+c} \nabla c \cdot \nabla w \leq \frac{1}{2} \int_{\Omega} p|\nabla w|^2 + C_1 \int_{\Omega} p|\nabla c|^2$$

for some $C_1 > 0$.

As $w_0 > 1 - \dfrac{1}{\rho}$, we can find some $\varepsilon_1 > 0$ such that $\rho(1 - w_0)_+ \leq 1 - \varepsilon_1$, where $(1 - w_0)_+ = \max\{0, 1 - w_0\}$. Hence from the w-equation in (3.1.4), it follows that

$$1 - w = (1 - w_0)e^{-\gamma \int_0^t p(s)ds}, \tag{3.5.20}$$

and thus

$$\int_\Omega (\rho^2 - 2\gamma\kappa + 2\rho\gamma\kappa(1-w))p|\nabla w|^2$$

$$\leq \int_\Omega (\rho^2 - 2\gamma\kappa + 2\rho\gamma\kappa(1-w_0)_+)p|\nabla w|^2$$

$$\leq \int_\Omega (\rho^2 - 2\gamma\kappa\varepsilon_1)p|\nabla w|^2$$

$$\leq -\int_\Omega p|\nabla w|^2$$

if we pick $\kappa > 0$ sufficiently large such that $\rho^2 - 2\gamma\kappa\varepsilon_1 < -1$.

Denote the lower order terms of $G(p, w, c)$ by $\theta(p, w)$, i.e.,

$$\theta(p, w) =: \lambda \int_\Omega p(1-p)\ln p + \lambda\rho \int_\Omega p(1-p)(w-1) + \gamma\rho \int_\Omega p^2(1-w)$$

$$+ 2\gamma^2\kappa \int_\Omega p^2(w-1)^2 - \lambda\gamma\kappa \int_\Omega p(1-p)(w-1)^2.$$

Since $s(1-s)\ln s \leq 0$ for $s \geq 0$, we get

$$\theta(p, w) \leq \lambda\rho \int_\Omega p(1-p)(w-1) + \gamma\rho \int_\Omega p^2(1-w)$$

$$+ 2\gamma^2\kappa \int_\Omega p^2(w-1)^2 - \lambda\gamma\kappa \int_\Omega p(1-p)(w-1)^2$$

$$\leq C_2(\|p\|_{L^\infty(\Omega)}, \|w-1\|_{L^\infty(\Omega)}) \int_\Omega p|w-1|,$$

which, along with $\|p(\cdot, t)\|_{L^\infty(\Omega)} \leq C$ from Theorem 3.4 and $\|w(\cdot, t) - 1\|_{L^\infty(\Omega)} \leq \|w_0 - 1\|_{L^\infty(\Omega)}$ from (3.5.20), yields

$$\theta(p, w) \leq C_3 \int_\Omega p|w-1|.$$

The desired result (3.5.19) then immediately follows.

Lemma 3.34 *If* $w_0 > 1 - \dfrac{1}{\rho}$, *then*

$$\sup_{t\geq 0} \int_\Omega |\nabla w(t)|^2 + \int_0^\infty \int_\Omega \frac{|\nabla p|^2}{p} + \int_0^\infty \int_\Omega p|\nabla w|^2 < \infty. \qquad (3.5.21)$$

.

Proof Combining Lemmas 3.30 and 3.31, we have

$$\frac{d}{dt} F(p(t), w(t)) + \frac{1}{2} \int_\Omega \frac{|\nabla p|^2}{p} + \frac{1}{2} \int_\Omega p|\nabla w|^2 \le C \int_\Omega p|w - 1| + C \int_\Omega p|\nabla c|^2.$$

Hence, (3.5.21) follows upon integration on the time variable, and using (3.5.15) and (3.5.16).

Lemma 3.35 *If $w_0 > 1 - \dfrac{1}{\rho}$, then for any $r \ge 2$*

$$\lim_{t \to \infty} \| p(\cdot, t) - \overline{p}(t) \|_{L^r(\Omega)} = 0, \tag{3.5.22}$$

$$\lim_{t \to \infty} |\overline{p}(t) - 1| = 0, \tag{3.5.23}$$

where $\overline{p}(t) = \frac{1}{|\Omega|} \int_\Omega p(\cdot, t)$, and

$$\lim_{t \to \infty} \| w(\cdot, t) - 1 \|_{L^r(\Omega)} = 0. \tag{3.5.24}$$

Proof The proofs of (3.5.22) and (3.5.23) are similar to those of Lemmas 5.9–5.11 of Morales-Rodrigo and Tello (2014), respectively. However, for the reader's convenience, we only give a brief sketch of (3.5.23). In fact, from (3.1.4) and the Poincaré–Wirtinger inequality, it follows that

$$\begin{aligned}
\overline{p}_t &= \lambda \left(\overline{p} - \overline{p}^2 - \frac{1}{|\Omega|} \int_\Omega (p - \overline{p})^2 \right) \\
&\ge \lambda \overline{p} \left(1 - \overline{p} - C_1 \int_\Omega \frac{|\nabla p|^2}{p} \right).
\end{aligned}$$

Hence by (3.5.21), we get

$$\begin{aligned}
\overline{p}(t) &\ge \overline{p}_0 \exp\{\lambda t - \lambda \int_0^t \overline{p}(s)ds - C_1 \lambda \int_0^\infty \int_\Omega \frac{|\nabla p|^2}{p} ds\} \\
&\ge C_2 \exp\{\lambda t - \lambda \int_0^t \overline{p}(s)ds\},
\end{aligned}$$

which means that (3.5.23) is valid due to either $\overline{p}(t) \to 1$ or $\overline{p}(t) \to 0$ in Lemma 5.10 of Morales-Rodrigo and Tello (2014). Indeed, suppose that $\overline{p}(t) \to 0$, then there exists $t_0 > 1$ such that $\overline{p}(t) \le \frac{1}{2}$ and thus $\int_0^t \overline{p}(s)ds \le \frac{t}{2} + \int_0^{t_0} \overline{p}(s)ds$ for all $t \ge t_0$. Therefore, we arrive at $\overline{p}(t) \ge C_3 e^{\frac{\lambda t}{2}}$ for all $t \ge t_0$, which contradicts $\overline{p}(t) \to 0$.

Now we turn to show (3.5.24). Invoking the Poincaré inequality in the form

$$\int_\Omega |\varphi(x) - \frac{1}{|\Omega|} \int_\Omega \varphi(y)dy|^2 dx \le C_p \int_\Omega |\nabla \varphi|^2 dx \quad \text{for all } \varphi \in W^{1,2}(\Omega)$$

for some $C_p > 0$, one can find that for all $j \in \mathbb{N}$

$$\int_{j}^{j+1} \|p(s) - \overline{p}(s)\|_{L^2(\Omega)}^2 ds \le C_p \int_{j}^{j+1} \|\nabla p(s)\|_{L^2(\Omega)}^2 ds$$

$$\le C_p \sup_{t \ge 0} \|p(t)\|_{L^\infty(\Omega)} \int_{j}^{j+1} \int_{\Omega} \frac{|\nabla p(s)|^2}{p(s)} ds,$$

$$(3.5.25)$$

which, along with (3.5.21) and Theorem 3.4, shows that

$$\int_{\Omega} \int_{j}^{j+1} |p(x,s) - \overline{p}(s)|^2 ds dx = \int_{j}^{j+1} \|p(s) - \overline{p}(s)\|_{L^2(\Omega)}^2 ds \to 0 \quad (3.5.26)$$

as $j \to \infty$.

Now defining $p_j(x) := \int_{j}^{j+1} |p(x,s) - \overline{p}(s)|^2 ds$, $x \in \Omega$, $j \in \mathbb{N}$, (3.5.26) tells us that $p_j \to 0$ in $L^1(\Omega)$ as $j \to \infty$. There exist a certain null set $Q \subseteq \Omega$ and a subsequence $(j_k)_{k \in \mathbb{N}} \subset \mathbb{N}$ such that $j_k \to \infty$ and $p_{j_k}(x) \to 0$ for every $x \in \Omega \setminus Q$ as $k \to \infty$. Restated in the original variable, this becomes

$$\int_{j_k}^{j_k+1} |p(x,s) - \overline{p}(s)|^2 ds \to 0 \qquad (3.5.27)$$

for every $x \in \Omega \setminus Q$ as $k \to \infty$.

Therefore, from (3.5.20) and $p(x,t) \ge 0$, it follows that for any $x \in \Omega \setminus Q$

$$|w(x,t) - 1|$$

$$\le \|w_0 - 1\|_{L^\infty(\Omega)} \exp\{-\gamma \int_{0}^{[t]} p(x,s) ds\}$$

$$\le \|w_0 - 1\|_{L^\infty(\Omega)} \exp\{-\gamma \sum_{k=0}^{m(t)} \int_{j_k}^{j_k+1} p(x,s) ds\}$$

$$\le \|w_0 - 1\|_{L^\infty(\Omega)} \exp\{\gamma \sum_{k=0}^{m(t)} \int_{j_k}^{j_k+1} |p(x,s) - \overline{p}(s)| ds - \gamma \sum_{k=0}^{m(t)} \int_{j_k}^{j_k+1} \overline{p}(s) ds\}$$

$$\le \|w_0 - 1\|_{L^\infty(\Omega)} \exp\{\gamma \sum_{k=0}^{m(t)} (\int_{j_k}^{j_k+1} |p(x,s) - \overline{p}(s)|^2 ds)^{\frac{1}{2}} - \gamma \sum_{k=0}^{m(t)} \int_{j_k}^{j_k+1} \overline{p}(s) ds\},$$

where $m(t) := \max_{k \in \mathbb{N}} \{j_k + 1, [t]\}$. Furthermore, by (3.5.23), there exists $k_0 \in \mathbb{N}$ such that $\int_{j_k}^{j_k+1} \overline{p}(s) ds \ge \frac{1}{2}$ for all $k \ge k_0$. Hence by the fact that $m(t) \to \infty$ as $t \to \infty$ and (3.5.27), we obtain that $w(x,t) - 1 \to 0$ almost everywhere in Ω as $t \to \infty$. On the other hand, as $|w(x,t) - 1| \le \|w_0 - 1\|_{L^\infty(\Omega)}$, the dominated convergence theorem ensures that (3.5.24) holds for any $r \in (2, \infty)$.

Remark 3.4 (1) It is observed that since $W^{1,2}(\Omega) \hookrightarrow L^\infty(\Omega)$ is invalid in the two-dimensional setting, $\|p(s) - \overline{p}(s)\|_{L^2(\Omega)}^2$ in (3.5.25) cannot be replaced by $\|p(s) -$

$\overline{p}(s)\|_{L^\infty(\Omega)}^2$, and thus we cannot infer that $\lim\limits_{t\to\infty}\|w(\cdot,t)-1\|_{L^\infty(\Omega)}=0$, even though we have established that all the related estimates of (p,c,w) in Morales-Rodrigo and Tello (2014) continue to hold under the milder condition imposed on the initial data w_0.

(2) Similar to the remark above, we note that, even though $\|w(\cdot,t)\|_{W^{1,n}(\Omega)}\leq C(T)$ for any $n\geq 2$ and $t\leq T$, we are not able to infer the global estimate $\sup_{t\geq 0}\int_\Omega|\nabla w(t)|^{2+\varepsilon}\leq C$. Otherwise, we would be able to apply regularity estimates for bounded solutions of semilinear parabolic equations (see Porzio and Vespri (1993) for instance) to obtain the Hölder estimates of $p(x,t)$ in $\Omega\times(1,\infty)$, and thereby conclude $\lim\limits_{t\to\infty}\|p(\cdot,t)-1\|_{L^\infty(\Omega)}=0$. As things stand at the moment, we are only able to infer convergence in L^r.

2. L^∞-convergence of solutions with exponential rate in one dimension

It is observed that the results, in particular Lemma 3.35, in the previous subsection are still valid in the one-dimensional case. Moreover, in the one-dimensional setting, the weak convergence result in Lemma 3.35 can be improved via a bootstrap argument. In fact, we shall derive some a priori estimates of (p,c,w) and thereby demonstrate that (p,c,w) converges to $(1,0,1)$ in $L^\infty(\Omega)$ as $t\to\infty$. Furthermore, by a regularity argument involving the variation-of-constants formula for p and smoothing L^p-L^q type estimates for the Neumann heat semigroup, we will show that $p(\cdot,t)-1$ decays exponentially in $L^\infty(\Omega)$.

As pointed out in the introduction, the main technical difficulty in the derivation of Theorem 3.5 stems from the coupling between p and w. Indeed, the lack of the regularization effect in the space variable in the w-equation and the presence of p there demand tedious estimates of the solution.

The following lemma plays a crucial role in establishing the uniform convergence of p as $t\to\infty$ (see Lemma 3.40). Though the proof thereof only involves elementary analysis, we give full proof here for the sake of the reader's convenience since we could not find a precise reference covering our situation.

Lemma 3.36 *Let $k(t)$ be a function satisfying*

$$k(t)\geq 0,\qquad \int_0^\infty k(t)dt<\infty.$$

If $k'(t)\leq h(t)$ for some $h(t)\in L^1(0,\infty)$, then $k(t)\to 0$ as $t\to\infty$.

Proof Supposing the contrary, then we can find $A>0$ and a sequence $(t_j)_{j\in\mathbb{N}}\subset(1,\infty)$ such that $t_j\geq t_{j-1}+2$, $t_j\to\infty$ as $j\to\infty$ and $k(t_j)\geq A$ for all $j\in\mathbb{N}$. On the other hand, by $k'(t)\leq h(t)$, we have

$$k(t_j-\tau)\geq k(t_j)-\int_{t_j-\tau}^{t_j}|h(s)|ds\geq k(t_j)-\int_{t_j-1}^{t_j}|h(s)|ds \qquad (3.5.28)$$

for all $\tau\in(0,1)$.

Since $h(t) \in L^1(0, \infty)$, we have $\int_{t_j-1}^{t_j} |h(s)| ds \to 0$ as $t_j \to \infty$ and thereby there exists $j_0 \in \mathbb{N}$ such that $\int_{t_j-1}^{t_j} |h(s)| ds \leq \frac{A}{2}$ for all $j \geq j_0$, which along with (3.5.28) implies that

$$k(t_j - \tau) \geq k(t_j) - \frac{A}{2} \geq \frac{A}{2} \tag{3.5.29}$$

for $j \geq j_0$ and $\tau \in (0, 1)$. It follows that $\int_{t_j-1}^{t_j} k(t) dt \geq \frac{A}{2}$ for all $j \geq j_0$, which contradicts $\int_0^\infty k(t) dt < \infty$ and thus completes the proof of the lemma.

Lemma 3.37 *If $w_0 > 1 - \frac{1}{\rho}$, then there exists a constant $C > 0$ such that*

$$\int_0^\infty \int_\Omega e^{\rho w} |z_x|^2 \leq C \tag{3.5.30}$$

where $z = pe^{-\rho w}$.

Proof We know that $z_x = e^{-\rho w} p_x - \rho z w_x$ and thus

$$e^{\rho w} |z_x|^2 \leq 4 e^{-\rho w} |p_x|^2 + 4 p^2 e^{-\rho w} \rho^2 |w_x|^2.$$

Integrating over $\Omega \times (0, \infty)$ and taking (3.5.5) into account, we have

$$\int_0^\infty \int_\Omega e^{\rho w} |z_x|^2 \leq 4 \int_0^\infty \int_\Omega |p_x|^2 + 4 \rho^2 \sup_{t \geq 0} \|p(t)\|_{L^\infty(\Omega)} \int_0^\infty \int_\Omega p |w_x|^2$$

$$\leq 4(1 + \rho^2) \sup_{t \geq 0} \|p(t)\|_{L^\infty(\Omega)} \left(\int_0^\infty \int_\Omega \frac{|p_x|^2}{p} + \int_0^\infty \int_\Omega p |w_x|^2 \right).$$

Hence by Theorem 3.4 and Lemma 3.34, we get (3.5.30).

Lemma 3.38 *If $w_0 > 1 - \frac{1}{\rho}$, then there exists a constant $C > 0$ such that*

$$\frac{d}{dt} \int_\Omega e^{\rho w} |z_x|^2 + \frac{1}{3} \int_\Omega e^{\rho w} z_t^2$$
$$\leq C \left(\int_\Omega e^{\rho w} |z_x|^2 + \int_\Omega p |w_x|^2 + \int_\Omega |c_{xx}|^2 + \int_\Omega p |c_x|^2 + \int_\Omega p |w - 1| + \int_\Omega p(p-1)^2 \right)$$

with $z = pe^{-\rho w}$.

Proof Note that z satisfies

$$z_t = e^{-\rho w} (e^{\rho w} z_x)_x - e^{-\rho w} \left(\frac{z_x e^{\rho w}}{1 + c} \nabla c \right)_x + \lambda z (1 - z e^{\rho w}) - \rho \gamma e^{\rho w} z^2 (1 - w).$$

Multiplying the above equation by $z_t e^{\rho w}$ and integrating in the spatial variable, we obtain

$$\int_\Omega e^{\rho w} z_t^2 + \int_\Omega e^{\rho w} z_x z_{xt}$$
$$= -\int_\Omega e^{\rho w} z_t \left(\frac{\alpha}{1+c} z_x c_x + \frac{\alpha z \rho}{1+c} w_x c_x - \frac{\alpha z}{(1+c)^2} |c_x|^2 + \frac{\alpha z}{1+c} c_{xx} \right) \quad (3.5.31)$$
$$+ \int_\Omega e^{\rho w} z_t (\lambda z (1 - z e^{\rho w}) - \rho \gamma e^{\rho w} z^2 (1 - w)).$$

Notice that

$$\int_\Omega e^{\rho w} z_x z_{xt} = \frac{1}{2} \frac{d}{dt} \int_\Omega e^{\rho w} |z_x|^2 - \frac{\gamma \rho}{2} \int_\Omega e^{2\rho w} z (1 - w) |z_x|^2$$
$$\geq \frac{1}{2} \frac{d}{dt} \int_\Omega e^{\rho w} |z_x|^2 - \frac{\gamma \rho}{2} \sup_{t \geq 0} \|p(t)\|_{L^\infty(\Omega)} \|1 - w_0\|_{L^\infty(\Omega)} \int_\Omega e^{\rho w} |z_x|^2,$$

$$-\int_\Omega z_t \frac{\alpha e^{\rho w}}{1+c} z_x c_x \leq \frac{1}{6} \int_\Omega e^{\rho w} z_t^2 + C_1 \sup_{t \geq 0} \|c_x(t)\|_{L^\infty(\Omega)}^2 \int_\Omega e^{\rho w} |z_x|^2,$$

$$\int_\Omega z_t \frac{\alpha z e^{\rho w} \rho}{1+c} w_x c_x \leq \frac{1}{6} \int_\Omega e^{\rho w} z_t^2 + C_1 \sup_{t \geq 0} \|c_x(t)\|_{L^\infty(\Omega)}^2 \sup_{t \geq 0} \|p(t)\|_{L^\infty(\Omega)} \int_\Omega p|w_x|^2,$$

$$\int_\Omega z_t \frac{\alpha z e^{\rho w}}{(1+c)^2} |c_x|^2 \leq \frac{1}{6} \int_\Omega e^{\rho w} z_t^2 + C_1 \sup_{t \geq 0} \|p(t)\|_{L^\infty(\Omega)} \sup_{t \geq 0} \|c_x(t)\|_{L^\infty(\Omega)}^2 \int_\Omega p|c_x|^2,$$

$$-\int_\Omega z_t \frac{\alpha z e^{\rho w}}{1+c} c_{xx} \leq \frac{1}{6} \int_\Omega e^{\rho w} z_t^2 + C_1 \sup_{t \geq 0} \|p(t)\|_{L^\infty(\Omega)}^2 \int_\Omega |c_{xx}|^2$$

and

$$\int_\Omega e^{\rho w} z_t (\lambda z (1 - z e^{\rho w}) - \rho \gamma e^{\rho w} z^2 (1 - w))$$
$$= \lambda \int_\Omega p z_t (1 - p) - \rho \gamma \int_\Omega z_t p^2 (1 - w))$$
$$\leq \frac{1}{6} \int_\Omega e^{\rho w} z_t^2 + C_1 \sup_{t \geq 0} \|p(t)\|_{L^\infty(\Omega)} \int_\Omega p(1 - p)^2$$
$$+ C_1 \sup_{t \geq 0} \|p(t)\|_{L^\infty(\Omega)}^3 \|1 - w_0\|_{L^\infty(\Omega)} \int_\Omega p|1 - w|.$$

Applying Theorem 3.4, (3.5.18) and inserting the above inequalities into (3.5.31), we obtain the desired inequality.

Now we focus our attention on the decay properties of the solutions. Indeed, we will show that $p(x, t)$ converges to 1 with respect to the norm in $L^\infty(\Omega)$ as $t \to \infty$. Subsequently, we will establish the exponential decay of $\|p(\cdot, t) - 1\|_{L^\infty(\Omega)}$ with explicit rate.

Lemma 3.39 *If $w_0 > 1 - \dfrac{1}{\rho}$, then*

$$\lim_{t\to\infty} \|w(\cdot,t) - 1\|_{L^\infty(\Omega)} = 0. \tag{3.5.32}$$

Proof From (3.5.20), it follows that for any $\epsilon > 0$

$$|w(x,t) - 1| \le \|w_0 - 1\|_{L^\infty(\Omega)} \exp\{-\gamma \int_0^t p(s)ds\} \tag{3.5.33}$$

$$\le \|w_0 - 1\|_{L^\infty(\Omega)} \exp\{\gamma \int_0^t \|p(s) - \overline{p}(s)\|_{L^\infty(\Omega)}ds - \gamma \int_0^t \overline{p}(s)ds\}$$

$$\le \|w_0 - 1\|_{L^\infty(\Omega)} \exp\{\frac{\gamma}{\epsilon} \int_0^t \|p(s) - \overline{p}(s)\|_{L^\infty(\Omega)}^2 ds + \epsilon\gamma t - \gamma \int_0^t \overline{p}(s)ds\},$$

where $\overline{p}(t) = \frac{1}{|\Omega|} \int_\Omega p(\cdot,t)$.

On the other hand, by the Poincaré–Wirtinger inequality, the Sobolev embedding theorem in one dimension and (3.5.21), we have

$$\int_0^t \|p(s) - \overline{p}(s)\|_{L^\infty(\Omega)}^2 ds \le C_1 \int_0^\infty \|p_x(s)\|_{L^2(\Omega)}^2 ds$$

$$\le C_1 \sup_{t \ge 0} \|p(t)\|_{L^\infty(\Omega)} \int_0^\infty \int_\Omega \frac{|p_x(s)|^2}{p(s)} ds \tag{3.5.34}$$

$$\le C_2$$

for some constant $C_2 > 0$. Combining (3.5.33) with (3.5.34) yields

$$\|w(t) - 1\|_{L^\infty(\Omega)} \le \|w_0 - 1\|_{L^\infty(\Omega)} \exp\{\frac{C_2\gamma}{\epsilon} + \epsilon\gamma t - \gamma \int_0^t \overline{p}(s)ds\}$$

for $t \ge 0$. The assertion now follows from the last inequality and the proof is complete.

Lemma 3.40 *If* $w_0 > 1 - \dfrac{1}{\rho}$, *then*

$$\lim_{t\to\infty} \|p(\cdot,t) - 1\|_{L^\infty(\Omega)} = 0. \tag{3.5.35}$$

Proof We first show that

$$\lim_{t\to\infty} \|z(\cdot,t) - \overline{z}(t)\|_{L^\infty(\Omega)} = 0 \tag{3.5.36}$$

where $\overline{z}(t) = \frac{1}{|\Omega|} \int_\Omega z(\cdot,t)$.

To this end, we consider the function $k(t) \ge 0$ defined by $k(t) = \int_\Omega e^{\rho w}|z_x|^2$ and prove that

$$\lim_{t\to\infty} k(t) = 0. \tag{3.5.37}$$

By Lemmas 3.36, 3.37 and 3.38, it is enough to prove that

$$h(t) := \int_{\Omega} e^{\rho w} |z_x|^2 + \int_{\Omega} p|w_x|^2 + \int_{\Omega} (|c_{xx}|^2 + p|c_x|^2) + \int_{\Omega} p|w-1| + \int_{\Omega} p(p-1)^2$$

$$\in L^1(0, \infty).$$

Noting (3.5.30), (3.5.21), (3.5.16), (3.5.15) and (3.5.4), it remains to estimate

$$\int_0^{\infty} \int_{\Omega} p(p-1)^2.$$

In fact, multiplying the p-equation in (3.1.4) by $p-1$, we have

$$\frac{1}{2} \frac{d}{dt} \int_{\Omega} (p-1)^2 = -\int_{\Omega} |p_x|^2 + \rho \int_{\Omega} p w_x p_x + \alpha \int_{\Omega} \frac{p}{1+c} c_x p_x - \lambda \int_{\Omega} p(p-1)^2$$

$$\leq -\frac{1}{2} \int_{\Omega} |p_x|^2 + C(\int_{\Omega} p^2 |w_x|^2 + \int_{\Omega} p^2 |c_x|^2) - \lambda \int_{\Omega} p(p-1)^2.$$

Hence, by the boundedness of p, (3.5.16) and (3.5.21), we easily infer that

$$\int_0^{\infty} \int_{\Omega} p(p-1)^2 \leq C.$$

Furthermore, by the Poincaré–Wirtinger inequality and the Sobolev embedding theorem in one dimension, we have

$$\|z(t) - \bar{z}(t)\|_{L^{\infty}(\Omega)} \leq C_p \|z_x(t)\|_{L^2(\Omega)},$$

which along with (3.5.37) yields (3.5.36).

On the other hand, for any $\{t_j\}_{j\in\mathbb{N}} \subset (1, \infty)$, there exists a subsequence along which $z(\cdot, t_j) - e^{-\rho} \to 0$ a.e. in Ω as $j \to \infty$ by Lemma 3.35. We apply the dominated convergence theorem along with the uniform majorization $|z(\cdot, t_j)| \leq \sup_{j\geq 1} \|z(t_j)\|_{L^{\infty}(\Omega)} \leq C$ to infer that

$$\lim_{t\to\infty} |\bar{z}(t) - e^{-\rho}| = 0. \tag{3.5.38}$$

Hence

$$\|p(\cdot, t) - 1\|_{L^{\infty}(\Omega)}$$
$$= \|e^{\rho w} z - 1\|_{L^{\infty}(\Omega)}$$
$$\leq e^{\rho(1+\|w_0\|_{\infty(\Omega)})} (\|z(\cdot, t) - \bar{z}(t)\|_{L^{\infty}(\Omega)} + |\bar{z}(t) - e^{-\rho}|) + \|e^{\rho w(\cdot, t)} - e^{\rho}\|_{L^{\infty}(\Omega)}$$
$$\leq e^{\rho(1+\|w_0\|_{\infty(\Omega)})} (\|z(\cdot, t) - \bar{z}(t)\|_{L^{\infty}(\Omega)} + |\bar{z}(t) - e^{-\rho}|) + C_1 \|w(\cdot, t) - 1\|_{L^{\infty}(\Omega)}$$

for some $C_1 > 0$, which together with (3.5.32), (3.5.36) and (3.5.38) yields the desired result.

Having established that $p(x, t)$ converges to 1 uniformly with respect to $x \in \Omega$ as $t \to \infty$, we now go on to establish an explicit exponential convergence rate. Using (3.5.35), we first look into the decay of $\int_\Omega |w_x(t)|^2$.

Lemma 3.41 *Let $w_0 > 1 - \dfrac{1}{\rho}$. Then for any $\epsilon > 0$, there exists $C(\epsilon) > 0$ such that*

$$\int_\Omega |w_x(t)|^2 \le C(\epsilon) e^{-2\gamma(1-\epsilon)t}. \tag{3.5.39}$$

Proof From (3.5.20), it follows that

$$|w_x(t)|^2 \le 4|w_{0x}|^2 e^{-2\gamma \int_0^t p(s)ds} + 4\gamma^2 |w_0 - 1|^2 e^{-2\gamma \int_0^t p(s)ds} \left(\int_0^t |p_x(s)|ds \right)^2$$

$$\le 4|w_{0x}|^2 e^{-2\gamma \int_0^t p(s)ds} + 4t\gamma^2 |w_0 - 1|^2 e^{-2\gamma \int_0^t p(s)ds} \int_0^t |p_x(s)|^2 ds.$$

Taking Lemma 3.40 into account, we know that for any $\epsilon > 0$, there exists $t_\epsilon > 1$ such that $p(x, t) > 1 - \epsilon$ for all $x \in \Omega, t > t_\epsilon$. Therefore, integrating the above inequality in the space variable yields

$$\int_\Omega |w_x(t)|^2$$

$$\le 4e^{-2\gamma(1-\epsilon)(t-t_\epsilon)} \int_\Omega |w_{0x}|^2 + 4t\gamma^2 \|w_0 - 1\|_{L^\infty(\Omega)}^2 e^{-2\gamma(1-\epsilon)(t-t_\epsilon)} \int_0^\infty \int_\Omega |p_x|^2$$

$$\le C_1(\epsilon)(1+t)e^{-2\gamma(1-\epsilon)t} \left(\int_\Omega |w_{0x}|^2 + \|w_0 - 1\|_{L^\infty(\Omega)}^2 \sup_{t\ge 0} \|p(t)\|_{L^\infty(\Omega)} \int_0^\infty \int_\Omega \frac{|p_x|^2}{p} \right),$$

for all $t > t_\epsilon$, which along with (3.5.21) implies (3.5.39).

Now we utilize the decay properties of $\int_\Omega |c_x(t)|^2$, $\int_\Omega |w_x(t)|^2$ and the uniform convergence of $|p(x, t) - 1|$ asserted by Lemma 3.40 to establish the decay property of $\|p(\cdot, t) - 1\|_{L^2(\Omega)}$.

Lemma 3.42 *Let $w_0 > 1 - \dfrac{1}{\rho}$. Then for any $\epsilon \in (0, \min\{1, \gamma, \lambda\})$, there exists $C(\epsilon) > 0$ such that*

$$\|p(\cdot, t) - 1\|_{L^2(\Omega)} \le C(\epsilon) e^{-(\min\{1, \gamma, \lambda\} - \epsilon)t}. \tag{3.5.40}$$

Proof By (3.5.35), we know that for any $\epsilon \in (0, \min\{1, \gamma, \lambda\})$, there exists $t_\epsilon > 1$ such that $p(x, t) > 1 - \epsilon$ for all $x \in \Omega, t > t_\epsilon$. Hence, we multiply the p-equation in (3.1.4) by $p - 1$ and integrate the result over Ω to get

$$
\frac{1}{2}\frac{d}{dt}\int_\Omega (p-1)^2 = -\int_\Omega |p_x|^2 + \rho\int_\Omega pw_x p_x + \alpha\int_\Omega \frac{p}{1+c}c_x p_x - \lambda\int_\Omega p(p-1)^2
$$

$$
\leq -\frac{1}{2}\int_\Omega |p_x|^2 + C_1\left(\int_\Omega |w_x|^2 + \int_\Omega |c_x|^2\right) - \lambda(1-\epsilon)\int_\Omega (p-1)^2
$$

for all $t > t_\epsilon$. Now, applying the Gronwall inequality, (3.5.18) and Lemma 3.43, we have

$$
\int_\Omega (p(t)-1)^2
$$

$$
\leq e^{-2\lambda(1-\epsilon)(t-t_\epsilon)}\int_\Omega (p(t_\epsilon)-1)^2 + C_1\int_0^t e^{-2\lambda(1-\epsilon)(t-s)}\left(\int_\Omega |w_x|^2 + \int_\Omega |c_x|^2\right)
$$

$$
\leq C_2(\epsilon)e^{-2\lambda(1-\epsilon)t} + C_3(\epsilon)\int_0^t e^{-2\lambda(1-\epsilon)(t-s)}\left(e^{-2\gamma(1-\epsilon)s} + e^{-2(1-\epsilon)s}\right)
$$

$$
\leq C_4(\epsilon)e^{-2\min\{\lambda,1,\gamma\}(1-\epsilon)t},
$$

where $c_i(\epsilon) > 0$ $(i = 2, 3, 4)$ are independent of time t. This completes the proof.

Moving forward, on the basis of Lemma 3.42, we come to establish the exponential decay of $\|p(\cdot, t) - \overline{p}(t)\|_{L^\infty(\Omega)}$ by means of a variation-of-constants representation of p, as follows.

Lemma 3.43 *Let* $w_0 > 1 - \dfrac{1}{\rho}$. *Then for any* $\epsilon \in (0, \min\{\lambda_1, 1, \gamma, \lambda\})$, *there exists* $C(\epsilon) > 0$ *such that*

$$
\|p(\cdot, t) - \overline{p}(t)\|_{L^\infty(\Omega)} \leq C(\epsilon)e^{-(\min\{\lambda_1, 1, \gamma, \lambda\}-\epsilon)t}. \tag{3.5.41}
$$

Proof By noting that $\overline{p}_t = \lambda\overline{p(1-p)}(t)$, applying the variation-of-constants formula to the p-equation in (3.1.4) yields

$$
p(\cdot, t) - \overline{p}(t) = e^{t\Delta}(p(\cdot, 0) - \overline{p}(0)) - \alpha\int_0^t e^{(t-s)\Delta}\left(\frac{p}{1+c}c_x\right)_x
$$

$$
- \rho\int_0^t e^{(t-s)\Delta}(pw_x)_x + \lambda\int_0^t e^{(t-s)\Delta}(p(1-p) - \overline{p(1-p)}).
$$

Together with (3.5.17), Lemmas 3.42 and 1.1, this gives

$$\|p(\cdot, t) - \overline{p}(t)\|_{L^\infty(\Omega)}$$

$$\leq \|e^{t\Delta}(p(\cdot, 0) - \overline{p}(0))\|_{L^\infty(\Omega)} + \alpha \int_0^t \|e^{(t-s)\Delta}(\frac{p}{1+c}c_x)_x\|_{L^\infty(\Omega)}$$

$$+ \rho \int_0^t \|e^{(t-s)\Delta}(pw_x)_x\|_{L^\infty(\Omega)} + \lambda \int_0^t \|e^{(t-s)\Delta}(p(1-p) - \overline{p(1-p)})\|_{L^\infty(\Omega)}$$

$$\leq c_1 e^{-\lambda_1 t}\|p(\cdot, 0) - \overline{p}(0)\|_{L^\infty(\Omega)} + C_1 \int_0^t (1 + (t-s)^{-\frac{3}{4}})e^{-\lambda_1(t-s)}\|w_x\|_{L^2(\Omega)}$$

$$+ C_1 \int_0^t (1 + (t-s)^{-\frac{3}{4}})e^{-\lambda_1(t-s)}\|c_x\|_{L^2(\Omega)}$$

$$+ C_1 \int_0^t (1 + (t-s)^{-\frac{3}{4}})e^{-\lambda_1(t-s)}\|p(1-p) - \overline{p(1-p)}\|_{L^2(\Omega)}$$

$$\leq c_1 e^{-\lambda_1 t}\|p(\cdot, 0) - \overline{p}(0)\|_{L^\infty(\Omega)} + C_2(\epsilon) \int_0^t (1 + (t-s)^{-\frac{3}{4}})e^{-\lambda_1(t-s)}e^{-\gamma(1-\epsilon)s}$$

$$+ C_2(\epsilon) \int_0^t (1 + (t-s)^{-\frac{3}{4}})e^{-\lambda_1(t-s)}e^{-(1-\epsilon)s}$$

$$+ C_1 \int_0^t (1 + (t-s)^{-\frac{3}{4}})e^{-\lambda_1(t-s)}\|p(1-p) - \overline{p(1-p)}\|_{L^2(\Omega)}.$$

It is observed that

$$\|p(1-p) - \overline{p(1-p)}\|_{L^2(\Omega)}^2 = \|p(1-p)\|_{L^2(\Omega)}^2 - |\Omega||\overline{p(1-p)}|^2$$
$$\leq \|p(1-p)\|_{L^2(\Omega)}^2.$$

Hence from (3.5.15) and Lemma 3.42, it follows that

$$\|p(\cdot, t) - \overline{p}(t)\|_{L^\infty(\Omega)}$$

$$\leq c_1 e^{-\lambda_1 t}\|p(\cdot, 0) - \overline{p}(0)\|_{L^\infty(\Omega)} + C_2(\epsilon) \int_0^t (1 + (t-s)^{-\frac{3}{4}})e^{-\lambda_1(t-s)}e^{-\gamma(1-\epsilon)s}$$

$$+ C_2(\epsilon) \int_0^t (1 + (t-s)^{-\frac{3}{4}})e^{-\lambda_1(t-s)}e^{-(1-\epsilon)s}$$

$$+ C_3(\epsilon) \int_0^t (1 + (t-s)^{-\frac{3}{4}})e^{-\lambda_1(t-s)}e^{-\min\{1,\gamma,\lambda\}(1-\epsilon)s}$$
$$\leq C_4(\epsilon)e^{-\min\{\lambda_1,1,\gamma,\lambda\}(1-\epsilon)t},$$

which implies (3.5.41).

Lemma 3.44 *Let $w_0 > 1 - \dfrac{1}{\rho}$. Then for any $\epsilon \in (0, \min\{\lambda_1, 1, \gamma, \lambda\})$, there exists $C(\epsilon) > 0$ such that*

$$|\overline{p}(t) - 1| \leq C(\epsilon)e^{-(\min\{2\lambda_1, 2, 2\gamma, \lambda\}-\epsilon)t}. \tag{3.5.42}$$

Proof We integrate the p-equation in the spatial variable over Ω to obtain

$$
\begin{aligned}
(\overline{p} - 1)_t &= \lambda(\overline{p} - \overline{p}^2 - \frac{1}{|\Omega|}\int_\Omega (p - \overline{p})^2) \\
&= -\lambda\overline{p}(\overline{p} - 1) - \frac{\lambda}{|\Omega|}\|p - \overline{p}\|_{L^2(\Omega)}^2.
\end{aligned}
\tag{3.5.43}
$$

By (3.5.23), there exists $t_\epsilon > 0$ such that $\overline{p}(t) \geq 1 - \epsilon$ for $t \geq t_\epsilon$. Hence by (3.5.41) and (3.5.43), solving the differential equation entails

$$
\begin{aligned}
|\overline{p}(t) - 1| &\leq |\overline{p}(t_\varepsilon) - 1|e^{-\lambda\int_{t_\varepsilon}^t \overline{p}(s)ds} + \frac{\lambda}{|\Omega|}\int_{t_\varepsilon}^t e^{-\lambda\int_s^t \overline{p}(\sigma)d\sigma}\|p(s) - \overline{p}(s)\|_{L^2(\Omega)}^2 \\
&\leq |\overline{p}(t_\varepsilon) - 1|e^{-\lambda(1-\epsilon)(t-t_\epsilon)} + \frac{\lambda}{|\Omega|}\int_{t_\varepsilon}^t e^{-\lambda(1-\epsilon)(t-s)}\|p(s) - \overline{p}(s)\|_{L^2(\Omega)}^2 \\
&\leq |\overline{p}(t_\varepsilon) - 1|e^{-\lambda(1-\epsilon)(t-t_\epsilon)} + C_1(\epsilon)\int_0^t e^{-\lambda(1-\epsilon)(t-s)}e^{-2\min\{\lambda_1,1,\gamma,\lambda\}(1-\epsilon)s} \\
&\leq |\overline{p}(t_\varepsilon) - 1|e^{-\lambda(1-\epsilon)(t-t_\epsilon)} + C_2(\epsilon)e^{-\min\{2\lambda_1,2,2\gamma,\lambda\}(1-\epsilon)t} \\
&\leq C_3(\epsilon)e^{-\min\{2\lambda_1,2,2\gamma,\lambda\}(1-\epsilon)t}
\end{aligned}
$$

for $t \geq t_\epsilon$, which proves (3.5.42).

Lemma 3.45 *Let $w_0 > 1 - \dfrac{1}{\rho}$. Then for any $\epsilon \in (0, \min\{\lambda_1, 1, \gamma, \lambda\})$, there exists $C(\epsilon) > 0$ such that*

$$\|p(\cdot, t) - 1\|_{L^\infty(\Omega)} \leq C(\epsilon)e^{-(\min\{\lambda_1, 1, \gamma, \lambda\}-\epsilon)t}. \tag{3.5.44}$$

Proof Combining above two lemmas, we have

$$\|p(\cdot, t) - 1\|_{L^\infty(\Omega)} \leq \|p(\cdot, t) - \overline{p}(t)\|_{L^\infty(\Omega)} + |\overline{p}(t) - 1| \leq C(\epsilon)e^{-(\min\{\lambda_1, 1, \gamma, \lambda\}-\epsilon)t}.$$

Proof of Theorem 3.5. (3.1.7) is a direct consequence of Lemma 3.35 in the previous subsection. As for (3.1.8)–(3.1.10), we only need to collect (3.5.19), (3.5.32) and (3.5.44).

Remark 3.5 In comparison with (3.5.44), by (3.5.21) and (3.5.37), we have

$$\sup_{t\geq 0}\int_{\Omega}|w_x(t)|^2+|z_x(t)|^2\leq C,$$

and thus $\sup_{t\geq 0}\|p(t)\|_{W^{1,2}(\Omega)}\leq C$. Hence, an interpolation by means of the Gagliardo–Nirenberg inequality in the one-dimensional setting provides

$$\|p(\cdot,t)-1\|_{L^{\infty}(\Omega)}\leq\|p(\cdot,t)\|_{W^{1,2}(\Omega)}^{\frac{1}{2}}\|p(\cdot,t)-1\|_{L^2(\Omega)}^{\frac{1}{2}}\leq C(\epsilon)e^{-(\frac{1}{2}\min\{1,\gamma,\lambda\}-\epsilon)t},$$

where we have used (3.5.40).

Chapter 4
Keller–Segel–(Navier–)Stokes System Modeling Coral Fertilization

4.1 Introduction

Chemotaxis, the directed movement caused by the concentration of certain chemicals, is ubiquitous in biology and ecology, and has a significant effect on pattern formation in numerous biological contexts (Hillen and Painter 2009; Maini et al. 1991). The first mathematically rigorous studies of chemotaxis were carried out by Patlak (1953) and Keller–Segel (1970). The latter work involves the derivation of a system of PDEs, now known as the Keller–Segel system, which, despite its simple structure, was proved to have a lasting impact as a theoretical framework describing the collective behavior of populations under the influence of a chemotactic signal produced by the populations themselves (Bellomo et al. 2015; Herrero and Velázquez 1997; Winkler 2013, 2014c). In contrast to this well-understood Keller–Segel system, there seem to be few theoretical results on nontrivial behavior in situations where the signal is not produced by the population, such as in oxygenotaxis processes of swimming aerobic bacteria (Tuval et al. 2005), or where the signal production occurs by indirect processes, such as in glycolysis reaction, tumor invasion and the spread of the mountain pine beetle (Chaplain and Lolas 2005; Dillon et al. 1994; Fujie and Senba 2017; Hu and Tao 2016; Painter et al. 2000; Tao and Winkler 2017b).

In this chapter, we study the decay property of the chemotaxis–fluid systems modeling coral fertilization. Section 4.3 is concerned with the following Keller–Segel–Stokes system

$$
\begin{cases}
\rho_t + u \cdot \nabla\rho = \Delta\rho - \nabla \cdot (\rho\mathscr{S}(x, \rho, c)\nabla c) - \rho m, & (x, t) \in \Omega \times (0, T), \\
m_t + u \cdot \nabla m = \Delta m - \rho m, & (x, t) \in \Omega \times (0, T), \\
c_t + u \cdot \nabla c = \Delta c - c + m, & (x, t) \in \Omega \times (0, T), \\
u_t = \Delta u - \nabla P + (\rho + m)\nabla\phi, \quad \nabla \cdot u = 0, & (x, t) \in \Omega \times (0, T), \\
(\nabla\rho - \rho\mathscr{S}(x, \rho, c)\nabla c) \cdot v = \nabla m \cdot v = \nabla c \cdot v = 0, u = 0, & (x, t) \in \partial\Omega \times (0, T), \\
\rho(x, 0) = \rho_0(x), \quad m(x, 0) = m_0(x), \quad c(x, 0) = c_0(x), & u(x, 0) = u_0(x), \; x \in \Omega,
\end{cases}
$$

$$\tag{4.1.1}$$

© The Author(s) 2022

Y. Ke et al., *Analysis of Reaction-Diffusion Models with the Taxis Mechanism*, Financial Mathematics and Fintech, https://doi.org/10.1007/978-981-19-3763-7_4

where $T \in (0, \infty]$, $\Omega \subset \mathbb{R}^3$ is a bounded domain with smooth boundary $\partial\Omega$, the chemotactic sensitivity tensor $\mathcal{S}(x, \rho, c) = (s_{ij}(x, \rho, c)) \in C^2(\overline{\Omega} \times [0, \infty)^2)$, $i, j \in \{1, 2, 3\}$, and $\phi \in W^{2,\infty}(\Omega)$.

This PDE system describes the phenomenon of coral broadcast spawning (Espejo and Suzuki 2017; Espejo and Winkler 2018; Kiselev and Ryzhik 2012a, b), where the sperm ρ chemotactically moves toward the higher concentration of the chemical c released by the egg m, while the egg m is merely affected by random diffusion, fluid transport and degradation upon contact with the sperm. Meanwhile, the fluid flow vector u, modeling the ambient ocean environment, satisfies a Stokes equation, where $P = P(x, t)$ represents the associated pressure, and the buoyancy effect of the sperm and egg on the velocity, mediated through a given gravitational potential ϕ, is taken into account. We note that the use of the Stokes equation instead of the Navier–Stokes equation is justified by the observation that the fluid flow is relatively slow compared with the movement of the sperm and egg. We further note that the sensitivity tensor $\mathcal{S}(x, \rho, c)$ may take values that are matrices possibly containing nontrivial off-diagonal entries, which reflects that the chemotactic migration may not necessarily be oriented along the gradient of the chemical signal, but may rather involve rotational flux components (see Xue and Othmer (2009); Xue (2015) for the detailed model derivation).

A two-component variant of (4.1.1) has been used in the mathematical study of coral broadcast spawning. Indeed, in Kiselev and Ryzhik (2012a, b), Kiselev and Ryzhik investigated the important effect of chemotaxis on the coral fertilization process via the Keller–Segel type system of the form

$$
\begin{cases}
\rho_t + u \cdot \nabla\rho = \Delta\rho - \chi\nabla \cdot (\rho\nabla c) - \mu\rho^q, \\
0 = \Delta c + \rho
\end{cases}
\tag{4.1.2}
$$

with a given regular solenoidal fluid flow vector u. This model implicitly assumes that the densities of sperm and egg gametes are identical, and that the Péclet number for the chemical concentration c is small which allows us to ignore the effects of convection on c. The authors showed that, for the Cauchy problem in $\mathbb{R}^2$, the total mass $\int_{\mathbb{R}^2} \rho(x, t)dx$ can become arbitrarily small with increasing χ in the case $q > 2$ of supercritical reaction, whereas in the critical case $q = 2$, a weaker but related effect within finite time intervals is observed. Recently, Ahn et al. (2017) established the global well-posedness of regular solutions for the variant model of (4.1.2) with $c_t + u \cdot \nabla c = \Delta c - c + \rho$ instead of $0 = \Delta c + \rho$. They also proved that $\int_{\mathbb{R}^d} \rho(x, t)dx$ ($d = 2, 3$) asymptotically approaches a strictly positive constant $C(\chi)$ which tends to 0 as $\chi \to \infty$.

In Espejo and Suzuki (2015), Espejo and Suzuki studied the three-component variant of (4.1.1)

$$\begin{cases} \rho_t + u \cdot \nabla \rho = \Delta \rho - \chi \nabla \cdot (\rho \mathscr{S}(x, \rho, c) \nabla c) - \mu \rho^2, \\ c_t + u \cdot \nabla c = \Delta c - c + \rho, \\ u_t + \kappa(u \cdot \nabla)u = \Delta u - \nabla P + \rho \nabla \phi, \\ \nabla \cdot u = 0 \end{cases} \qquad (4.1.3)$$

in the modeling of broadcast spawning when the interaction of chemotactic movement of the gametes and the surrounding fluid is not negligible. Here the coefficient $\kappa \in \mathbb{R}$ is related to the strength of nonlinear convection. In particular, when the fluid flow is slow, we can use the Stokes instead of the Navier–Stokes equation, i.e., assume $\kappa = 0$ (see Difrancesco et al. (2010); Lorz (2010)). It should be mentioned that the chemotaxis–fluid model with $c_t + u \cdot \nabla c = \Delta c - c\rho$ replacing the second equation in (4.1.3) has also been used to describe the behavior of bacteria of the species Bacillus subtilis suspended in sessile water drops (Tuval et al. 2005). From the viewpoint of mathematical analysis, this chemotaxis–fluid system compounds the known difficulties in the study of fluid dynamics with the typical intricacies in the study of chemotaxis systems. It has also been observed that when $\mathscr{S} = \mathscr{S}(x, \rho, c)$ is a tensor, the corresponding chemotaxis–fluid system loses some energy-like structure, which plays a key role in the analysis of the scalar-valued case. Despite these challenges, some comprehensive results on the global boundedness and large time behavior of solutions are available in the literature (see Cao and Lankeit (2016); Li et al. (2019a); Liu and Wang (2017); Tao and Winkler (2015b); Wang and Xiang (2016); Winkler (2012, 2017b, 2018c, e) for example). It has been shown that when $\mathscr{S} = \mathscr{S}(x, \rho, c)$ is a tensor fulfilling

$$|\mathscr{S}(x, \rho, c)| \leq \frac{C_{\mathscr{S}}}{(1 + \rho)^\alpha} \quad \text{for some } \alpha > 0 \text{ and } C_{\mathscr{S}} > 0, \qquad (4.1.4)$$

the three-dimensional system (4.1.3) with $\mu = 0$, $\kappa = 0$ admits globally bounded weak solutions for $\alpha > 1/2$ (Wang and Xiang 2016), which is slightly stronger than the corresponding subcritical assumption $\alpha > 1/3$ for the fluid-free system. As for $\alpha \geq 0$, when the suitably regular initial data (ρ_0, c_0, u_0) fulfill a smallness condition, (4.1.3) with $\mu = 0$, $\kappa = 1$ possesses a global classical solution which decays to $(\bar{\rho}_0, \bar{\rho}_0, 0)$ exponentially with $\bar{\rho}_0 = \frac{1}{|\Omega|} \int_\Omega \rho_0(x)dx$ (Yu et al. 2018). Removing the presupposition that the densities of the sperm and egg coincide at each point, Espejo and Suzuki (2017) looked at a simplified version of (4.1.1) in two dimensions, namely

$$\begin{cases} \rho_t + u \cdot \nabla \rho = \Delta \rho - \chi \nabla \cdot (\rho \nabla c) - \rho m, \\ m_t + u \cdot \nabla m = \Delta m - \rho m, \\ 0 = \Delta c + k_0(m - \frac{1}{|\Omega|} \int_\Omega m dx) \text{ with } \int_\Omega c dx = 0, \end{cases} \qquad (4.1.5)$$

and showed that $\int_\Omega \rho_0(x)dx \geq \int_\Omega m_0(x)dx$ implies that $m(x,t)$ vanishes asymptotically, while $\int_\Omega \rho(x,t)dx \to \frac{1}{|\Omega|}(\int_\Omega \rho_0(x)dx - \int_\Omega m_0(x)dx)$ as $t \to \infty$, provided that χ is small enough and u is low. In two dimensions, Espejo and Winkler (2018) have recently considered the Navier–Stokes version of (4.1.1)

$$\begin{cases} \rho_t + u \cdot \nabla\rho = \Delta\rho - \nabla \cdot (\rho\nabla c) - \rho m, \\ m_t + u \cdot \nabla m = \Delta m - \rho m, \\ c_t + u \cdot \nabla c = \Delta c - c + m, \\ u_t + \kappa(u \cdot \nabla) = \Delta u - \nabla P + (\rho + m)\nabla\phi, \quad \nabla \cdot u = 0, \end{cases} \tag{4.1.6}$$

and established the global existence of classical solutions to the associated initial-boundary value problem, which tend toward a spatially homogeneous equilibrium in the large time limit.

In Sect. 4.3, motivated by the above works, we shall consider the properties of solutions to (4.1.1). In particular, we shall show that the corresponding solutions converge to a spatially homogeneous equilibrium exponentially as $t \to \infty$ as well.

Throughout the rest of this part, we shall assume that

$$\begin{cases} \rho_0 \in C^0(\overline{\Omega}), \ \rho_0 \geq 0 \text{ and } \rho_0 \not\equiv 0, \\ m_0 \in C^0(\overline{\Omega}), \ m_0 \geq 0 \text{ and } m_0 \not\equiv 0, \\ c_0 \in W^{1,\infty}(\Omega), \ c_0 \geq 0 \text{ and } c_0 \not\equiv 0, \\ u_0 \in D(A^\beta) \text{ for all } \beta \in (\frac{3}{4}, 1), \end{cases} \tag{4.1.7}$$

where A denotes the realization of the Stokes operator in $L^2(\Omega)$. Under these assumptions, we shall first establish the existence of global bounded classical solutions to (4.1.1):

Theorem 4.1 *Suppose that (4.1.4), (4.1.7) hold with $\alpha > \frac{1}{3}$. Then the system (4.1.1) admits a global classical solution (ρ, m, c, u, P), which is uniformly bounded in the sense that for any $\beta \in (\frac{3}{4}, 1)$, there exists $K > 0$ such that for all $t \in (0, \infty)$*

$$\|\rho(\cdot,t)\|_{L^\infty(\Omega)} + \|m(\cdot,t)\|_{L^\infty(\Omega)} + \|c(\cdot,t)\|_{W^{1,\infty}(\Omega)} + \|A^\beta u(\cdot,t)\|_{L^2(\Omega)} \leq K. \tag{4.1.8}$$

Then, we establish the large time behavior of these solutions as follows:

Theorem 4.2 *Under the assumptions of Theorem 4.1, the solutions given by Theorem 4.1 satisfy*

$$\rho(\cdot,t) \to \rho_\infty, \ m(\cdot,t) \to m_\infty, \ c(\cdot,t) \to m_\infty, \ u(\cdot,t) \to 0 \text{ in } L^\infty(\Omega) \text{ as } t \to \infty.$$

Furthermore, when $\int_\Omega \rho_0 \neq \int_\Omega m_0$, there exist $K > 0$ and $\delta > 0$ such that

$$\|\rho(\cdot, t) - \rho_\infty\|_{L^2(\Omega)} \leq K e^{-\delta t}, \tag{4.1.9}$$

$$\|m(\cdot, t) - m_\infty\|_{L^\infty(\Omega)} \leq K e^{-\delta t}, \tag{4.1.10}$$

$$\|c(\cdot, t) - m_\infty\|_{L^\infty(\Omega)} \leq K e^{-\delta t}, \tag{4.1.11}$$

$$\|u(\cdot, t)\|_{L^\infty(\Omega)} \leq K e^{-\delta t}, \tag{4.1.12}$$

where $\rho_\infty = \frac{1}{|\Omega|} \left\{ \int_\Omega \rho_0 - \int_\Omega m_0 \right\}_+$, $m_\infty = \frac{1}{|\Omega|} \left\{ \int_\Omega m_0 - \int_\Omega \rho_0 \right\}_+$.

According to the result for the related fluid–free system, the subcritical restriction $\alpha > \frac{1}{3}$ seems to be necessary for the existence of global bounded solutions. However, for $\alpha \leq \frac{1}{3}$, inspired by Cao and Lankeit (2016); Yu et al. (2018), we investigate the existence of global bounded classical solutions and their large time behavior under a smallness assumption imposed on the initial data, which can be stated as follows Li et al. (2019b):

Theorem 4.3 *Suppose that (4.1.4) hold with $\alpha = 0$ and $\int_\Omega \rho_0 \neq \int_\Omega m_0$. Further, let $N = 3$ and $p_0 \in (\frac{N}{2}, \infty)$, $q_0 \in (N, \infty)$ if $\int_\Omega \rho_0 > \int_\Omega m_0$; and $p_0 \in (\frac{2N}{3}, \infty)$, $q_0 \in (N, \infty)$ if $\int_\Omega \rho_0 < \int_\Omega m_0$. Then there exists $\varepsilon > 0$ such that for any initial data (ρ_0, m_0, c_0, u_0) fulfilling (4.1.7) as well as*

$$\|\rho_0 - \rho_\infty\|_{L^{p_0}(\Omega)} \leq \varepsilon, \quad \|m_0 - m_\infty\|_{L^{q_0}(\Omega)} \leq \varepsilon, \quad \|\nabla c_0\|_{L^N(\Omega)} \leq \varepsilon, \quad \|u_0\|_{L^N(\Omega)} \leq \varepsilon,$$

(4.1.1) possesses a global classical solution (ρ, m, c, u, P). Moreover, for any $\alpha_1 \in (0, \min\{\lambda_1, m_\infty + \rho_\infty\})$, $\alpha_2 \in (0, \min\{\alpha_1, \lambda_1', 1\})$, there exist constants K_i, $i = 1, 2, 3, 4$, such that for all $t \geq 1$,

$$\|m(\cdot, t) - m_\infty\|_{L^\infty(\Omega)} \leq K_1 e^{-\alpha_1 t}, \quad \|\rho(\cdot, t) - \rho_\infty\|_{L^\infty(\Omega)} \leq K_2 e^{-\alpha_1 t},$$

$$\|c(\cdot, t) - m_\infty\|_{W^{1,\infty}(\Omega)} \leq K_3 e^{-\alpha_2 t}, \quad \|u(\cdot, t)\|_{L^\infty(\Omega)} \leq K_4 e^{-\alpha_2 t}.$$

Here λ_1' is the first eigenvalue of A, and λ_1 is the first nonzero eigenvalue of $-\Delta$ on Ω under the Neumann boundary condition.

Remark 4.1 In Theorem 4.3, we have excluded the case $\int_\Omega \rho_0 = \int_\Omega m_0$. Indeed, in this case, some results of Cao and Winkler (2018) suggest that exponential decay of solutions may not hold.

Remark 4.2 It is observed that the similar result to Theorem 4.3 is also valid for the Navier–Stokes version of (4.1.1) upon slight modification of the definition of T in (4.3.60) and (4.3.94).

As mentioned above, compared with the scalar sensitivity $\mathscr{S}$, the system (4.1.1) with rotational tensor loses a favorable quasi-energy structure. For example, we note that the integral

$$\int_{\Omega} \rho ln\rho + a \int_{\Omega} |\nabla c|^2 + b \int_{\Omega} |u|^2$$

with appropriate positive constants a and b plays a favorable entropy-like functional in deriving the bounds of solution to (4.1.6). However, this will no longer be available in the present situation (see Espejo and Winkler (2018)). To overcome this difficulty, our approach underlying the derivation of Theorem 4.1 will be based on the estimate of the functional

$$\|\rho(\cdot, t)\|_{L^2(\Omega)}^2 + \|u(\cdot, t)\|_{W^{1,2}(\Omega)}^2 + \|\nabla c(\cdot, t)\|_{L^2(\Omega)}^2.$$

In addition, the proof of the exponential decay results in Theorem 4.2 relies on careful analysis of the functional

$$G(t) := \int_{\Omega} (\rho - \overline{\rho})^2 + a \int_{\Omega} (m - \overline{m})^2 + b \int_{\Omega} (c - \overline{c})^2 + c \int_{\Omega} \rho m$$

with suitable parameters a, b, $c > 0$. Indeed, it can be seen that $G(t)$ satisfies the ODE $G'(t) + \delta_1 G(t) \leq 0$ for some $\delta_1 > 0$, and thereby the convergence rate of solutions in $L^2(\Omega)$ is established. At the same time, in comparison with the chemotaxis–fluid system considered in Cao and Lankeit (2016); Yu et al. (2018), due to

$$\|e^{t\Delta}\omega\|_{L^p(\Omega)} \leq c_1 \left(1 + t^{-\frac{N}{2}(\frac{1}{q} - \frac{1}{p})}\right) e^{-\lambda_1 t} \|\omega\|_{L^q(\Omega)}$$

for all $\omega \in L^q(\Omega)$ with $\int_{\Omega} \omega = 0$, $-\rho m$ in the first equation of (4.1.1) gives rise to some difficulty in mathematical analysis despite its dissipative feature. Accordingly, it requires a nontrivial application of the mass conservation of $\rho(x, t) - m(x, t)$.

In Sect. 4.4, we are concerned with the asymptotic behavior of classical solutions of the three-dimensional Keller–Segal–Navier–Stokes system

$$\begin{cases} \rho_t + u \cdot \nabla\rho = \Delta\rho - \nabla \cdot (\rho \mathscr{S}(x, \rho, c)\nabla c) - \rho m, & (x, t) \in \Omega \times (0, T), \\ m_t + u \cdot \nabla m = \Delta m - \rho m, & (x, t) \in \Omega \times (0, T), \\ c_t + u \cdot \nabla c = \Delta c - c + m, & (x, t) \in \Omega \times (0, T), \\ u_t + (u \cdot \nabla)u = \Delta u - \nabla P + (\rho + m)\nabla\phi, \quad \nabla \cdot u = 0, & (x, t) \in \Omega \times (0, T), \\ (\nabla\rho - \rho\mathscr{S}(x, \rho, c)\nabla c) \cdot v = \nabla m \cdot v = \nabla c \cdot v = 0, u = 0, & (x, t) \in \partial\Omega \times (0, T), \\ \rho(x, 0) = \rho_0(x), \quad m(x, 0) = m_0(x), \quad c(x, 0) = c_0(x), & u(x, 0) = u_0(x), \ x \in \Omega. \end{cases}$$
$$(4.1.13)$$

In this coral fertilization model, the sperm ρ chemotactically moves toward the higher concentration of the chemical c released by the egg m, while the egg m is merely affected by random diffusion, fluid transport and degradation upon contact with the sperm. We assume that the tensor-valued chemotactic sensitivity $\mathscr{S} = \mathscr{S}(x, \rho, c)$ satisfies

$$|\mathscr{S}(x, \rho, c)| \leq C_{\mathscr{S}} \quad \text{for some } C_{\mathscr{S}} > 0, \tag{4.1.14}$$

and the initial data satisfy

$$\begin{cases} \rho_0 \in C^0(\overline{\Omega}), \ \rho_0 \geq 0 \text{ and } \rho_0 \not\equiv 0, \\ m_0 \in C^0(\overline{\Omega}), \ m_0 \geq 0 \text{ and } m_0 \not\equiv 0, \\ c_0 \in W^{1,\infty}(\Omega), \ c_0 \geq 0 \text{ and } c_0 \not\equiv 0, \\ u_0 \in D(A^\beta) \text{ for all } \beta \in (\frac{3}{4}, 1), \end{cases} \tag{4.1.15}$$

where A denotes the realization of the Stokes operator in $L^2(\Omega)$.

Under these assumptions, our main result can be stated as follows Myowin et al. (2020):

Theorem 4.4 *Suppose that (4.1.14) hold and $\int_\Omega \rho_0 > \int_\Omega m_0$. Let $p_0 \in (\frac{3}{2}, 3)$, $q_0 \in (3, \frac{3p_0}{3-p_0})$. Then, there exists $\varepsilon > 0$ such that for any initial data (ρ_0, m_0, c_0, u_0) fulfilling (4.1.15) as well as*

$$\|\rho_0 - \rho_\infty\|_{L^{p_0}(\Omega)} < \varepsilon, \quad \|m_0\|_{L^{q_0}(\Omega)} < \varepsilon, \quad \|c_0\|_{L^\infty(\Omega)} < \varepsilon, \quad \|u_0\|_{L^3(\Omega)} < \varepsilon,$$

(4.1.13) admits a global classical solution (ρ, m, c, u, P). In particular, for any $\alpha_1 \in (0, \min\{\lambda_1, \rho_\infty\})$, $\alpha_2 \in (0, \min\{\alpha_1, \lambda_1', 1\})$, there exist constants K_i, $i = 1, 2, 3, 4$, such that for all $t \geq 1$

$$\|m(\cdot, t)\|_{L^\infty(\Omega)} \leq K_1 e^{-\alpha_1 t},$$
$$\|\rho(\cdot, t) - \rho_\infty\|_{L^\infty(\Omega)} \leq K_2 e^{-\alpha_1 t},$$
$$\|c(\cdot, t)\|_{W^{1,\infty}(\Omega)} \leq K_3 e^{-\alpha_2 t},$$
$$\|u(\cdot, t)\|_{L^\infty(\Omega)} \leq K_4 e^{-\alpha_2 t}.$$

Here, λ_1 is the first nonzero eigenvalue of $-\Delta$ on Ω under the Neumann boundary condition, $\rho_\infty = \frac{1}{|\Omega|}(\int_\Omega \rho_0 - \int_\Omega m_0)$. and λ_1' is the first eigenvalue of A.

As for the case $\int_\Omega \rho_0 < \int_\Omega m_0$, i.e., $m_\infty = \frac{1}{|\Omega|}(\int_\Omega m_0 - \int_\Omega \rho_0) > 0$, we have Myowin et al. (2020)

Theorem 4.5 *Assume that (4.1.14) and $\int_\Omega \rho_0 < \int_\Omega m_0$ hold, and let $p_0 \in (2, 3)$, $q_0 \in (3, \frac{3p_0}{2(3-p_0)})$. Then there exists $\varepsilon > 0$ such that for any initial data (ρ_0, m_0, c_0, u_0) fulfilling (4.1.15) as well as*

$$\|\rho_0\|_{L^{p_0}(\Omega)} \leq \varepsilon, \quad \|m_0 - m_\infty\|_{L^{q_0}(\Omega)} \leq \varepsilon, \quad \|\nabla c_0\|_{L^3(\Omega)} \leq \varepsilon, \quad \|u_0\|_{L^3(\Omega)} \leq \varepsilon,$$

(4.1.13) admits a global classical solution (ρ, m, c, u, P). Furthermore, for any $\alpha_1 \in (0, \min\{\lambda_1, m_\infty, 1\})$, $\alpha_2 \in (0, \min\{\alpha_1, \lambda_1'\})$, there exist constants $K_i > 0$, $i = 1, 2, 3, 4$, such that

$$\|m(\cdot,t) - m_\infty\|_{L^\infty(\Omega)} \leq K_1 e^{-\alpha_1 t},$$
$$\|\rho(\cdot,t)\|_{L^\infty(\Omega)} \leq K_2 e^{-\alpha_1 t},$$
$$\|c(\cdot,t) - m_\infty\|_{W^{1,\infty}(\Omega)} \leq K_3 e^{-\alpha_2 t},$$
$$\|u(\cdot,t)\|_{L^\infty(\Omega)} \leq K_4 e^{-\alpha_2 t}.$$

Remark 4.3 In our results, we have excluded the case $\int_\Omega \rho_0 = \int_\Omega m_0$. Indeed, in light of results of Cao and Winkler (2018); Htwe and Wang (2019), algebraic decay rather than exponential decay of the solutions is expected in this case.

It is noted that the nonlinear convection $(u \cdot \nabla)u$ in the three-dimensional Navier–Stokes equation may lead to the spontaneous emergence of singularities, resulting in a blow-up with respect to the norm of $L^\infty(\Omega)$. Hence, we subject the study of classical solutions of (4.1.13) to small initial data. We further note the substantial difference between dimensions two and three, and acknowledge results on global boundedness in two dimensions obtained by Espejo (2018) in the case of scalar-valued sensitivity and by Li (2019) in the case of tensor-valued sensitivity with saturation effect or suitably small initial data.

Section 4.5 is devoted to the large time behavior in a chemotaxis-Stokes system modeling coral fertilization with arbitrarily slow porous medium diffusion. In accordance with the phenomena observed from experiments (Coll et al. 1994, 1995; Miller 1979, 1985), oriented motions may occur to sperms in response to some chemical signal secreted by eggs during the period of coral fertilization. In order to describe this in mathematics, a model appearing as

$$\begin{cases} n_t + u \cdot \nabla n = \nabla \cdot (D(n)\nabla n) - \nabla \cdot (n\nabla c) - nv, \\ c_t + u \cdot \nabla c = \Delta c - c + v, \\ v_t + u \cdot \nabla v = \Delta v - vn, \\ u_t = \Delta u + \nabla P + (n + v)\nabla\Phi, \end{cases} \tag{4.1.16}$$

was proposed under the assumptions that sperms and eggs enjoy different densities n and v, respectively, that P and Φ separately stand for the liquid pressure and the gravitational potential with

$$\Phi \in W^{2,\infty}(\Omega), \tag{4.1.17}$$

and that the fluid velocity u is an unknown function (Espejo and Winkler 2018).

For simplified versions of (4.1.16), such as $D \equiv 1$ together with $n \equiv v$ or with a given fluid field u, related analytical results on global dynamic behaviors of the solution can be found in Espejo and Suzuki (2015, 2017). Whereas for more complex situations, during the past years, a number of analytic approaches have been developed to explore global dynamics in (4.1.16) and the variants thereof.

In particular, under the interaction of linear diffusion, i.e., $D \equiv 1$, with proper saturation effects of cells, by constructing appropriate weighted functions g and whereafter detecting the evolution of

$$\int_{\Omega} n^p g(c) \tag{4.1.18}$$

with any $p > 1$, system (4.1.16) coupled with (Navier–)Stokes-fluid is proved to be globally solvable in the classical sense (Li 2019) or in the weak sense (Zheng 2021). Moreover, arguments based on L^p-L^q estimates for Neumann heat semigroup further show exponential decay features of the corresponding classical solutions under suitable smallness assumptions on initial data (Htwe et al. 2020; Li et al. 2019b).

As a more frequently used method, energy-based arguments, which start from constructions of proper energy functionals, play a crucial role in the whole study of systems related to (4.1.16). More precisely, as shown in Espejo and Winkler (2018), an analysis of a suitably established entropy-like functional

$$\int_{\Omega} n \ln n + k_1 \int_{\Omega} |\nabla c|^2 + k_2 \int_{\Omega} |u|^2 \tag{4.1.19}$$

with $k_1 > 0$ and $k_2 > 0$ underlies the derivation of global boundedness and stabilization of the unique classical solution to the Navier–Stokes version of system (4.1.16) with $D \equiv 1$ in spatially two-dimensional setting. In cases when saturation influence of cells is accounted for in the cross-diffusion term of n-equation, the construction of a similar but different functional as compared to (4.1.19) is also viewed as the fundament in deriving global solvability of system (4.1.16) with $D \equiv 1$, both in the Stokes-fluid context (Li et al. 2019b) and in the Navier–Stokes-fluid setting (Liu et al. 2020). Apart from that, when cell mobility depends on gradients of some unknown quantity, such as p-Laplacian cell diffusion, the pursuance of global solvability involves an analysis of a functional with more complex structure (Liu 2020).

Actually, whether by establishing weighted estimates as (4.1.18) or by constructing energy functionals of different types, the core of the analysis is to derive a uniform L^p bound of component n for any $p > 1$. Taking a recent work (Liu 2020) as an example, in the presence of a porous medium type diffusion, namely D in (4.1.16) is chosen to generalize the prototypical case

$$D(s) = s^{m-1}, \quad s > 0 \tag{4.1.20}$$

with some $m > 1$, the condition

$$m > \frac{37}{33} \tag{4.1.21}$$

therein reflects an explicitly quantitative requirement for the strength of nonlinear diffusion in the derivation of temporally independent L^p estimates for n. However, since complementary results on possibly emerging explosion phenomena are rather barren, it is still unknown that corresponding uniform L^p bounds could be achieved for smaller values of m or even for the optimal restriction $m \geq 1$.

In the present work, we attempt to make use of a different method, by which conditional estimates for u and c subject to some uniform L^p norms of n are established, to explore how far the porous medium type diffusion of sperms can prevent the occurrence of singularity formation phenomena.

For precisely formulating our main results, let us close the considered problem involving system (4.1.16) with the following initial-boundary conditions

$$n(x, 0) = n_0(x), c(x, 0) = c_0(x), v(x, 0) = v_0(x) \text{ and } u(x, 0) = u_0(x), \quad x \in \Omega \tag{4.1.22}$$

as well as

$$D(n)\frac{\partial n}{\partial v} = \frac{\partial c}{\partial v} = \frac{\partial v}{\partial v} = 0 \text{ and } u = 0, \quad x \in \partial\Omega, \ t > 0, \tag{4.1.23}$$

where $\Omega \subset \mathbb{R}^3$ is a bounded domain with smooth boundary, where the function D fulfills

$$D \in C^\mu_{loc}([0, \infty)) \bigcap C^2_{loc}((0, \infty)) \text{ and } D(s) \geq C_D s^{m-1} \text{ for any } s \geq 0 \tag{4.1.24}$$

with certain $\mu \in (0, 1)$, $C_D > 0$ and $m \geq 1$, and where the initial data satisfies

$$\begin{cases} n_0 \in C^\nu(\bar{\Omega}) \text{ for some } \nu > 0 \text{ with } n_0 \geq 0 \text{ in } \Omega \text{ and } n_0 \not\equiv 0, \\ c_0 \in W^{1,\infty}(\Omega) \text{ with } c_0 \geq 0 \text{ in } \Omega, \\ v_0 \in W^{1,\infty}(\Omega) \text{ with } v_0 \geq 0 \text{ in } \Omega, \text{ and} \\ u_0 \in D(A^\alpha) \text{ for certain } \alpha \in \left(\frac{3}{4}, 1\right) \end{cases} \tag{4.1.25}$$

with A representing the realization of the Stokes operator with its domain defined as $D(A) := W^{2,2}(\Omega; \mathbb{R}^3) \bigcap W_0^{1,2}(\Omega; \mathbb{R}^3) \bigcap L_\sigma^2(\Omega)$ with $L_\sigma^2(\Omega) := \{\omega \in L^2(\Omega; \mathbb{R}^3)| \nabla \cdot \omega = 0\}$ (Sohr 2001).

Within this framework, our main results can be read as follows (Wang and Liu 2022).

Theorem 4.6 *Assume that $\Omega \subset \mathbb{R}^3$ is a bounded domain with smooth boundary. Let (4.1.17) be satisfied, and let (4.1.24) hold with*

$$m > 1. \tag{4.1.26}$$

Then for each (n_0, c_0, v_0, u_0) complying with (4.1.25), there exist functions n, c, v and u fulfilling

$$\begin{cases} n \in L^\infty(\Omega \times (0, \infty)) \cap C^0([0, \infty); (W_0^{2,2}(\Omega))^*), \\ c \in \cap_{r>3} L^\infty((0, \infty); W^{1,r}(\Omega)) \cap C^0(\bar{\Omega} \times [0, \infty)) \cap C^{1,0}(\bar{\Omega} \times (0, \infty)), \\ v \in \cap_{r>3} L^\infty((0, \infty); W^{1,r}(\Omega)) \cap C^0(\bar{\Omega} \times [0, \infty)) \cap C^{1,0}(\bar{\Omega} \times (0, \infty)), \\ u \in L^\infty(\Omega \times (0, \infty); \mathbb{R}^3) \cap L_{loc}^2([0, \infty); W_0^{1,2}(\Omega; \mathbb{R}^3) \cap L_\sigma^2(\Omega)) \cap C^0(\bar{\Omega} \times [0, \infty); \mathbb{R}^3), \end{cases} \tag{4.1.27}$$

such that $n \geq 0$, $c \geq 0$ and $v \geq 0$, and that along with certain $P \in C^0$ ($\Omega \times (0, \infty)$) the quintuple (n, c, v, u, P) becomes a global weak solution of the problem (4.1.16), (4.1.22) and (4.1.23) in the sense of Definition 4.1 below, and has the stabilization features that

$$\|n(\cdot, t) - n_\infty\|_{L^p(\Omega)} + \|c(\cdot, t) - v_\infty\|_{W^{1,\infty}(\Omega)} + \|v(\cdot, t) - v_\infty\|_{W^{1,\infty}(\Omega)} + \|u(\cdot, t)\|_{L^\infty(\Omega)} \to 0$$
$$(4.1.28)$$

for any $p \geq 1$ as $t \to \infty$ with

$$n_\infty := \frac{1}{|\Omega|} \left\{ \int_\Omega n_0 - \int_\Omega v_0 \right\}_+ \quad and \quad v_\infty := \frac{1}{|\Omega|} \left\{ \int_\Omega v_0 - \int_\Omega n_0 \right\}_+.$$

From (4.1.26), which shows the values that m could be taken herein for successfully establishing temporally independent L^p bounds of n, one can see that an apparent relaxation is realized in comparison to the previously derived range of m, i.e., (4.1.21). In fact, for introduced approximated problems of (4.1.16), (4.1.22) and (4.1.23), which is verified to be locally solvable with an extensible blow-up criterion, the hypothesis (4.1.26) allows for an application of a standard testing procedure to derive the uniform L^p estimates of $(n_\varepsilon)_{\varepsilon \in (0,1)}$ with the aids of conditionally uniform L^∞ estimates of $(\nabla c_\varepsilon)_{\varepsilon \in (0,1)}$ which are established by utilizing L^p-L^q estimates for fractional powers of a sectorial operator on the basis of basic estimates implied in the regularized problems and of some well-established conditional estimates of $(u_\varepsilon)_{\varepsilon \in (0,1)}$ (see Sects. 4.3–4.4). The derivation of (4.1.28) is essentially based on the dissipative effect of the considered consumption process, as shown in Espejo and Winkler (2018) for two-dimensional Navier–Stokes version of (4.1.16) with $m = 1$, or in Winkler (2015b, 2018c) and Winkler (2014b, 2017b, 2021a) for simplified oxygen-consumption type chemotaxis-fluid models with $m > 1$ and $m = 1$, respectively. More precisely, the absorptive contribution $-nv$ to the third equation in (4.1.16) implies time-independently uniform bounds of spatio-temporal integrals for nv and for the square of the gradients of both v and c, which underlies the achievement of the convergence of n, c and v in (4.1.28). Thanks to the convergence of n and v in (4.1.28), the large time behavior of u can be detected by means of a combination of variation-of-constants formula with regularity properties of analytic semigroup.

4.2 Preliminaries

In this subsection, we provide some preliminary results that will be used in the subsequent sections.

Next we introduce the Stokes operator and recall estimates for the corresponding semigroup. With $L_\sigma^p(\Omega) := \{\varphi \in L^p(\Omega) | \nabla \cdot \varphi = 0\}$ and $\mathscr{P}$ representing the Helmholtz projection of $L^p(\Omega)$ onto $L_\sigma^p(\Omega)$, the Stokes operator on $L_\sigma^p(\Omega)$ is defined as $A_p = -\mathscr{P}\Delta$ with domain $D(A_p) := W^{2,p}(\Omega) \cap W_0^{2,p}(\Omega) \cap L_\sigma^p(\Omega)$. Since A_{p_1}

and A_{p_2} coincide on the intersection of their domains for p_1, $p_2 \in (1, \infty)$, we will drop the index in the following.

Lemma 4.1 (Lemma 4.2 of Cao and Lankeit (2016)) *The Stokes operator A generates the analytic semigroup $(e^{-tA})_{t>0}$ in $L^r_\sigma(\Omega)$. Its spectrum satisfies $\lambda'_1 = \inf Re\sigma(A) > 0$ and we fix $\mu \in (0, \lambda'_1)$. For any such μ, we have*

(i) For any $p \in (1, \infty)$ and $\gamma \geq 0$, there is $c_5(p, \gamma) > 0$ such that for all $\phi \in L^p_\sigma(\Omega)$,

$$\|A^\gamma e^{-tA}\phi\|_{L^p(\Omega)} \leq c_5(p, \gamma)t^{-\gamma}e^{-\mu t}\|\phi\|_{L^p(\Omega)};$$

(ii) For any p, q with $1 < p \leq q < \infty$, there is $c_6(p, q) > 0$ such that for all $\phi \in L^p_\sigma(\Omega)$,

$$\|e^{-tA}\phi\|_{L^q(\Omega)} \leq c_6(p, q)t^{-\frac{N}{2}\left(\frac{1}{p}-\frac{1}{q}\right)}e^{-\mu t}\|\phi\|_{L^p(\Omega)};$$

(iii) For any p, q with $1 < p \leq q < \infty$, there is $c_7(p, q) > 0$ such that for all $\phi \in L^p_\sigma(\Omega)$,

$$\|\nabla e^{-tA}\phi\|_{L^q(\Omega)} \leq c_7(p, q)t^{-\frac{1}{2}-\frac{N}{2}\left(\frac{1}{p}-\frac{1}{q}\right)}e^{-\mu t}\|\phi\|_{L^p(\Omega)};$$

(iv) If $\gamma \geq 0$ and $1 < p < q < \infty$ satisfy $2\gamma - \frac{N}{q} \geq 1 - \frac{N}{p}$, there is $c_8(\gamma, p, q) > 0$ such that for all $\phi \in D(A^\gamma_q)$,

$$\|\phi\|_{W^{1,p}(\Omega)} \leq c_8(\gamma, p, q)\|A^\gamma\phi\|_{L^q(\Omega)}.$$

Lemma 4.2 (Theorem 1 and Theorem 2 of Fujiwara and Morimoto (1977)) *The Helmholtz projection $\mathscr{P}$ defines a bounded linear operator $\mathscr{P}: L^p(\Omega) \to L^p_\sigma(\Omega)$; in particular, for any $p \in (1, \infty)$, there exists $c_9(p) > 0$ such that $\|\mathscr{P}\omega\|_{L^p(\Omega)} \leq c_9(p)\|\omega\|_{L^p(\Omega)}$ for every $\omega \in L^p(\omega)$.*

The following elementary lemma provides some useful information on both the short time and the large time behavior of certain integrals, which is used in the proof of Theorem 4.3.

Lemma 4.3 (Lemma 1.2 of Winkler (2010)) *Let $\alpha < 1$, $\beta < 1$, and γ, δ be positive constants such that $\gamma \neq \delta$. Then there exists $c_{10}(\alpha, \beta, \gamma, \delta) > 0$ such that*

$$\int_0^t (1 + s^{-\alpha})(1 + (t - s)^{-\beta})e^{-\gamma s}e^{-\delta(t-s)}ds$$

$$\leq c_{10}(\alpha, \beta, \gamma, \delta)\left(1 + t^{\min\{0, 1-\alpha-\beta\}}\right)e^{-\min\{\gamma, \delta\}t}.$$

4.3 Global Boundedness and Decay Property of Solutions to a 3D Coral Fertilization Model

4.3.1 A Convenient Extensibility Criterion

At the beginning, we recall the result of the local existence of classical solutions, which can be proved by a straightforward adaptation of a well-known fixed point argument (see Winkler (2012) for example).

Lemma 4.4 *Suppose that* (4.1.4), (4.1.7) *and*

$$\mathscr{S}(x, \rho, c) = 0, \quad (x, \rho, c) \in \partial\Omega \times [0, \infty) \times [0, \infty) \tag{4.3.1}$$

hold. Then there exist $T_{max} \in (0, \infty]$ and a classical solution (ρ, m, c, u, P) of (4.1.1) *on* $(0, T_{max})$. *Moreover, ρ, m, c are nonnegative in $\Omega \times (0, T_{max})$, and if $T_{max} < \infty$, then for $\beta \in (\frac{3}{4}, 1)$,*

$$\lim_{t \to T_{max}} \left(\|\rho(\cdot, t)\|_{L^\infty(\Omega)} + \|m(\cdot, t)\|_{L^\infty(\Omega)} + \|c(\cdot, t)\|_{W^{1,\infty}(\Omega)} + \|A^\beta u(\cdot, t)\|_{L^2(\Omega)} \right) = \infty.$$

This solution is unique, up to addition of constants to P.

The following elementary properties of the solutions in Lemma 4.4 are immediate consequences of the integration of the first and second equations in (4.1.1), as well as an application of the maximum principle to the second and third equations.

Lemma 4.5 *Suppose that* (4.1.4), (4.1.7) *and* (4.3.1) *hold. Then for all $t \in (0, T_{max})$, the solution of* (4.1.1) *from Lemma 4.4 satisfies*

$$\|\rho(\cdot, t)\|_{L^1(\Omega)} \leq \|\rho_0\|_{L^1(\Omega)}, \quad \|m(\cdot, t)\|_{L^1(\Omega)} \leq \|m_0\|_{L^1(\Omega)}, \tag{4.3.2}$$

$$\int_0^t \|\rho(\cdot, s)m(\cdot, s)\|_{L^1(\Omega)}ds \leq \min\{\|\rho_0\|_{L^1(\Omega)}, \|m_0\|_{L^1(\Omega)}\}, \tag{4.3.3}$$

$$\|\rho(\cdot, t)\|_{L^1(\Omega)} - \|m(\cdot, t)\|_{L^1(\Omega)} = \|\rho_0\|_{L^1(\Omega)} - \|m_0\|_{L^1(\Omega)}, \tag{4.3.4}$$

$$\|m(\cdot, t)\|_{L^2(\Omega)}^2 + 2\int_0^t \|\nabla m(\cdot, s)\|_{L^2(\Omega)}^2 ds \leq \|m_0\|_{L^2(\Omega)}^2, \tag{4.3.5}$$

$$\|m(\cdot, t)\|_{L^\infty(\Omega)} \leq \|m_0\|_{L^\infty(\Omega)}, \tag{4.3.6}$$

$$\|c(\cdot, t)\|_{L^\infty(\Omega)} \leq \max\{\|m_0\|_{L^\infty(\Omega)}, \|c_0\|_{L^\infty(\Omega)}\}. \tag{4.3.7}$$

4.3.2 Global Boundedness and Decay for $\mathscr{S} = 0$ on $\partial\Omega$

In this subsection, we shall consider the case in which besides (4.1.4), the sensitivity satisfies $\mathscr{S} = 0$ on $\partial\Omega$. Under this hypothesis, the boundary condition for ρ in (4.1.1) actually reduces to the homogeneous Neumann condition $\nabla\rho \cdot \nu = 0$.

1. Global boundedness for $\mathscr{S} = 0$ on $\partial\Omega$

Lemma 4.6 *Suppose that* (4.1.4), (4.1.7), (4.3.1) *hold with* $\alpha > \frac{1}{3}$. *Then for any* $\varepsilon > 0$, *there exists* $K(\varepsilon) > 0$ *such that, for all* $t \in (0, T_{max})$, *the solution of* (4.1.1) *satisfies*

$$\frac{d}{dt}\|\rho(\cdot, t)\|^2_{L^2(\Omega)} + \frac{1}{2}\|\nabla\rho(\cdot, t)\|^2_{L^2(\Omega)} \leq \varepsilon\|\Delta c(\cdot, t)\|^2_{L^2(\Omega)} + K(\varepsilon). \quad (4.3.8)$$

Proof Multiplying the first equation of (4.1.1) by ρ, we obtain

$$\frac{1}{2}\frac{d}{dt}\int_\Omega \rho^2 + \int_\Omega |\nabla\rho|^2$$

$$= \int_\Omega \rho\mathscr{S}(x, \rho, c)\nabla\rho\nabla c - \int_\Omega \rho^2 m \quad (4.3.9)$$

$$\leq \frac{1}{2}\int_\Omega |\nabla\rho|^2 + \frac{C_S^2}{2}\int_\Omega \frac{\rho^2}{(1+\rho)^{2\alpha}}|\nabla c|^2.$$

Now we estimate the term $\frac{C_S^2}{2}\int_\Omega \frac{\rho^2}{(1+\rho)^{2\alpha}}|\nabla c|^2$ on the right-hand side of (4.3.9). In fact, if $\alpha \geq \frac{3}{4}$,

$$\frac{C_S^2}{2}\int_\Omega \frac{\rho^2}{(1+\rho)^{2\alpha}}|\nabla c|^2 \leq \varepsilon\int_\Omega |\nabla c|^4 + K(\varepsilon), \quad (4.3.10)$$

while for $\alpha \in \left(\frac{1}{3}, \frac{3}{4}\right)$,

$$\frac{C_S^2}{2}\int_\Omega \frac{\rho^2}{(1+\rho)^{2\alpha}}|\nabla c|^2 \leq \frac{C_S^2}{2}\int_\Omega \rho^{2-2\alpha}|\nabla c|^2$$

$$\leq \frac{C_S^4}{16\varepsilon}\int_\Omega \rho^{4-4\alpha} + \varepsilon\int_\Omega |\nabla c|^4. \quad (4.3.11)$$

On the other hand, by Lemma 4.5 and the Gagliardo–Nirenberg inequality, we get

$$\int_\Omega |\nabla c|^4 \leq C_{GN}\left\{\|\Delta c\|^2_{L^2(\Omega)}\|c\|^2_{L^\infty(\Omega)} + \|c\|^4_{L^\infty(\Omega)}\right\} \quad (4.3.12)$$

$$\leq C'_{GN}(\|\Delta c\|^2_{L^2(\Omega)} + 1)$$

and

$$\int_\Omega |\rho|^{4-4\alpha} = \|\rho\|^{4-4\alpha}_{L^{4-4\alpha}(\Omega)} \leq C_{GN}\left\{\|\nabla\rho\|^{(4-4\alpha)\lambda_2}_{L^2(\Omega)}\|\rho\|^{(4-4\alpha)(1-\lambda_2)}_{L^1(\Omega)} + \|\rho\|^{4-4\alpha}_{L^1(\Omega)}\right\}$$

with $\lambda_2 = \frac{6(3-4\alpha)}{5(4-4\alpha)}$. Due to $\alpha \in \left(\frac{1}{3}, \frac{3}{4}\right)$, we have $(4-4\alpha)\lambda_2 < 2$ and thus

$$\frac{C_S^4}{16\varepsilon} \int_\Omega |\rho|^{4-4\alpha} \le \frac{1}{4} \int_\Omega |\nabla\rho|^2 + K_1 \tag{4.3.13}$$

by the Young inequality. Combining (4.3.9)–(4.3.13), we readily have (4.3.8).

Lemma 4.7 *Under the assumptions of Lemma 4.6, there exists a positive constant* $C = C(m_0, c_0)$ *such that for all* $t \in (0, T_{max})$*, the solution of (4.1.1) satisfies*

$$\frac{d}{dt}\|\nabla c(\cdot, t)\|_{L^2(\Omega)}^2 + 2\|\nabla c(\cdot, t)\|_{L^2(\Omega)}^2 + \|\Delta c(\cdot, t)\|_{L^2(\Omega)}^2$$
$$\le K(\|\nabla u(\cdot, t)\|_{L^2(\Omega)}^2 + 1). \tag{4.3.14}$$

Proof Multiplying the c-equation of (4.1.1) by $-\Delta c$ and by the Young inequality, we obtain

$$\frac{1}{2}\frac{d}{dt}\int_\Omega |\nabla c|^2 + \int_\Omega |\Delta c|^2 + \int_\Omega |\nabla c|^2$$
$$\le -\int_\Omega m\Delta c + \int_\Omega (u \cdot \nabla c)\Delta c \tag{4.3.15}$$
$$= -\int_\Omega m\Delta c - \int_\Omega \nabla c \cdot (\nabla u \cdot \nabla c) - \int_\Omega \nabla c \cdot (D^2 c \cdot u)$$
$$= -\int_\Omega m\Delta c - \int_\Omega \nabla c \cdot (\nabla u \cdot \nabla c)$$
$$\le \int_\Omega |m|^2 + \frac{1}{4}\int_\Omega |\Delta c|^2 + \left(\int_\Omega |\nabla c|^4\right)^{\frac{1}{2}}\left(\int_\Omega |\nabla u|^2\right)^{\frac{1}{2}}$$
$$\le \|m\|_{L^2(\Omega)}^2 + \frac{1}{4}\|\Delta c\|_{L^2(\Omega)}^2 + \frac{1}{2\varepsilon}\|\nabla u\|_{L^2(\Omega)}^2 + \frac{\varepsilon}{2}\|\nabla c\|_{L^4(\Omega)}^4,$$

where the fact that u is solenoidal and vanishes on $\partial\Omega$ is used to ensure $\int_\Omega \nabla c \cdot (D^2 c \cdot u) = 0$.

By (4.3.12) and taking $\varepsilon = \frac{1}{2C_{GN}'}$ in the above inequality, we have

$$\frac{1}{2}\frac{d}{dt}\int_\Omega |\nabla c|^2 + \frac{1}{2}\int_\Omega |\Delta c|^2 + \int_\Omega |\nabla c|^2 \le \|m\|_{L^2(\Omega)}^2 + C_{GN}'\|\nabla u\|_{L^2(\Omega)}^2 + \frac{1}{4},$$

which along with (4.3.5) readily ensures the validity of (4.3.14).

Lemma 4.8 *Under the assumptions of Lemma 4.6, the solution of (4.1.1) satisfies*

$$\frac{d}{dt}\|u(\cdot, t)\|_{L^2(\Omega)}^2 + \|\nabla u(\cdot, t)\|_{L^2(\Omega)}^2 \le K\left(\|\rho(\cdot, t)\|_{L^2(\Omega)}^2 + 1\right), \tag{4.3.16}$$
$$\frac{d}{dt}\|\nabla u(\cdot, t)\|_{L^2(\Omega)}^2 + \|Au(\cdot, t)\|_{L^2(\Omega)}^2 \le K\left(\|\rho(\cdot, t)\|_{L^2(\Omega)}^2 + 1\right) \tag{4.3.17}$$

for all $t \in (0, T_{max})$ *for a positive constant* K.

Proof Testing the u-equation in (4.1.1) by u, using the Hölder inequality and Poincaré inequality, we can get

$$\frac{1}{2}\frac{d}{dt}\int_\Omega |u|^2 + \int_\Omega |\nabla u|^2 = \int_\Omega (\rho + m)\nabla\phi \cdot u$$
$$\leq \|\nabla\phi\|_{L^\infty(\Omega)}\|\rho + m\|_{L^2(\Omega)}\|u\|_{L^2(\Omega)}$$
$$\leq \frac{1}{2}\|\nabla u\|_{L^2(\Omega)}^2 + K_1(\|\rho\|_{L^2(\Omega)}^2 + \|m\|_{L^2(\Omega)}^2),$$

which together with (4.3.5) yields (4.3.16). Applying the Helmholtz projection $\mathscr{P}$ to the fourth equation in (4.1.1), testing the resulting identity by Au and using the Young inequality, we have

$$\frac{1}{2}\frac{d}{dt}\int_\Omega |A^{\frac{1}{2}}u|^2 + \int_\Omega |Au|^2 = -\int_\Omega \mathscr{P}[(\rho + m)\nabla\phi] \cdot Au$$
$$\leq \frac{1}{2}\int_\Omega |Au|^2 + K_2(\int_\Omega \rho^2 + \int_\Omega m^2),$$

which yields (4.3.17), due to (4.3.5) and the fact that $\int_\Omega |\nabla u|^2 = \int_\Omega |A^{\frac{1}{2}}u|^2$.

Lemma 4.9 *Under the assumptions of Lemma 4.6, one can find $C > 0$ such that for all $t \in (0, T_{max})$, the solution of (4.1.1) satisfies*

$$\|\rho(\cdot, t)\|_{L^2(\Omega)}^2 + \|\nabla c(\cdot, t)\|_{L^2(\Omega)}^2 + \|u(\cdot, t)\|_{W^{1,2}(\Omega)}^2 \leq K.$$

Proof By the Gagliardo–Nirenberg inequality

$$\|\rho\|_{L^2(\Omega)} \leq C_{GN}\left(\|\nabla\rho\|_{L^2(\Omega)}^{\frac{3}{5}}\|\rho\|_{L^1(\Omega)}^{\frac{2}{5}} + \|\rho\|_{L^1(\Omega)}\right)$$

and (4.3.8), for any $\varepsilon > 0$, there exists $K(\varepsilon) > 0$ such that

$$\frac{d}{dt}\|\rho\|_{L^2(\Omega)}^2 + \|\rho\|_{L^2(\Omega)}^2 + \frac{1}{4}\|\nabla\rho\|_{L^2(\Omega)}^2 \leq \varepsilon\|\Delta c\|_{L^2(\Omega)}^2 + K_1(\varepsilon). \qquad (4.3.18)$$

Adding (4.3.16) and (4.3.17), and by the Poincaré inequality, one can find constants $K_i > 0$, $i = 2, 3, 4$, such that

$$\frac{d}{dt}(\|u\|_{L^2(\Omega)}^2 + \|\nabla u\|_{L^2(\Omega)}^2) + K_2(\|u\|_{L^2(\Omega)}^2 + \|\nabla u\|_{L^2(\Omega)}^2)$$
$$\leq K_3\left(\|\rho\|_{L^2(\Omega)}^2 + 1\right) \qquad (4.3.19)$$
$$\leq \frac{1}{8}\|\nabla\rho\|_{L^2(\Omega)}^2 + K_4.$$

Recalling (4.3.14), we get

$$\frac{d}{dt}\|\nabla c\|_{L^2(\Omega)}^2 + 2\|\nabla c\|_{L^2(\Omega)}^2 + \|\Delta c\|_{L^2(\Omega)}^2 \le K_5\left(\|\nabla u\|_{L^2(\Omega)}^2 + 1\right). \quad (4.3.20)$$

Now combining the above inequalities and choosing $\varepsilon = \frac{K_2}{2K_5}$, one can see that there exists some constant $K_6 > 0$ such that

$$Y(t) := \|\rho(\cdot,t)\|_{L^2(\Omega)}^2 + \|u(\cdot,t)\|_{W^{1,2}(\Omega)}^2 + \varepsilon\|\nabla c(\cdot,t)\|_{L^2(\Omega)}^2$$

satisfies $Y'(t) + \delta Y(t) \le K_6$, where $\delta = \min\{1, \frac{K_2}{2}\}$. Hence, by an ODE comparison argument, we obtain $Y(t) \le K_7$ for some constant $K_7 > 0$ and thereby complete the proof.

With all of the above estimates at hand, we can now establish the global existence result in the case $\mathscr{S} = 0$ on $\partial\Omega$.

Proof of Theorem 4.1 *in the case* $\mathscr{S} = 0$ *on* $\partial\Omega$.   To establish the existence of globally bounded classical solution, by the extensibility criterion in Lemma 4.4, we only need to show that

$$\|\rho(\cdot,t)\|_{L^\infty(\Omega)} + \|m(\cdot,t)\|_{L^\infty(\Omega)} + \|c(\cdot,t)\|_{W^{1,\infty}(\Omega)} + \|A^\beta u(\cdot,t)\|_{L^2(\Omega)} \le K_1$$
$$(4.3.21)$$

for all $t \in (0, T_{max})$ with some positive constant K_1 independent of T_{max}. To this end, by the estimate of Stokes operator (Corollary 3.4 of Winkler (2015b)), we first get

$$\|u\|_{L^\infty(\Omega)} \le K_2\|u\|_{W^{1,5}(\Omega)} \le K_3 \quad (4.3.22)$$

with positive constant $K_3 > 0$ independent of T_{max}, due to $\|\rho\|_{L^2(\Omega)} \le K_4$ and $\|m\|_{L^\infty(\Omega)} \le K_4$ from Lemma 4.9 and Lemma 4.5, respectively.

By Lemma 1.1, Lemma 4.9 and the Young inequality, we have

$$\sup_{t\in(0,T_{max})}\|\nabla c\|_{L^\infty(\Omega)} \le K_5(1 + \sup_{t\in(0,T_{max})}\|m - u\cdot\nabla c\|_{L^4(\Omega)})$$

$$\le K_5(1 + \sup_{t\in(0,T_{max})}(\|m\|_{L^4(\Omega)} + \|u\|_{L^6(\Omega)}\|\nabla c\|_{L^{12}(\Omega)}))$$

$$\le K_5(1 + \sup_{t\in(0,T_{max})}(\|m\|_{L^4(\Omega)} + \|u\|_{L^6(\Omega)}\|\nabla c\|_{L^2(\Omega)}^{\frac{1}{6}}\|\nabla c\|_{L^\infty(\Omega)}^{\frac{5}{6}}))$$

$$\le K_6(1 + \sup_{t\in(0,T_{max})}\|\nabla c\|_{L^\infty(\Omega)}^{\frac{5}{6}}),$$

which implies that $\sup_{t\in(0,T_{max})}\|\nabla c(\cdot,t)\|_{L^\infty(\Omega)} \le K_7$. Along with (4.3.7) this implies $\|c(\cdot,t)\|_{W^{1,\infty}(\Omega)} \le K_8$. Furthermore, applying the variation-of-constants formula to the ρ-equation in (4.1.1), the maximum principle, Lemma 1.1(iv) and Lemma 4.9, we get

$$\|\rho\|_{L^\infty(\Omega)} \leq \|e^{t\Delta}\rho_0\|_{L^\infty(\Omega)} + \int_0^t \|e^{(t-s)\Delta}\nabla \cdot (\rho\mathscr{S}\nabla c + \rho u)\|_{L^\infty(\Omega)}ds$$

$$\leq \|\rho_0\|_{L^\infty(\Omega)} + c_4 \int_0^t (1 + (t-s)^{-\frac{7}{8}})e^{-\lambda_1(t-s)}\|\rho\mathscr{S}\nabla c + \rho u\|_{L^4(\Omega)}ds$$

$$\leq \|\rho_0\|_{L^\infty(\Omega)} + K_9 \int_0^t (1 + (t-s)^{-\frac{7}{8}})e^{-\lambda_1(t-s)}\|\rho\|_{L^4(\Omega)}ds$$

$$\leq \|\rho_0\|_{L^\infty(\Omega)} + K_9 \int_0^t (1 + (t-s)^{-\frac{7}{8}})e^{-\lambda_1(t-s)}\|\rho\|_{L^\infty(\Omega)}^{\frac{1}{2}}\|\rho\|_{L^2(\Omega)}^{\frac{1}{2}}ds$$

$$\leq \|\rho_0\|_{L^\infty(\Omega)} + K_{10} \sup_{s\in(0,T_{max})} \|\rho\|_{L^\infty(\Omega)}^{\frac{1}{2}}$$

with $K_{10} = K_9 \sup_{t\in(0,T_{max})} \|\rho\|_{L^2(\Omega)}^{\frac{1}{2}} \int_0^\infty (1 + s^{-\frac{7}{8}})e^{-\lambda_1 s}ds$, where we have used $\nabla \cdot u = 0$. Taking supremum on the left-hand side of the above inequality over $(0, T_{max})$, we obtain

$$\sup_{t\in(0,T_{max})} \|\rho\|_{L^\infty(\Omega)} \leq \|\rho_0\|_{L^\infty(\Omega)} + K_{10} \sup_{t\in(0,T_{max})} \|\rho\|_{L^\infty(\Omega)}^{\frac{1}{2}},$$

and thereby $\sup_{t\in(0,T_{max})} \|\rho\|_{L^\infty(\Omega)} \leq K_{11}$ by the Young inequality. Finally, by a straightforward argument (see [Espejo and Winkler (2018), Lemma 3.1] or [Tuval et al. (2005), p. 340]), one can find $K_{12} > 0$ such that $\sup_{t\in(0,T_{max})} \|A^\beta u\|_{L^2(\Omega)} \leq K_{12}$. The boundedness estimate (4.3.21) is now a direct consequence of the above inequalities and this completes the proof.

2. Large time behavior for $\mathscr{S} = 0$ on $\partial\Omega$

This subsection is devoted to showing the large time behavior of global solutions to (4.1.1) obtained in the above subsection. In order to derive the convergence properties of the solution with respect to the norm in $L^2(\Omega)$, we shall make use of the following lemma. In the sequel, we denote $\overline{f} = \frac{1}{|\Omega|}\int_\Omega f(x)dx$.

Lemma 4.10 (Lemma 4.6 of Espejo and Winkler (2018)) *Let $\lambda > 0$, $C > 0$, and suppose that $y \in C^1([0, \infty))$ and $h \in C^0([0, \infty))$ are nonnegative functions satisfying $y'(t) + \lambda y(t) \leq h(t)$ for some $\lambda > 0$ and all $t > 0$. Then if $\int_0^\infty h(s)ds \leq C$, we have $y(t) \to 0$ as $t \to \infty$.*

By means of the testing procedure and the Young inequality, we have

$$\frac{d}{dt}\int_\Omega (\rho - \overline{\rho})^2 = 2\int_\Omega (\rho - \overline{\rho})(\Delta\rho - \nabla(\rho\mathscr{S}(x, \rho, c)\nabla c) - u \cdot \nabla\rho - \rho m + \overline{\rho m})$$

$$= -2\int_\Omega |\nabla\rho|^2 + 2\int_\Omega \rho\mathscr{S}(x, \rho, c)\nabla c \cdot \nabla\rho - 2\int_\Omega (\rho - \overline{\rho})(\rho m - \overline{\rho m})$$

$$\leq -\int_\Omega |\nabla\rho|^2 + K_1\int_\Omega |\nabla c|^2 - 2\int_\Omega (\rho - \overline{\rho})\rho m, \qquad (4.3.23)$$

$$\frac{d}{dt}\int_\Omega (m-\overline{m})^2 = 2\int_\Omega (m-\overline{m})(\Delta m - u\cdot\nabla m - \rho m + \overline{\rho m}) \tag{4.3.24}$$

$$= 2\int_\Omega m(\Delta m - u\cdot\nabla m) - 2\int_\Omega (m-\overline{m})(\rho m - \overline{\rho m})$$

$$\leq -2\int_\Omega |\nabla m|^2 - 2\int_\Omega (m-\overline{m})\rho m,$$

$$\frac{d}{dt}\int_\Omega (c-\overline{c})^2 = 2\int_\Omega (c-\overline{c})(\Delta c - u\cdot\nabla c - (c-\overline{c}) + (m-\overline{m})) \tag{4.3.25}$$

$$= 2\int_\Omega c(\Delta c - u\cdot\nabla c) - 2\int_\Omega (c-\overline{c})^2 + 2\int_\Omega (c-\overline{c})(m-\overline{m})$$

$$\leq -2\int_\Omega |\nabla c|^2 - \int_\Omega (c-\overline{c})^2 + \int_\Omega (m-\overline{m})^2,$$

$$\frac{d}{dt}\int_\Omega |u|^2 = -2\int_\Omega |\nabla u|^2 + 2\int_\Omega (\rho+m)\nabla\phi\cdot u - 2\int_\Omega \nabla P\cdot u \tag{4.3.26}$$

$$= -2\int_\Omega |\nabla u|^2 + 2\int_\Omega (\rho-\overline{\rho}+m-\overline{m})\nabla\phi\cdot u$$

$$\leq -2\int_\Omega |\nabla u|^2 + K_2\left(\int_\Omega |\rho-\overline{\rho}+m-\overline{m}|^2\right)^{\frac{1}{2}}\left(\int_\Omega |u|^2\right)^{\frac{1}{2}}$$

$$\leq -\int_\Omega |\nabla u|^2 + K_3\left(\int_\Omega |\rho-\overline{\rho}|^2 + \int_\Omega |m-\overline{m}|^2\right),$$

where $\nabla\cdot u = 0$, $u\,|_{\partial\Omega}= 0$ and the boundedness of u, $\nabla\phi$ and $\mathscr{S}$ are used.

Lemma 4.11 *Under the assumptions of Lemma 4.6,*

$$\|(\rho-\overline{\rho})(\cdot,t)\|_{L^\infty(\Omega)} \to 0 \quad as\ t\to\infty,$$
$$\|(m-\overline{m})(\cdot,t)\|_{L^\infty(\Omega)} \to 0 \quad as\ t\to\infty,$$
$$\|(c-\overline{c})(\cdot,t)\|_{L^\infty(\Omega)} \to 0 \quad as\ t\to\infty,$$
$$\|u(\cdot,t)\|_{L^\infty(\Omega)} \to 0 \quad as\ t\to\infty.$$

Proof From (4.3.23)–(4.3.26), it follows that

$$\frac{d}{dt}\int_\Omega (\rho-\overline{\rho})^2 \leq -\int_\Omega |\nabla\rho|^2 + K_1\int_\Omega |\nabla c|^2 + 2\overline{\rho}\int_\Omega \rho m, \tag{4.3.27}$$

$$\frac{d}{dt}\int_\Omega (m-\overline{m})^2 \leq -2\int_\Omega |\nabla m|^2 + 2\overline{m}\int_\Omega \rho m, \tag{4.3.28}$$

$$\frac{d}{dt}\int_\Omega (c-\overline{c})^2 \leq -2\int_\Omega |\nabla c|^2 - \int_\Omega (c-\overline{c})^2 + \int_\Omega (m-\overline{m})^2, \tag{4.3.29}$$

$$\frac{d}{dt}\int_\Omega |u|^2 \leq -\int_\Omega |\nabla u|^2 + K_3\left(\int_\Omega |\rho-\overline{\rho}|^2 + \int_\Omega |m-\overline{m}|^2\right). \tag{4.3.30}$$

Since $\int_\Omega |m - \overline{m}|^2 \leq C_p \|\nabla m\|_{L^2(\Omega)}^2$ and $\int_0^\infty \int_\Omega \rho m \leq K_4$ by (4.3.3), an application of Lemma 4.10 to (4.3.28) yields

$$\|m(\cdot, t) - \overline{m}(t)\|_{L^2(\Omega)} \to 0 \quad \text{as } t \to \infty. \tag{4.3.31}$$

Since

$$\int_0^\infty \int_\Omega |(m - \overline{m})|^2 ds \leq C_p \int_0^\infty \|\nabla m\|_{L^2(\Omega)}^2 ds \leq K_5, \tag{4.3.32}$$

the application of Lemma 4.10 to (4.3.29) also yields

$$\|c(\cdot, t) - \overline{c}(t)\|_{L^2(\Omega)} \to 0 \quad \text{as } t \to \infty \tag{4.3.33}$$

and

$$\int_0^\infty \|\nabla c\|_{L^2(\Omega)}^2 \leq \int_0^\infty \int_\Omega |m - \overline{m}|^2 + \int_\Omega |c_0 - \overline{c_0}|^2 \leq K_6. \tag{4.3.34}$$

Furthermore, by (4.3.34), $\int_\Omega |\rho - \overline{\rho}|^2 \leq C_p \|\nabla \rho\|_{L^2(\Omega)}^2$ and $\int_0^\infty \int_\Omega \rho m \leq K_4$, Lemma 4.10 implies that

$$\|\rho(\cdot, t) - \overline{\rho}(t)\|_{L^2(\Omega)} \to 0 \quad \text{as } t \to \infty, \tag{4.3.35}$$

$$\int_0^\infty \|\rho - \overline{\rho}\|_{L^2(\Omega)}^2 \leq C_p \int_0^\infty \|\nabla \rho\|_{L^2(\Omega)}^2 \leq K_7. \tag{4.3.36}$$

Hence, from (4.3.32), (4.3.36), $\int_\Omega |u|^2 \leq C_p \|\nabla u\|_{L^2(\Omega)}^2$ and Lemma 4.10, it follows that

$$\|u(\cdot, t)\|_{L^2(\Omega)} \to 0 \quad \text{as } t \to \infty \tag{4.3.37}$$

as well as $\int_0^\infty \|\nabla u\|_{L^2(\Omega)}^2 \leq K_8$.

Now we turn the above convergence in $L^2(\Omega)$ into $L^\infty(\Omega)$ with the help of the higher regularity of the solutions. Indeed, similar to the proof of $\|c(\cdot, t)\|_{W^{1,\infty}(\Omega)} \leq K$ in Theorem 4.1 in the case $\mathscr{S} = 0$ on $\partial\Omega$, $\|m(\cdot, t)\|_{W^{1,\infty}(\Omega)} \leq K_{10}$ can be proved since $\|\rho(\cdot, t)\|_{L^\infty(\Omega)} + \|m(\cdot, t)\|_{L^\infty(\Omega)} \leq K_9$ for all $t > 0$ in (4.3.21). Hence, from (4.3.21), there exists a constant $K_{11} > 0$, such that $\|m(\cdot, t) - \overline{m}(t)\|_{W^{1,\infty}(\Omega)} \leq K_{11}$, $\|c(\cdot, t) - \overline{c}(t)\|_{W^{1,\infty}(\Omega)} \leq K_{11}$, $\|u(\cdot, t)\|_{W^{1,5}(\Omega)} \leq K_{11}$ for all $t > 1$. Therefore, by (4.3.31), (4.3.33) and (4.3.37), the application of the interpolation inequality yields as $t \to \infty$,

$$\|m - \overline{m}\|_{L^\infty(\Omega)} \leq C \left(\|m - \overline{m}\|_{W^{1,\infty}(\Omega)}^{\frac{3}{5}} \|m - \overline{m}\|_{L^2(\Omega)}^{\frac{2}{5}} + \|m - \overline{m}\|_{L^2(\Omega)} \right) \to 0,$$

$$\|c(\cdot, t) - \overline{c}(t)\|_{L^\infty(\Omega)} \to 0, \quad \|u(\cdot, t)\|_{L^\infty(\Omega)} \to 0.$$

In addition, similar to Lemma 4.4 in Espejo and Winkler (2018) or Lemma 5.2 in Cao and Lankeit (2016), there exist $\vartheta \in (0, 1)$ and constant $K_{12} > 0$ such that $\|\rho\|_{C^{\vartheta, \frac{\vartheta}{2}}(\overline{\Omega} \times [t, t+1])} \leq K_{12}$ for all $t > 1$, which along with (4.3.35) implies that $\|\rho(\cdot, t) - \overline{\rho}(t)\|_{C_{loc}(\overline{\Omega})} \to 0$ as $t \to \infty$ and then by the finite covering theorem, $\|\rho(\cdot, t) - \overline{\rho}(t)\|_{L^\infty(\Omega)} \to 0$ as $t \to \infty$.

By a very similar argument as in Lemma 4.2 of Espejo and Winkler (2018), we have

Lemma 4.12 *Under the assumptions of Lemma 4.6,*

$$\overline{\rho}(t) \to \rho_\infty, \quad \overline{m}(t) \to m_\infty, \quad \overline{c}(t) \to m_\infty \ \ as \ t \to \infty$$

with $\rho_\infty = \{\overline{\rho_0} - \overline{m_0}\}_+$ and $m_\infty = \{\overline{m_0} - \overline{\rho_0}\}_+$.

Proof From (4.3.3) and (4.3.5), we have

$$\int_{t-1}^t \|\rho m\|_{L^1(\Omega)} \to 0 \ \ as \ t \to \infty, \tag{4.3.38}$$

$$\int_{t-1}^t \|\nabla m\|_{L^2(\Omega)}^2 \to 0 \ \ as \ t \to \infty. \tag{4.3.39}$$

On the other hand,

$$\begin{aligned}
\int_{t-1}^t \|\rho m\|_{L^1(\Omega)} &= \int_{t-1}^t \int_\Omega \rho(m - \overline{m}) + \int_{t-1}^t \int_\Omega \rho\overline{m} \\
&\geq -\int_{t-1}^t \|\rho(\cdot, s)\|_{L^2(\Omega)} \|m - \overline{m}\|_{L^2(\Omega)} + |\Omega| \int_{t-1}^t \overline{\rho} \cdot \overline{m} \\
&\geq -K \int_{t-1}^t \|\nabla m\|_{L^2(\Omega)} + |\Omega| \int_{t-1}^t \overline{\rho} \cdot \overline{m} \\
&\geq -K \left(\int_{t-1}^t \|\nabla m\|_{L^2(\Omega)}^2 \right)^{\frac{1}{2}} + |\Omega| \int_{t-1}^t \overline{\rho} \cdot \overline{m}.
\end{aligned}$$

Inserting (4.3.38) and (4.3.39) into the above inequality, we obtain

$$\int_{t-1}^t \overline{\rho} \cdot \overline{m} \to 0 \ \ as \ t \to \infty. \tag{4.3.40}$$

Now if $\overline{\rho_0} - \overline{m_0} \geq 0$, (4.3.4) warrants that $\overline{\rho} - \overline{m} \geq 0$, which along with (4.3.40) implies that

$$\int_{t-1}^t \overline{m}^2(s)ds \to 0 \ \ as \ t \to \infty. \tag{4.3.41}$$

Noticing that $\overline{m}(s) \geq \overline{m}(t)$ for all $t \geq s$, we have $0 \leq \overline{m}(t)^2 \leq \int_{t-1}^{t} \overline{m}^2(s)ds \to$ 0 as $t \to \infty$, and thus $\overline{\rho} \to \rho_\infty$ as $t \to \infty$ due to (4.3.4). By very similar argument, one can see that $\overline{\rho} \to 0$ as $t \to \infty$ and $\overline{m} \to m_\infty$ as $t \to \infty$ in the case of $\overline{\rho_0} - \overline{m_0} < 0$. Finally, it is observed that $c(\cdot, t) \to m_\infty$ in $L^2(\Omega)$ as $t \to \infty$ is also valid (see Lemma 4.7 of Espejo and Winkler (2018) for example) and thus $\overline{c}(t) \to m_\infty$ as $t \to \infty$ by the Hölder inequality.

Combining Lemma 4.11 with Lemma 4.12, we have

Lemma 4.13 *Under the assumptions of Lemma 4.6, we have*

$$\rho(\cdot, t) \to \rho_\infty, \ m(\cdot, t) \to m_\infty, \ c(\cdot, t) \to m_\infty, \ u(\cdot, t) \to 0 \ in \ L^\infty(\Omega) \ as \ t \to \infty.$$

Now we proceed to estimate the decay rate of $\|\rho(\cdot, t) - \rho_\infty\|_{L^\infty(\Omega)}$, $\|m(\cdot, t) - m_\infty\|_{L^\infty(\Omega)}$, $\|c(\cdot, t) - c_\infty\|_{L^\infty(\Omega)}$, and $\|u(\cdot, t)\|_{L^\infty(\Omega)}$ when $\int_\Omega \rho_0 \neq \int_\Omega m_0$. To this end, we first consider its decay rate in $L^2(\Omega)$ based on a differential inequality.

Lemma 4.14 *Under the assumptions of Lemma 4.6 and $\int_\Omega \rho_0 \neq \int_\Omega m_0$, for any $\varepsilon > 0$, there exist constants $K(\varepsilon) > 0$ and $t_\varepsilon > 0$ such that for $t > t_\varepsilon$,*

$$|\overline{\rho}(t) - \rho_\infty| + |\overline{m}(t) - m_\infty| \leq \qquad K(\varepsilon)e^{-(\rho_\infty + m_\infty - \varepsilon)t}, \qquad (4.3.42)$$

$$|\overline{c}(t) - m_\infty| \leq \qquad K(\varepsilon)e^{-\min\{1, (\rho_\infty + m_\infty - \varepsilon)\}t}. \qquad (4.3.43)$$

Proof For the case $\int_\Omega \rho_0 > \int_\Omega m_0$, we have $\rho_\infty > 0$ and $m_\infty = 0$. By Lemma 4.13, there exists $t_\varepsilon > 0$ such that $\rho(x, t) \geq \rho_\infty - \varepsilon$ for $t > t_\varepsilon$ and $x \in \Omega$, and thereby $\frac{d}{dt} \int_\Omega m = -\int_\Omega \rho m \leq -(\rho_\infty - \varepsilon) \int_\Omega m$ for $t > t_\varepsilon$, which implies that $\overline{m}(t) \leq \overline{m_0}e^{-(\rho_\infty - \varepsilon)(t - t_\varepsilon)}$ for $t > t_\varepsilon$. Moreover, due to $\overline{\rho} = \overline{m} + \rho_\infty$ by (4.3.4), we have $|\overline{\rho}(t) - \rho_\infty| = \overline{m}(t) \leq \overline{m_0}e^{-(\rho_\infty - \varepsilon)(t - t_\varepsilon)}$ for $t > t_\varepsilon$. As for the case $\int_\Omega \rho_0 < \int_\Omega m_0$, similarly we can prove that $|\overline{m}(t) - m_\infty| = \overline{\rho} \leq \overline{\rho_0}e^{-(m_\infty - \varepsilon)(t - t_\varepsilon)}$ for $t > t_\varepsilon$. Furthermore, by the third equation of (4.1.1), we have $\frac{d}{dt} \int_\Omega (c - m_\infty) = \int_\Omega (m - m_\infty) - \int_\Omega (c - m_\infty)$, and thereby $|\overline{c}(t) - m_\infty| \leq K(\varepsilon)e^{-\min\{1, \rho_\infty + m_\infty - \varepsilon\}t}$.

Proof of Theorem 4.2 *in the case $\mathscr{S} = 0$ on $\partial\Omega$.* By Lemmas 4.11 and 4.13, as $t \to \infty$, we have

$$\rho(\cdot, t) - \overline{\rho}(t) \to 0, \ m(\cdot, t) - \overline{m}(t) \to 0, \ \rho(\cdot, t) \to \rho_\infty, \ m(\cdot, t) \to m_\infty \ in \ L^\infty(\Omega),$$

which implies that for any $\varepsilon \in (0, \frac{\rho_\infty + m_\infty}{2})$, there exists $t_\varepsilon > 0$ such that $|\rho(\cdot, t) - \overline{\rho}(t)| < \varepsilon$, $|m(\cdot, t) - \overline{m}(t)| < \varepsilon$, $\rho(\cdot, t) + m(\cdot, t) \geq \rho_\infty + m_\infty - \varepsilon$ for all $t > t_\varepsilon$ and $x \in \Omega$. Hence, from (4.3.23)–(4.3.26), we have

$$\frac{d}{dt} \int_{\Omega} (\rho - \overline{\rho})^2 + \int_{\Omega} |\nabla \rho|^2 \le K_1 \int_{\Omega} |\nabla c|^2 + 2\varepsilon \int_{\Omega} \rho m, \qquad (4.3.44)$$

$$\frac{d}{dt} \int_{\Omega} (m - \overline{m})^2 + 2 \int_{\Omega} |\nabla m|^2 \le 2\varepsilon \int_{\Omega} \rho m, \qquad (4.3.45)$$

$$\frac{d}{dt} \int_{\Omega} (c - \overline{c})^2 + 2 \int_{\Omega} |\nabla c|^2 + \int_{\Omega} (c - \overline{c})^2 \le \int_{\Omega} (m - \overline{m})^2, \qquad (4.3.46)$$

$$\frac{d}{dt} \int_{\Omega} |u|^2 + \int_{\Omega} |\nabla u|^2 \le K_3 \left(\int_{\Omega} (\rho - \overline{\rho})^2 + \int_{\Omega} (m - \overline{m})^2 \right) \qquad (4.3.47)$$

for $t > t_\varepsilon$, as well as

$$\frac{d}{dt} \int_{\Omega} \rho m \qquad (4.3.48)$$

$$= \int_{\Omega} [\rho(\Delta m - u \cdot \nabla m - \rho m) + m(\Delta \rho - \nabla(\rho S(x, \rho, c) \nabla c) - u \cdot \nabla \rho - \rho m)]$$

$$= -2 \int_{\Omega} \nabla \rho \nabla m - \int_{\Omega} (\rho u \cdot \nabla m + m u \cdot \nabla \rho) + \int_{\Omega} \rho S(x, \rho, c) \nabla c \cdot \nabla m - \int_{\Omega} \rho m^2$$

$$- \int_{\Omega} \rho^2 m$$

$$\le \int_{\Omega} |\nabla \rho|^2 + 2 \int_{\Omega} |\nabla m|^2 - \int_{\Omega} u \cdot \nabla(\rho m) + K_3 \int_{\Omega} |\nabla c|^2 - \int_{\Omega} \rho m(\rho + m)$$

$$\le \int_{\Omega} |\nabla \rho|^2 + 2 \int_{\Omega} |\nabla m|^2 + K_3 \int_{\Omega} |\nabla c|^2 - \frac{1}{2} (\rho_\infty + m_\infty) \int_{\Omega} \rho m,$$

where $\nabla \cdot u = 0$, $u |_{\partial \Omega} = 0$ and the boundedness of ρ are used.

On the other hand, by Poincare's inequality, there exists $C_P > 0$, such that

$$\int_{\Omega} |\nabla \rho|^2 \ge C_P \int_{\Omega} (\rho - \overline{\rho})^2, \quad \int_{\Omega} |\nabla m|^2 \ge C_P \int_{\Omega} (m - \overline{m})^2,$$

$$\int_{\Omega} |\nabla c|^2 \ge C_P \int_{\Omega} (c - \overline{c})^2, \quad \int_{\Omega} |\nabla u|^2 \ge C_P \int_{\Omega} (u - \overline{u})^2.$$

Therefore, combining the above inequalities, and taking $\varepsilon < \frac{a(\rho_\infty + m_\infty) C_P}{8(K_1 + C_P)}$ with $a = \min\{\frac{1}{2}, \frac{K_1}{4C_P}, \frac{K_1}{K_3}\}$, the functional $G(t) := \int_{\Omega} (\rho - \overline{\rho})^2 + \frac{K_1}{C_P} \int_{\Omega} (m - \overline{m})^2 + K_1 \int_{\Omega} (c - \overline{c})^2 + a \int_{\Omega} \rho m$ satisfies the ordinary differential inequality $\frac{d}{dt} G(t) + \delta_1 G(t) \le 0$ with $\delta_1 = \min\{\frac{C_P}{2}, 1, \frac{\rho_\infty + m_\infty}{4}\}$, which implies that

$$\|\rho(\cdot, t) - \overline{\rho}\|_{L^2(\Omega)} + \|m(\cdot, t) - \overline{m}\|_{L^2(\Omega)} + \|c(\cdot, t) - \overline{c}\|_{L^2(\Omega)} \le Ce^{-\frac{\delta_1}{2} t}. \quad (4.3.49)$$

Moreover, by (4.3.49) and (4.3.47), $\|u(\cdot, t)\|_{L^2(\Omega)} \le Ce^{-\delta_2 t}$ for some $\delta_2 > 0$. At this position, combining (4.3.49) with Lemma 4.14, we can find $\delta_3 > 0$ such that

$$\|\rho(\cdot,t) - \rho_\infty\|_{L^2(\Omega)} + \|m(\cdot,t) - m_\infty\|_{L^2(\Omega)} + \|c(\cdot,t) - m_\infty\|_{L^2(\Omega)} \le Ce^{-\delta_3 t}.$$
(4.3.50)

Hence, as in the proof of Lemma 4.11, we can obtain the decay estimates (4.1.9)–(4.1.12) by an application of the interpolation inequality, and thus the proof is complete.

3. Exponential decay under smallness condition

In this subsection, we give the proof of Theorem 4.3 under the assumption that $\mathscr{S} = 0$ on $\partial\Omega$. The proof thereof is divided into two cases (Propositions 4.1 and 4.2).

(1) The case $\int_\Omega \rho_0 > \int_\Omega m_0$

In this subsection, we consider the case $\int_\Omega \rho_0 > \int_\Omega m_0$, i.e., $\rho_\infty > 0, m_\infty = 0$.

Proposition 4.1 *Suppose that* (4.1.4) *hold with* $\alpha = 0$ *and* $\int_\Omega \rho_0 > \int_\Omega m_0$. *Let* $N = 3$, $p_0 \in (\frac{N}{2}, N)$, $q_0 \in (N, \frac{Np_0}{N-p_0})$. *There exists* $\varepsilon > 0$ *such that for any initial data* (ρ_0, m_0, c_0, u_0) *fulfilling* (4.1.7) *as well as*

$$\|\rho_0 - \rho_\infty\|_{L^{p_0}(\Omega)} \le \varepsilon, \quad \|m_0\|_{L^{q_0}(\Omega)} \le \varepsilon, \quad \|\nabla c_0\|_{L^N(\Omega)} \le \varepsilon, \quad \|u_0\|_{L^N(\Omega)} \le \varepsilon,$$

(4.1.1) *admits a global classical solution* (ρ, m, c, u, P). *In particular, for any* $\alpha_1 \in (0, \min\{\lambda_1, \rho_\infty\})$, $\alpha_2 \in (0, \min\{\alpha_1, \lambda_1', 1\})$, *there exist constants* K_i, $i = 1, 2, 3, 4$, *such that for all* $t \ge 1$

$$\|m(\cdot,t)\|_{L^\infty(\Omega)} \le K_1 e^{-\alpha_1 t}, \tag{4.3.51}$$

$$\|\rho(\cdot,t) - \rho_\infty\|_{L^\infty(\Omega)} \le K_2 e^{-\alpha_1 t}, \tag{4.3.52}$$

$$\|c(\cdot,t)\|_{W^{1,\infty}(\Omega)} \le K_3 e^{-\alpha_2 t}, \tag{4.3.53}$$

$$\|u(\cdot,t)\|_{L^\infty(\Omega)} \le K_4 e^{-\alpha_2 t}. \tag{4.3.54}$$

Proposition 4.1 is the consequence of the following lemmas. In the proof of these lemmas, the constants $c_i > 0$, $i = 1, \ldots, 10$, refer to those in Lemmas 1.1, 4.1–4.3, respectively. We first collect some easily verifiable observations in the following lemma:

Lemma 4.15 *Under the assumptions of Proposition 4.1 and*

$$\sigma = \int_0^\infty \left(1 + s^{-\frac{N}{2p_0}}\right) e^{-\alpha_1 s} ds,$$

there exist $M_1 > 0$, $M_2 > 0$ *and* $\varepsilon > 0$, *such that*

$$c_3 + 2c_2c_{10}e^{(1+c_1+c_1|\Omega|^{\frac{1}{p_0}-\frac{1}{q_0}})\sigma} \leq \frac{M_2}{4}, \quad M_1\varepsilon < 1, \tag{4.3.55}$$

$$12c_2c_{10}(c_6 + 4c_6c_9c_{10}\|\nabla\phi\|_{L^\infty(\Omega)}(M_1 + c_1 + c_1|\Omega|^{\frac{1}{p_0}-\frac{1}{q_0}}$$

$$+ 4e^{(1+c_1+c_1|\Omega|^{\frac{1}{p_0}-\frac{1}{q_0}})\sigma}))\varepsilon < 1, \tag{4.3.56}$$

$$c_4c_{10}C_SM_2(e^{(1+c_1+c_1|\Omega|^{\frac{1}{p_0}-\frac{1}{q_0}})\sigma} + \rho_\infty|\Omega|^{\frac{1}{q_0}}) \leq \frac{M_1}{8}, \tag{4.3.57}$$

$$3c_{10}c_4C_S(M_1 + c_1 + c_1|\Omega|^{\frac{1}{p_0}-\frac{1}{q_0}})M_2\varepsilon \leq \frac{M_1}{8}, \tag{4.3.58}$$

$$3c_{10}c_7c_6(M_1 + c_1 + c_1|\Omega|^{\frac{1}{p_0}-\frac{1}{q_0}})(1 + 2c_9c_{10}\|\nabla\phi\|_{L^\infty(\Omega)}$$

$$\cdot (M_1 + c_1 + c_1|\Omega|^{\frac{1}{p_0}-\frac{1}{q_0}} + 4e^{(1+c_1+c_1|\Omega|^{\frac{1}{p_0}-\frac{1}{q_0}})\sigma}))\varepsilon \leq \frac{M_1}{4}. \tag{4.3.59}$$

Let

$$T \triangleq \sup\left\{\widetilde{T} \in (0, T_{max}) \,\middle|\, \begin{array}{l} \|(\rho-m)(\cdot,t)-e^{t\Delta}(\rho_0-m_0)\|_{L^\theta(\Omega)} \\ \leq M_1\varepsilon(1+t^{-\frac{N}{2}(\frac{1}{p_0}-\frac{1}{\theta})})e^{-\alpha_1 t} \text{ for all } \theta \in [q_0, \infty], \, t \in [0, \widetilde{T}); \\ \|\nabla c(\cdot,t)\|_{L^\infty(\Omega)} \leq M_2\varepsilon(1+t^{-\frac{1}{2}})e^{-\alpha_1 t} \text{ for all } t \in [0, \widetilde{T}). \end{array}\right\} \tag{4.3.60}$$

By (4.1.7) and Lemma 4.4, $T > 0$ is well-defined. We first show $T = T_{max}$. To this end, we will show that all of the estimates mentioned in (4.3.60) is valid with even smaller coefficients on the right-hand side. The derivation of these estimates will mainly rely on $L^p - L^q$ estimates for the Neumann heat semigroup and the fact that the classical solutions on $(0, T_{max})$ can be represented as

$$(\rho - m)(\cdot, t) = e^{t\Delta}(\rho_0 - m_0)$$

$$- \int_0^t e^{(t-s)\Delta}(\nabla \cdot (\rho\mathscr{S}(x, \rho, c)\nabla c) + u \cdot \nabla(\rho - m))(\cdot, s)ds, \tag{4.3.61}$$

$$m(\cdot, t) = e^{t\Delta}m_0 - \int_0^t e^{(t-s)\Delta}(\rho m + u \cdot \nabla m)(\cdot, s)ds, \tag{4.3.62}$$

$$c(\cdot, t) = e^{t(\Delta-1)}c_0 + \int_0^t e^{(t-s)(\Delta-1)}(m - u \cdot \nabla c)(\cdot, s)ds, \tag{4.3.63}$$

$$u(\cdot, t) = e^{-tA}u_0 + \int_0^t e^{-(t-s)A}\mathscr{P}((\rho + m)\nabla\phi)(\cdot, s)ds \tag{4.3.64}$$

for all $t \in (0, T_{max})$ as per the variation-of-constants formula.

Lemma 4.16 *Under the assumptions of Proposition 4.1, for all $t \in (0, T)$ and $\theta \in [q_0, \infty]$,*

$$\|(\rho - m)(\cdot, t) - \rho_\infty\|_{L^\theta(\Omega)} \le M_3 \varepsilon (1 + t^{-\frac{N}{2}(\frac{1}{p_0} - \frac{1}{\theta})}) e^{-\alpha_1 t}.$$

Proof Since $e^{t\Delta} \rho_\infty = \rho_\infty$ and $\int_\Omega (\rho_0 - m_0 - \rho_\infty) = 0$, the definition of T and Lemma 1.1(i) show that

$$
\begin{aligned}
&\|(\rho - m)(\cdot, t) - \rho_\infty\|_{L^\theta(\Omega)} \\
\le& \|(\rho - m)(\cdot, t) - e^{t\Delta}(\rho_0 - m_0)\|_{L^\theta(\Omega)} + \|e^{t\Delta}(\rho_0 - m_0 - \rho_\infty)\|_{L^\theta(\Omega)} \\
\le& M_1 \varepsilon (1 + t^{-\frac{N}{2}(\frac{1}{p_0} - \frac{1}{\theta})}) e^{-\alpha_1 t} + c_1 (1 + t^{-\frac{N}{2}(\frac{1}{p_0} - \frac{1}{\theta})})(\|\rho_0 - \rho_\infty\|_{L^{p_0}(\Omega)} \\
&+ \|m_0\|_{L^{p_0}(\Omega)}) e^{-\lambda_1 t} \\
\le& M_3 \varepsilon (1 + t^{-\frac{N}{2}(\frac{1}{p_0} - \frac{1}{\theta})}) e^{-\alpha_1 t}
\end{aligned}
$$

for all $t \in (0, T)$ and $\theta \in [q_0, \infty]$, where $M_3 = M_1 + c_1 + c_1 |\Omega|^{\frac{1}{p_0} - \frac{1}{q_0}}$.

Lemma 4.17 *Under the assumptions of Proposition 4.1, for any $k > 1$,*

$$\|m(\cdot, t)\|_{L^k(\Omega)} \le M_4 \|m_0\|_{L^k(\Omega)} e^{-\rho_\infty t} \quad \text{for all } t \in (0, T) \tag{4.3.65}$$

with $\sigma = \int_0^\infty (1 + s^{-\frac{N}{2p_0}}) e^{-\alpha_1 s} ds$ and $M_4 = e^{M_3 \sigma \varepsilon}$.

Proof Multiplying the m-equation in (4.1.1) by km^{k-1} and integrating the result over Ω, we get $\frac{d}{dt} \int_\Omega m^k \le -k \int_\Omega \rho m^k$ on $(0, T)$. Since

$$-\rho \le |\rho - m - \rho_\infty| - m - \rho_\infty \le -\rho_\infty + |\rho - m - \rho_\infty|,$$

Lemma 4.16 yields

$$
\begin{aligned}
\frac{d}{dt} \int_\Omega m^k &\le -k\rho_\infty \int_\Omega m^k + k \int_\Omega m^k |\rho - m - \rho_\infty| \\
&\le -k\rho_\infty \int_\Omega m^k + k\|\rho - m - \rho_\infty\|_{L^\infty(\Omega)} \int_\Omega m^k \\
&\le -k\rho_\infty \int_\Omega m^k + kM_3 \varepsilon \left(1 + t^{-\frac{N}{2p_0}}\right) e^{-\alpha_1 t} \int_\Omega m^k
\end{aligned}
$$

and thus

$$
\begin{aligned}
\int_\Omega m^k &\le \int_\Omega m_0^k \exp\{-k\rho_\infty t + kM_3 \varepsilon \int_0^t (1 + s^{-\frac{N}{2p_0}}) e^{-\alpha_1 s} ds\} \\
&\le \|m_0\|_{L^k(\Omega)}^k e^{k(M_3 \sigma \varepsilon - \rho_\infty t)}.
\end{aligned}
$$

The assertion (4.3.65) follows immediately.

Lemma 4.18 *Under the assumptions of Proposition 4.1, there exists $M_3 > 0$, such that $\|u(\cdot, t)\|_{L^{q_0}(\Omega)} \le M_5 \varepsilon \left(1 + t^{-\frac{1}{2} + \frac{N}{2q_0}}\right) e^{-\alpha_2 t}$ for all $t \in (0, T)$.*

Proof For any given $\alpha_2 < \lambda_1'$, we fix $\mu \in (\alpha_2, \lambda_1')$. By (4.3.64), Lemmas 4.1 and 4.2, we obtain

$$\|u(\cdot, t)\|_{L^{q_0}(\Omega)}$$

$$\leq c_6 t^{-\frac{N}{2}\left(\frac{1}{N}-\frac{1}{q_0}\right)} e^{-\mu t} \|u_0\|_{L^N(\Omega)} + \int_0^t \|e^{-(t-s)A}\mathscr{P}((\rho+m)\nabla\phi)(\cdot, s)\|_{L^{q_0}(\Omega)} ds$$

$$\leq c_6 t^{-\frac{N}{2}\left(\frac{1}{N}-\frac{1}{q_0}\right)} e^{-\mu t} \|u_0\|_{L^N(\Omega)} \qquad (4.3.66)$$

$$+ c_6 \int_0^t e^{-\mu(t-s)} \|\mathscr{P}((\rho+m-\overline{\rho+m})\nabla\phi)(\cdot, s)\|_{L^{q_0}(\Omega)} ds$$

$$\leq c_6 t^{-\frac{1}{2}+\frac{N}{2q_0}} e^{-\mu t} \|u_0\|_{L^N(\Omega)}$$

$$+ c_6 c_9 \|\nabla\phi\|_{L^\infty(\Omega)} \int_0^t e^{-\mu(t-s)} \|(\rho+m-\overline{\rho+m})(\cdot, s)\|_{L^{q_0}(\Omega)} ds,$$

where $\mathscr{P}(\overline{\rho+m}\nabla\phi) = \overline{\rho+m}\mathscr{P}(\nabla\phi) = 0$ is used. On the other hand, due to $\alpha_1 < \rho_\infty$, Lemmas 4.16 and 4.17 show that

$$\|(\rho+m-\overline{\rho+m})(\cdot, s)\|_{L^{q_0}(\Omega)}$$

$$=\|(\rho-m-\overline{\rho-m})(\cdot, s) + 2(m-\overline{m})(\cdot, s)\|_{L^{q_0}(\Omega)} \qquad (4.3.67)$$

$$\leq \|(\rho-m-\rho_\infty)(\cdot, s)\|_{L^{q_0}(\Omega)} + 2\|(m-\overline{m})(\cdot, s)\|_{L^{q_0}(\Omega)}$$

$$\leq M_5'\varepsilon(1 + s^{-\frac{N}{2}\left(\frac{1}{p_0}-\frac{1}{q_0}\right)})e^{-\alpha_1 s}$$

with $M_5' = M_3 + 4e^{M_3\sigma\varepsilon}$. Combining (4.3.66) with (4.3.67) and applying Lemma 4.2, we have

$$\|u(\cdot, t)\|_{L^{q_0}(\Omega)}$$

$$\leq c_6 t^{-\frac{1}{2}+\frac{N}{2q_0}} e^{-\mu t} \|u_0\|_{L^N(\Omega)}$$

$$+ c_6 c_9 \|\nabla\phi\|_{L^\infty(\Omega)} M_5'\varepsilon \int_0^t (1 + s^{-\frac{N}{2}\left(\frac{1}{p_0}-\frac{1}{q_0}\right)})e^{-\alpha_1 s} e^{-\mu(t-s)} ds$$

$$\leq c_6 t^{-\frac{1}{2}+\frac{N}{2q_0}} e^{-\mu t} \|u_0\|_{L^N(\Omega)} + c_6 c_9 c_{10} \|\nabla\phi\|_{L^\infty(\Omega)} M_5'\varepsilon(1 + t^{\min\{0, 1-\frac{N}{2}\left(\frac{1}{p_0}-\frac{1}{q_0}\right)\}})e^{-\alpha_2 t}$$

$$\leq c_6 t^{-\frac{1}{2}+\frac{N}{2q_0}} e^{-\mu t}\varepsilon + 2c_6 c_9 c_{10} \|\nabla\phi\|_{L^\infty(\Omega)} M_5'\varepsilon e^{-\alpha_2 t}$$

$$\leq M_5\varepsilon(1 + t^{-\frac{1}{2}+\frac{N}{2q_0}})e^{-\alpha_2 t},$$

where $M_5 = c_6 + 2c_6 c_9 c_{10} \|\nabla\phi\|_{L^\infty(\Omega)} M_5'$ and $\frac{N}{2}\left(\frac{1}{p_0} - \frac{1}{q_0}\right) < 1$ is used.

Lemma 4.19 *Under the assumptions of Proposition 4.1, for all $t \in (0, T)$,*

$$\|\nabla c(\cdot, t)\|_{L^\infty(\Omega)} \leq \frac{M_2}{2}\varepsilon(1 + t^{-\frac{1}{2}})e^{-\alpha_1 t}.$$

Proof By (4.3.63) and Lemma 1.1(iii), we have

$$\|\nabla c(\cdot, t)\|_{L^\infty(\Omega)}$$

$$\leq \|e^{t(\Delta-1)}\nabla c_0\|_{L^\infty(\Omega)} + \int_0^t \|\nabla e^{(t-s)(\Delta-1)}(m - u \cdot \nabla c)(\cdot, s)\|_{L^\infty(\Omega)}ds \quad (4.3.68)$$

$$\leq c_3(1 + t^{-\frac{1}{2}})e^{-(\lambda_1+1)t}\|\nabla c_0\|_{L^N(\Omega)} + \int_0^t \|\nabla e^{(t-s)(\Delta-1)}m(\cdot, s)\|_{L^\infty(\Omega)}ds$$

$$+ \int_0^t \|\nabla e^{(t-s)(\Delta-1)}u \cdot \nabla c(\cdot, s)\|_{L^\infty(\Omega)}ds.$$

Now we estimate the last two integrals on the right-hand side of the above inequality. From Lemmas 1.1(ii), 4.3, 4.17 with $k = q_0$ and the fact that $q_0 > N$, it follows that

$$\int_0^t \|\nabla e^{(t-s)(\Delta-1)}m\|_{L^\infty(\Omega)}ds$$

$$\leq c_2 \int_0^t (1 + (t-s)^{-\frac{1}{2}-\frac{N}{2q_0}})e^{-(\lambda_1+1)(t-s)}\|m\|_{L^{q_0}(\Omega)}ds \quad (4.3.69)$$

$$\leq c_2 M_4\varepsilon \int_0^t (1 + (t-s)^{-\frac{1}{2}-\frac{N}{2q_0}})e^{-(\lambda_1+1)(t-s)}e^{-\rho_\infty s}ds$$

$$\leq c_2 c_{10} M_4(1 + t^{\min\{0,\frac{1}{2}-\frac{N}{2q_0}\}})\varepsilon e^{-\alpha_1 t}$$

$$\leq 2c_2 c_{10} M_4(1 + t^{-\frac{1}{2}})\varepsilon e^{-\alpha_1 t}.$$

On the other hand, by Lemmas 4.3, 4.18 and the definition of T, we obtain

$$\int_0^t \|\nabla e^{(t-s)(\Delta-1)}u \cdot \nabla c\|_{L^\infty(\Omega)}ds$$

$$\leq c_2 \int_0^t (1 + (t-s)^{-\frac{1}{2}-\frac{N}{2q_0}})e^{-(\lambda_1+1)(t-s)}\|u \cdot \nabla c\|_{L^{q_0}(\Omega)}ds \quad (4.3.70)$$

$$\leq c_2 \int_0^t (1 + (t-s)^{-\frac{1}{2}-\frac{N}{2q_0}})e^{-(\lambda_1+1)(t-s)}\|u\|_{L^{q_0}(\Omega)}\|\nabla c\|_{L^\infty(\Omega)}ds$$

$$\leq c_2 M_5 M_2\varepsilon^2 \int_0^t (1 + (t-s)^{-\frac{1}{2}-\frac{N}{2q_0}})e^{-(\lambda_1+1)(t-s)}(1 + s^{-\frac{1}{2}+\frac{N}{2q_0}})(1 + s^{-\frac{1}{2}})e^{-(\alpha_1+\alpha_2)s}ds$$

$$\leq 3c_2 M_5 M_2\varepsilon^2 \int_0^t e^{-(\lambda_1+1)(t-s)}e^{-(\alpha_1+\alpha_2)s}(1 + (t-s)^{-\frac{1}{2}-\frac{N}{2q_0}})(1 + s^{-1+\frac{N}{2q_0}})ds$$

$$\leq 3c_2 c_{10} M_2 M_5\varepsilon^2(1 + t^{-\frac{1}{2}})e^{-\alpha_1 t}.$$

From (4.3.68)–(4.3.70), it follows that

$$\|\nabla c\|_{L^{\infty}(\Omega)} \le (c_3 + 2c_2c_{10}M_4 + 3c_2c_{10}M_2M_5\varepsilon)(1 + t^{-\frac{1}{2}})\varepsilon e^{-\alpha_1 t}$$
$$\le \frac{M_2}{2}(1 + t^{-\frac{1}{2}})\varepsilon e^{-\alpha_1 t},$$

due to the choice of M_1, M_2 and ε satisfying (4.3.55), (4.3.56), and thereby completes the proof.

Lemma 4.20 *Under the assumptions of Proposition 4.1, for all $\theta \in [q_0, \infty]$ and $t \in (0, T)$,*

$$\|(\rho - m)(\cdot, t) - e^{t\Delta}(\rho_0 - m_0)\|_{L^{\theta}(\Omega)} \le \frac{M_1}{2}\varepsilon(1 + t^{-\frac{N}{2}(\frac{1}{p_0} - \frac{1}{\theta})})e^{-\alpha_1 t}.$$

Proof According to (4.3.61), Lemmas 1.1(iv) and 4.1, we have

$$\|(\rho - m)(\cdot, t) - e^{t\Delta}(\rho_0 - m_0)\|_{L^{\theta}(\Omega)}$$
$$\le \int_0^t \|e^{(t-s)\Delta}(\nabla \cdot (\rho\mathscr{S}(x, \rho, c)\nabla c) + u \cdot \nabla(\rho - m))(\cdot, s)\|_{L^{\theta}(\Omega)}ds$$
$$\le \int_0^t \|e^{(t-s)\Delta}\nabla \cdot (\rho\mathscr{S}(x, \rho, c)\nabla c)(\cdot, s)\|_{L^{\theta}(\Omega)}ds$$
$$\quad + \int_0^t \|e^{(t-s)\Delta}\nabla \cdot ((\rho - m - \rho_{\infty})u)(\cdot, s)\|_{L^{\theta}(\Omega)}ds$$
$$\le c_4 C_{\mathscr{S}} \int_0^t (1 + (t-s)^{-\frac{1}{2} - \frac{N}{2}(\frac{1}{q_0} - \frac{1}{\theta})})e^{-\lambda_1(t-s)}\|\rho(\cdot, s)\|_{L^{q_0}(\Omega)}\|\nabla c(\cdot, s)\|_{L^{\infty}(\Omega)}ds$$
$$\quad + c_7 \int_0^t (1 + (t-s)^{-\frac{1}{2} - \frac{N}{2}(\frac{1}{q_0} - \frac{1}{\theta})})e^{-\mu(t-s)}\|u(\rho - m - \rho_{\infty})(\cdot, s)\|_{L^{q_0}(\Omega)}ds$$
$$= I_1 + I_2.$$

Now we need to estimate I_1 and I_2. Firstly, from Lemmas 4.16 and 4.17, we obtain

$$\|\rho(\cdot, s)\|_{L^{q_0}(\Omega)} \le \|(\rho - m - \rho_{\infty})(\cdot, s)\|_{L^{q_0}(\Omega)} + \|m(\cdot, s)\|_{L^{q_0}(\Omega)} + \|\rho_{\infty}\|_{L^{q_0}(\Omega)}$$
$$\le M_3\varepsilon(1 + s^{-\frac{N}{2}(\frac{1}{p_0} - \frac{1}{q_0})})e^{-\alpha_1 s} + M_6 \tag{4.3.71}$$

with $M_6 = e^{(1 + c_1 + c_1|\Omega|^{\frac{1}{p_0} - \frac{1}{q_0}})\sigma} + \rho_{\infty}|\Omega|^{\frac{1}{q_0}}$, which together with Lemmas 4.19 and 1.1 implies that

$$I_1 \leq c_4 C_S M_6 \int_0^t (1 + (t-s)^{-\frac{1}{2} - \frac{N}{2}(\frac{1}{q_0} - \frac{1}{\theta})}) e^{-\lambda_1(t-s)} \|\nabla c\|_{L^\infty(\Omega)} ds \qquad (4.3.72)$$

$$+ M_7 \varepsilon \int_0^t (1 + (t-s)^{-\frac{1}{2} - \frac{N}{2}(\frac{1}{q_0} - \frac{1}{\theta})})(1 + s^{-\frac{N}{2}(\frac{1}{p_0} - \frac{1}{q_0})}) e^{-\alpha_1 s} e^{-\lambda_1(t-s)} \|\nabla c\|_{L^\infty(\Omega)} ds$$

$$\leq c_4 C_S M_6 M_2 \varepsilon \int_0^t (1 + (t-s)^{-\frac{1}{2} - \frac{N}{2}(\frac{1}{q_0} - \frac{1}{\theta})}) e^{-\lambda_1(t-s)}(1 + s^{-\frac{1}{2}}) e^{-\alpha_1 s} ds$$

$$+ 3 M_7 M_2 \varepsilon^2 \int_0^t (1 + (t-s)^{-\frac{1}{2} - \frac{N}{2}(\frac{1}{q_0} - \frac{1}{\theta})})(1 + s^{-\frac{1}{2} - \frac{N}{2}(\frac{1}{p_0} - \frac{1}{q_0})}) e^{-2\alpha_1 s} e^{-\lambda_1(t-s)} ds$$

$$\leq c_{10}(c_4 C_S M_6 M_2 + 3 M_7 M_2 \varepsilon)(1 + t^{-\frac{N}{2}(\frac{1}{p_0} - \frac{1}{\theta})}) \varepsilon e^{-\alpha_1 t}$$

$$\leq \frac{M_1}{4} \varepsilon (1 + t^{-\frac{N}{2}(\frac{1}{p_0} - \frac{1}{\theta})}) e^{-\alpha_1 t}$$

with $M_7 := c_4 C_S M_3$, where we have used (4.3.57) and (4.3.58) and $\frac{1}{p_0} - \frac{1}{q_0} < \frac{1}{N}$. On the other hand, from Lemmas 4.16 and 4.18, it follows that

$$I_2 = c_7 \int_0^t (1 + (t-s)^{-\frac{1}{2} - \frac{N}{2}(\frac{1}{q_0} - \frac{1}{\theta})}) e^{-\mu(t-s)} \|\rho - m - \rho_\infty\|_{L^\infty(\Omega)} \|u\|_{L^{q_0}(\Omega)} ds$$

$$\leq 3 c_7 M_3 M_5 \varepsilon^2 \int_0^t (1 + (t-s)^{-\frac{1}{2} - \frac{N}{2}(\frac{1}{q_0} - \frac{1}{\theta})}) e^{-\mu(t-s)}(1 + s^{-\frac{1}{2} + \frac{N}{2q_0} - \frac{N}{2p_0}}) e^{-(\alpha_1 + \alpha_2)s} ds$$

$$\leq 3 c_7 M_3 M_5 c_{10} \varepsilon^2 (1 + t^{\min\{0, \frac{N}{2}(\frac{1}{\theta} - \frac{1}{p_0})\}}) e^{-\min\{\mu, \alpha_1 + \alpha_2\} t}$$

$$\leq \frac{M_1}{4} \varepsilon (1 + t^{-\frac{N}{2}(\frac{1}{p_0} - \frac{1}{\theta})}) e^{-\alpha_1 t}, \qquad (4.3.73)$$

where we have used (4.3.59) and $\frac{1}{p_0} - \frac{1}{q_0} < \frac{1}{N}$. Hence, combining the above inequalities leads to our conclusion immediately.

Proof of Theorem 4.3 *in the case* $\mathscr{S} = 0$ *on* $\partial\Omega$, *part 1 (Proposition* 4.1*)*. First we claim that $T = T_{max}$. In fact, if $T < T_{max}$, then by Lemmas 4.19 and 4.20, we have $\|\nabla c(\cdot, t)\|_{L^\infty(\Omega)} \leq \frac{M_2}{2} \varepsilon (1 + t^{-\frac{1}{2}}) e^{-\alpha_1 t}$ and

$$\|(\rho - m)(\cdot, t) - e^{t\Delta}(\rho_0 - m_0)\|_{L^\theta(\Omega)} \leq \frac{M_1}{2} \varepsilon (1 + t^{-\frac{N}{2}(\frac{1}{p_0} - \frac{1}{\theta})}) e^{-\alpha_1 t}$$

for all $\theta \in [q_0, \infty]$ and $t \in (0, T)$, which contradicts the definition of T in (4.3.60). Next, we show that $T_{max} = \infty$. In fact, if $T_{max} < \infty$, we only need to show that as $t \to T_{max}$,

$$\|\rho(\cdot, t)\|_{L^\infty(\Omega)} + \|m(\cdot, t)\|_{L^\infty(\Omega)} + \|c(\cdot, t)\|_{W^{1,\infty}(\Omega)} + \|A^\beta u(\cdot, t)\|_{L^2(\Omega)} \to \infty$$

according to the extensibility criterion in Lemma 4.4.

Let $t_0 := \min\{1, \frac{T_{max}}{3}\}$. Then from Lemma 4.17, there exists $K_1 > 0$ such that for $t \in (t_0, T_{max})$,

$$\|m(\cdot, t)\|_{L^\infty(\Omega)} \leq K_1 e^{-\rho_\infty t}. \qquad (4.3.74)$$

Moreover, from Lemma 4.16 and the fact that

$$\|\rho(\cdot, t) - \rho_\infty\|_{L^\infty(\Omega)} \leq \|(\rho - m)(\cdot, t) - \rho_\infty\|_{L^\infty(\Omega)} + \|m(\cdot, t)\|_{L^\infty(\Omega)},$$

it follows that for all $t \in (t_0, T_{max})$ and some constant $K_2 > 0$,

$$\|\rho(\cdot, t) - \rho_\infty\|_{L^\infty(\Omega)} \leq K_2 e^{-\alpha_1 t}. \tag{4.3.75}$$

Furthermore, Lemma 4.19 implies that there exists $K_3' > 0$, such that

$$\|\nabla c(\cdot, t)\|_{L^\infty(\Omega)} \leq K_3' e^{-\alpha_2 t} \quad \text{for all } t \in (t_0, T_{max}). \tag{4.3.76}$$

On the other hand, we can conclude that $\|c(\cdot, t)\|_{L^\infty(\Omega)} + \|A^\beta u(\cdot, t)\|_{L^2(\Omega)} \leq C$ for $t \in (t_0, T_{max})$. In fact, we first show that there exists a constant $M_9 > 0$, such that

$$\|A^\beta u(\cdot, t)\|_{L^2(\Omega)} \leq M_9 e^{-\alpha_2 t} \tag{4.3.77}$$

for $t_0 < t < T_{max}$. By (4.3.64), we have

$$\|A^\beta u(\cdot, t)\|_{L^2(\Omega)}$$
$$\leq \|A^\beta e^{-tA} u_0\|_{L^2(\Omega)} + \int_0^t \|A^\beta e^{-(t-s)A} \mathcal{P}((\rho + m - \rho_\infty)\nabla\phi)(\cdot, s)\|_{L^2(\Omega)} ds.$$

According to Lemma 4.1, $\|A^\beta e^{-tA} u_0\|_{L^2(\Omega)} \leq c_5 e^{-\mu t} \|A^\beta u_0\|_{L^2(\Omega)}$ for all $t \in (0, T_{max})$. On the other hand, from Lemmas 4.1, 4.2, and 4.16, it follows that there exists $\hat{M} > 1$, such that

$$\int_0^t \|A^\beta e^{-(t-s)A} \mathcal{P}((\rho + m - \rho_\infty)\nabla\phi)(\cdot, s)\|_{L^2(\Omega)} ds$$

$$\leq c_9 c_5 \|\nabla\phi\|_{L^\infty(\Omega)} |\Omega|^{\frac{q_0-2}{2q_0}} \int_0^t e^{-\mu(t-s)} (t-s)^{-\beta} (\|(\rho - m - \rho_\infty)(\cdot, s)\|_{L^{q_0}(\Omega)}$$
$$+ 2\|m(\cdot, s)\|_{L^{q_0}(\Omega)}) ds$$

$$\leq c_9 c_5 \|\nabla\phi\|_{L^\infty(\Omega)} |\Omega|^{\frac{q_0-2}{2q_0}} \hat{M} \int_0^t e^{-\mu(t-s)} (t-s)^{-\beta} (1 + s^{-\frac{N}{2}(\frac{1}{p_0}-\frac{1}{q_0})}) e^{-\alpha_1 s} ds$$

$$\leq c_5 c_9 c_{10} \|\nabla\phi\|_{L^\infty(\Omega)} |\Omega|^{\frac{q_0-2}{2q_0}} \hat{M} e^{-\alpha_2 t} (1 + t^{\min\{0, 1-\beta-\frac{N}{2}(\frac{1}{p_0}-\frac{1}{q_0})\}})$$

$$\leq c_5 c_9 c_{10} \|\nabla\phi\|_{L^\infty(\Omega)} |\Omega|^{\frac{q_0-2}{2q_0}} \hat{M} e^{-\alpha_2 t} (1 + t_0^{\min\{0, 1-\beta-\frac{N}{2}(\frac{1}{p_0}-\frac{1}{q_0})\}})$$

for $t_0 < t < T_{max}$. Hence, combining the above inequalities, we arrive at (4.3.77).

Since $D(A^\beta) \hookrightarrow L^\infty(\Omega)$ with $\beta \in (\frac{N}{4}, 1)$, we have

$$\|u(\cdot, t)\|_{L^\infty(\Omega)} \leq K_4 e^{-\alpha_2 t} \quad \text{for some } K_4 > 0 \text{ and } t \in (0, T_{max}). \tag{4.3.78}$$

Now we turn to show that there exists $K_3'' > 0$, such that

$$\|c(\cdot, t)\|_{L^\infty(\Omega)} \le K_3'' e^{-\alpha_2 t} \quad \text{for all } t \in (0, T_{max}). \tag{4.3.79}$$

Indeed, from (4.3.63), it follows that

$$\begin{aligned}
\|c\|_{L^\infty(\Omega)} &\le \|e^{t(\Delta - 1)} c_0\|_{L^\infty(\Omega)} + \int_0^t \|e^{(t-s)(\Delta - 1)}(m - u \cdot \nabla c)\|_{L^\infty(\Omega)} ds \\
&\le e^{-t} \|c_0\|_{L^\infty(\Omega)} + \int_0^t \|e^{(t-s)(\Delta - 1)} m(\cdot, s)\|_{L^\infty(\Omega)} ds \\
&\quad + \int_0^t \|e^{(t-s)(\Delta - 1)} u \cdot \nabla c(\cdot, s)\|_{L^\infty(\Omega)} ds.
\end{aligned} \tag{4.3.80}$$

An application of (4.3.65) with $k = \infty$ yields

$$\begin{aligned}
\int_0^t \|e^{(t-s)(\Delta - 1)} m(\cdot, s)\|_{L^\infty(\Omega)} ds &\le \int_0^t e^{-(t-s)} \|m(\cdot, s)\|_{L^\infty(\Omega)} ds \tag{4.3.81} \\
&\le \|m_0\|_{L^\infty(\Omega)} M_4 \int_0^t e^{-(t-s)} e^{-\rho_\infty s} ds \\
&\le M_4 c_{10} e^{-\alpha_2 t}.
\end{aligned}$$

On the other hand, from (4.3.78) and (4.3.76), we can see that

$$\begin{aligned}
\int_0^t \|e^{(t-s)(\Delta - 1)} u \cdot \nabla c\|_{L^\infty(\Omega)} ds &\le \int_0^t e^{-(t-s)} \|u\|_{L^\infty(\Omega)} \|\nabla c\|_{L^\infty(\Omega)} ds \tag{4.3.82} \\
&\le K_3' K_4 \int_0^t e^{-2\alpha_2 s} e^{-(t-s)} ds \\
&\le K_3' K_4 c_{10} e^{-\alpha_2 t}.
\end{aligned}$$

Hence, inserting (4.3.81), (4.3.82) into (4.3.80), we arrive at the conclusion (4.3.79). Therefore, we have $T_{max} = \infty$, and the decay estimates in (4.3.51)–(4.3.54) follow from (4.3.74)–(4.3.79), respectively.

(2) The case $\int_\Omega \rho_0 < \int_\Omega m_0$

In this subsection, we consider the case $\int_\Omega \rho_0 < \int_\Omega m_0$, i.e., $m_\infty > 0$, $\rho_\infty = 0$.

Proposition 4.2 *Suppose that (4.1.4) hold with $\alpha = 0$ and $\int_\Omega \rho_0 < \int_\Omega m_0$. Let $N = 3$, $p_0 \in (\frac{2N}{3}, N)$, $q_0 \in (N, \frac{Np_0}{2(N - p_0)})$. Then there exists $\varepsilon > 0$ such that for any initial data (ρ_0, m_0, c_0, u_0) fulfilling (4.1.7) as well as*

$$\|\rho_0\|_{L^{p_0}(\Omega)} \le \varepsilon, \quad \|m_0 - m_\infty\|_{L^{q_0}(\Omega)} \le \varepsilon, \quad \|\nabla c_0\|_{L^N(\Omega)} \le \varepsilon, \quad \|u_0\|_{L^N(\Omega)} \le \varepsilon,$$

(4.1.1) *admits a global classical solution* (ρ, m, c, u, P). *Furthermore, for any* $\alpha_1 \in$ $(0, \min\{\lambda_1, m_\infty\})$, $\alpha_2 \in (0, \min\{\alpha_1, \lambda_1', 1\})$, *there exist constants* $K_i > 0$, $i = 1, 2,$ $3, 4$, *such that*

$$\|m(\cdot, t) - m_\infty\|_{L^\infty(\Omega)} \le K_1 e^{-\alpha_1 t}, \tag{4.3.83}$$

$$\|\rho(\cdot, t)\|_{L^\infty(\Omega)} \le K_2 e^{-\alpha_1 t}, \tag{4.3.84}$$

$$\|c(\cdot, t) - m_\infty\|_{W^{1,\infty}(\Omega)} \le K_3 e^{-\alpha_2 t}, \tag{4.3.85}$$

$$\|u(\cdot, t)\|_{L^\infty(\Omega)} \le K_4 e^{-\alpha_2 t}. \tag{4.3.86}$$

The proof of Proposition 4.2 proceeds in a parallel fashion to that of Proposition 4.1. However, due to differences in the properties of ρ and m, there are significant differences in the details of their proofs. Thus, for the convenience of the reader, we will give the full proof of Proposition 4.2. The following can be verified easily:

Lemma 4.21 *Under the assumptions of Proposition 4.2, it is possible to choose* $M_1 > 0$, $M_2 > 0$ *and* $\varepsilon > 0$, *such that*

$$c_3 \le \frac{M_2}{6}, \quad c_2 c_{10}(1 + c_1 + c_1|\Omega|^{\frac{1}{p_0} - \frac{1}{q_0}} + M_1) \le \frac{M_2}{6}, \tag{4.3.87}$$

$$18 c_2 c_6 c_{10}(1 + 2 c_9 c_{10}(1 + c_1 + c_1|\Omega|^{\frac{1}{p_0} - \frac{1}{q_0}} + 2M_1))\|\nabla\phi\|_{L^\infty(\Omega)})\varepsilon \le 1, \tag{4.3.88}$$

$$2c_1 + (\min\{1, |\Omega|\})^{-\frac{1}{p_0}} \le \frac{M_1}{8}, \quad 24 c_4 C_S c_{10} M_2 \varepsilon < 1, \tag{4.3.89}$$

$$24 c_4 c_{10} c_6(1 + 2 c_9 c_{10}(1 + c_1 + c_1|\Omega|^{\frac{1}{p_0} - \frac{1}{q_0}} + 2M_1))\|\nabla\phi\|_{L^\infty(\Omega)})\varepsilon < 1, \tag{4.3.90}$$

$$24 c_4 c_{10}(1 + c_1 + c_1|\Omega|^{\frac{1}{p_0} - \frac{1}{q_0}} + M_1)\varepsilon < 1, \tag{4.3.91}$$

$$12 c_4 C_S c_{10} M_1 M_2 \varepsilon < 1, \tag{4.3.92}$$

$$c_{10} c_6 c_4(1 + c_1 + c_1|\Omega|^{\frac{1}{p_0} - \frac{1}{q_0}})(1 + 2 c_9 c_{10}(1 + c_1 + c_1|\Omega|^{\frac{1}{p_0} - \frac{1}{q_0}} + 2M_1))\|\nabla\phi\|_{L^\infty(\Omega)})\varepsilon$$
$$< \frac{1}{24}. \tag{4.3.93}$$

Similar to the proof of Proposition 4.1, we define

$$T \triangleq \sup\left\{\widetilde{T} \in (0, T_{max}) \left| \begin{array}{l} \|(m-\rho)(\cdot, t) - e^{t\Delta}(m_0 - \rho_0)\|_{L^\theta(\Omega)} \le \varepsilon(1 + t^{-\frac{N}{2}(\frac{1}{p_0} - \frac{1}{\theta})})e^{-\alpha_1 t}, \\ \|\rho(\cdot, t)\|_{L^\theta(\Omega)} \le M_1\varepsilon(1 + t^{-\frac{N}{2}(\frac{1}{p_0} - \frac{1}{\theta})})e^{-\alpha_1 t}, \ \forall\theta \in [q_0, \infty], \\ \|\nabla c(\cdot, t)\|_{L^\infty(\Omega)} \le M_2\varepsilon(1 + t^{-\frac{1}{2}})e^{-\alpha_1 t}, \ \forall t \in [0, \widetilde{T}). \end{array} \right. \right\} \tag{4.3.94}$$

By Lemma 4.3.7 and (4.1.7), $T > 0$ is well-defined. As in the previous subsection, we first show $T = T_{max}$, and then $T_{max} = \infty$. To this end, we will show that all of the estimates mentioned in (4.3.94) are valid with even smaller coefficients on the right-hand side than appearing in (4.3.94). The derivation of these estimates will mainly rely on $L^p - L^q$ estimates for the Neumann heat semigroup and the corresponding semigroup for Stokes operator, and the fact that the classical solutions of (4.1.1) on $(0, T)$ can be represented as

$$(m - \rho)(\cdot, t) = e^{t\Delta}(m_0 - \rho_0)$$
$$+ \int_0^t e^{(t-s)\Delta}(\nabla \cdot (\rho \mathscr{S}(x, \rho, c)\nabla c) - u \cdot \nabla(m - \rho))(\cdot, s)ds,$$

$$\text{(4.3.95)}$$

$$\rho(\cdot, t) = e^{t\Delta}\rho_0 - \int_0^t e^{(t-s)\Delta}(\nabla \cdot (\rho \mathscr{S}(x, \rho, c)\nabla c) + u \cdot \nabla\rho + \rho m)(\cdot, s)ds,$$

$$\text{(4.3.96)}$$

$$c(\cdot, t) = e^{t(\Delta - 1)}c_0 + \int_0^t e^{(t-s)(\Delta - 1)}(m - u \cdot \nabla c)(\cdot, s)ds,$$

$$\text{(4.3.97)}$$

$$u(\cdot, t) = e^{-tA}u_0 + \int_0^t e^{-(t-s)A}\mathscr{P}((\rho + m)\nabla\phi)(\cdot, s)ds.$$

$$\text{(4.3.98)}$$

Lemma 4.22 *Under the assumptions of Proposition 4.2, we have*

$$\|(m - \rho)(\cdot, t) - m_\infty\|_{L^\theta(\Omega)} \leq M_3\varepsilon(1 + t^{-\frac{N}{2}(\frac{1}{p_0} - \frac{1}{\theta})})e^{-\alpha_1 t}$$

for all $t \in (0, T)$ and $\theta \in [q_0, \infty]$.

Proof Since $e^{t\Delta}(\overline{m}_0 - \overline{\rho}_0) = m_\infty$ and $\int_\Omega(m_0 - \rho_0 - m_\infty) = 0$, from the Definition of T and Lemma 1.1(i), we get

$$\|(m - \rho)(\cdot, t) - m_\infty\|_{L^\theta(\Omega)}$$
$$\leq \|(m - \rho)(\cdot, t) - e^{t\Delta}(m_0 - \rho_0)\|_{L^\theta(\Omega)} + \|e^{t\Delta}(m_0 - \rho_0) - e^{t\Delta}m_\infty\|_{L^\theta(\Omega)}$$
$$\leq \varepsilon(1 + t^{-\frac{N}{2}(\frac{1}{p_0} - \frac{1}{\theta})})e^{-\alpha_1 t} + c_1(1 + t^{-\frac{N}{2}(\frac{1}{p_0} - \frac{1}{\theta})})(\|\rho_0\|_{L^{p_0}(\Omega)} + \|m_0 - m_\infty\|_{L^{p_0}(\Omega)})e^{-\lambda_1 t}$$
$$\leq (1 + c_1 + c_1|\Omega|^{\frac{1}{p_0} - \frac{1}{q_0}})\varepsilon(1 + t^{-\frac{N}{2}(\frac{1}{p_0} - \frac{1}{\theta})})e^{-\alpha_1 t}$$

for all $t \in (0, T)$ and $\theta \in [q_0, \infty]$. This lemma is proved for

$$M_3 = 1 + c_1 + c_1|\Omega|^{\frac{1}{p_0} - \frac{1}{q_0}}.$$

Lemma 4.23 *Under the assumptions of Proposition 4.2, we have*

$$\|m(\cdot, t) - m_\infty\|_{L^\theta(\Omega)} \leq M_4\varepsilon(1 + t^{-\frac{N}{2}(\frac{1}{p_0} - \frac{1}{\theta})})e^{-\alpha_1 t} \quad \text{for all } t \in (0, T), \theta \in [q_0, \infty].$$

Proof From Lemma 4.22 and the definition of T, it follows that

$$\|m(\cdot, t) - m_\infty\|_{L^\theta(\Omega)} \leq \|(m - \rho - m_\infty)(\cdot, t)\|_{L^\theta(\Omega)} + \|\rho(\cdot, t)\|_{L^\theta(\Omega)}$$

$$\leq (M_3 + M_1)\varepsilon(1 + t^{-\frac{N}{2}(\frac{1}{p_0} - \frac{1}{\theta})})e^{-\alpha_1 t}.$$

The lemma is proved for $M_4 = M_3 + M_1$.

Lemma 4.24 *Under the assumptions of Proposition 4.2, there exists $M_5 > 0$, such that*

$$\|u(\cdot, t)\|_{L^{q_0}(\Omega)} \leq M_5\varepsilon(1 + t^{-\frac{1}{2} + \frac{N}{2q_0}})e^{-\alpha_2 t} \quad \text{for all } t \in (0, T).$$

Proof For any given $\alpha_2 < \lambda_1'$, we can fix $\mu \in (\alpha_2, \lambda_1')$. By (4.3.98), Lemmas 4.1, 4.2 and $\mathscr{P}(\nabla\phi) = 0$, we obtain that

$$\|u(\cdot, t)\|_{L^{q_0}(\Omega)}$$

$$\leq c_6 t^{-\frac{N}{2}(\frac{1}{N} - \frac{1}{q_0})}e^{-\mu t}\|u_0\|_{L^N(\Omega)} + \int_0^t \|e^{-(t-s)A}\mathscr{P}((\rho + m)\nabla\phi)(\cdot, s)\|_{L^{q_0}(\Omega)}ds$$

$$\leq c_6 t^{-\frac{N}{2}(\frac{1}{N} - \frac{1}{q_0})}e^{-\mu t}\|u_0\|_{L^N(\Omega)} \tag{4.3.99}$$

$$+ c_6 c_9 \int_0^t e^{-\mu(t-s)}\|(\rho + m - m_\infty)(\cdot, s)\|_{L^{q_0}(\Omega)}\|\nabla\phi\|_{L^\infty(\Omega)}ds$$

$$\leq c_6 t^{-\frac{1}{2} + \frac{N}{2q_0}}e^{-\mu t}\|u_0\|_{L^N(\Omega)}$$

$$+ c_6 c_9 \|\nabla\phi\|_{L^\infty(\Omega)}\int_0^t e^{-\mu(t-s)}\|(\rho + m - m_\infty)(\cdot, s)\|_{L^{q_0}(\Omega)}ds.$$

By Lemma 4.23 and the definition of T, we get

$$\|(\rho + m - m_\infty)(\cdot, s)\|_{L^{q_0}(\Omega)} = \|(m - m_\infty)(\cdot, s)\|_{L^{q_0}(\Omega)} + \|\rho(\cdot, s)\|_{L^{q_0}(\Omega)} \tag{4.3.100}$$

$$\leq (M_4 + M_1)\varepsilon(1 + s^{-\frac{N}{2}(\frac{1}{p_0} - \frac{1}{q_0})})e^{-\alpha_1 s}.$$

Inserting (4.3.100) into (4.3.99), and noting $\frac{N}{2}(\frac{1}{p_0} - \frac{1}{q_0}) < 1$, we have

$$\|u(\cdot, t)\|_{L^{q_0}(\Omega)}$$

$$\leq c_6 t^{-\frac{1}{2} + \frac{N}{2q_0}}e^{-\mu t}\|u_0\|_{L^N(\Omega)}$$

$$+ c_6 c_9 (M_4 + M_1)\|\nabla\phi\|_{L^\infty(\Omega)}\varepsilon\int_0^t (1 + s^{-\frac{N}{2}(\frac{1}{p_0} - \frac{1}{q_0})})e^{-\alpha_1 s}e^{-\mu(t-s)}ds$$

$$\leq c_6 t^{-\frac{1}{2} + \frac{N}{2q_0}}e^{-\mu t}\|u_0\|_{L^N(\Omega)}$$

$$+ c_6 c_9 c_{10}(M_4 + M_1)\|\nabla\phi\|_{L^\infty(\Omega)}\varepsilon(1 + t^{\min\{0, 1 - \frac{N}{2}(\frac{1}{p_0} - \frac{1}{q_0})\}})e^{-\alpha_2 t}$$

$$\leq c_6 t^{-\frac{1}{2}+\frac{N}{2q_0}}\varepsilon e^{-\mu t} + 2c_6 c_9 c_{10}(M_4 + M_1)\|\nabla\phi\|_{L^\infty(\Omega)}\varepsilon e^{-\alpha_2 t}$$
$$= M_5 \varepsilon (1 + t^{-\frac{1}{2}+\frac{N}{2q_0}})e^{-\alpha_2 t}$$

with $M_5 = c_6 + 2c_6 c_9 c_{10}(M_4 + M_1)\|\nabla\phi\|_{L^\infty(\Omega)}$.

Lemma 4.25 *Under the assumptions of Proposition 4.2, we have*

$$\|\nabla c(\cdot, t)\|_{L^\infty(\Omega)} \leq \frac{M_2}{2}\varepsilon(1 + t^{-\frac{1}{2}})e^{-\alpha_1 t} \quad \text{for all } t \in (0, T).$$

Proof From (4.3.97) and Lemma 1.1(iii), we have

$$\|\nabla c(\cdot, t)\|_{L^\infty(\Omega)}$$
$$\leq \|e^{t(\Delta-1)}\nabla c_0\|_{L^\infty(\Omega)} + \int_0^t \|\nabla e^{(t-s)(\Delta-1)}(m - u\cdot\nabla c)(\cdot, s)\|_{L^\infty(\Omega)}ds \quad (4.3.101)$$
$$\leq c_3(1 + t^{-\frac{1}{2}})e^{-(\lambda_1+1)t}\|\nabla c_0\|_{L^N(\Omega)} + \int_0^t \|\nabla e^{(t-s)(\Delta-1)}(m - m_\infty)(\cdot, s)\|_{L^\infty(\Omega)}ds$$
$$+ \int_0^t \|\nabla e^{(t-s)(\Delta-1)}u\cdot\nabla c(\cdot, s)\|_{L^\infty(\Omega)}ds.$$

In the second inequality, we have used $\nabla e^{(t-s)(\Delta-1)}m_\infty = 0$.

From Lemmas 1.1, 4.3 and 4.23, it follows that

$$\int_0^t \|\nabla e^{(t-s)(\Delta-1)}(m - m_\infty)(\cdot, s)\|_{L^\infty(\Omega)}ds$$
$$\leq c_2 \int_0^t (1 + (t-s))^{-\frac{1}{2}-\frac{N}{2q_0}}e^{-(\lambda_1+1)(t-s)}\|(m - m_\infty)(\cdot, s)\|_{L^{q_0}(\Omega)}ds \quad (4.3.102)$$
$$\leq c_2 M_4 \varepsilon \int_0^t (1 + (t-s))^{-\frac{1}{2}-\frac{N}{2q_0}}e^{-(\lambda_1+1)(t-s)}(1 + s^{-\frac{N}{2}(\frac{1}{p_0}-\frac{1}{q_0})})e^{-\alpha_1 s}ds$$
$$\leq c_2 c_{10} M_4 \varepsilon (1 + t^{\min\{0,\frac{1}{2}-\frac{N}{2p_0}\}})e^{-\min\{\alpha_1,\lambda_1+1\}t}$$
$$\leq c_2 c_{10} M_4 \varepsilon (1 + t^{-\frac{1}{2}})e^{-\alpha_1 t}.$$

On the other hand, by Lemmas 1.1(ii), 4.3 and the definition of T, we obtain

$$\int_0^t \|\nabla e^{(t-s)(\Delta-1)} u \cdot \nabla c(\cdot, s)\|_{L^\infty(\Omega)} ds$$

$$\leq c_2 \int_0^t (1 + (t-s)^{-\frac{1}{2} - \frac{N}{2q_0}}) e^{-(\lambda_1+1)(t-s)} \|u \cdot \nabla c(\cdot, s)\|_{L^{q_0}(\Omega)} ds \qquad (4.3.103)$$

$$\leq c_2 \int_0^t (1 + (t-s)^{-\frac{1}{2} - \frac{N}{2q_0}}) e^{-(\lambda_1+1)(t-s)} \|u(\cdot, s)\|_{L^{q_0}(\Omega)} \|\nabla c(\cdot, s)\|_{L^\infty(\Omega)} ds$$

$$\leq c_2 M_5 M_2 \varepsilon^2 \int_0^t (1 + (t-s)^{-\frac{1}{2} - \frac{N}{2q_0}}) e^{-(\lambda_1+1)(t-s)} (1 + s^{-\frac{1}{2} + \frac{N}{2q_0}})(1 + s^{-\frac{1}{2}}) e^{-(\alpha_1+\alpha_2)s} ds$$

$$\leq 3c_2 M_5 M_2 \varepsilon^2 \int_0^t e^{-(\lambda_1+1)(t-s)} e^{-(\alpha_1+\alpha_2)s} (1 + (t-s)^{-\frac{1}{2} - \frac{N}{2q_0}})(1 + s^{-1 + \frac{N}{2q_0}}) ds$$

$$\leq 3c_2 M_5 M_2 c_{10} \varepsilon^2 (1 + t^{-\frac{1}{2}}) e^{-\min\{\lambda_1+1, \alpha_1+\alpha_2\} t}$$

$$\leq 3c_2 M_5 M_2 c_{10} \varepsilon^2 (1 + t^{-\frac{1}{2}}) e^{-\alpha_1 t}.$$

Hence, combining above inequalities with (4.3.87) and (4.3.88), we arrive at the conclusion.

Lemma 4.26 *Under the assumptions of Proposition 4.2, we have*

$$\|\rho(\cdot, t)\|_{L^\theta(\Omega)} \leq \frac{M_1}{2} \varepsilon (1 + t^{-\frac{N}{2}(\frac{1}{p_0} - \frac{1}{\theta})}) e^{-\alpha_1 t} \quad \text{for all } t \in (0, T), \ \theta \in [q_0, \infty].$$

Proof By the variation-of-constants formula, we have

$$\rho(\cdot, t) = e^{t(\Delta - m_\infty)} \rho_0 - \int_0^t e^{(t-s)(\Delta - m_\infty)} (\nabla \cdot (\rho \mathscr{S}(\cdot, \rho, c) \nabla c) - u \cdot \nabla \rho)(\cdot, s) ds$$

$$+ \int_0^t e^{(t-s)(\Delta - m_\infty)} \rho(m_\infty - m)(\cdot, s) ds.$$

By Lemma 1.1, the result in Sect. 2 of Horstmann and Winkler (2005) and $\alpha_1 < \min\{\lambda_1, m_\infty\}$, we obtain

$$\|\rho(\cdot, t)\|_{L^\theta(\Omega)}$$
$$\leq e^{-m_\infty t} (\|e^{t\Delta}(\rho_0 - \overline{\rho}_0)\|_{L^\theta(\Omega)} + \|\overline{\rho}_0\|_{L^\theta(\Omega)})$$
$$+ \int_0^t \|e^{(t-s)(\Delta - m_\infty)} \nabla \cdot (\rho \mathscr{S}(\cdot, \rho, c) \nabla c)(\cdot, s)\|_{L^\theta(\Omega)} ds$$
$$+ \int_0^t \|e^{(t-s)(\Delta - m_\infty)} (u \cdot \nabla \rho)(\cdot, s)\|_{L^\theta(\Omega)} ds$$
$$+ \int_0^t \|e^{(t-s)(\Delta - m_\infty)} \rho(m_\infty - m)(\cdot, s)\|_{L^\theta(\Omega)} ds$$
$$\leq c_1 (1 + t^{-\frac{N}{2}(\frac{1}{p_0} - \frac{1}{\theta})}) e^{-(\lambda_1 + m_\infty) t} \|\rho_0 - \overline{\rho}_0\|_{L^{p_0}(\Omega)}$$

$$+ (\min\{1, |\Omega|\})^{-\frac{1}{p_0}} e^{-m_\infty t} \varepsilon$$

$$+ c_4 C_S \int_0^t (1 + (t-s))^{-\frac{1}{2} - \frac{N}{2}(\frac{1}{q_0} - \frac{1}{\theta})} e^{-(\lambda_1 + m_\infty)(t-s)} \|\rho\|_{L^{q_0}(\Omega)} \|\nabla c\|_{L^\infty(\Omega)} ds$$

$$+ \int_0^t \|e^{(t-s)(\Delta - m_\infty)} \nabla \cdot (\rho u)(\cdot, s)\|_{L^\theta(\Omega)} ds$$

$$+ \int_0^t \|e^{(t-s)(\Delta - m_\infty)} \rho(m_\infty - m)(\cdot, s)\|_{L^\theta(\Omega)} ds$$

$$\leq (2c_1 + (\min\{1, |\Omega|\})^{-\frac{1}{p_0}})(1 + t^{-\frac{N}{2}(\frac{1}{p_0} - \frac{1}{\theta})}) \varepsilon e^{-\alpha_1 t}$$

$$+ c_4 C_S \int_0^t (1 + (t-s))^{-\frac{1}{2} - \frac{N}{2}(\frac{1}{q_0} - \frac{1}{\theta})} e^{-(\lambda_1 + m_\infty)(t-s)} \|\rho\|_{L^{q_0}(\Omega)} \|\nabla c\|_{L^\infty(\Omega)} ds$$

$$+ c_4 \int_0^t (1 + (t-s))^{-\frac{1}{2} - \frac{N}{2}(\frac{1}{q_0} - \frac{1}{\theta})} e^{-(\lambda_1 + m_\infty)(t-s)} \|\rho\|_{L^\infty(\Omega)} \|u\|_{L^{q_0}(\Omega)} ds$$

$$+ c_1 \int_0^t (1 + (t-s))^{-\frac{N}{2}(\frac{1}{q_0} - \frac{1}{\theta})} e^{-m_\infty(t-s)} \|\rho\|_{L^{q_0}(\Omega)} \|m - m_\infty\|_{L^\infty(\Omega)} ds$$

$$= (2c_1 + (\min\{1, |\Omega|\})^{-\frac{1}{p_0}})(1 + t^{-\frac{N}{2}(\frac{1}{p_0} - \frac{1}{\theta})}) \varepsilon e^{-\alpha_1 t} + I_1 + I_2 + I_3.$$

By the definition of T, Lemmas 4.25, 4.3 and (4.3.89), we get

$$I_1 \leq 3c_4 C_S M_1 M_2 \varepsilon^2 \int_0^t (1 + (t-s))^{-\frac{1}{2} - \frac{N}{2}(\frac{1}{q_0} - \frac{1}{\theta})} e^{-\lambda_1(t-s)} e^{-2\alpha_1 s}$$

$$\cdot (1 + s^{-\frac{1}{2} - \frac{N}{2}(\frac{1}{p_0} - \frac{1}{q_0})}) ds$$

$$\leq 3c_4 C_S c_{10} M_1 M_2 \varepsilon^2 (1 + t^{\min\{0, -\frac{N}{2}(\frac{1}{p_0} - \frac{1}{\theta})\}}) e^{-\min\{\lambda_1, 2\alpha_1\} t}$$

$$\leq \frac{M_1}{8} \varepsilon (1 + t^{-\frac{N}{2}(\frac{1}{p_0} - \frac{1}{\theta})}) e^{-\alpha_1 t}.$$

Similarly, by (4.3.91) and (4.3.92), we can also get

$$I_2 \leq 3c_4 M_1 M_5 \varepsilon^2 \int_0^t (1 + (t-s))^{-\frac{1}{2} - \frac{N}{2}(\frac{1}{q_0} - \frac{1}{\theta})} e^{-\lambda_1(t-s)} e^{-2\alpha_1 s} (1 + s^{-\frac{1}{2} - \frac{N}{2}(\frac{1}{p_0} - \frac{1}{q_0})}) ds$$

$$\leq 3c_4 c_{10} M_5 M_1 \varepsilon^2 (1 + t^{\min\{0, -\frac{N}{2}(\frac{1}{p_0} - \frac{1}{\theta})\}}) e^{-\min\{\lambda_1, 2\alpha_1\} t}$$

$$\leq \frac{M_1}{8} \varepsilon (1 + t^{-\frac{N}{2}(\frac{1}{p_0} - \frac{1}{\theta})}) e^{-\alpha_1 t},$$

$$I_3 \leq 3c_4 M_1 M_4 \varepsilon^2 \int_0^t (1 + (t-s))^{-\frac{1}{2} - \frac{N}{2}(\frac{1}{q_0} - \frac{1}{\theta})} e^{-m_\infty (t-s)} e^{-2\alpha_1 s}(1 + s^{-\frac{N}{p_0} + \frac{N}{2q_0}})ds$$

$$\leq 3c_4 c_{10} M_1 M_4 \varepsilon^2 (1 + t^{\min\{0, -\frac{N}{2}(\frac{1}{p_0} - \frac{1}{\theta})\}}) e^{-\min\{m_\infty, 2\alpha_1\}t}$$

$$\leq \frac{M_1}{8}\varepsilon(1 + t^{-\frac{N}{2}(\frac{1}{p_0} - \frac{1}{\theta})})e^{-\alpha_1 t},$$

respectively, where the fact that $q_0 \in (N, \frac{Np_0}{2(N-p_0)})$ warrants $-\frac{N}{p_0} + \frac{N}{2q_0} > -1$ is used. Hence, the combination of the above inequalities yields

$$\|\rho(\cdot, t)\|_{L^\theta(\Omega)} \leq \frac{M_1}{2}\varepsilon(1 + t^{-\frac{N}{2}(\frac{1}{p_0} - \frac{1}{\theta})})e^{-\alpha_1 t}.$$

Lemma 4.27 *Under the assumptions of Proposition 4.2, we have*

$$\|(m - \rho)(\cdot, t) - e^{t\Delta}(m_0 - \rho_0)\|_{L^\theta(\Omega)} \leq \frac{\varepsilon}{2}(1 + t^{-\frac{N}{2}(\frac{1}{p_0} - \frac{1}{\theta})})e^{-\alpha_1 t}$$

for $\theta \in [q_0, \infty]$, $t \in (0, T)$.

Proof From (4.3.95) and Lemma 1.1(iv), it follows that

$$\|(m - \rho)(\cdot, t) - e^{t\Delta}(m_0 - \rho_0)\|_{L^\theta(\Omega)}$$

$$\leq \int_0^t \|e^{(t-s)\Delta}(\nabla \cdot (\rho \mathscr{S}(\cdot, \rho, c)\nabla c) - u \cdot \nabla(m - \rho))(\cdot, s)\|_{L^\theta(\Omega)}ds$$

$$\leq \int_0^t \|e^{(t-s)\Delta}\nabla \cdot (\rho \mathscr{S}(\cdot, \rho, c)\nabla c)(\cdot, s)\|_{L^\theta(\Omega)}ds$$

$$+ \int_0^t \|e^{(t-s)\Delta}\nabla \cdot ((m - \rho - m_\infty)u)(\cdot, s)\|_{L^\theta(\Omega)}ds$$

$$\leq c_4 C_S \int_0^t (1 + (t-s))^{-\frac{1}{2} - \frac{N}{2}(\frac{1}{q_0} - \frac{1}{\theta})} e^{-\lambda_1(t-s)} \|\rho(\cdot, s)\|_{L^{q_0}(\Omega)} \|\nabla c(\cdot, s)\|_{L^\infty(\Omega)}ds$$

$$+ c_4 \int_0^t (1 + (t-s))^{-\frac{1}{2} - \frac{N}{2}(\frac{1}{q_0} - \frac{1}{\theta})} e^{-\lambda_1(t-s)} \|u(m - \rho - m_\infty)(\cdot, s)\|_{L^{q_0}(\Omega)}ds$$

$$= I_1 + I_2.$$

From the definition of T and (4.3.93), we have

$$I_1 \leq c_4 C_S M_1 M_2 \varepsilon^2 \int_0^t (1 + (t-s))^{-\frac{1}{2} - \frac{N}{2}(\frac{1}{q_0} - \frac{1}{\theta})} e^{-\lambda_1(t-s)}(1 + s^{-\frac{1}{2} - \frac{N}{2}(\frac{1}{p_0} - \frac{1}{q_0})})e^{-2\alpha_1 s}ds$$

$$\leq 3c_4 C_S c_{10} M_1 M_2 \varepsilon^2 (1 + t^{\min\{0, -\frac{N}{2}(\frac{1}{p_0} - \frac{1}{\theta})\}}) e^{-\min\{\lambda_1, 2\alpha_1\}t}$$

$$\leq \frac{\varepsilon}{4}(1 + t^{-\frac{N}{2}(\frac{1}{p_0} - \frac{1}{\theta})})e^{-\alpha_1 t}.$$

On the other hand, from Lemmas 4.22, 4.24 and (4.3.94), it follows that

$$I_2 = c_4 \int_0^t (1 + (t-s))^{-\frac{1}{2} - \frac{N}{2}(\frac{1}{q_0} - \frac{1}{\theta})} e^{-\lambda_1(t-s)} \|m - \rho - m_\infty\|_{L^\infty(\Omega)} \|u\|_{L^{q_0}(\Omega)} ds$$

$$\leq 2c_4 M_3 M_5 \varepsilon^2 \int_0^t (1 + (t-s))^{-\frac{1}{2} - \frac{N}{2}(\frac{1}{q_0} - \frac{1}{\theta})} e^{-\lambda_1(t-s)}$$

$$\cdot (1 + s^{-\frac{N}{2p_0}}) e^{-\alpha_1 s} (1 + s^{-\frac{1}{2} + \frac{N}{2q_0}}) e^{-\alpha_2 s} ds$$

$$\leq 6c_4 M_3 M_5 \varepsilon^2 \int_0^t (1 + (t-s))^{-\frac{1}{2} - \frac{N}{2}(\frac{1}{q_0} - \frac{1}{\theta})} (1 + s^{-\frac{1}{2} + \frac{N}{2}(\frac{1}{q_0} - \frac{1}{p_0})})$$

$$\cdot e^{-\lambda_1(t-s)} e^{-(\alpha_1 + \alpha_2)s} ds$$

$$\leq 6c_{10} c_4 M_3 M_5 \varepsilon^2 e^{-\min\{\lambda_1, \alpha_1 + \alpha_2\} t} (1 + t^{\min\{0, \frac{N}{2}(\frac{1}{\theta} - \frac{1}{p_0})\}})$$

$$\leq \frac{\varepsilon}{4} (1 + t^{-\frac{N}{2}(\frac{1}{p_0} - \frac{1}{\theta})}) e^{-\alpha_1 t}.$$

Combining the above inequalities, we arrive at

$$\|(\rho - m)(\cdot, t) - e^{t\Delta}(\rho_0 - m_0)\|_{L^\theta(\Omega)} \leq \frac{\varepsilon}{2} (1 + t^{-\frac{N}{2}(\frac{1}{p_0} - \frac{1}{\theta})}) e^{-\alpha_1 t},$$

and thus complete the proof of this lemma.

By the above lemmas, we can claim that $T = T_{max}$. Indeed, if $T < T_{max}$, by Lemmas 4.27, 4.26 and 4.25, we have

$$\|(m - \rho)(\cdot, t) - e^{t\Delta}(m_0 - \rho_0)\|_{L^\theta(\Omega)} \leq \frac{\varepsilon}{2} (1 + t^{-\frac{N}{2}(\frac{1}{p_0} - \frac{1}{\theta})}) e^{-\alpha_1 t},$$

$$\|\rho(\cdot, t)\|_{L^\theta(\Omega)} \leq \frac{M_1}{2} \varepsilon (1 + t^{-\frac{N}{2}(\frac{1}{p_0} - \frac{1}{\theta})}) e^{-\alpha_1 t}$$

as well as

$$\|\nabla c(\cdot, t)\|_{L^\infty(\Omega)} \leq \frac{M_2}{2} \varepsilon \left(1 + t^{-\frac{1}{2}}\right) e^{-\alpha_1 t}$$

for all $\theta \in [q_0, \infty]$ and $t \in (0, T)$, which contradict the definition of T in (4.3.94). Next, the further estimates of solutions are established to ensure $T_{max} = \infty$.

Lemma 4.28 *Under the assumptions of Proposition 4.2, there exists $M_6 > 0$ such that*

$$\|A^\beta u(\cdot, t)\|_{L^2(\Omega)} \leq \varepsilon M_6 e^{-\alpha_2 t} \quad \text{for } t \in (t_0, T_{max}) \text{ with } t_0 = \min\{\frac{T_{max}}{6}, 1\}.$$

Proof For any given $\alpha_2 < \lambda_1'$, we can fix $\mu \in (\alpha_2, \lambda_1')$. From (4.3.98), it follows that

$$\|A^\beta u(\cdot, t)\|_{L^2(\Omega)}$$

$$\leq \|A^\beta e^{-tA} u_0\|_{L^2(\Omega)} + \int_0^t \|A^\beta e^{-(t-s)A} \mathscr{P}((\rho + m - m_\infty)\nabla\phi)(\cdot, s)\|_{L^2(\Omega)} ds.$$

In the first integral, we apply Lemma 4.1, which gives

$$\|A^\beta e^{-tA} u_0\|_{L^2(\Omega)} \leq c_5 |\Omega|^{\frac{N-2}{2N}} t^{-\beta} e^{-\alpha_2 t} \|u_0\|_{L^N(\Omega)} \leq c_5 |\Omega|^{\frac{N-2}{2N}} t^{-\beta} e^{-\alpha_2 t} \varepsilon$$

for all $t \in (0, T)$. Next by Lemmas 4.2, 4.22 and 4.26, we have

$$\int_0^t \|A^\beta e^{-(t-s)A} \mathscr{P}((\rho + m - m_\infty)\nabla\phi)(\cdot, s)\|_{L^2(\Omega)} ds$$

$$\leq c_9 c_5 \|\nabla\phi\|_{L^\infty(\Omega)} |\Omega|^{\frac{q_0-2}{2q_0}} \int_0^t e^{-\mu(t-s)} (t-s)^{-\beta}$$

$$\cdot (\|m(\cdot, s) - \rho(\cdot, s) - m_\infty\|_{L^{q_0}(\Omega)} + 2\|\rho(\cdot, s)\|_{L^{q_0}(\Omega)}) ds$$

$$\leq M_6' \varepsilon \int_0^t e^{-\mu(t-s)} (t-s)^{-\beta} (1 + s^{-\frac{N}{2}(\frac{1}{p_0} - \frac{1}{q_0})}) e^{-\alpha_1 s} ds$$

$$\leq M_6' \varepsilon c_{10} (1 + t^{-1}) e^{-\alpha_2 t},$$

where $M_6' = (M_3 + M_1) c_9 c_5 \|\nabla\phi\|_{L^\infty(\Omega)} |\Omega|^{\frac{q_0-2}{2q_0}}$. Therefore there exists $M_6 > 0$ such that $\|A^\beta u(\cdot, t)\|_{L^2(\Omega)} \leq \varepsilon M_6 e^{-\alpha_2 t}$ for $t \in (t_0, T_{max})$.

Lemma 4.29 *Under the assumptions of Proposition 4.2, there exists $M_7 > 0$, such that $\|c(\cdot, t) - m_\infty\|_{L^\infty(\Omega)} \leq M_7 e^{-\alpha_2 t}$ for all (t_0, T_{max}) with $t_0 = \min\{\frac{T_{max}}{6}, 1\}$.*

Proof From (4.3.97) and Lemma 1.1, we have

$$\|(c - m_\infty)(\cdot, t)\|_{L^\infty(\Omega)}$$

$$\leq c_1 e^{-t} \|c_0 - m_\infty\|_{L^\infty(\Omega)} + \int_0^t \|e^{(t-s)(\Delta-1)} (m - m_\infty)(\cdot, s)\|_{L^\infty(\Omega)} ds$$

$$+ \int_0^t \|e^{(t-s)(\Delta-1)} u \cdot \nabla c(\cdot, s)\|_{L^\infty(\Omega)} ds. \tag{4.3.104}$$

By Lemmas 4.3 and 4.23, we obtain

$$\int_0^t \|e^{(t-s)(\Delta-1)} (m - m_\infty)(\cdot, s)\|_{L^\infty(\Omega)} ds$$

$$\leq c_1 \int_0^t (1 + (t-s)^{-\frac{N}{2q_0}}) e^{-(t-s)} \|(m - m_\infty)(\cdot, s)\|_{L^{q_0}(\Omega)} ds \tag{4.3.105}$$

$$\leq c_1 c_{10} M_4 \varepsilon e^{-\alpha_2 t}.$$

On the other hand, by Lemmas 4.3, 4.24 and 4.25, we get

$$\int_0^t \|e^{(t-s)(\Delta-1)} u \cdot \nabla c(\cdot, s)\|_{L^\infty(\Omega)} ds$$

$$\leq c_1 \int_0^t (1 + (t-s)^{-\frac{N}{2q_0}}) e^{-(t-s)} \|u \cdot \nabla c(\cdot, s)\|_{L^{q_0}(\Omega)} ds$$

$$\leq c_1 \int_0^t (1 + (t-s)^{-\frac{N}{2q_0}}) e^{-(t-s)} \|u(\cdot, s)\|_{L^{q_0}(\Omega)} \|\nabla c(\cdot, s)\|_{L^\infty(\Omega)} ds$$

$$\leq 6 c_1 M_5 M_2 c_{10} \varepsilon^2 e^{-\alpha_2 t}. \tag{4.3.106}$$

Therefore combining the above equalities, we arrive at the desired result.

Proof of Theorem 4.3 *in the case* $\mathscr{S} = 0$ *on* $\partial\Omega$, *part 2 (Proposition 4.2).* We now come to the final step to show that $T_{max} = \infty$. According to the extensibility criterion in Lemma 4.4, it remains to show that there exists $C > 0$ such that for $t_0 := \min\{\frac{T_{max}}{6}, 1\} < t < T_{max}$

$$\|\rho(\cdot, t)\|_{L^\infty(\Omega)} + \|m(\cdot, t)\|_{L^\infty(\Omega)} + \|c(\cdot, t)\|_{W^{1,\infty}(\Omega)} + \|A^\beta u(\cdot, t)\|_{L^2(\Omega)} < C.$$

From Lemmas 4.23 and 4.26, there exists $K_i > 0$, $i = 1, 2, 3$, such that

$$\|m(\cdot, t) - m_\infty\|_{L^\infty(\Omega)} \leq K_1 e^{-\alpha_1 t},$$
$$\|\rho(\cdot, t)\|_{L^\infty(\Omega)} \leq K_2 e^{-\alpha_1 t},$$
$$\|\nabla c(\cdot, t)\|_{L^\infty(\Omega)} \leq K_3 e^{-\alpha_1 t}$$

for $t \in (t_0, T_{max})$. Furthermore, Lemma 4.29 implies that $\|c(\cdot, t) - m_\infty\|_{W^{1,\infty}(\Omega)} \leq K_3' e^{-\alpha_2 t}$ with some $K_3' > 0$ for all $t \in (t_0, T_{max})$. Since $D(A^\beta) \hookrightarrow L^\infty(\Omega)$ with $\beta \in (\frac{N}{4}, 1)$, it follows from Lemma 4.28 that $\|u(\cdot, t)\|_{L^\infty(\Omega)} \leq K_4 e^{-\alpha_2 t}$ for some $K_4 > 0$ for all $t \in (t_0, T_{max})$. This completes the proof of Proposition 4.2.

Before we move to the next section, we remark that the following result is also valid by suitably adjusting $\varepsilon > 0$ for the larger values of p_0 or q_0.

Corollary 4.1 *Let* $N = 3$ *and* $\int_\Omega \rho_0 \neq \int_\Omega m_0$. *Further, let* $p_0 \in (\frac{N}{2}, \infty)$, $q_0 \in (N, \infty)$ *if* $\int_\Omega \rho_0 > \int_\Omega m_0$, *and* $p_0 \in (\frac{2N}{3}, \infty)$, $q_0 \in (N, \infty)$ *if* $\int_\Omega \rho_0 < \int_\Omega m_0$. *There exists* $\varepsilon > 0$ *such that for any initial data* (ρ_0, m_0, c_0, u_0) *fulfilling* (4.1.7) *as well as*

$$\|\rho_0 - \rho_\infty\|_{L^{p_0}(\Omega)} \leq \varepsilon, \quad \|m_0 - m_\infty\|_{L^{q_0}(\Omega)} \leq \varepsilon, \quad \|\nabla c_0\|_{L^N(\Omega)} \leq \varepsilon, \quad \|u_0\|_{L^N(\Omega)} \leq \varepsilon,$$

(4.1.1) *admits a global classical solution* (ρ, m, c, u, P). *Moreover, for any* $\alpha_1 \in (0, \min\{\lambda_1, m_\infty + \rho_\infty\})$, $\alpha_2 \in (0, \min\{\alpha_1, \lambda_1', 1\})$, *there exist constants* K_i $i = 1, 2, 3, 4$, *such that for all* $t \geq 1$

$$\|m(\cdot, t) - m_\infty\|_{L^\infty(\Omega)} \leq K_1 e^{-\alpha_1 t}, \quad \|\rho(\cdot, t) - \rho_\infty\|_{L^\infty(\Omega)} \leq K_2 e^{-\alpha_1 t},$$
$$\|c(\cdot, t) - m_\infty\|_{W^{1,\infty}(\Omega)} \leq K_3 e^{-\alpha_2 t}, \quad \|u(\cdot, t)\|_{L^\infty(\Omega)} \leq K_4 e^{-\alpha_2 t}.$$

4.3.3 Global Boundedness and Decay for General $\mathscr{S}$

In this subsection, we give the proof of our results for the general matrix-valued $\mathscr{S}$. This is accomplished by an approximation procedure. In order to make the previous results applicable, we introduce a family of smooth functions $\rho_\eta \in C_0^\infty(\Omega)$ and $0 \leq \rho_\eta(x) \leq 1$ for $\eta \in (0, 1)$, $\lim_{\eta \to 0} \rho_\eta(x) = 1$ and let $\mathscr{S}_\eta(x, \rho, c) = \rho_\eta(x)\mathscr{S}(x, \rho, c)$. Using this definition, we regularize (4.1.1) as follows:

$$\begin{cases} (\rho_\eta)_t + u_\eta \cdot \nabla \rho_\eta = \Delta \rho_\eta - \nabla \cdot (\rho_\eta \mathscr{S}_\eta(x, \rho_\eta, c_\eta)\nabla c_\eta) - \rho_\eta m_\eta, \\ (m_\eta)_t + u_\eta \cdot \nabla m_\eta = \Delta m_\eta - \rho_\eta m_\eta, \\ (c_\eta)_t + u_\eta \cdot \nabla c_\eta = \Delta c_\eta - c_\eta + m_\eta, \\ (u_\eta)_t = \Delta u_\eta - \nabla P_\eta + (\rho_\eta + m_\eta)\nabla\phi, \quad \nabla \cdot u_\eta = 0, \\ \dfrac{\partial \rho_\eta}{\partial \nu} = \dfrac{\partial m_\eta}{\partial \nu} = \dfrac{\partial c_\eta}{\partial \nu} = 0, \ u_\eta = 0 \end{cases} \quad (4.3.107)$$

with the initial data

$$\rho_\eta(x, 0) = \rho_0(x), \ m_\eta(x, 0) = m_0(x), \ c(x, 0) = c_0(x), \ u_\eta(x, 0) = u_0(x), \ x \in \Omega. \quad (4.3.108)$$

It is observed that $\mathscr{S}_\eta$ satisfies the additional condition $\mathscr{S} = 0$ on $\partial\Omega$. Therefore, based on the discussion in Sect. 4.3.2, under the assumptions of Theorem 4.1 and Theorem 4.3, the problem (4.3.107)–(4.3.108) admits a global classical solution $(\rho_\eta, m_\eta, c_\eta, u_\eta, P_\eta)$ that satisfies

$$\|m_\eta(\cdot, t) - m_\infty\|_{L^\infty(\Omega)} \leq K_1 e^{-\alpha_1 t}, \quad \|\rho_\eta(\cdot, t) - \rho_\infty\|_{L^\infty(\Omega)} \leq K_2 e^{-\alpha_1 t},$$
$$\|c_\eta(\cdot, t) - m_\infty\|_{W^{1,\infty}(\Omega)} \leq K_3 e^{-\alpha_2 t}, \quad \|u_\eta(\cdot, t)\|_{L^\infty(\Omega)} \leq K_4 e^{-\alpha_2 t}.$$

for some constants $K_i, i = 1, 2, 3, 4$, and $t \geq 0$. Applying a standard procedure such as in Lemmas 5.2 and 5.6 of Cao and Lankeit (2016), one can obtain a subsequence of $\{\eta_j\}_{j \in \mathbb{N}}$ with $\eta_j \to 0$ as $j \to \infty$, such that $\rho_{\eta_j} \to \rho$, $m_{\eta_j} \to m$, $c_{\eta_j} \to c$, $u_{\eta_j} \to u$ in $C_{loc}^{\vartheta, \frac{\vartheta}{2}}(\overline{\Omega} \times (0, \infty))$ as $j \to \infty$ for some $\vartheta \in (0, 1)$. Moreover, by the arguments as in Lemmas 5.7, 5.8 of Cao and Lankeit (2016), one can also show that (ρ, m, c, u, P) is a classical solution of (4.1.1) with the decay properties asserted in Theorems 4.2 and 4.3. The proofs of Theorems 4.1–4.3 are thus complete.

4.4 Asymptotic Behavior of Solutions to a Coral Fertilization Model

4.4.1 A Convenient Extensibility Criterion

Firstly, we recall the result of the local existence of classical solutions, which can be proved by a straightforward adaptation of a well-known fixed point argument (see Winkler (2012) for example).

Lemma 4.30 *Suppose that* (4.1.14), (4.1.15) *and*

$$\mathscr{S}(x, \rho, c) = 0, \quad (x, \rho, c) \in \partial\Omega \times [0, \infty) \times [0, \infty) \tag{4.4.1}$$

hold. Then there exist $T_{max} \in (0, \infty]$ and a classical solution (ρ, m, c, u, P) of (4.1.13) *on* $(0, T_{max})$. *Moreover, ρ, m, c are nonnegative in $\Omega \times (0, T_{max})$, and if $T_{max} < \infty$, then for $\beta \in (\frac{3}{4}, 1)$, as $t \to T_{max}$*

$$\|\rho(\cdot, t)\|_{L^\infty(\Omega)} + \|m(\cdot, t)\|_{L^\infty(\Omega)} + \|c(\cdot, t)\|_{W^{1,\infty}(\Omega)} + \|A^\beta u(\cdot, t)\|_{L^2(\Omega)} \to \infty.$$

This solution is unique, up to addition of constants to P.

The following elementary properties of the solutions in Lemma 4.30 are immediate consequences of the integration of the first and second equations in (4.1.13), as well as an application of the maximum principle to the second and third equations.

Lemma 4.31 *Suppose that* (4.1.14), (4.1.15) *and* (4.4.1) *hold. Then for all $t \in (0, T_{max})$, the solution of* (4.1.13) *from Lemma 4.30 satisfies*

$$\|\rho(\cdot, t)\|_{L^1(\Omega)} \le \|\rho_0\|_{L^1(\Omega)}, \quad \|m(\cdot, t)\|_{L^1(\Omega)} \le \|m_0\|_{L^1(\Omega)}, \tag{4.4.2}$$

$$\int_0^t \|\rho(\cdot, s) m(\cdot, s)\|_{L^1(\Omega)} ds \le \min\{\|\rho_0\|_{L^1(\Omega)}, \|m_0\|_{L^1(\Omega)}\}, \tag{4.4.3}$$

$$\|\rho(\cdot, t)\|_{L^1(\Omega)} - \|m(\cdot, t)\|_{L^1(\Omega)} = \|\rho_0\|_{L^1(\Omega)} - \|m_0\|_{L^1(\Omega)}, \tag{4.4.4}$$

$$\|m(\cdot, t)\|_{L^2(\Omega)}^2 + 2 \int_0^t \|\nabla m(\cdot, s)\|_{L^2(\Omega)}^2 ds \le \|m_0\|_{L^2(\Omega)}^2, \tag{4.4.5}$$

$$\|m(\cdot, t)\|_{L^\infty(\Omega)} \le \|m_0\|_{L^\infty(\Omega)}, \tag{4.4.6}$$

$$\|c(\cdot, t)\|_{L^\infty(\Omega)} \le \max\{\|m_0\|_{L^\infty(\Omega)}, \|c_0\|_{L^\infty(\Omega)}\}. \tag{4.4.7}$$

4.4.2 Global Boundedness and Decay for $\mathscr{S} = 0$ on $\partial\Omega$

Throughout this section, we assume that $\mathscr{S} = 0$ on $\partial\Omega$. We note that, under this assumption, the boundary condition for ρ in (4.1.13) reduces to the homogeneous Neumann condition $\nabla\rho \cdot \nu = 0$.

In the case $\int_\Omega \rho_0 > \int_\Omega m_0$, i.e., $\rho_\infty > 0$, $m_\infty = 0$, Theorem 4.4 reduces to:

Proposition 4.3 *Suppose that (4.1.14) hold and $\int_\Omega \rho_0 > \int_\Omega m_0$. Let $p_0 \in (\frac{3}{2}, 3)$, $q_0 \in (3, \frac{3p_0}{3-p_0})$. There exists $\varepsilon > 0$, such that for any initial data (ρ_0, m_0, c_0, u_0) fulfilling (4.1.15) as well as*

$$\|\rho_0 - \rho_\infty\|_{L^{p_0}(\Omega)} < \varepsilon, \quad \|m_0\|_{L^{q_0}(\Omega)} < \varepsilon, \quad \|c_0\|_{L^\infty(\Omega)} < \varepsilon, \quad \|u_0\|_{L^3(\Omega)} < \varepsilon,$$

(4.1.13) admits a global classical solution (ρ, m, c, u, P). In particular, for any $\alpha_1 \in (0, \min\{\lambda_1, \rho_\infty\})$, $\alpha_2 \in (0, \min\{\alpha_1, \lambda_1', 1\})$, there exist constants K_i, $i = 1, 2, 3, 4$, such that for all $t \geq 1$

$$\|m(\cdot, t)\|_{L^\infty(\Omega)} \leq K_1 e^{-\alpha_1 t}, \tag{4.4.8}$$

$$\|\rho(\cdot, t) - \rho_\infty\|_{L^\infty(\Omega)} \leq K_2 e^{-\alpha_1 t}, \tag{4.4.9}$$

$$\|c(\cdot, t)\|_{W^{1,\infty}(\Omega)} \leq K_3 e^{-\alpha_2 t}, \tag{4.4.10}$$

$$\|u(\cdot, t)\|_{L^\infty(\Omega)} \leq K_4 e^{-\alpha_2 t}. \tag{4.4.11}$$

Proposition 4.3 is the consequence of the following lemmas. In the proofs thereof, the constants c_i, $i = 1, 2, 3, 4$ refer to those in Lemma 1.1, $c_i > 0$, $i = 5, \ldots, 10$, refer to those in Lemmas 4.1–4.3.

Lemma 4.32 *Under the assumptions of Proposition 4.3 and*

$$\sigma = \int_0^\infty \left(1 + s^{-\frac{3}{2p_0}}\right) e^{-\alpha_1 s} ds,$$

there exist $M_1 > 0$, $M_2 > 0$ and $\varepsilon \in (0, 1)$, such that

$$c_2 + 2c_2 c_{10} e^{(1+c_1+c_1|\Omega|^{\frac{1}{p_0}-\frac{1}{q_0}})\sigma} \leq \frac{M_2}{4}, \tag{4.4.12}$$

$$c_4 c_{10} C_S M_2 (e^{(1+c_1+c_1|\Omega|^{\frac{1}{p_0}-\frac{1}{q_0}})\sigma} + \rho_\infty |\Omega|^{\frac{1}{q_0}}) \leq \frac{M_1}{8}, \tag{4.4.13}$$

$$c_6 + 2c_6 c_9 c_{10} \|\nabla\phi\|_{L^\infty(\Omega)} (M_1 + c_1 + c_1|\Omega|^{\frac{1}{p_0}-\frac{1}{q_0}} + 4e^{(1+c_1+c_1|\Omega|^{\frac{1}{p_0}-\frac{1}{q_0}})\sigma}) < \frac{M_3}{4}, \tag{4.4.14}$$

$$c_7 + 2c_7 c_9 c_{10} \|\nabla\phi\|_{L^\infty(\Omega)} |\Omega|^{\frac{1}{3}-\frac{1}{q_0}} (M_1 + c_1 + c_1|\Omega|^{\frac{1}{p_0}-\frac{1}{q_0}} + 4e^{(1+c_1+c_1|\Omega|^{\frac{1}{p_0}-\frac{1}{q_0}})\sigma})$$
$$< \frac{M_4}{4}, \tag{4.4.15}$$

$$3c_{10} c_4 C_S (M_1 + c_1 + c_1|\Omega|^{\frac{1}{p_0}-\frac{1}{q_0}}) M_2 \varepsilon \leq \frac{M_1}{8}, \tag{4.4.16}$$

$$3c_{10} c_4 (M_1 + c_1 + c_1|\Omega|^{\frac{1}{p_0}-\frac{1}{q_0}}) M_3 \varepsilon \leq \frac{M_1}{4}, \tag{4.4.17}$$

$$12 c_2 c_{10} M_3 \varepsilon < 1, \quad 12 c_7 c_9 c_{10} M_3 \varepsilon \leq 1, \quad 12 c_6 c_9 c_{10} M_4 \varepsilon \leq 1. \tag{4.4.18}$$

Let

$$
T \triangleq \sup \left\{ \widetilde{T} \in (0, T_{max}) \left|
\begin{aligned}
&\left\| (\rho - m)(\cdot, t) - e^{t\Delta}(\rho_0 - m_0) \right\|_{L^\theta(\Omega)} \\
&\quad \leq M_1 \varepsilon (1 + t^{-\frac{3}{2}(\frac{1}{p_0} - \frac{1}{\theta})}) e^{-\alpha_1 t} \quad \forall \theta \in [q_0, \infty], \ t \in [0, \widetilde{T}); \\
&\| \nabla c(\cdot, t) \|_{L^\infty(\Omega)} \leq M_2 \varepsilon (1 + t^{-\frac{1}{2}}) e^{-\alpha_1 t} \quad \forall t \in [0, \widetilde{T}); \\
&\| u(\cdot, t) \|_{L^{q_0}(\Omega)} \leq M_3 \varepsilon (1 + t^{-\frac{1}{2} + \frac{3}{2q_0}}) e^{-\alpha_2 t} \quad \forall t \in [0, \widetilde{T}); \\
&\| \nabla u(\cdot, t) \|_{L^3(\Omega)} \leq M_4 \varepsilon (1 + t^{-\frac{1}{2}}) e^{-\alpha_2 t} \quad \forall t \in [0, \widetilde{T}).
\end{aligned}
\right. \right\}.
$$

$$(4.4.19)$$

Then $T > 0$ is well-defined by Lemma 4.30 and (4.1.15). Now we claim that $T = T_{max} = \infty$ if ε is sufficiently small. To this end, by the contradiction argument, it suffices to verify that all of the estimates mentioned in (4.4.19) still hold for even smaller coefficients on the right-hand side. This mainly relies on $L^p - L^q$ estimates for the Neumann heat semigroup and the fact that the classical solution on $(0, T_{max})$ can be written as

$$
\begin{aligned}
&(\rho - m)(\cdot, t) \\
&= e^{t\Delta}(\rho_0 - m_0) - \int_0^t e^{(t-s)\Delta}(\nabla \cdot (\rho \mathscr{S}(x, \rho, c)\nabla c) + u \cdot \nabla(\rho - m))(\cdot, s)ds,
\end{aligned}
$$

$$(4.4.20)$$

$$
m(\cdot, t) = e^{t\Delta}m_0 - \int_0^t e^{(t-s)\Delta}(\rho m + u \cdot \nabla m)(\cdot, s)ds, \tag{4.4.21}
$$

$$
c(\cdot, t) = e^{t(\Delta-1)}c_0 + \int_0^t e^{(t-s)(\Delta-1)}(m - u \cdot \nabla c)(\cdot, s)ds, \tag{4.4.22}
$$

$$
u(\cdot, t) = e^{-tA}u_0 + \int_0^t e^{-(t-s)A}\mathscr{P}((\rho + m)\nabla\phi - (u \cdot \nabla)u)(\cdot, s)ds \tag{4.4.23}
$$

for all $t \in (0, T_{max})$ according to the variation-of-constants formula.

Although the proofs of Lemmas 4.33 and 4.34 below are similar to those of Lemmas 3.11 and 3.12 in Li et al. (2019b), respectively, we provide their proofs for the convenience of the interested reader.

Lemma 4.33 *Under the assumptions of Proposition 4.3, for all $t \in (0, T)$ and $\theta \in [q_0, \infty]$, there exists constant $M_5 > 0$, such that*

$$
\| (\rho - m)(\cdot, t) - \rho_\infty \|_{L^\theta(\Omega)} \leq M_5 \varepsilon (1 + t^{-\frac{3}{2}(\frac{1}{p_0} - \frac{1}{\theta})}) e^{-\alpha_1 t}.
$$

Proof Due to $e^{t\Delta}\rho_\infty = \rho_\infty$ and $\int_\Omega (\rho_0 - m_0 - \rho_\infty) = 0$, the definition of T and Lemma 1.1(i) show that for all $t \in (0, T)$ and $\theta \in [q_0, \infty]$,

$$\|(\rho - m)(\cdot, t) - \rho_\infty\|_{L^\theta(\Omega)}$$

$$\leq \|(\rho - m)(\cdot, t) - e^{t\Delta}(\rho_0 - m_0)\|_{L^\theta(\Omega)} + \|e^{t\Delta}(\rho_0 - m_0 - \rho_\infty)\|_{L^\theta(\Omega)}$$

$$\leq M_1\varepsilon(1 + t^{-\frac{3}{2}(\frac{1}{p_0} - \frac{1}{\theta})})e^{-\alpha_1 t}$$

$$\quad + c_1(1 + t^{-\frac{3}{2}(\frac{1}{p_0} - \frac{1}{\theta})})(\|\rho_0 - \rho_\infty\|_{L^{p_0}(\Omega)} + \|m_0\|_{L^{p_0}(\Omega)})e^{-\lambda_1 t}$$

$$\leq M_5\varepsilon(1 + t^{-\frac{3}{2}(\frac{1}{p_0} - \frac{1}{\theta})})e^{-\alpha_1 t},$$

where $M_5 = M_1 + c_1 + c_1|\Omega|^{\frac{1}{p_0} - \frac{1}{q_0}}$.

Lemma 4.34 *Under the assumptions of Proposition 4.3, for any $k > 1$,*

$$\|m(\cdot, t)\|_{L^k(\Omega)} \leq M_6\|m_0\|_{L^k(\Omega)}e^{-\rho_\infty t} \quad \text{for all } t \in (0, T) \tag{4.4.24}$$

with $\sigma = \int_0^\infty (1 + s^{-\frac{3}{2p_0}})e^{-\alpha_1 s}ds$ and $M_6 = e^{M_5\sigma\varepsilon}$.

Proof Testing the first equation in (4.1.13) with m^{k-1} ($k > 1$) and integrating by parts, we have

$$\frac{d}{dt}\int_\Omega m^k \leq -k\int_\Omega \rho m^k \quad \text{on } (0, T).$$

In view of $-\rho \leq |\rho - m - \rho_\infty| - m - \rho_\infty \leq -\rho_\infty + |\rho - m - \rho_\infty|$, Lemma 4.33 yields

$$\frac{d}{dt}\int_\Omega m^k \leq -k\rho_\infty\int_\Omega m^k + k\int_\Omega m^k|\rho - m - \rho_\infty|$$

$$\leq -k\rho_\infty\int_\Omega m^k + k\|\rho - m - \rho_\infty\|_{L^\infty(\Omega)}\int_\Omega m^k$$

$$\leq -k\rho_\infty\int_\Omega m^k + kM_5\varepsilon(1 + t^{-\frac{3}{2p_0}})e^{-\alpha_1 t}\int_\Omega m^k$$

and thus

$$\int_\Omega m^k \leq \int_\Omega m_0^k \exp\{-k\rho_\infty t + kM_5\varepsilon\int_0^t (1 + s^{-\frac{3}{2p_0}})e^{-\alpha_1 s}ds\} \leq \|m_0\|_{L^k(\Omega)}^k e^{k(M_5\sigma\varepsilon - \rho_\infty t)},$$

from which (4.4.24) follows immediately.

Lemma 4.35 *Under the assumptions of Proposition 4.3, we have*

$$\|u(\cdot, t)\|_{L^{q_0}(\Omega)} \leq \frac{M_3}{2}\varepsilon\left(1 + t^{-\frac{1}{2} + \frac{3}{2q_0}}\right)e^{-\alpha_2 t} \quad \text{for all } t \in (0, T).$$

Proof For $\alpha_2 < \lambda_1'$, we fix $\mu \in (\alpha_2, \lambda_1')$. According to (4.4.23), Lemmas 4.1(ii) and 4.2, we infer that

$$\|u(\cdot, t)\|_{L^{q_0}(\Omega)}$$

$$\leq c_6 t^{-\frac{3}{2}\left(\frac{1}{3}-\frac{1}{q_0}\right)} e^{-\mu t} \|u_0\|_{L^3(\Omega)}$$

$$+ \int_0^t \|e^{-(t-s)A}\mathcal{P}((\rho + m)\nabla\phi - (u \cdot \nabla)u)(\cdot, s)\|_{L^{q_0}(\Omega)}ds$$

$$\leq c_6 t^{-\frac{3}{2}\left(\frac{1}{3}-\frac{1}{q_0}\right)} e^{-\mu t} \|u_0\|_{L^3(\Omega)}$$

$$+ c_6 \int_0^t e^{-\mu(t-s)} \|\mathcal{P}((\rho + m - \overline{\rho + m})\nabla\phi)(\cdot, s)\|_{L^{q_0}(\Omega)}ds$$

$$+ c_6 \int_0^t e^{-\mu(t-s)} \|\mathcal{P}((u \cdot \nabla)u)(\cdot, s)\|_{L^{q_0}(\Omega)}ds \qquad (4.4.25)$$

$$\leq c_6 t^{-\frac{1}{2}+\frac{3}{2q_0}} e^{-\mu t} \|u_0\|_{L^3(\Omega)}$$

$$+ c_6 c_9 \|\nabla\phi\|_{L^\infty(\Omega)} \int_0^t e^{-\mu(t-s)} \|(\rho + m - \overline{\rho + m})(\cdot, s)\|_{L^{q_0}(\Omega)}ds$$

$$+ c_6 c_9 \int_0^t (t-s)^{-\frac{1}{2}} e^{-\mu(t-s)} \|(u \cdot \nabla)u(\cdot, s)\|_{L^{\frac{1}{\frac{1}{3}+\frac{1}{q_0}}}(\Omega)} ds$$

$$=: c_6 t^{-\frac{1}{2}+\frac{3}{2q_0}} e^{-\mu t} \|u_0\|_{L^3(\Omega)} + J_1 + J_2,$$

where $\mathcal{P}(\overline{\rho + m}\nabla\phi) = \overline{\rho + m}\mathcal{P}(\nabla\phi) = 0$ is used.

Due to $\alpha_1 < \rho_\infty$, an application of Lemmas 4.33 and 4.34 shows that

$$J_1 \leq c_6 c_9 \|\nabla\phi\|_{L^\infty(\Omega)} \int_0^t e^{-\mu(t-s)} \|(\rho - m - \overline{\rho - m})(\cdot, s) + 2(m - \overline{m})(\cdot, s)\|_{L^{q_0}(\Omega)}ds$$

$$\leq c_6 c_9 \|\nabla\phi\|_{L^\infty(\Omega)} \int_0^t e^{-\mu(t-s)} (\|(\rho - m - \rho_\infty)(\cdot, s)\|_{L^{q_0}(\Omega)} + 2\|(m - \overline{m})(\cdot, s)\|_{L^{q_0}(\Omega)})ds$$

$$\leq c_6 c_9 \|\nabla\phi\|_{L^\infty(\Omega)} M_7' \varepsilon \int_0^t e^{-\mu(t-s)} (1 + s^{-\frac{3}{2}\left(\frac{1}{p_0}-\frac{1}{q_0}\right)}) e^{-\alpha_1 s} ds \qquad (4.4.26)$$

with $M_7' = M_5 + 4e^{M_5\sigma\varepsilon}$.

On the other hand, by the Hölder inequality and definition of T, we have

$$J_2 \leq c_6 c_9 \int_0^t (t-s)^{-\frac{1}{2}} e^{-\mu(t-s)} \|u(\cdot, s)\|_{L^{q_0}(\Omega)} \|\nabla u(\cdot, s)\|_{L^3(\Omega)} ds$$

$$\leq 3 c_6 c_9 M_3 M_4 \varepsilon^2 \int_0^t (t-s)^{-\frac{1}{2}} e^{-\mu(t-s)} (1 + s^{-1+\frac{3}{2q_0}}) e^{-2\alpha_2 s} ds. \qquad (4.4.27)$$

Now, plugging (4.4.26), (4.4.27) into (4.4.25) and applying Lemma 4.3, we end up with

$$\|u(\cdot, t)\|_{L^{q_0}(\Omega)}$$

$$\leq c_6 t^{-\frac{1}{2}+\frac{3}{2q_0}} e^{-\mu t} \|u_0\|_{L^3(\Omega)} + c_6 c_9 c_{10} \|\nabla\phi\|_{L^\infty(\Omega)} M_7' \varepsilon (1 + t^{\min\{0, 1-\frac{3}{2}(\frac{1}{p_0}-\frac{1}{q_0})\}}) e^{-\alpha_2 t}$$

$$+ 3 c_6 c_9 c_{10} M_3 M_4 \varepsilon^2 (1 + t^{-\frac{1}{2}+\frac{3}{2q_0}}) e^{-\alpha_2 t}$$

$$\leq c_6 t^{-\frac{1}{2}+\frac{3}{2q_0}} e^{-\mu t} \varepsilon + 2 c_6 c_9 c_{10} \|\nabla\phi\|_{L^\infty(\Omega)} M_7' \varepsilon e^{-\alpha_2 t}$$

$$+ 3 c_6 c_9 c_{10} M_3 M_4 \varepsilon^2 (1 + t^{-\frac{1}{2}+\frac{3}{2q_0}}) e^{-\alpha_2 t}$$

$$\leq \frac{M_3}{2} \varepsilon (1 + t^{-\frac{1}{2}+\frac{3}{2q_0}}) e^{-\alpha_2 t},$$

where (4.4.14), (4.4.18) and the fact that $\frac{3}{2}(\frac{1}{p_0} - \frac{1}{q_0}) < 1$ are used.

In the next lemma, we show that the estimate for the gradient is also preserved.

Lemma 4.36 *Under the assumptions of Proposition 4.3, we have*

$$\|\nabla u(\cdot, t)\|_{L^3(\Omega)} \leq \frac{M_4}{2} \varepsilon (1 + t^{-\frac{1}{2}}) e^{-\alpha_2 t} \text{ for all } t \in (0, T).$$

Proof According to (4.4.23), we have

$$\nabla u(\cdot, t) = \nabla e^{-tA} u_0 + \int_0^t \nabla e^{-(t-s)A} (\mathscr{P}((\rho + m)\nabla\phi) - \mathscr{P}((u \cdot \nabla)u))(\cdot, s) ds.$$

Applying Lemmas 4.1(iii), 4.2 and the Hölder inequality, we arrive at

$$\|\nabla u(\cdot, t)\|_{L^3(\Omega)}$$

$$\leq c_7 t^{-\frac{1}{2}} e^{-\mu t} \|u_0\|_{L^3(\Omega)} + \int_0^t \|\nabla e^{-(t-s)A} \mathscr{P}((\rho + m)\nabla\phi - (u \cdot \nabla)u)(\cdot, s)\|_{L^3(\Omega)} ds$$

$$\leq c_7 t^{-\frac{1}{2}} e^{-\mu t} \varepsilon + c_7 \int_0^t (t - s)^{-\frac{1}{2}} e^{-\mu(t-s)} \|\mathscr{P}((\rho + m - \overline{\rho + m})\nabla\phi)(\cdot, s)\|_{L^3(\Omega)} ds$$

$$+ c_7 \int_0^t (t - s)^{-\frac{1}{2}-\frac{3}{2q_0}} e^{-\mu(t-s)} \|\mathscr{P}((u \cdot \nabla)u)(\cdot, s)\|_{L^{\frac{3q_0}{3+q_0}}(\Omega)} ds \qquad (4.4.28)$$

$$\leq c_7 t^{-\frac{1}{2}} e^{-\mu t} \varepsilon$$

$$+ c_7 c_9 \|\nabla\phi\|_{L^\infty(\Omega)} |\Omega|^{\frac{1}{3}-\frac{1}{q_0}} \int_0^t (t - s)^{-\frac{1}{2}} e^{-\mu(t-s)} \|(\rho + m - \overline{\rho + m})(\cdot, s)\|_{L^{q_0}(\Omega)} ds$$

$$+ c_7 c_9 \int_0^t (t - s)^{-\frac{1}{2}-\frac{3}{2q_0}} e^{-\mu(t-s)} \|\nabla u(\cdot, s)\|_{L^3(\Omega)} \|u(\cdot, s)\|_{L^{q_0}(\Omega)} ds$$

$$=: c_7 t^{-\frac{1}{2}} e^{-\mu t} \varepsilon + \tau_1 + \tau_2,$$

where $\mathscr{P}(\overline{\rho + m}\nabla\phi) = \overline{\rho + m}\mathscr{P}(\nabla\phi) = 0$ is used.

Due to $\alpha_1 < \rho_\infty$, an application of Lemmas 4.33 and 4.34 shows that

$$\tau_1 \leq c_7 c_9 \|\nabla \phi\|_{L^\infty(\Omega)} |\Omega|^{\frac{1}{3}-\frac{1}{q_0}} \int_0^t (t-s)^{-\frac{1}{2}} e^{-\mu(t-s)}$$

$$\|(\rho - m - \overline{\rho - m})(\cdot, s) + 2(m - \overline{m})(\cdot, s)\|_{L^{q_0}(\Omega)} ds \tag{4.4.29}$$

$$\leq c_7 c_9 \|\nabla \phi\|_{L^\infty(\Omega)} |\Omega|^{\frac{1}{3}-\frac{1}{q_0}} \int_0^t (t-s)^{-\frac{1}{2}} e^{-\mu(t-s)}$$

$$(\|(\rho - m - \rho_\infty)(\cdot, s)\|_{L^{q_0}(\Omega)} + 2\|(m - \overline{m})(\cdot, s)\|_{L^{q_0}(\Omega)}) ds$$

$$\leq c_7 c_9 \|\nabla \phi\|_{L^\infty(\Omega)} |\Omega|^{\frac{1}{3}-\frac{1}{q_0}} M_7' \varepsilon \int_0^t e^{-\mu(t-s)}(1 + s^{-\frac{3}{2}(\frac{1}{p_0}-\frac{1}{q_0})})(t-s)^{-\frac{1}{2}} e^{-\alpha_1 s} ds.$$

On the other hand, from the Hölder inequality and definition of T, it follows that

$$\tau_2 \leq 3c_7 c_9 M_3 M_4 \varepsilon^2 \int_0^t (t-s)^{-\frac{1}{2}-\frac{3}{2q_0}} e^{-\mu(t-s)}(1 + s^{-1+\frac{3}{2q_0}}) e^{-2\alpha_2 s} ds. \tag{4.4.30}$$

Therefore, inserting (4.4.30), (4.4.29) into (4.4.28) and applying Lemma 4.3, we get

$$\|\nabla u(\cdot, t)\|_{L^{q_0}(\Omega)}$$

$$\leq c_7 t^{-\frac{1}{2}} e^{-\mu t} \varepsilon + c_7 c_9 c_{10} \|\nabla \phi\|_{L^\infty(\Omega)} |\Omega|^{\frac{1}{3}-\frac{1}{q_0}} M_7' \varepsilon (1 + t^{\min\{0, \frac{1}{2}-\frac{3}{2}(\frac{1}{p_0}-\frac{1}{q_0})\}}) e^{-\alpha_2 t}$$

$$+ 3c_7 c_9 c_{10} M_3 M_4 \varepsilon^2 (1 + t^{-\frac{1}{2}}) e^{-\alpha_2 t}$$

$$\leq c_7 t^{-\frac{1}{2}} e^{-\mu t} \varepsilon + 2c_7 c_9 c_{10} \|\nabla \phi\|_{L^\infty(\Omega)} |\Omega|^{\frac{1}{3}-\frac{1}{q_0}} M_7' \varepsilon e^{-\alpha_2 t}$$

$$+ 3c_7 c_9 c_{10} M_3 M_4 \varepsilon^2 (1 + t^{-\frac{1}{2}}) e^{-\alpha_2 t}$$

$$\leq \frac{M_4}{2} \varepsilon (1 + t^{-\frac{1}{2}}) e^{-\alpha_2 t},$$

where (4.4.15), (4.4.18) and the fact that $q_0 \in (3, \frac{3p_0}{3-p_0})$, $p_0 \in (\frac{3}{2}, 3)$ are used.

Lemma 4.37 *Under the assumptions of Proposition 4.3, we have*

$$\|\nabla c(\cdot, t)\|_{L^\infty(\Omega)} \leq \frac{M_2}{2} \varepsilon (1 + t^{-\frac{1}{2}}) e^{-\alpha_1 t} \ \text{for all} \ t \in (0, T).$$

Proof By (4.4.22) and Lemma 1.1(ii), we have

$$\|\nabla c(\cdot, t)\|_{L^\infty(\Omega)}$$

$$\leq \|e^{t(\Delta-1)} \nabla c_0\|_{L^\infty(\Omega)} + \int_0^t \|\nabla e^{(t-s)(\Delta-1)}(m - u \cdot \nabla c)(\cdot, s)\|_{L^\infty(\Omega)} ds$$

$$\leq c_2 (1 + t^{-\frac{1}{2}}) e^{-(\lambda_1+1)t} \|c_0\|_{L^\infty(\Omega)} + \int_0^t \|\nabla e^{(t-s)(\Delta-1)} m(\cdot, s)\|_{L^\infty(\Omega)} ds$$

$$+ \int_0^t \|\nabla e^{(t-s)(\Delta-1)} u \cdot \nabla c(\cdot, s)\|_{L^\infty(\Omega)} ds. \tag{4.4.31}$$

Now we estimate the last two integrals on the right-hand side of the above inequality. From Lemmas 1.1(ii), 4.3, 4.34 with $k = q_0$ and the fact that $q_0 > 3$, it follows that

$$\int_0^t \|\nabla e^{(t-s)(\Delta-1)}m\|_{L^\infty(\Omega)}ds \leq c_2 \int_0^t (1 + (t-s)^{-\frac{1}{2}-\frac{3}{2q_0}})e^{-(\lambda_1+1)(t-s)}\|m\|_{L^{q_0}(\Omega)}ds$$

$$\leq c_2 M_6\varepsilon \int_0^t (1 + (t-s)^{-\frac{1}{2}-\frac{3}{2q_0}})e^{-(\lambda_1+1)(t-s)}e^{-\rho_\infty s}ds$$

$$\leq c_2 c_{10} M_6(1 + t^{\min\{0,\frac{1}{2}-\frac{3}{2q_0}\}})\varepsilon e^{-\alpha_1 t} \qquad (4.4.32)$$

$$\leq 2c_2 c_{10} M_6\varepsilon e^{-\alpha_1 t}.$$

On the other hand, by Lemmas 1.1(ii), 4.3, 4.35 and the definition of T, we obtain

$$\int_0^t \|\nabla e^{(t-s)(\Delta-1)}u \cdot \nabla c\|_{L^\infty(\Omega)}ds$$

$$\leq c_2 \int_0^t (1 + (t-s)^{-\frac{1}{2}-\frac{3}{2q_0}})e^{-(\lambda_1+1)(t-s)}\|u \cdot \nabla c\|_{L^{q_0}(\Omega)}ds \qquad (4.4.33)$$

$$\leq c_2 \int_0^t (1 + (t-s)^{-\frac{1}{2}-\frac{3}{2q_0}})e^{-(\lambda_1+1)(t-s)}\|u\|_{L^{q_0}(\Omega)}\|\nabla c\|_{L^\infty(\Omega)}ds$$

$$\leq c_2 M_3 M_2\varepsilon^2 \int_0^t (1 + (t-s)^{-\frac{1}{2}-\frac{3}{2q_0}})e^{-(\lambda_1+1)(t-s)}(1 + s^{-\frac{1}{2}+\frac{3}{2q_0}})(1 + s^{-\frac{1}{2}})e^{-(\alpha_1+\alpha_2)s}ds$$

$$\leq 3c_2 M_3 M_2\varepsilon^2 \int_0^t e^{-(\lambda_1+1)(t-s)}e^{-(\alpha_1+\alpha_2)s}(1 + (t-s)^{-\frac{1}{2}-\frac{3}{2q_0}})(1 + s^{-1+\frac{3}{2q_0}})ds$$

$$\leq 3c_2 c_{10} M_2 M_3\varepsilon^2(1 + t^{-\frac{1}{2}})e^{-\alpha_1 t}.$$

From (4.4.31)–(4.4.33), it follows that

$$\|\nabla c\|_{L^\infty(\Omega)} \leq (c_2 + 2c_2 c_{10} M_6 + 3c_2 c_{10} M_2 M_3\varepsilon)(1 + t^{-\frac{1}{2}})\varepsilon e^{-\alpha_1 t}$$

$$\leq \frac{M_2}{2}(1 + t^{-\frac{1}{2}})\varepsilon e^{-\alpha_1 t},$$

due to the choice of M_2, M_3 and ε in (4.4.12) and (4.4.18), and thereby completes the proof.

Lemma 4.38 *Under the assumptions of Proposition 4.3, for all $\theta \in [q_0, \infty]$ and $t \in (0, T)$,*

$$\|(\rho - m)(\cdot, t) - e^{t\Delta}(\rho_0 - m_0)\|_{L^\theta(\Omega)} \leq \frac{M_1}{2}\varepsilon(1 + t^{-\frac{3}{2}(\frac{1}{p_0}-\frac{1}{\theta})})e^{-\alpha_1 t}.$$

Proof According to (4.4.20), Lemma 1.1(iv), we have

$$\|(\rho - m)(\cdot, t) - e^{t\Delta}(\rho_0 - m_0)\|_{L^{\theta}(\Omega)}$$

$$\leq \int_0^t \|e^{(t-s)\Delta}(\nabla \cdot (\rho \mathscr{S}(x, \rho, c)\nabla c) + u \cdot \nabla(\rho - m))(\cdot, s)\|_{L^{\theta}(\Omega)} ds$$

$$\leq \int_0^t \|e^{(t-s)\Delta}\nabla \cdot (\rho \mathscr{S}(x, \rho, c)\nabla c)(\cdot, s)\|_{L^{\theta}(\Omega)} ds$$

$$+ \int_0^t \|e^{(t-s)\Delta}\nabla \cdot ((\rho - m - \rho_{\infty})u)(\cdot, s)\|_{L^{\theta}(\Omega)} ds$$

$$\leq c_4 C_S \int_0^t (1 + (t-s))^{-\frac{1}{2} - \frac{3}{2}(\frac{1}{q_0} - \frac{1}{\theta})} e^{-\lambda_1(t-s)} \|\rho(\cdot, s)\|_{L^{q_0}(\Omega)} \|\nabla c(\cdot, s)\|_{L^{\infty}(\Omega)} ds$$

$$+ c_4 \int_0^t (1 + (t-s))^{-\frac{1}{2} - \frac{3}{2}(\frac{1}{q_0} - \frac{1}{\theta})} e^{-\lambda_1(t-s)} \|u(\rho - m - \rho_{\infty})(\cdot, s)\|_{L^{q_0}(\Omega)} ds$$

$$=: I_1 + I_2.$$

Now we need to estimate I_1 and I_2. Firstly, from Lemmas 4.33 and 4.34, we obtain

$$\|\rho(\cdot, s)\|_{L^{q_0}(\Omega)} \leq \|(\rho - m - \rho_{\infty})(\cdot, s)\|_{L^{q_0}(\Omega)} + \|m(\cdot, s)\|_{L^{q_0}(\Omega)} + \|\rho_{\infty}\|_{L^{q_0}(\Omega)}$$

$$\leq M_5 \varepsilon (1 + s^{-\frac{3}{2}\left(\frac{1}{p_0} - \frac{1}{q_0}\right)}) e^{-\alpha_1 s} + M_8 \tag{4.4.34}$$

with $M_8 = e^{(1 + c_1 + c_1 |\Omega|^{\frac{1}{p_0} - \frac{1}{q_0}})\sigma} + \rho_{\infty} |\Omega|^{\frac{1}{q_0}}$, which along with Lemmas 4.37 and 1.1 implies that

$$I_1 \leq c_4 C_S M_8 \int_0^t (1 + (t-s))^{-\frac{1}{2} - \frac{3}{2}(\frac{1}{q_0} - \frac{1}{\theta})} e^{-\lambda_1(t-s)} \|\nabla c\|_{L^{\infty}(\Omega)} ds$$

$$+ c_4 C_S M_5 \varepsilon \int_0^t (1 + (t-s))^{-\frac{1}{2} - \frac{3}{2}(\frac{1}{q_0} - \frac{1}{\theta})} (1 + s^{-\frac{3}{2}(\frac{1}{p_0} - \frac{1}{q_0})}) e^{-\alpha_1 s} e^{-\lambda_1(t-s)}$$

$$\cdot \|\nabla c\|_{L^{\infty}(\Omega)} ds \tag{4.4.35}$$

$$\leq c_4 C_S M_8 M_2 \varepsilon \int_0^t (1 + (t-s))^{-\frac{1}{2} - \frac{3}{2}(\frac{1}{q_0} - \frac{1}{\theta})} e^{-\lambda_1(t-s)} (1 + s^{-\frac{1}{2}}) e^{-\alpha_1 s} ds$$

$$+ 3 c_4 C_S M_5 M_2 \varepsilon^2 \int_0^t (1 + (t-s))^{-\frac{1}{2} - \frac{3}{2}(\frac{1}{q_0} - \frac{1}{\theta})} (1 + s^{-\frac{1}{2} - \frac{3}{2}(\frac{1}{p_0} - \frac{1}{q_0})}) e^{-2\alpha_1 s}$$

$$\cdot e^{-\lambda_1(t-s)} ds$$

$$\leq c_{10} c_4 C_S (M_8 M_2 + 3 M_5 M_2 \varepsilon)(1 + t^{-\frac{3}{2}(\frac{1}{p_0} - \frac{1}{\theta})}) \varepsilon e^{-\alpha_1 t}$$

$$\leq \frac{M_1}{4}(1 + t^{-\frac{3}{2}(\frac{1}{p_0} - \frac{1}{\theta})}) \varepsilon e^{-\alpha_1 t},$$

where we have used (4.4.13) and (4.4.16) and $\frac{1}{p_0} - \frac{1}{q_0} < \frac{1}{3}$.

On the other hand, from Lemmas 4.33 and 4.35, it follows that

$$
\begin{aligned}
I_2 &= c_4 \int_0^t (1 + (t-s))^{-\frac{1}{2} - \frac{3}{2}(\frac{1}{q_0} - \frac{1}{\theta})} e^{-\alpha_1(t-s)} \|\rho - m - \rho_\infty\|_{L^\infty(\Omega)} \|u\|_{L^{q_0}(\Omega)} ds \\
&\le 3c_4 M_3 M_5 \varepsilon^2 \int_0^t (1 + (t-s))^{-\frac{1}{2} - \frac{3}{2}(\frac{1}{q_0} - \frac{1}{\theta})} e^{-\alpha_1(t-s)} (1 + s^{-\frac{1}{2} + \frac{3}{2q_0} - \frac{3}{2p_0}}) \\
&\quad \cdot e^{-(\alpha_1 + \alpha_2)s} ds \qquad\qquad\qquad\qquad\qquad\qquad\qquad\qquad (4.4.36) \\
&\le 3c_4 c_{10} M_3 M_5 \varepsilon^2 (1 + t^{\min\{0, \frac{3}{2}(\frac{1}{\theta} - \frac{1}{p_0})\}}) e^{-\alpha_1 t} \\
&\le \frac{M_1}{4} \varepsilon (1 + t^{-\frac{3}{2}(\frac{1}{p_0} - \frac{1}{\theta})}) e^{-\alpha_1 t},
\end{aligned}
$$

where we have used (4.4.17) and $\frac{1}{p_0} - \frac{1}{q_0} < \frac{1}{3}$. Hence, combining the above inequalities leads to our conclusion immediately.

Now we are ready to complete the proof of Proposition 4.3.

Proof of Proposition 4.3. First from Lemmas 4.35–4.38 and Definition (4.4.19), it follows that $T = T_{max}$. It remains to show that $T_{max} = \infty$ and to establish convergence result asserted in Proposition 4.3.

Supposed that $T_{max} < \infty$. We only need to show that for all $t \le T_{max}$,

$$
\|\rho(\cdot, t)\|_{L^\infty(\Omega)} + \|m(\cdot, t)\|_{L^\infty(\Omega)} + \|c(\cdot, t)\|_{W^{1,\infty}(\Omega)} + \|A^\beta u(\cdot, t)\|_{L^2(\Omega)} < \infty
$$

with $\beta \in (\frac{3}{4}, 1)$ according to the extensibility criterion in Lemma 4.30.

Let $t_0 := \min\{1, \frac{T_{max}}{3}\}$. Then from Lemma 4.34, there exists $K_1 > 0$, such that for $t \in (t_0, T_{max})$,

$$
\|m(\cdot, t)\|_{L^\infty(\Omega)} \le K_1 e^{-\rho_\infty t}. \qquad (4.4.37)
$$

Moreover, from Lemma 4.33 and the fact that

$$
\|\rho(\cdot, t) - \rho_\infty\|_{L^\infty(\Omega)} \le \|(\rho - m)(\cdot, t) - \rho_\infty\|_{L^\infty(\Omega)} + \|m(\cdot, t)\|_{L^\infty(\Omega)},
$$

it follows that for all $t \in (t_0, T_{max})$ and some constant $K_2 > 0$,

$$
\|\rho(\cdot, t) - \rho_\infty\|_{L^\infty(\Omega)} \le K_2 e^{-\alpha_1 t}. \qquad (4.4.38)
$$

Furthermore, Lemma 4.37 implies that there exists $K_3' > 0$, such that

$$
\|\nabla c(\cdot, t)\|_{L^\infty(\Omega)} \le K_3' e^{-\alpha_1 t} \quad \text{for all } t \in (t_0, T_{max}) \qquad (4.4.39)
$$

Hence, it only remains to show that

$$
\|c(\cdot, t)\|_{L^\infty(\Omega)} + \|A^\beta u(\cdot, t)\|_{L^2(\Omega)} \le C \quad \text{for all } t \in (t_0, T_{max}).
$$

for some constant $C > 0$. In fact, we will show that

$$\|A^\beta u(\cdot, t)\|_{L^2(\Omega)} \leq Ce^{-\alpha_2 t} \tag{4.4.40}$$

for $t_0 < t < T_{max}$ with some constant $C > 0$.

By (4.4.23), we have

$$\|A^\beta u(\cdot, t)\|_{L^2(\Omega)} \tag{4.4.41}$$

$$\leq \|A^\beta e^{-tA} u_0\|_{L^2(\Omega)} + \int_0^t \|A^\beta e^{-(t-s)A} \mathscr{P}((\rho + m - \rho_\infty)\nabla\phi)(\cdot, s)\|_{L^2(\Omega)} ds$$

$$+ \int_0^t \|A^\beta e^{-(t-s)A} \mathscr{P}((u \cdot \nabla)u)(\cdot, s)\|_{L^2(\Omega)} ds.$$

According to Lemma 4.1,

$$\|A^\beta e^{-tA} u_0\|_{L^2(\Omega)} \leq c_5 e^{-\mu t} \|A^\beta u_0\|_{L^2(\Omega)} \quad \text{for all } t \in (0, T_{max}).$$

From Lemmas 4.1, 4.2, 4.33 and the Hölder inequality, it follows that there exists $l_1 > 0$, such that

$$\int_0^t \|A^\beta e^{-(t-s)A} \mathscr{P}((\rho + m - \rho_\infty)\nabla\phi)(\cdot, s)\|_{L^2(\Omega)} ds$$

$$\leq c_5 c_9 \|\nabla\phi\|_{L^\infty(\Omega)} |\Omega|^{\frac{q_0-2}{2q_0}} \int_0^t (\|(\rho - m - \rho_\infty)(\cdot, s)\|_{L^{q_0}(\Omega)}$$

$$+ 2\|m(\cdot, s)\|_{L^{q_0}(\Omega)})(t - s)^{-\beta} e^{-\mu(t-s)} ds$$

$$\leq c_5 c_9 \|\nabla\phi\|_{L^\infty(\Omega)} |\Omega|^{\frac{q_0-2}{2q_0}} l_1 \int_0^t e^{-\mu(t-s)}(t - s)^{-\beta}(1 + s^{-\frac{3}{2}(\frac{1}{p_0} - \frac{1}{q_0})}) e^{-\alpha_1 s} ds$$

$$\leq c_5 c_9 c_{10} \|\nabla\phi\|_{L^\infty(\Omega)} |\Omega|^{\frac{q_0-2}{2q_0}} l_1 e^{-\alpha_2 t}(1 + t^{\min\{0, 1-\beta-\frac{3}{2}(\frac{1}{p_0} - \frac{1}{q_0})\}}).$$

On the other hand, let $M(t) := e^{-\alpha_2 t} \|A^\beta u(\cdot, t)\|_{L^2(\Omega)}$ for $0 < t < T_{max}$. By Lemmas 4.1(iv) and the Gagliardo–Nirenberg type inequality, one can see that

$$\|(u \cdot \nabla)u(\cdot, s)\|_{L^2(\Omega)} \leq |\Omega|^{\frac{1}{6}} \|u(\cdot, s)\|_{L^\infty(\Omega)} \|\nabla u(\cdot, s)\|_{L^3(\Omega)}$$

$$\leq l_2 \|A^\beta u(\cdot, s)\|_{L^2(\Omega)}^{\vartheta} \|u(\cdot, s)\|_{L^{q_0}(\Omega)}^{1-\vartheta} \|\nabla u(\cdot, s)\|_{L^3(\Omega)}$$

for some $l_2 > 0$ with $\vartheta = \frac{1}{q_0}/(\frac{1}{q_0} - \frac{1}{2} + \frac{2\beta}{3})$, and thereby an application of Lemmas 2.2, 4.2, 4.35 and 4.36 gives

$$\int_0^t \|A^\beta e^{-(t-s)A} \mathscr{P}((u \cdot \nabla)u)(\cdot, s)\|_{L^2(\Omega)} ds$$

$$\leq c_5 c_9 l_2 \int_0^t \|A^\beta u(\cdot, s)\|_{L^2(\Omega)}^\vartheta \|u(\cdot, s)\|_{L^{q_0}(\Omega)}^{1-\vartheta} \|\nabla u(\cdot, s)\|_{L^3(\Omega)}$$

$$\leq l_3 (\max_{0 \leq s < T_{max}} M(s))^\vartheta \int_0^t e^{-\mu(t-s)}(t-s)^{-\beta}(1 + s^{-\frac{1}{2}+(-\frac{1}{2}+\frac{3}{2q_0})(1-\vartheta)})e^{-2\alpha_2 s} ds$$

$$\leq c_{10} l_3 (\max_{0 \leq s < T_{max}} M(s))^\vartheta (1 + t^{\min\{0, \frac{1}{2}-\beta+(\frac{3}{2q_0}-\frac{1}{2})(1-\vartheta)\}})e^{-\alpha_2 t}$$

for some $l_3 > 0$. Now inserting the above inequalities into (4.4.41), we arrive at

$$M(t) \leq c_5 \|A^\beta u_0\|_{L^2(\Omega)} + c_5 c_9 c_{10} \|\nabla \phi\|_{L^\infty(\Omega)} |\Omega|^{\frac{q_0-2}{2q_0}} l_1 (1 + t^{\min\{0, 1-\beta-\frac{3}{2}(\frac{1}{p_0}-\frac{1}{q_0})\}})$$

$$+ c_{10} l_3 (\max_{0 \leq s < T_{max}} M(s))^\vartheta (1 + t^{\min\{0, \frac{1}{2}-\beta+(\frac{3}{2q_0}-\frac{1}{2})(1-\vartheta)\}}),$$

which implies that for some $l_4 > 0$ depending on t_0, we have

$$\max_{t_0 \leq t < T_{max}} M(t) \leq l_4 + l_4 (\max_{0 \leq t < T_{max}} M(t))^\vartheta.$$

On the other hand, from Lemma 4.30, $\max_{0 \leq t \leq t_0} M(t) \leq l_5$. Therefore, we get

$$\max_{0 \leq t < T_{max}} M(t) \leq l_4 + l_5 + l_4 (\max_{0 \leq t < T_{max}} M(t))^\vartheta.$$

As $\vartheta < 1$, we infer that $M(t) \leq l_6$ for all $t \in (0, T_{max})$ for some $l_6 > 0$ independent of T_{max} and hence arrive at (4.4.40).

Furthermore, due to $D(A^\beta) \hookrightarrow L^\infty(\Omega)$ with $\beta \in (\frac{3}{4}, 1)$ and Lemma 4.35, we get

$$\|u(\cdot, t)\|_{L^\infty(\Omega)} \leq K_4 e^{-\alpha_2 t} \quad \text{for some } K_4 > 0 \text{ and } t \in (0, T_{max}). \quad (4.4.42)$$

Now we turn to showing that there exists $K_3'' > 0$, such that

$$\|c(\cdot, t)\|_{L^\infty(\Omega)} \leq K_3'' e^{-\alpha_2 t} \quad \text{for all } t \in (0, T_{max}). \quad (4.4.43)$$

From (4.4.22), it follows that

$$\|c\|_{L^\infty(\Omega)} \leq \|e^{t(\Delta-1)} c_0\|_{L^\infty(\Omega)} + \int_0^t \|e^{(t-s)(\Delta-1)}(m - u \cdot \nabla c)\|_{L^\infty(\Omega)} ds$$

$$\leq e^{-t} \|c_0\|_{L^\infty(\Omega)} + \int_0^t \|e^{(t-s)(\Delta-1)} m(\cdot, s)\|_{L^\infty(\Omega)} ds \quad (4.4.44)$$

$$+ \int_0^t \|e^{(t-s)(\Delta-1)} u \cdot \nabla c(\cdot, s)\|_{L^\infty(\Omega)} ds.$$

An application of (4.4.24) with $k = \infty$ yields

$$\int_0^t \|e^{(t-s)(\Delta-1)}m(\cdot, s)\|_{L^\infty(\Omega)}ds \le \int_0^t e^{-(t-s)}\|m(\cdot, s)\|_{L^\infty(\Omega)}ds \qquad (4.4.45)$$

$$\le \|m_0\|_{L^\infty(\Omega)}M_6\int_0^t e^{-(t-s)}e^{-\rho_\infty s}ds$$

$$\le \|m_0\|_{L^\infty(\Omega)}M_6c_{10}e^{-\alpha_2 t}.$$

On the other hand, from (4.4.42) and (4.4.39), we can see that

$$\int_0^t \|e^{(t-s)(\Delta-1)}u \cdot \nabla c\|_{L^\infty(\Omega)}ds \le \int_0^t e^{-(t-s)}\|u\|_{L^\infty(\Omega)}\|\nabla c\|_{L^\infty(\Omega)}ds \qquad (4.4.46)$$

$$\le K_3' K_4 \int_0^t e^{-(\alpha_1+\alpha_2)s}e^{-(t-s)}ds$$

$$\le K_3' K_4 c_{10}e^{-\alpha_2 t}.$$

Inserting (4.4.45), (4.4.46) into (4.4.44), we arrive at the conclusion (4.4.43). We have thus established that $T_{max} = \infty$, and the decay estimates in (4.4.8)–(4.4.11) follow from (4.4.37)–(4.4.40) and (4.4.43), respectively.

As for the case $\int_\Omega \rho_0 < \int_\Omega m_0$, i.e., $m_\infty > 0$, $\rho_\infty = 0$, Theorem 4.5 reduces to

Proposition 4.4 *Assume that (4.1.14) and $\int_\Omega \rho_0 < \int_\Omega m_0$ hold, and let $p_0 \in (2, 3)$, $q_0 \in (3, \frac{3p_0}{2(3-p_0)})$. Then there exists $\varepsilon > 0$, such that for any initial data (ρ_0, m_0, c_0, u_0) fulfilling (4.1.15) as well as*

$$\|\rho_0\|_{L^{p_0}(\Omega)} \le \varepsilon, \quad \|m_0 - m_\infty\|_{L^{q_0}(\Omega)} \le \varepsilon, \quad \|\nabla c_0\|_{L^3(\Omega)} \le \varepsilon, \quad \|u_0\|_{L^3(\Omega)} \le \varepsilon,$$

(4.1.13) admits a global classical solution (ρ, m, c, u, P). Furthermore, for any $\alpha_1 \in (0, \min\{\lambda_1, m_\infty, 1\})$, $\alpha_2 \in (0, \min\{\alpha_1, \lambda_1'\})$, there exist constants $K_i > 0$, $i = 1, 2, 3, 4$, such that

$$\|m(\cdot, t) - m_\infty\|_{L^\infty(\Omega)} \le K_1 e^{-\alpha_1 t}, \qquad (4.4.47)$$

$$\|\rho(\cdot, t)\|_{L^\infty(\Omega)} \le K_2 e^{-\alpha_1 t}, \qquad (4.4.48)$$

$$\|c(\cdot, t) - m_\infty\|_{W^{1,\infty}(\Omega)} \le K_3 e^{-\alpha_2 t}, \qquad (4.4.49)$$

$$\|u(\cdot, t)\|_{L^\infty(\Omega)} \le K_4 e^{-\alpha_2 t}. \qquad (4.4.50)$$

The basic strategy of the proof of Proposition 4.4 parallels that of Proposition 4.3 to a certain extent. However, due to differences in the properties of ρ and m, there are significant differences in the details of their proofs. Thus, for the convenience of the reader, we will sketch the proof of Proposition 4.4.

The following elementary observations can be verified easily:

Lemma 4.39 *Under the assumptions of Proposition 4.4, it is possible to choose $M_1 > 0$, $M_2 > 0$ and $\varepsilon > 0$, such that*

$$c_3 + c_2 c_{10}(1 + c_1 + c_1|\Omega|^{\frac{1}{p_0} - \frac{1}{q_0}} + M_1) \le \frac{M_2}{4}, \tag{4.4.51}$$

$$c_6 + 2c_6 c_9 c_{10}(M_1 + 2 + 2c_1 + 2c_1|\Omega|^{\frac{1}{p_0} - \frac{1}{q_0}})\|\nabla\phi\|_{L^\infty(\Omega)} < \frac{M_3}{4} \tag{4.4.52}$$

$$c_7 + 2c_7 c_9 c_{10}(M_1 + 2 + 2c_1 + 2c_1|\Omega|^{\frac{1}{p_0} - \frac{1}{q_0}})\|\nabla\phi\|_{L^\infty(\Omega)}|\Omega|^{\frac{1}{3} - \frac{1}{q_0}} < \frac{M_4}{4} \tag{4.4.53}$$

$$12 c_2 c_{10} M_3 \varepsilon \le 1, \tag{4.4.54}$$

$$2c_1 + (\min\{1, |\Omega|\})^{-\frac{1}{p_0}} \le \frac{M_1}{8}, \quad 12 c_6 c_9 c_{10} M_4 \varepsilon < 1, \tag{4.4.55}$$

$$24 c_4 C_S c_{10} M_2 \varepsilon < 1, \tag{4.4.56}$$

$$12 c_7 c_9 c_{10} M_3 \varepsilon < 1, \tag{4.4.57}$$

$$12 c_4 c_{10} C_S M_1 M_2 \varepsilon < 1, \tag{4.4.58}$$

$$24 c_1 c_{10}(1 + c_1 + c_1|\Omega|^{\frac{1}{p_0} - \frac{1}{q_0}} + M_1)\varepsilon < 1, \tag{4.4.59}$$

$$18 c_4 c_{10} M_3 \varepsilon < 1. \tag{4.4.60}$$

$$12 c_{10} c_4 M_3(1 + c_1 + c_1|\Omega|^{\frac{1}{p_0} - \frac{1}{q_0}})\varepsilon < 1. \tag{4.4.61}$$

Define

$$T := \sup \left\{ \widetilde{T} \in (0, T_{max}) \left| \begin{array}{l} \|(m-\rho)(\cdot, t) - e^{t\Delta}(m_0 - \rho_0)\|_{L^\theta(\Omega)} \le \varepsilon(1 + t^{-\frac{3}{2}(\frac{1}{p_0} - \frac{1}{\theta})})e^{-\alpha_1 t}; \\[2mm] \|\rho(\cdot, t)\|_{L^\theta(\Omega)} \le M_1 \varepsilon(1 + t^{-\frac{3}{2}(\frac{1}{p_0} - \frac{1}{\theta})})e^{-\alpha_1 t}, \ \forall \theta \in [q_0, \infty]; \\[2mm] \|\nabla c(\cdot, t)\|_{L^\infty(\Omega)} \le M_2 \varepsilon(1 + t^{-\frac{1}{2}})e^{-\alpha_1 t} \ \text{for all } t \in [0, \widetilde{T}); \\[2mm] \|u(\cdot, t)\|_{L^{q_0}(\Omega)} \le M_3 \varepsilon \left(1 + t^{-\frac{1}{2} + \frac{3}{2q_0}}\right)e^{-\alpha_2 t} \ \text{for all } t \in [0, \widetilde{T}); \\[2mm] \|\nabla u(\cdot, t)\|_{L^3(\Omega)} \le M_4 \varepsilon \left(1 + t^{-\frac{1}{2}}\right)e^{-\alpha_2 t} \ \text{for all } t \in [0, \widetilde{T}). \end{array} \right. \right\} \tag{4.4.62}$$

By Lemma 4.30 and (4.1.15), $T > 0$ is well-defined. As in the proof of Proposition 4.3, we first show $T = T_{max}$, and then $T_{max} = \infty$. To this end, we will show that all of the estimates mentioned in (4.4.62) are still valid with even smaller coefficients on the right-hand side. The derivation of these estimates will mainly rely on $L^p - L^q$ estimates for the Neumann heat semigroup and the corresponding semigroup for the Stokes operator, and the fact that the classical solution of (4.1.13) on $(0, T)$ can be represented as

$$(m - \rho)(\cdot, t) = e^{t\Delta}(m_0 - \rho_0)$$

$$+ \int_0^t e^{(t-s)\Delta}(\nabla \cdot (\rho \mathscr{S}(x, \rho, c)\nabla c) - u \cdot \nabla(m - \rho))(\cdot, s)ds, \quad (4.4.63)$$

$$\rho(\cdot, t) = e^{t\Delta}\rho_0 - \int_0^t e^{(t-s)\Delta}(\nabla \cdot (\rho \mathscr{S}(x, \rho, c)\nabla c) + u \cdot \nabla\rho + \rho m)(\cdot, s)ds,$$

$$(4.4.64)$$

$$c(\cdot, t) = e^{t(\Delta - 1)}c_0 + \int_0^t e^{(t-s)(\Delta - 1)}(m - u \cdot \nabla c)(\cdot, s)ds, \quad (4.4.65)$$

$$u(\cdot, t) = e^{-tA}u_0 + \int_0^t e^{-(t-s)A}\mathscr{P}((\rho + m)\nabla\phi - (u \cdot \nabla)u)(\cdot, s)ds. \quad (4.4.66)$$

Lemma 4.40 (Lemma 3.17 in Li et al. (2019b)) *Under the assumptions of Proposition 4.4,*

$$\|(m - \rho)(\cdot, t) - m_\infty\|_{L^\theta(\Omega)} \le M_5 \varepsilon(1 + t^{-\frac{3}{2}(\frac{1}{p_0} - \frac{1}{\theta})})e^{-\alpha_1 t}$$

for all $t \in (0, T)$ and $\theta \in [q_0, \infty]$ with $M_5 = 1 + c_1 + c_1|\Omega|^{\frac{1}{p_0} - \frac{1}{q_0}}$.

Lemma 4.41 (Lemma 3.18 in Li et al. (2019b)) *Under the assumptions of Proposition 4.4,*

$$\|m(\cdot, t) - m_\infty\|_{L^\theta(\Omega)} \le (M_5 + M_1)\varepsilon(1 + t^{-\frac{3}{2}(\frac{1}{p_0} - \frac{1}{\theta})})e^{-\alpha_1 t}$$

for all $t \in (0, T)$, $\theta \in [q_0, \infty]$.

Lemma 4.42 *Under the assumptions of Proposition 4.4, we have*

$$\|u(\cdot, t)\|_{L^{q_0}(\Omega)} \le \frac{M_3}{2}\varepsilon(1 + t^{-\frac{1}{2} + \frac{3}{2q_0}})e^{-\alpha_2 t} \quad \text{for all } t \in (0, T).$$

Proof For any given $\alpha_2 < \lambda_1'$, we can fix $\mu \in (\alpha_2, \lambda_1')$. By (4.4.66), Lemmas 4.1, 4.2 and $\mathscr{P}(\nabla\phi) = 0$, we obtain that

$$\|u(\cdot, t)\|_{L^{q_0}(\Omega)}$$

$$\le c_6 t^{-\frac{3}{2}(\frac{1}{3} - \frac{1}{q_0})}e^{-\mu t}\|u_0\|_{L^3(\Omega)}$$

$$+ \int_0^t \|e^{-(t-s)A}\mathscr{P}((\rho + m)\nabla\phi - (u \cdot \nabla)u)(\cdot, s)\|_{L^{q_0}(\Omega)}ds \quad (4.4.67)$$

$$\le c_6 t^{-\frac{1}{2} + \frac{3}{2q_0}}e^{-\mu t}\varepsilon + c_6 c_9\|\nabla\phi\|_{L^\infty(\Omega)}\int_0^t e^{-\mu(t-s)}\|(\rho + m - m_\infty)(\cdot, s)\|_{L^{q_0}(\Omega)}ds$$

$$+ c_6 c_9 \int_0^t (t - s)^{-\frac{1}{2}}e^{-\mu(t-s)}\|(u \cdot \nabla)u(\cdot, s)\|_{L^{\frac{1}{\frac{1}{3} + \frac{1}{q_0}}}(\Omega)}ds.$$

By Lemma 4.41 and the definition of T, we get

$$\|(\rho + m - m_\infty)(\cdot, s)\|_{L^{q_0}(\Omega)} = \|(m - m_\infty)(\cdot, s)\|_{L^{q_0}(\Omega)} + \|\rho(\cdot, s)\|_{L^{q_0}(\Omega)} \quad (4.4.68)$$
$$\leq (2M_5 + M_1)\varepsilon(1 + s^{-\frac{3}{2}(\frac{1}{p_0} - \frac{1}{q_0})})e^{-\alpha_1 s}.$$

Inserting (4.4.68) into (4.4.67), by the definition of T and noting that $\frac{3}{2}(\frac{1}{p_0} - \frac{1}{q_0}) < 1$, we have

$$\|u(\cdot, t)\|_{L^{q_0}(\Omega)}$$
$$\leq c_6 t^{-\frac{1}{2} + \frac{3}{2q_0}} e^{-\mu t} \varepsilon$$
$$+ c_6 c_9 (2M_5 + M_1)\|\nabla\phi\|_{L^\infty(\Omega)}\varepsilon \int_0^t (1 + s^{-\frac{3}{2}(\frac{1}{p_0} - \frac{1}{q_0})})e^{-\alpha_1 s}e^{-\mu(t-s)}ds$$
$$+ c_6 c_9 \int_0^t (t - s)^{-\frac{1}{2}}e^{-\mu(t-s)}\|\nabla u(\cdot, s)\|_{L^3(\Omega)}\|u(\cdot, s)\|_{L^{q_0}(\Omega)}ds$$
$$\leq c_6 t^{-\frac{1}{2} + \frac{3}{2q_0}} e^{-\mu t} \varepsilon + c_6 c_9 c_{10}(2M_5 + M_1)\|\nabla\phi\|_{L^\infty(\Omega)}\varepsilon(1 + t^{\min\{0, 1 - \frac{3}{2}(\frac{1}{p_0} - \frac{1}{q_0})\}})e^{-\alpha_2 t}$$
$$+ 3c_6 c_9 M_3 M_4 \varepsilon^2 \int_0^t (t - s)^{-\frac{1}{2}}e^{-\mu(t-s)}(1 + s^{-1 + \frac{3}{2q_0}})e^{-2\alpha_2 s}ds$$
$$\leq c_6 t^{-\frac{1}{2} + \frac{3}{2q_0}} \varepsilon e^{-\mu t} + 2c_6 c_9 c_{10}(2M_5 + M_1)\|\nabla\phi\|_{L^\infty(\Omega)}\varepsilon e^{-\alpha_2 t}$$
$$+ 3c_6 c_9 c_{10} M_3 M_4 (1 + t^{-\frac{1}{2} + \frac{3}{2q_0}})\varepsilon^2 e^{-\alpha_2 t}$$
$$\leq \frac{M_3}{2}\varepsilon(1 + t^{-\frac{1}{2} + \frac{3}{2q_0}})e^{-\alpha_2 t},$$

where we have used (4.4.52) and (4.4.55).

Lemma 4.43 *Under the assumptions of Proposition 4.4, we have*

$$\|\nabla u(\cdot, t)\|_{L^3(\Omega)} \leq \frac{M_4}{2}\varepsilon(1 + t^{-\frac{1}{2}})e^{-\alpha_2 t} \text{ for all } t \in (0, T).$$

Proof According to (4.4.66), and applying Lemmas 4.1(iii) and 4.2, we arrive at

$$\|\nabla u(\cdot, t)\|_{L^3(\Omega)}$$
$$\leq c_7 t^{-\frac{1}{2}}e^{-\mu t}\|u_0\|_{L^3(\Omega)} + \int_0^t \|\nabla e^{-(t-s)A}\mathcal{P}((\rho + m)\nabla\phi - (u \cdot \nabla)u)(\cdot, s)\|_{L^3(\Omega)}ds$$
$$\leq c_7 t^{-\frac{1}{2}}e^{-\mu t}\varepsilon + c_7|\Omega|^{\frac{1}{3} - \frac{1}{q_0}}\int_0^t (t - s)^{-\frac{1}{2}}e^{-\mu(t-s)}\|\mathcal{P}((\rho + m - m_\infty)\nabla\phi)(\cdot, s)\|_{L^{q_0}(\Omega)}ds$$
$$+ c_7 \int_0^t (t - s)^{-\frac{1}{2} - \frac{3}{2q_0}}e^{-\mu(t-s)}\|\mathcal{P}((u \cdot \nabla)u)(\cdot, s)\|_{L^{\frac{3q_0}{3+q_0}}(\Omega)}ds \quad (4.4.69)$$
$$\leq c_7 t^{-\frac{1}{2}}e^{-\mu t}\varepsilon$$
$$+ c_7 c_9 \|\nabla\phi\|_{L^\infty(\Omega)}|\Omega|^{\frac{1}{3} - \frac{1}{q_0}}\int_0^t (t - s)^{-\frac{1}{2}}e^{-\mu(t-s)}\|(\rho + m - m_\infty)(\cdot, s)\|_{L^{q_0}(\Omega)}ds$$

$$+ c_7 c_9 \int_0^t (t-s)^{-\frac{1}{2}-\frac{3}{2q_0}} e^{-\mu(t-s)} \|\nabla u(\cdot, s)\|_{L^3(\Omega)} \|u(\cdot, s)\|_{L^{q_0}(\Omega)} ds,$$

where $\mathscr{P}(m_\infty \nabla \phi) = m_\infty \mathscr{P}(\nabla \phi) = 0$ is used.

From (4.4.68), it follows that

$$\int_0^t (t-s)^{-\frac{1}{2}} e^{-\mu(t-s)} \|(\rho + m - m_\infty)(\cdot, s)\|_{L^{q_0}(\Omega)} ds \qquad (4.4.70)$$

$$\leq (2M_5 + M_1)\varepsilon \int_0^t (t-s)^{-\frac{1}{2}} e^{-\mu(t-s)} (1 + s^{-\frac{3}{2}(\frac{1}{p_0}-\frac{1}{q_0})}) e^{-\alpha_1 s} ds.$$

In addition, an application of the Hölder inequality and definition of T shows that

$$\int_0^t (t-s)^{-\frac{1}{2}-\frac{3}{2q_0}} e^{-\mu(t-s)} \|u(\cdot, s)\|_{L^{q_0}(\Omega)} \|\nabla u(\cdot, s)\|_{L^3(\Omega)} ds$$

$$\leq 3M_3 M_4 \varepsilon^2 \int_0^t (t-s)^{-\frac{1}{2}-\frac{3}{2q_0}} e^{-\mu(t-s)} (1 + s^{-1+\frac{3}{2q_0}}) e^{-2\alpha_2 s} ds. \qquad (4.4.71)$$

Therefore, inserting (4.4.71), (4.4.70) into (4.4.69) and applying Lemma 4.3, we get

$$\|\nabla u(\cdot, t)\|_{L^3(\Omega)}$$

$$\leq c_7 t^{-\frac{1}{2}} e^{-\mu t} \varepsilon$$

$$+ c_7 c_9 c_{10} \|\nabla \phi\|_{L^\infty(\Omega)} |\Omega|^{\frac{1}{3}-\frac{1}{q_0}} (2M_5 + M_1)\varepsilon (1 + t^{\min\{0, \frac{1}{2}-\frac{3}{2}(\frac{1}{p_0}-\frac{1}{q_0})\}}) e^{-\alpha_2 t}$$

$$+ 3c_7 c_9 c_{10} M_3 M_4 \varepsilon^2 (1 + t^{-\frac{1}{2}}) e^{-\alpha_2 t}$$

$$\leq \frac{M_4}{2} \varepsilon (1 + t^{-\frac{1}{2}}) e^{-\alpha_2 t},$$

where (4.4.53), (4.4.57) are used.

Lemma 4.44 *Under the assumptions of Proposition 4.4, we have*

$$\|\nabla c(\cdot, t)\|_{L^\infty(\Omega)} \leq \frac{M_2}{2} \varepsilon (1 + t^{-\frac{1}{2}}) e^{-\alpha_1 t} \quad \text{for all } t \in (0, T).$$

Proof From (4.4.65) and the standard regularization properties of the Neumann heat semigroup $(e^{\tau\Delta})_{\tau>0}$ in Winkler (2010), one can conclude that

$$\|\nabla c(\cdot, t)\|_{L^\infty(\Omega)}$$

$$\leq e^{-t}\|\nabla e^{t\Delta}c_0\|_{L^\infty(\Omega)} + \int_0^t \|\nabla e^{(t-s)(\Delta-1)}(m - u \cdot \nabla c)(\cdot, s)\|_{L^\infty(\Omega)}ds$$

$$\leq c_3(1 + t^{-\frac{1}{2}})e^{-t}\|\nabla c_0\|_{L^3(\Omega)} + \int_0^t \|\nabla e^{(t-s)(\Delta-1)}(m - m_\infty)(\cdot, s)\|_{L^\infty(\Omega)}ds$$

$$+ \int_0^t \|\nabla e^{(t-s)(\Delta-1)}u \cdot \nabla c(\cdot, s)\|_{L^\infty(\Omega)}ds. \tag{4.4.72}$$

In the second inequality, we have used $\nabla e^{(t-s)(\Delta-1)}m_\infty = 0$.

From Lemmas 1.1(ii), 4.41 and 4.3, it follows that

$$\int_0^t \|\nabla e^{(t-s)(\Delta-1)}(m - m_\infty)(\cdot, s)\|_{L^\infty(\Omega)}ds$$

$$\leq c_2 \int_0^t (1 + (t - s)^{-\frac{1}{2}-\frac{3}{2q_0}})e^{-(\lambda_1+1)(t-s)}\|(m - m_\infty)(\cdot, s)\|_{L^{q_0}(\Omega)}ds \tag{4.4.73}$$

$$\leq c_2(M_5 + M_1)\varepsilon \int_0^t (1 + (t - s)^{-\frac{1}{2}-\frac{3}{2q_0}})e^{-(\lambda_1+1)(t-s)}(1 + s^{-\frac{3}{2}(\frac{1}{p_0}-\frac{1}{q_0})})e^{-\alpha_1 s}ds$$

$$\leq c_2 c_{10}(M_5 + M_1)\varepsilon(1 + t^{\min\{0,\frac{1}{2}-\frac{3}{2p_0}\}})e^{-\min\{\alpha_1, \lambda_1+1\}t}$$

$$\leq c_2 c_{10}(M_5 + M_1)\varepsilon(1 + t^{-\frac{1}{2}})e^{-\alpha_1 t}.$$

On the other hand, by Lemmas 1.1(ii), 4.3 and the definition of T, we obtain

$$\int_0^t \|\nabla e^{(t-s)(\Delta-1)}u \cdot \nabla c(\cdot, s)\|_{L^\infty(\Omega)}ds$$

$$\leq c_2 \int_0^t (1 + (t - s)^{-\frac{1}{2}-\frac{3}{2q_0}})e^{-(\lambda_1+1)(t-s)}\|u \cdot \nabla c(\cdot, s)\|_{L^{q_0}(\Omega)}ds \tag{4.4.74}$$

$$\leq c_2 \int_0^t (1 + (t - s)^{-\frac{1}{2}-\frac{3}{2q_0}})e^{-(\lambda_1+1)(t-s)}\|u(\cdot, s)\|_{L^{q_0}(\Omega)}\|\nabla c(\cdot, s)\|_{L^\infty(\Omega)}ds$$

$$\leq c_2 M_3 M_2 \varepsilon^2 \int_0^t (1 + (t - s)^{-\frac{1}{2}-\frac{3}{2q_0}})e^{-(\lambda_1+1)(t-s)}(1 + s^{-\frac{1}{2}+\frac{3}{2q_0}})(1 + s^{-\frac{1}{2}})e^{-(\alpha_1+\alpha_2)s}ds$$

$$\leq 3c_2 M_3 M_2 \varepsilon^2 \int_0^t e^{-(\lambda_1+1)(t-s)}e^{-(\alpha_1+\alpha_2)s}(1 + (t - s)^{-\frac{1}{2}-\frac{3}{2q_0}})(1 + s^{-1+\frac{3}{2q_0}})ds$$

$$\leq 3c_2 M_3 M_2 c_{10}\varepsilon^2(1 + t^{-\frac{1}{2}})e^{-\alpha_1 t}.$$

Hence, combining the above inequalities and applying (4.4.51) and (4.4.54), we arrive at the desired conclusion.

Lemma 4.45 *Under the assumptions of Proposition 4.4, we have*

$$\|\rho(\cdot, t)\|_{L^\theta(\Omega)} \leq \frac{M_1}{2}\varepsilon(1 + t^{-\frac{3}{2}(\frac{1}{p_0}-\frac{1}{\theta})})e^{-\alpha_1 t} \quad \text{for all } t \in (0, T), \ \theta \in [q_0, \infty].$$

Proof From (4.4.64), we have

$$\rho(\cdot, t) = e^{t(\Delta - m_\infty)}\rho_0 - \int_0^t e^{(t-s)(\Delta - m_\infty)}(\nabla \cdot (\rho \mathscr{S}(\cdot, \rho, c)\nabla c) - u \cdot \nabla \rho)(\cdot, s)ds$$

$$+ \int_0^t e^{(t-s)(\Delta - m_\infty)}\rho(m_\infty - m)(\cdot, s)ds.$$

By Lemma 1.1, the result in Sect. 2 of Winkler (2010) and $\alpha_1 < \min\{\lambda_1, m_\infty\}$, we obtain

$$\|\rho(\cdot, t)\|_{L^\theta(\Omega)}$$

$$\leq e^{-m_\infty t}(\|e^{t\Delta}(\rho_0 - \overline{\rho}_0)\|_{L^\theta(\Omega)} + \|\overline{\rho}_0\|_{L^\theta(\Omega)})$$

$$+ \int_0^t \|e^{(t-s)(\Delta - m_\infty)}\nabla \cdot (\rho \mathscr{S}(\cdot, \rho, c)\nabla c)(\cdot, s)\|_{L^\theta(\Omega)}ds$$

$$+ \int_0^t \|e^{(t-s)(\Delta - m_\infty)}(u \cdot \nabla \rho)(\cdot, s)\|_{L^\theta(\Omega)}ds$$

$$+ \int_0^t \|e^{(t-s)(\Delta - m_\infty)}\rho(m_\infty - m)(\cdot, s)\|_{L^\theta(\Omega)}ds$$

$$\leq c_1(1 + t^{-\frac{3}{2}(\frac{1}{p_0} - \frac{1}{\theta})})e^{-(\lambda_1 + m_\infty)t}\|\rho_0 - \overline{\rho}_0\|_{L^{p_0}(\Omega)} + (\min\{1, |\Omega|\})^{-\frac{1}{p_0}}e^{-m_\infty t}\varepsilon$$

$$+ c_4 C_S \int_0^t (1 + (t-s)^{-\frac{1}{2} - \frac{3}{2}(\frac{1}{q_0} - \frac{1}{\theta})})e^{-(\lambda_1 + m_\infty)(t-s)}\|\rho\|_{L^{q_0}(\Omega)}\|\nabla c\|_{L^\infty(\Omega)}ds$$

$$+ \int_0^t \|e^{(t-s)(\Delta - m_\infty)}\nabla \cdot (\rho u)(\cdot, s)\|_{L^\theta(\Omega)}ds$$

$$+ \int_0^t \|e^{(t-s)(\Delta - m_\infty)}\rho(m_\infty - m)(\cdot, s)\|_{L^\theta(\Omega)}ds$$

$$\leq (2c_1 + (\min\{1, |\Omega|\})^{-\frac{1}{p_0}})(1 + t^{-\frac{3}{2}(\frac{1}{p_0} - \frac{1}{\theta})})\varepsilon e^{-\alpha_1 t}$$

$$+ c_4 C_S \int_0^t (1 + (t-s)^{-\frac{1}{2} - \frac{3}{2}(\frac{1}{q_0} - \frac{1}{\theta})})e^{-(\lambda_1 + m_\infty)(t-s)}\|\rho\|_{L^{q_0}(\Omega)}\|\nabla c\|_{L^\infty(\Omega)}ds$$

$$+ c_4 \int_0^t (1 + (t-s)^{-\frac{1}{2} - \frac{3}{2}(\frac{1}{q_0} - \frac{1}{\theta})})e^{-(\lambda_1 + m_\infty)(t-s)}\|\rho\|_{L^\infty(\Omega)}\|u\|_{L^{q_0}(\Omega)}ds$$

$$+ c_1 \int_0^t (1 + (t-s)^{-\frac{3}{2}(\frac{1}{q_0} - \frac{1}{\theta})})e^{-m_\infty(t-s)}\|\rho\|_{L^{q_0}(\Omega)}\|m - m_\infty\|_{L^\infty(\Omega)}ds.$$

According to the definition of T, Lemmas 4.44 and 4.3, this shows that

$$\int_0^t (1 + (t - s))^{-\frac{1}{2} - \frac{3}{2}(\frac{1}{q_0} - \frac{1}{\theta})} e^{-(\lambda_1 + m_\infty)(t-s)} \|\rho\|_{L^{q_0}(\Omega)} \|\nabla c\|_{L^\infty(\Omega)} ds$$

$$\leq 3M_1 M_2 \varepsilon^2 \int_0^t (1 + (t - s))^{-\frac{1}{2} - \frac{3}{2}(\frac{1}{q_0} - \frac{1}{\theta})} e^{-\lambda_1(t-s)} e^{-2\alpha_1 s} (1 + s^{-\frac{1}{2} - \frac{3}{2}(\frac{1}{p_0} - \frac{1}{q_0})}) ds$$

$$\leq 3c_{10} M_1 M_2 \varepsilon^2 (1 + t^{\min\{0, -\frac{3}{2}(\frac{1}{p_0} - \frac{1}{\theta})\}}) e^{-\min\{\lambda_1, 2\alpha_1\} t}.$$

Similarly, we can also get

$$\int_0^t (1 + (t - s))^{-\frac{1}{2} - \frac{3}{2}(\frac{1}{q_0} - \frac{1}{\theta})} e^{-(\lambda_1 + m_\infty)(t-s)} \|\rho\|_{L^\infty(\Omega)} \|u\|_{L^{q_0}(\Omega)} ds$$

$$\leq 3M_1 M_3 \varepsilon^2 \int_0^t (1 + (t - s))^{-\frac{1}{2} - \frac{3}{2}(\frac{1}{q_0} - \frac{1}{\theta})} e^{-\lambda_1(t-s)} e^{-2\alpha_1 s} (1 + s^{-\frac{1}{2} - \frac{3}{2}(\frac{1}{p_0} - \frac{1}{q_0})}) ds$$

$$\leq 3c_{10} M_3 M_1 \varepsilon^2 (1 + t^{\min\{0, -\frac{3}{2}(\frac{1}{p_0} - \frac{1}{\theta})\}}) e^{-\min\{\lambda_1, 2\alpha_1\} t}$$

and

$$\int_0^t (1 + (t - s))^{-\frac{3}{2}(\frac{1}{q_0} - \frac{1}{\theta})} e^{-m_\infty(t-s)} \|\rho\|_{L^{q_0}(\Omega)} \|m - m_\infty\|_{L^\infty(\Omega)} ds$$

$$\leq 3M_1 (M_5 + M_1) \varepsilon^2 \int_0^t (1 + (t - s))^{-\frac{1}{2} - \frac{3}{2}(\frac{1}{q_0} - \frac{1}{\theta})} e^{-m_\infty(t-s)} e^{-2\alpha_1 s} (1 + s^{-\frac{3}{p_0} + \frac{3}{2q_0}}) ds$$

$$\leq 3c_{10} M_1 (M_5 + M_1) \varepsilon^2 (1 + t^{\min\{0, -\frac{3}{2}(\frac{1}{p_0} - \frac{1}{\theta})\}}) e^{-\min\{m_\infty, 2\alpha_1\} t},$$

where the fact that $q_0 \in (3, \frac{3p_0}{2(3 - p_0)})$ warrants $-\frac{3}{p_0} + \frac{3}{2q_0} > -1$ is used. Hence, the combination of the above inequalities yields

$$\|\rho(\cdot, t)\|_{L^\theta(\Omega)} \leq \frac{M_1}{2} \varepsilon (1 + t^{-\frac{3}{2}(\frac{1}{p_0} - \frac{1}{\theta})}) e^{-\alpha_1 t},$$

thanks to (4.4.60), (4.4.59) and (4.4.56).

Lemma 4.46 *Under the assumptions of Proposition 4.4, we have*

$$\|(m - \rho)(\cdot, t) - e^{t\Delta}(m_0 - \rho_0)\|_{L^\theta(\Omega)} \leq \frac{\varepsilon}{2} (1 + t^{-\frac{3}{2}(\frac{1}{p_0} - \frac{1}{\theta})}) e^{-\alpha_1 t}$$

for $\theta \in [q_0, \infty]$, $t \in (0, T)$.

Proof From (4.4.63) and Lemma 1.1(iv), it follows that

$$\|(m - \rho)(\cdot, t) - e^{t\Delta}(m_0 - \rho_0)\|_{L^\theta(\Omega)}$$

$$\leq \int_0^t \|e^{(t-s)\Delta}(\nabla \cdot (\rho \mathscr{S}(\cdot, \rho, c)\nabla c) - u \cdot \nabla(m - \rho))(\cdot, s)\|_{L^\theta(\Omega)}ds$$

$$\leq \int_0^t \|e^{(t-s)\Delta}\nabla \cdot (\rho \mathscr{S}(\cdot, \rho, c)\nabla c)(\cdot, s)\|_{L^\theta(\Omega)}ds$$

$$+ \int_0^t \|e^{(t-s)\Delta}\nabla \cdot ((m - \rho - m_\infty)u)(\cdot, s)\|_{L^\theta(\Omega)}ds$$

$$\leq c_4 C_S \int_0^t (1 + (t - s)^{-\frac{1}{2} - \frac{3}{2}(\frac{1}{q_0} - \frac{1}{\theta})})e^{-\lambda_1(t-s)}\|\rho(\cdot, s)\|_{L^{q_0}(\Omega)}\|\nabla c(\cdot, s)\|_{L^\infty(\Omega)}ds$$

$$+ c_4 \int_0^t (1 + (t - s)^{-\frac{1}{2} - \frac{3}{2}(\frac{1}{q_0} - \frac{1}{\theta})})e^{-\lambda_1(t-s)}\|u(m - \rho - m_\infty)(\cdot, s)\|_{L^{q_0}(\Omega)}ds$$

$$=: I_1 + I_2.$$

From the definition of T and (4.4.58), we have

$$I_1 \leq 3c_4 C_S M_1 M_2 \varepsilon^2 \int_0^t (1 + (t - s)^{-\frac{1}{2} - \frac{3}{2}(\frac{1}{q_0} - \frac{1}{\theta})})e^{-\lambda_1(t-s)}(1 + s^{-\frac{1}{2} - \frac{3}{2}(\frac{1}{p_0} - \frac{1}{q_0})})e^{-2\alpha_1 s}ds$$

$$\leq 3c_4 C_S c_{10} M_1 M_2 \varepsilon^2 (1 + t^{\min\{0, -\frac{3}{2}(\frac{1}{p_0} - \frac{1}{\theta})\}})e^{-\min\{\lambda_1, 2\alpha_1\}t}$$

$$\leq \frac{\varepsilon}{4}(1 + t^{-\frac{3}{2}(\frac{1}{p_0} - \frac{1}{\theta})})e^{-\alpha_1 t}.$$

From Lemmas 4.40, 4.42 and (4.4.61), it follows that

$$I_2 = c_4 \int_0^t (1 + (t - s)^{-\frac{1}{2} - \frac{3}{2}(\frac{1}{q_0} - \frac{1}{\theta})})e^{-\lambda_1(t-s)}\|m - \rho - m_\infty\|_{L^\infty(\Omega)}\|u\|_{L^{q_0}(\Omega)}ds$$

$$\leq c_4 M_3 M_5 \varepsilon^2 \int_0^t (1 + (t - s)^{-\frac{1}{2} - \frac{3}{2}(\frac{1}{q_0} - \frac{1}{\theta})})e^{-\lambda_1(t-s)}(1 + s^{-\frac{3}{2p_0}})e^{-\alpha_1 s}$$

$$\cdot (1 + s^{-\frac{1}{2} + \frac{3}{2q_0}})e^{-\alpha_2 s}ds$$

$$\leq 3c_4 M_3 M_5 \varepsilon^2 \int_0^t (1 + (t - s)^{-\frac{1}{2} - \frac{3}{2}(\frac{1}{q_0} - \frac{1}{\theta})})(1 + s^{-\frac{1}{2} + \frac{3}{2}(\frac{1}{q_0} - \frac{1}{p_0})})$$

$$\cdot e^{-\lambda_1(t-s)}e^{-(\alpha_1 + \alpha_2)s}ds$$

$$\leq 3c_{10}c_4 M_3 M_5 \varepsilon^2 e^{-\min\{\lambda_1, \alpha_1 + \alpha_2\}t}(1 + t^{\min\{0, \frac{3}{2}(\frac{1}{\theta} - \frac{1}{p_0})\}})$$

$$\leq \frac{\varepsilon}{4}(1 + t^{-\frac{3}{2}(\frac{1}{p_0} - \frac{1}{\theta})})e^{-\alpha_1 t}.$$

Combining the above inequalities, we arrive at

$$\|(\rho - m)(\cdot, t) - e^{t\Delta}(\rho_0 - m_0)\|_{L^\theta(\Omega)} \leq \frac{\varepsilon}{2}(1 + t^{-\frac{3}{2}(\frac{1}{p_0} - \frac{1}{\theta})})e^{-\alpha_1 t}$$

and thus complete the proof of this lemma.

By the above lemmas, one can see that $T = T_{max}$. We will need two more estimates to show that $T_{max} = \infty$.

Lemma 4.47 *Under the assumptions of Proposition 4.4, for all $\beta \in (\frac{3}{4}, \min\{\frac{5}{4} - \frac{3}{2q_0}, 1\})$ there exists $M_6 > 0$, such that*

$$\|A^\beta u(\cdot, t)\|_{L^2(\Omega)} \le \varepsilon M_6 e^{-\alpha_2 t} \quad \text{for } t \in (t_0, T_{max}) \text{ with } t_0 = \min\{\frac{T_{max}}{6}, 1\}.$$

Proof The proof is similar to that of (4.4.40), and thus is omitted here.

Lemma 4.48 *Under the assumptions of Proposition 4.4, there exists $M_7 > 0$, such that $\|c(\cdot, t) - m_\infty\|_{L^\infty(\Omega)} \le M_7 e^{-\alpha_2 t}$ for all (t_0, T_{max}) with $t_0 = \min\{\frac{T_{max}}{6}, 1\}$.*

Proof We refer the readers to the proof of Lemma 3.24 in Li et al. (2019b).

Proof of Proposition 4.4. We first show that the solution is global, i.e., $T_{max} = \infty$. To this end, according to the extensibility criterion in Lemma 4.30, it suffices to show that there exists $C > 0$, such that for all $t_0 < t < T_{max}$

$$\|\rho(\cdot, t)\|_{L^\infty(\Omega)} + \|m(\cdot, t)\|_{L^\infty(\Omega)} + \|c(\cdot, t)\|_{W^{1,\infty}(\Omega)} + \|A^\beta u(\cdot, t)\|_{L^2(\Omega)} < C.$$

From Lemmas 4.41, 4.45 and 4.47, there exist $K_i > 0$, $i = 1, 2, 3, 4$, such that

$$\|m(\cdot, t) - m_\infty\|_{L^\infty(\Omega)} \le K_1 e^{-\alpha_1 t}, \quad \|\rho(\cdot, t)\|_{L^\infty(\Omega)} \le K_2 e^{-\alpha_1 t},$$

$$\|\nabla c(\cdot, t)\|_{L^\infty(\Omega)} \le K_3 e^{-\alpha_1 t}, \quad \|A^\beta u(\cdot, t)\|_{L^2(\Omega)} \le K_4 e^{-\alpha_2 t}$$

for $t \in (t_0, T_{max})$. Furthermore, Lemma 4.48 implies that $\|c(\cdot, t) - m_\infty\|_{W^{1,\infty}(\Omega)} \le K_3' e^{-\alpha_2 t}$ with some $K_3' > 0$ for all $t \in (t_0, T_{max})$. Since $D(A^\beta) \hookrightarrow L^\infty(\Omega)$ with $\beta \in (\frac{3}{4}, 1)$, it follows from Lemma 4.47 that $\|u(\cdot, t)\|_{L^\infty(\Omega)} \le K_4 e^{-\alpha_2 t}$ for some $K_4 > 0$ for all $t \in (t_0, T_{max})$. This completes the proof of Proposition 4.4.

4.4.3 Global Boundedness and Decay for General $\mathscr{S}$

Proof of Theorems for general $\mathscr{S}$. We complete the proofs of our theorems by an approximation procedure (see Cao and Lankeit (2016) for example). In order to make the previous results applicable, we introduce a family of smooth functions $\rho_\eta \in C_0^\infty(\Omega)$ with $0 \le \rho_\eta(x) \le 1$ for $\eta \in (0, 1)$, and $\lim_{\eta \to 0} \rho_\eta(x) = 1$, and let $\mathscr{S}_\eta(x, \rho, c) = \rho_\eta(x)\mathscr{S}(x, \rho, c)$.

Using this definition, we regularize (4.1.13) as follows:

$$\begin{cases} (\rho_\eta)_t + u_\eta \cdot \nabla \rho_\eta = \Delta \rho_\eta - \nabla \cdot (\rho_\eta \mathscr{S}_\eta(x, \rho_\eta, c_\eta) \nabla c_\eta) - \rho_\eta m_\eta, \\ (m_\eta)_t + u_\eta \cdot \nabla m_\eta = \Delta m_\eta - \rho_\eta m_\eta, \\ (c_\eta)_t + u_\eta \cdot \nabla c_\eta = \Delta c_\eta - c_\eta + m_\eta, \\ (u_\eta)_t + (u_\eta \cdot \nabla) u_\eta = \Delta u_\eta - \nabla P_\eta + (\rho_\eta + m_\eta) \nabla \phi, \quad \nabla \cdot u_\eta = 0, \\ \dfrac{\partial \rho_\eta}{\partial \nu} = \dfrac{\partial m_\eta}{\partial \nu} = \dfrac{\partial c_\eta}{\partial \nu} = 0, \ u_\eta = 0 \end{cases} \quad (4.4.75)$$

with the initial data

$$\rho_\eta(x, 0) = \rho_0(x), \ m_\eta(x, 0) = m_0(x), \ c(x, 0) = c_0(x), \ u_\eta(x, 0) = u_0(x), \quad x \in \Omega. \quad (4.4.76)$$

It is observed that $\mathscr{S}_\eta$ satisfies the additional condition $\mathscr{S} = 0$ on $\partial\Omega$. Therefore based on the discussion in Sect. 4.4.2, under the assumptions of Theorem 4.4 and Theorem 4.5, the problem (4.4.75)–(4.4.76) admits a global classical solution $(\rho_\eta, m_\eta, c_\eta, u_\eta, P_\eta)$ that satisfies

$$\|m_\eta(\cdot, t) - m_\infty\|_{L^\infty(\Omega)} \le K_1 e^{-\alpha_1 t}, \quad \|\rho_\eta(\cdot, t) - \rho_\infty\|_{L^\infty(\Omega)} \le K_2 e^{-\alpha_1 t},$$
$$\|c_\eta(\cdot, t) - m_\infty\|_{W^{1,\infty}(\Omega)} \le K_3 e^{-\alpha_2 t}, \quad \|A^\beta u_\eta(\cdot, t)\|_{L^2(\Omega)} \le K_4 e^{-\alpha_2 t}$$

for some constants K_i, $i = 1, 2, 3, 4$, and all $t \ge 0$. Applying a standard procedure such as in Lemmas 5.2 and 5.6 of Cao and Lankeit (2016), one can obtain a subsequence of $\{\eta_j\}_{j \in \mathbb{N}}$ with $\eta_j \to 0$ as $j \to \infty$ such that $\rho_{\eta_j} \to \rho$, $m_{\eta_j} \to m$, $c_{\eta_j} \to c$, $u_{\eta_j} \to u$ in $C_{loc}^{\nu, \frac{\nu}{2}}(\overline{\Omega} \times (0, \infty))$ as $j \to \infty$ for some $\nu \in (0, 1)$. Moreover, by the arguments as in Lemmas 5.7 and 5.8 of Cao and Lankeit (2016), one can also show that (ρ, m, c, u, P) is a classical solution of (4.1.13) with the decay properties asserted in Theorems 4.4 and 4.5, respectively. The proof of our main results is thus complete.

4.5 Large Time Behavior of Solutions to a Coral Fertilization Model with Nonlinear Diffusion

4.5.1 Regularized Problems

At first, we present a natural notion of weak solvability to (4.1.16), (4.1.22) and (4.1.23).

Definition 4.1 For a quadruple of functions (n, c, v, u), we call it a global weak solution of (4.1.16), (4.1.22) and (4.1.23), if it fulfills

$$\begin{cases} n \in L^1_{loc}(\bar{\Omega} \times [0, \infty)), \\ c \in L^\infty_{loc}(\bar{\Omega} \times [0, \infty)) \bigcap L^1_{loc}([0, \infty); W^{1,1}(\Omega)), \\ v \in L^\infty_{loc}(\bar{\Omega} \times [0, \infty)) \bigcap L^1_{loc}([0, \infty); W^{1,1}(\Omega)), \\ u \in L^1_{loc}([0, \infty); W^{1,1}_0(\Omega; \mathbb{R}^3)), \end{cases}$$ (4.5.1)

with $n \geq 0, c \geq 0, v \geq 0$ in $\Omega \times (0, \infty)$, and

$$E(n), \ n|\nabla c|, \ n|u|, \ c|u| \text{ and } v|u| \text{ belong to } L^1_{loc}(\bar{\Omega} \times [0, \infty)), \quad (4.5.2)$$

where $E(s) := \int_0^s D(\sigma)d\sigma$, if $\nabla \cdot u = 0$ in the distributional sense, if

$$-\int_0^\infty \int_\Omega n\varphi_t - \int_\Omega n_0\varphi(\cdot, 0) = -\int_0^\infty \int_\Omega E(n)\Delta\varphi + \int_0^\infty \int_\Omega n\nabla c \cdot \nabla\varphi \\ + \int_0^\infty \int_\Omega nu \cdot \nabla\varphi - \int_0^\infty \int_\Omega nv\varphi$$ (4.5.3)

for any $\varphi \in C_0^\infty(\bar{\Omega} \times [0, \infty))$ satisfying $\frac{\partial\varphi}{\partial v} = 0$, if

$$-\int_0^\infty \int_\Omega c\varphi_t - \int_\Omega c_0\varphi(\cdot, 0) = -\int_0^\infty \int_\Omega \nabla c \cdot \nabla\varphi - \int_0^\infty \int_\Omega c\varphi + \int_0^\infty \int_\Omega v\varphi \\ + \int_0^\infty \int_\Omega cu \cdot \nabla\varphi$$

(4.5.4)

and

$$-\int_0^\infty \int_\Omega v\varphi_t - \int_\Omega v_0\varphi(\cdot, 0) = -\int_0^\infty \int_\Omega \nabla v \cdot \nabla\varphi - \int_0^\infty \int_\Omega vn\varphi + \int_0^\infty \int_\Omega vu \cdot \nabla\varphi$$ (4.5.5)

for any $\varphi \in C_0^\infty(\bar{\Omega} \times [0, \infty))$, as well as if

$$-\int_0^\infty \int_\Omega u \cdot \varphi_t - \int_\Omega u_0\varphi(\cdot, 0) = -\int_0^\infty \int_\Omega \nabla u \cdot \nabla\varphi + \int_0^\infty \int_\Omega (n + v)\nabla\Phi \cdot \varphi$$ (4.5.6)

for any $\varphi \in C_0^\infty(\bar{\Omega} \times [0, \infty); \mathbb{R}^3)$ fulfilling $\nabla \cdot \varphi \equiv 0$.

Now, in line with the analysis in closely related settings (Winkler 2015b, 2018c), let us introduce a family of regularized problems of (4.1.16), (4.1.22) and (4.1.23) through a standard approximation procedure. Thereupon, the corresponding approximated problems appear as

$$\begin{cases} n_{\varepsilon t} + u_\varepsilon \cdot \nabla n_\varepsilon = \nabla \cdot (D_\varepsilon(n_\varepsilon)\nabla n_\varepsilon) - \nabla \cdot (n_\varepsilon F'_\varepsilon(n_\varepsilon)\nabla c_\varepsilon) - F_\varepsilon(n_\varepsilon)v_\varepsilon, \quad x \in \Omega, \ t > 0, \\ c_{\varepsilon t} + u_\varepsilon \cdot \nabla c_\varepsilon = \Delta c_\varepsilon - c_\varepsilon + v_\varepsilon, \quad x \in \Omega, \ t > 0, \\ v_{\varepsilon t} + u_\varepsilon \cdot \nabla v_\varepsilon = \Delta v_\varepsilon - v_\varepsilon F_\varepsilon(n_\varepsilon), \quad x \in \Omega, \ t > 0, \\ u_{\varepsilon t} = \Delta u_\varepsilon + \nabla P_\varepsilon + (n_\varepsilon + v_\varepsilon)\nabla \Phi, \quad \nabla \cdot u_\varepsilon = 0, \quad x \in \Omega, \ t > 0, \\ \dfrac{\partial n_\varepsilon}{\partial v} = \dfrac{\partial c_\varepsilon}{\partial v} = \dfrac{\partial v_\varepsilon}{\partial v} = 0, \ u_\varepsilon = 0, \quad x \in \partial\Omega, \ t > 0, \\ n_\varepsilon(x,0) = n_0(x), \ c_\varepsilon(x,0) = c_0(x), \ v_\varepsilon(x,0) = v_0(x), \ u_\varepsilon(x,0) = u_0(x), \quad x \in \Omega, \end{cases}$$
(4.5.7)

where for each $\varepsilon \in (0,1)$ $(D_\varepsilon)_{\varepsilon \in (0,1)} \in C^2([0,\infty))$ fulfills

$$D_\varepsilon(s) \geq \varepsilon \ \text{ and } \ D(s) \leq D_\varepsilon(s) \leq D(s) + 2\varepsilon, \ \ s \geq 0, \tag{4.5.8}$$

and where

$$F_\varepsilon(s) := \int_0^s \rho_\varepsilon(\sigma)d\sigma, \ \ s \geq 0 \ \text{ for all } \ \varepsilon \in (0,1) \tag{4.5.9}$$

with $(\rho_\varepsilon)_{\varepsilon \in (0,1)} \subset C_0^\infty([0,\infty))$ having the properties that for each $\varepsilon \in (0,1)$

$$0 \leq \rho_\varepsilon \leq 1 \ \text{ in } [0,\infty), \ \ \rho_\varepsilon \equiv 1 \ \text{ in } [0,\tfrac{1}{\varepsilon}] \ \text{ and } \ \rho_\varepsilon \equiv 0 \ \text{ in } [\tfrac{2}{\varepsilon},\infty), \tag{4.5.10}$$

from which and (4.5.9) one can infer that for each $\varepsilon \in (0,1)$

$$F_\varepsilon \in C^\infty([0,\infty)), \ \ 0 \leq F_\varepsilon(s) \leq s \ \text{ and } \ 0 \leq F'_\varepsilon(s) \leq 1, \ \ s \geq 0 \tag{4.5.11}$$

as well as

$$F_\varepsilon(s) \to s \ \text{ and } \ F'_\varepsilon(s) \to 1 \ \text{ for any } \ s > 0 \ \text{ as } \ \varepsilon \to 0. \tag{4.5.12}$$

Actually, the local solvability of (4.5.7) can be verified through a suitable adaptation of standard fixed point arguments as proceeding in Winkler (2012, Lemma 2.1), so here we merely present the associated assertions.

Lemma 4.49 *Let (4.1.17) be fulfilled. Then for any (n_0, c_0, v_0, u_0) complying with (4.1.25) and each $\varepsilon \in (0,1)$, one can find $T_{\max,\varepsilon} \in (0,+\infty]$ and functions*

$$\begin{cases} n_\varepsilon \in C^0(\bar\Omega \times [0, T_{\max,\varepsilon})) \bigcap C^{2,1}(\bar\Omega \times (0, T_{\max,\varepsilon})), \\ c_\varepsilon \in \bigcap_{r>3} C^0([0, T_{\max,\varepsilon}); W^{1,r}(\Omega)) \bigcap C^{2,1}(\bar\Omega \times (0, T_{\max,\varepsilon})), \\ v_\varepsilon \in \bigcap_{r>3} C^0([0, T_{\max,\varepsilon}); W^{1,r}(\Omega)) \bigcap C^{2,1}(\bar\Omega \times (0, T_{\max,\varepsilon})), \\ u_\varepsilon \in C^0(\bar\Omega \times [0, T_{\max,\varepsilon}); \mathbb{R}^3) \bigcap C^{2,1}(\bar\Omega \times (0, T_{\max,\varepsilon}); \mathbb{R}^3), \end{cases}$$
(4.5.13)

such that $n_\varepsilon \geq 0$, $c_\varepsilon \geq 0$ and $v_\varepsilon \geq 0$ in $\Omega \times [0, T_{\max,\varepsilon})$, that with $P_\varepsilon \in C^{1,0}(\Omega \times (0, T_{\max,\varepsilon}))$ the quintuple of functions $(n_\varepsilon, c_\varepsilon, v_\varepsilon, u_\varepsilon, P_\varepsilon)$ forms a classical solution of (4.5.7) in $\Omega \times [0, T_{\max,\varepsilon})$, and that with some $\alpha \in (\frac{3}{4}, 1)$ either $T_{\max,\varepsilon} < \infty$ or

$$\limsup_{t \nearrow T_{\max,\varepsilon}} \left\{ \|n_\varepsilon(\cdot, t)\|_{L^\infty(\Omega)} + \|c_\varepsilon(\cdot, t)\|_{W^{1,r}(\Omega)} + \|v_\varepsilon(\cdot, t)\|_{W^{1,r}(\Omega)} + \|A^\alpha u_\varepsilon(\cdot, t)\|_{L^2(\Omega)} \right\}$$
$$= \infty$$

(4.5.14)

holds.

Thanks to the consumption interaction between n and v, (4.5.7) implies following basic estimates.

Lemma 4.50 *Let $M_0 := \max\{\int_\Omega n_0, \int_\Omega v_0, \|c_0\|_{L^\infty(\Omega)}, \|v_0\|_{L^\infty(\Omega)}\}$. Then the solutions constructed in Lemma 4.49 satisfy*

$$\int_\Omega n_\varepsilon(\cdot, t) \leq M_0 \quad \text{and} \quad \int_\Omega v_\varepsilon(\cdot, t) \leq M_0 \qquad (4.5.15)$$

for all $t \in (0, T_{\max,\varepsilon})$ and $\varepsilon \in (0, 1)$, as well as

$$\|v_\varepsilon(\cdot, t)\|_{L^\infty(\Omega)} \leq M_0 \quad \text{and} \quad \|c_\varepsilon(\cdot, t)\|_{L^\infty(\Omega)} \leq M_0 \qquad (4.5.16)$$

for all $t \in (0, T_{\max,\varepsilon})$ and $\varepsilon \in (0, 1)$. Moreover, there exists some $C > 0$ such that

$$\int_0^\infty \int_\Omega F(n_\varepsilon)v_\varepsilon \leq C \quad \text{for all } \varepsilon \in (0, 1) \qquad (4.5.17)$$

and

$$\int_0^\infty \int_\Omega |\nabla v_\varepsilon|^2 \leq C \quad \text{for all } \varepsilon \in (0, 1) \qquad (4.5.18)$$

as well as

$$\int_0^\infty \int_\Omega |\nabla c_\varepsilon|^2 \leq C \quad \text{for all } \varepsilon \in (0, 1). \qquad (4.5.19)$$

Proof The detailed process of the derivation thereof can be found in Espejo and Winkler (2018); Liu (2020).

In the final, we provide some conditional estimates of $(u_\varepsilon)_{\varepsilon \in (0,1)}$, which reveal the relationships between temporally independent estimates of $(u_\varepsilon)_{\varepsilon \in (0,1)}$ and uniform L^p $(p > 1)$ norms of $(n_\varepsilon)_{\varepsilon \in (0,1)}$.

Lemma 4.51 *uppose $(n_\varepsilon, c_\varepsilon, v_\varepsilon, u_\varepsilon)_{\varepsilon \in (0,1)}$ are solutions as established in Lemma 4.49. Let $p \geq 2, l > 3$ and $\alpha \in (\frac{3}{4}, 1)$. Then for any $\iota > 0$, one can find $M_1 = M_1(p, l, \iota, M_0) > 0$ and $M_2 = M_2(\alpha, p, \iota, M_0) > 0$ such that*

$$\|u_\varepsilon(\cdot, t)\|_{L^l(\Omega)} \le M_1 \cdot \left\{ 1 + \sup_{\tau \in (0,t)} \|n_\varepsilon(\cdot, \tau)\|_{L^p(\Omega)} \right\}^{\frac{p}{p-1} \cdot \left(\frac{l-3}{3l} + \iota\right)} \tag{4.5.20}$$

for each $t \in (0, T_{\max,\varepsilon})$ and all $\varepsilon \in (0, 1)$, and that

$$\|A^\alpha u_\varepsilon(\cdot, t)\|_{L^2(\Omega)} \le M_2 \cdot \left\{ 1 + \sup_{\tau \in (0,t)} \|n_\varepsilon(\cdot, \tau)\|_{L^p(\Omega)} \right\}^{\frac{p}{p-1} \cdot \left(\frac{4\alpha-1}{6} + \iota\right)} \tag{4.5.21}$$

for each $t \in (0, T_{\max,\varepsilon})$ and all $\varepsilon \in (0, 1)$.

Proof Recalling (Winkler 2021b, Corollary 2.1), we infer from u_ε-equation in (4.5.7) that

$$\|u_\varepsilon(\cdot, t)\|_{L^l(\Omega)} \le C_1 \cdot \left\{ 1 + \sup_{\tau \in (0,t)} \|n_\varepsilon(\cdot, \tau) + v_\varepsilon(\cdot, \tau)\|_{L^p(\Omega)} \right\}^{\frac{p}{p-1} \cdot \left(\frac{l-3}{3l} + \iota\right)} \tag{4.5.22}$$

with some $C_1 = C_1(p, l, \iota, M_0) > 0$ for each $t \in (0, T_{\max,\varepsilon})$ and all $\varepsilon \in (0, 1)$, where due to (4.5.16),

$$\begin{aligned}\|n_\varepsilon(\cdot, \tau) + v_\varepsilon(\cdot, \tau)\|_{L^p(\Omega)} &\le \|n_\varepsilon(\cdot, \tau)\|_{L^p(\Omega)} + \|v_\varepsilon(\cdot, \tau)\|_{L^p(\Omega)} \\ &\le \|n_\varepsilon(\cdot, \tau)\|_{L^p(\Omega)} + M_0|\Omega|^{\frac{1}{p}} \end{aligned} \tag{4.5.23}$$

for any $\tau \in (0, t)$ with each $t \in (0, T_{\max,\varepsilon})$ for all $\varepsilon \in (0, 1)$. Thereupon, we can rewrite (4.5.22) as

$$\|u_\varepsilon(\cdot, t)\|_{L^l(\Omega)} \le C_1 \cdot \left\{ 1 + M_0|\Omega|^{\frac{1}{p}} + \sup_{\tau \in (0,t)} \|n_\varepsilon(\cdot, \tau)\|_{L^p(\Omega)} \right\}^{\frac{p}{p-1} \cdot \left(\frac{l-3}{3l} + \iota\right)} \tag{4.5.24}$$

for each $t \in (0, T_{\max,\varepsilon})$ and all $\varepsilon \in (0, 1)$. With the choice of $M_1 := C_1 \cdot (1 + M_0|\Omega|^{\frac{1}{p}})$, (4.5.20) is implied by (4.5.24). In a flavor quite similar to the reasoning of (4.5.20), (4.5.21) follows from a combination of Winkler (2021b, Proposition 1.1) with (4.5.23). $\blacksquare$

4.5.2 Conditional Uniform Bounds for $(\nabla c_\varepsilon)_{\varepsilon \in (0,1)}$

In fact, for the derivation of uniform L^p bounds of $(n_\varepsilon)_{\varepsilon \in (0,1)}$, besides the ε-independent conditional estimates of $(u_\varepsilon)_{\varepsilon \in (0,1)}$ as given by Lemma 4.51, it is also essential to gain similar uniform estimates for signal gradients with respect to the temporally independent L^p norms of $(n_\varepsilon)_{\varepsilon \in (0,1)}$ in accordance with the recursive frameworks established in Winkler (2021b). For convenience in expressions, we make use of the following abbreviations:

$$I_{p,\varepsilon}(t) := 1 + \sup_{\tau \in (0,t)} \|n_\varepsilon(\cdot, \tau)\|_{L^p(\Omega)}, \quad t \in (0, T_{\max,\varepsilon}) \text{ for all } \varepsilon \in (0, 1) \quad (4.5.25)$$

and

$$K_{q,\theta,\varepsilon}(t) := 1 + \sup_{\tau \in (0,t)} \left\| B^\theta \left(c_\varepsilon(\cdot, \tau) - e^{-sB} c_0 \right) \right\|_{L^q(\Omega)}, \quad t \in (0, T_{\max,\varepsilon}) \quad (4.5.26)$$

for all $\varepsilon \in (0, 1)$.

Lemma 4.52 *Let $\theta \in (\frac{1}{2}, 1)$ and $q > 3$. Then for any $\iota > 0$, one can find some $C = C(\theta, q, \iota) > 0$ satisfying*

$$\left\| \nabla(c_\varepsilon(\cdot, t) - e^{-\tau B} c_0) \right\|_{L^\infty(\Omega)} \leq C \cdot \left\{ 1 + \sup_{\tau \in (0,t)} \left\| B^\theta (c_\varepsilon(\cdot, \tau) - e^{-\tau B} c_0) \right\|_{L^q(\Omega)} \right\}^{\frac{q+3}{2\theta q} + \iota} \quad (4.5.27)$$

for each $t \in (0, T_{\max,\varepsilon})$ and all $\varepsilon \in (0, 1)$.

Proof Since $\theta \in (\frac{1}{2}, 1)$ enables us to choose $q > 3$ large enough such that $1 - \frac{q+3}{2\theta q} > 0$, this ensures the existence of $\iota > 0$ sufficiently small such that $\iota < 1 - \frac{q+3}{2\theta q}$, which allows for the following choice of ϑ, namely

$$\vartheta(\iota) := \frac{q+3}{2q} + \iota\theta < \theta. \quad (4.5.28)$$

From the interpolation inequality provided by Friedman (1969, Theorem 2.14.1) for fractional powers of sectorial operators, it follows that

$$\begin{aligned}
&\left\| B^\vartheta (c_\varepsilon(\cdot, t) - e^{-\tau B} c_0) \right\|_{L^q(\Omega)} \\
&\leq C_1 \left\| B^\theta (c_\varepsilon(\cdot, t) - e^{-\tau B} c_0) \right\|_{L^q(\Omega)}^{\frac{\vartheta}{\theta}} \left\| c_\varepsilon(\cdot, t) - e^{-\tau B} c_0 \right\|_{L^q(\Omega)}^{\frac{\theta - \vartheta}{\theta}} \\
&\leq C_1 \left\{ 2M_0 |\Omega|^{\frac{1}{q}} \right\}^{1 - \iota - \frac{q+3}{2\theta q}} \left\| B^\theta (c_\varepsilon(\cdot, t) - e^{-\tau B} c_0) \right\|_{L^q(\Omega)}^{\frac{q+3}{2\theta q} + \iota}
\end{aligned} \quad (4.5.29)$$

with some $C_1 = C_1(\theta, q, \iota) > 0$ and $M_0 > 0$ as taken in Lemma 4.50 for each $t \in (0, T_{\max,\varepsilon})$ and all $\varepsilon \in (0, 1)$. Combining with the embedding $D(B^\vartheta) \hookrightarrow W^{1,\infty}(\Omega)$ (Henry 1981), we obtain from (4.5.29) that

$$\begin{aligned}
&\left\| \nabla(c_\varepsilon(\cdot, t) - e^{-\tau B} c_0) \right\|_{L^\infty(\Omega)} \\
&\leq C_2 \left\| B^\vartheta (c_\varepsilon(\cdot, t) - e^{-\tau B} c_0) \right\|_{L^q(\Omega)} \\
&\leq C_2 C_1 \left\{ 2M_0 |\Omega|^{\frac{1}{q}} \right\}^{1 - \iota - \frac{q+3}{2\theta q}} \left\| B^\theta (c_\varepsilon(\cdot, t) - e^{-\tau B} c_0) \right\|_{L^q(\Omega)}^{\frac{q+3}{2\theta q} + \iota}
\end{aligned}$$

with $C_2 = C_2(\theta, q, \iota) > 0$ for each $t \in (0, T_{\max,\varepsilon})$ and all $\varepsilon \in (0, 1)$, and whereafter
(4.5.27) holds with $C := C_2 C_1 \left\{ 2M_0 |\Omega|^{\frac{1}{q}} \right\}^{1-\iota-\frac{q+3}{2\theta q}}$.

With the aid of Lemma 4.52, the following conditional estimates can be established by means of the L^p-L^q estimates for fractional powers of sectorial operators (Horstmann and Winkler 2005, (3)).

Lemma 4.53 *Let $\theta \in (\frac{1}{2}, 1)$, and let $q > 3$, $p \geq 2$. Then for each $\iota > 0$ there exists $C = C(\theta, q, p, \iota) > 0$ such that*

$$\left\| B^\theta (c_\varepsilon(\cdot, \tau) - e^{-\tau B} c_0) \right\|_{L^q(\Omega)} \leq C \cdot \left\{ 1 + \sup_{\tau \in (0,t)} \|n_\varepsilon(\cdot, \tau)\|_{L^p(\Omega)} \right\}^{\frac{p}{p-1} \cdot \left(\frac{2\theta}{3} + \iota \right)}$$

(4.5.30)

for all $t \in (0, T_{\max,\varepsilon})$ and $\varepsilon \in (0, 1)$.

Proof Picking $\iota > 0$ sufficiently small such that

$$\iota < \min \left\{ 1 - \frac{q+3}{2\theta q}, 2q(1-\theta) \right\},$$

(4.5.31)

and then letting

$$l := \frac{3q}{3 + 2q(1-\theta) - \iota},$$

(4.5.32)

one can observe from $\iota < 2q(1-\theta)$ and $q > 3 + 2q - 2q\theta + 2q\theta\iota > 3 + 2q - 2q\theta$ implied by (4.5.31) and (4.5.32), respectively, that

$$q = \frac{3q}{3 + 2q - 2q\theta - 2q(1-\theta)} > l = \frac{3q}{3 + 2q - 2q\theta - \iota} > \frac{3q}{3 + 2q - 2q\theta} > 3.$$

(4.5.33)

Next, we apply B^θ to the following variation-of-constants representation

$$c_\varepsilon(\cdot, t) - e^{-tB} c_0 = \int_0^t e^{-B(t-\tau)} \{v_\varepsilon(\cdot, \tau) - u_\varepsilon(\cdot, \tau) \nabla c_\varepsilon(\cdot, \tau)\} d\tau$$

for all $t \in (0, T_{\max,\varepsilon})$ and $\varepsilon \in (0, 1)$ to have

$$\left\| B^\theta (c_\varepsilon(\cdot, t) - e^{-\tau B} c_0) \right\|_{L^q(\Omega)} \leq \int_0^t \left\| B^\theta e^{-B(t-\tau)} v_\varepsilon(\cdot, \tau) \right\|_{L^q(\Omega)} d\tau$$

$$+ \int_0^t \left\| B^\theta e^{-B(t-\tau)} u_\varepsilon(\cdot, \tau) \nabla c_\varepsilon(\cdot, \tau) \right\|_{L^q(\Omega)} d\tau$$

(4.5.34)

for all $t \in (0, T_{\max,\varepsilon})$ and $\varepsilon \in (0, 1)$. Recalling the L^p-L^q estimates for fractional powers of sectorial operators (Horstmann and Winkler, 2005, (3)) and the following regularity features of the Neumman heat semigroup (Henry 1981; Winkler 2010), namely

$$\left\| \nabla e^{-tB} c_0 \right\|_{L^\infty(\Omega)} \le C_1 \| \nabla c_0 \|_{L^\infty(\Omega)} \tag{4.5.35}$$

with some $C_1 > 0$, we gain from (4.5.16), (4.5.20), (4.5.25), (4.5.26) and (4.5.27) that

$$\int_0^t \left\| B^\theta e^{-B(t-\tau)} v_\varepsilon(\cdot, \tau) \right\|_{L^q(\Omega)} d\tau \le C_2 \int_0^t \left(1 + (t-\tau)^{-\theta} \right) e^{-(t-\tau)} \| v_\varepsilon(\cdot, \tau) \|_{L^q(\Omega)} d\tau$$

$$\le C_2 M_0 |\Omega|^{\frac{1}{q}} \int_0^t \left(1 + (t-\tau)^{-\theta} \right) e^{-(t-\tau)} d\tau \le C_3 \tag{4.5.36}$$

for all $t \in (0, T_{\max,\varepsilon})$ and $\varepsilon \in (0, 1)$ with $C_2 > 0$ and

$$C_3 := C_2 M_0 |\Omega|^{\frac{1}{q}} \int_0^\infty \left(1 + \sigma^{-\theta} \right) e^{-\sigma} d\sigma < \infty$$

thanks to $\theta \in \left(\frac{1}{2}, 1 \right)$, and that

$$\int_0^t \left\| B^\theta e^{-B(t-\tau)} u_\varepsilon(\cdot, \tau) \nabla c_\varepsilon(\cdot, \tau) \right\|_{L^q(\Omega)} d\tau$$

$$\le C_4 \int_0^t \left(1 + (t-\tau)^{-\theta - \frac{3}{2}(\frac{1}{l} - \frac{1}{q})} \right) e^{-(t-\tau)} \| u_\varepsilon(\cdot, \tau) \nabla c_\varepsilon(\cdot, \tau) \|_{L^l(\Omega)} d\tau$$

$$\le C_4 \int_0^t \left(1 + (t-\tau)^{-\theta - \frac{3}{2}(\frac{1}{l} - \frac{1}{q})} \right) e^{-(t-\tau)} \| u_\varepsilon(\cdot, \tau) \|_{L^l(\Omega)} \| \nabla c_\varepsilon(\cdot, \tau) \|_{L^\infty(\Omega)} d\tau$$

$$\le C_4 \int_0^t \left(1 + (t-\tau)^{-\theta - \frac{3}{2}(\frac{1}{l} - \frac{1}{q})} \right) e^{-(t-\tau)} \| u_\varepsilon(\cdot, \tau) \|_{L^l(\Omega)}$$

$$\cdot \left\{ \| \nabla(c_\varepsilon(\cdot, \tau) - e^{-tB} c_0) \|_{L^\infty(\Omega)} + \| \nabla e^{-tB} c_0 \|_{L^\infty(\Omega)} \right\} d\tau$$

$$\le C_4 M_1 \int_0^t \left(1 + (t-\tau)^{-\theta - \frac{3}{2}(\frac{1}{l} - \frac{1}{q})} \right) e^{-(t-\tau)} d\tau \cdot I_{p,\varepsilon}^{\frac{p}{p-1} \cdot \left(\frac{l-3}{3l} + \iota \right)}(t)$$

$$\cdot \left\{ C_5 K_{q,\theta,\varepsilon}^{\frac{q+3}{2\theta q} + \iota}(t) + C_1 \| \nabla c_0 \|_{L^\infty(\Omega)} \right\}$$

$$\le C_6 I_{p,\varepsilon}^{\frac{p}{p-1} \cdot \left(\frac{l-3}{3l} + \iota \right)}(t) \cdot K_{q,\theta,\varepsilon}^{\frac{q+3}{2\theta q} + \iota}(t) \tag{4.5.37}$$

for all $t \in (0, T_{\max,\varepsilon})$ and $\varepsilon \in (0, 1)$, where C_4, C_5 are positive constants and $C_6 := C_4 M_1 (C_5 + C_1 M_0) \int_0^\infty \left(1 + \sigma^{-\theta - \frac{3}{2}(\frac{1}{l} - \frac{1}{q})} \right) e^{-\sigma} d\sigma < \infty$ due to (4.5.33). Inserting (4.5.36) and (4.5.37) into (4.5.34) entails

$$\left\| B^\theta (c_\varepsilon(\cdot, t) - e^{-\tau B} c_0) \right\|_{L^q(\Omega)} \le C_3 + C_6 I_{p,\varepsilon}^{\frac{p}{p-1} \cdot \left(\frac{l-3}{3l} + \iota \right)}(t) \cdot K_{q,\theta,\varepsilon}^{\frac{q+3}{2\theta q} + \iota}(t) \tag{4.5.38}$$

for all $t \in (0, T_{\max,\varepsilon})$ and $\varepsilon \in (0, 1)$. It can be readily seen from (4.5.31) that

$$\iota + \frac{q+3}{2\theta q} < 1,$$

which allows for an application of Young's inequality to attain

$$\left\| B^{\theta}(c_{\varepsilon}(\cdot, t) - e^{-\tau B}c_0) \right\|_{L^q(\Omega)} \leq C_3 + \frac{1}{2}K_{q,\theta,\varepsilon}(t) + C_7 I_{p,\varepsilon}^{\frac{p}{p-1} \cdot \left(\frac{l-3}{3l}+\iota\right) \cdot \frac{2\theta q}{2\theta q - q - 3 - 2\theta q\iota}}(t)$$

with certain $C_7 = C_7(\theta, q, p, l, \iota)$ for all $t \in (0, T_{\max,\varepsilon})$ and $\varepsilon \in (0, 1)$. We thus combine with (4.5.26) to have

$$K_{q,\theta,\varepsilon}(t) \leq 1 + C_3 + \frac{1}{2}K_{q,\theta,\varepsilon}(t) + C_7 I_{p,\varepsilon}^{\frac{p}{p-1} \cdot \left(\frac{l-3}{3l}+\iota\right) \cdot \frac{2\theta q}{2\theta q - q - 3 - 2\theta q\iota}}(t),$$

namely

$$K_{q,\theta,\varepsilon}(t) \leq 2(1 + C_3) + 2C_7 I_{p,\varepsilon}^{\frac{p}{p-1} \cdot \left(\frac{l-3}{3l}+\iota\right) \cdot \frac{2\theta q}{2\theta q - q - 3 - 2\theta q\iota}}(t) \qquad (4.5.39)$$

for all $t \in (0, T_{\max,\varepsilon})$ and $\varepsilon \in (0, 1)$. Setting

$$\psi(\tilde{\iota}) := \frac{p}{p-1} \cdot \left(\frac{2\theta q - q - 3 + \tilde{\iota}}{3q} + \iota\right) \cdot \frac{2\theta q}{2\theta q - q - 3 - 2\theta q\tilde{\iota}},$$

we note that $\psi(\tilde{\iota}) \searrow \frac{p}{p-1} \cdot \frac{2\theta}{3}$ as $\tilde{\iota} \searrow 0$, whereupon for arbitrarily small $\iota > 0$ one can pick $\iota' \in \left(0, \min\left\{1 - \frac{q+3}{2\theta q}, 2q(1 - \theta)\right\}\right)$ such that

$$\psi(\iota') \leq \frac{p}{p-1} \cdot \frac{2\theta}{3} + \iota.$$

An elementary calculation along with (4.5.32) thus shows

$$\frac{p}{p-1} \cdot \left(\frac{l-3}{3l} + \iota'\right) \cdot \frac{2\theta q}{2\theta q - q - 3 - 2\theta q\iota'}$$

$$= \frac{p}{p-1} \cdot \left(\frac{2\theta q - q - 3 + \iota'}{3q} + \iota'\right) \cdot \frac{2\theta q}{2\theta q - q - 3 - 2\theta q\iota'}$$

$$= \psi(\iota') \leq \frac{p}{p-1} \cdot \frac{2\theta}{3} + \iota,$$

which in conjunction with (4.5.39), (4.5.25) and (4.5.26) yields (4.5.30). $\qquad \square$

With Lemmas 4.52–4.53 at hand, we are in the position to derive the desired conditional uniform L^{∞} estimates for $(\nabla c_{\varepsilon})_{\varepsilon \in (0,1)}$ from a well-known continuous embedding.

Lemma 4.54 *Suppose that $p \geq 2$. Then for any $\iota > 0$, one can find $C(p, \iota) > 0$ fulfilling*

$$\|\nabla c_\varepsilon(\cdot, t)\|_{L^\infty(\Omega)} \leq C \cdot \left\{ 1 + \sup_{\tau \in (0,t)} \|n_\varepsilon(\cdot, \tau)\|_{L^p(\Omega)} \right\}^{\frac{p}{p-1} \cdot \left(\frac{1}{3} + \iota\right)} \tag{4.5.40}$$

for any $t \in (0, T_{\max,\varepsilon})$ and all $\varepsilon \in (0, 1)$.

Proof For given $\iota > 0$, there exists $q > 3$ sufficiently large satisfying $\frac{1}{q} < \iota$, which shows

$$\frac{q+3}{3q} < \frac{1}{3} + \iota.$$

Define

$$\phi(\tilde{\iota}) := \left(\frac{q+3}{2\theta q} + \tilde{\iota} \right) \cdot \left(\frac{2\theta}{3} + \tilde{\iota} \right), \qquad \tilde{\iota} > 0.$$

We can readily see that

$$\phi(\tilde{\iota}) \searrow \frac{q+3}{2\theta q} \cdot \frac{2\theta}{3} = \frac{q+3}{3q} < \frac{1}{3} + \iota \quad \text{as} \quad \tilde{\iota} \searrow 0.$$

This enables us to pick some $\iota'' = \iota''(\iota) > 0$ such that

$$\phi(\iota'') \leq \frac{1}{3} + \iota. \tag{4.5.41}$$

Now, from Lemmas 4.52–4.53, we are able to find certain $C_1 = C_1(p, q, \theta, \iota'') > 0$ fulfilling

$$\left\| \nabla(c_\varepsilon(\cdot, t) - e^{-\tau B} c_0) \right\|_{L^\infty(\Omega)} \leq C_1 I_{p,\varepsilon}^{\frac{p}{p-1} \cdot \left(\frac{2\theta}{3} + \iota''\right) \cdot \left(\frac{q+3}{2\theta q} + \iota''\right)} \tag{4.5.42}$$

for any $t \in (0, T_{\max,\varepsilon})$ and all $\varepsilon \in (0, 1)$. Apart from that, (4.5.35) provides some $C_2 > 0$ satisfying

$$\left\| \nabla e^{-tB} c_0 \right\|_{L^\infty(\Omega)} \leq C_2 \|\nabla c_0\|_{L^\infty(\Omega)}. \tag{4.5.43}$$

Thereupon, it can be deduced from (4.5.25), (4.5.41), (4.5.42) and (4.5.43) that

$$\|\nabla c_\varepsilon(\cdot,t)\|_{L^\infty(\Omega)} \le \left\|\nabla(c_\varepsilon(\cdot,t) - e^{-\tau B}c_0)\right\|_{L^\infty(\Omega)} + \left\|\nabla e^{-\tau B}c_0\right\|_{L^\infty(\Omega)}$$

$$\le C_1 I_{p,\varepsilon}^{\frac{p}{p-1}\cdot\left(\frac{2\theta}{3}+\iota''\right)\cdot\left(\frac{q+3}{2\theta q}+\iota''\right)}(t) + C_2\|\nabla c_0\|_{L^\infty(\Omega)}$$

$$\le C_3 I_{p,\varepsilon}^{\frac{p}{p-1}\cdot\left(\frac{2\theta}{3}+\iota''\right)\cdot\left(\frac{q+3}{2\theta q}+\iota''\right)}(t)$$

$$= C_3 I_{p,\varepsilon}^{\frac{p}{p-1}\cdot\phi(\iota'')}(t)$$

$$\le C_3 I_{p,\varepsilon}^{\frac{p}{p-1}\cdot\left(\frac{1}{3}+\iota\right)}(t)$$

with $C_3 := C_1 + C_2\|\nabla c_0\|_{L^\infty(\Omega)}$ for any $t \in (0, T_{\max,\varepsilon})$ and all $\varepsilon \in (0,1)$, as claimed.

4.5.3 A Prior Estimates

Relying on the basic estimates and the conditional estimates obtained in previous sections, we can achieve the boundedness of $(n_\varepsilon)_{\varepsilon\in(0,1)}$ in temporally independent L^p-topology under a milder assumption on m as compared to that imposed in Liu (2020).

Lemma 4.55 *Let $m > 1$. Then for any $p > 1$ there exists $C = C(p) > 0$ such that*

$$\|n_\varepsilon(\cdot,t)\|_{L^p(\Omega)} \le C \tag{4.5.44}$$

for each $t \in (0, T_{\max,\varepsilon})$ and all $\varepsilon \in (0,1)$. In particular, for $p = m$, one can find $C_* > 0$ fulfilling

$$\int_0^T \int_\Omega n_\varepsilon^{2m-3}|\nabla n_\varepsilon|^2 \le C_*(T+1) \tag{4.5.45}$$

for any $T \in (0, T_{\max,\varepsilon})$.

Proof Thanks to the arbitrariness of $p > 1$, herein without loss of generality, we let

$$p > m. \tag{4.5.46}$$

In addition, since $m > 1$, it is possible to choose $\iota > 0$ sufficiently small such that

$$\lambda := \frac{1+3\iota}{3m-2} < 1. \tag{4.5.47}$$

In view of (4.1.24), (4.5.11), $\nabla \cdot u_\varepsilon = 0$ and the nonnegativity of n_ε and v_ε, we test n_ε-equation in (4.5.7) by pn_ε^{p-1} and invoke Young's inequality to have

$$\frac{d}{dt} \int_\Omega n_\varepsilon^p = -p(p-1) \int_\Omega D_\varepsilon(n_\varepsilon) n_\varepsilon^{p-2} |\nabla n_\varepsilon|^2 + p(p-1) \int_\Omega n_\varepsilon^{p-1} F_\varepsilon'(n_\varepsilon) \nabla n_\varepsilon \cdot \nabla c_\varepsilon$$

$$- p \int_\Omega n_\varepsilon^{p-1} F_\varepsilon(n_\varepsilon) v_\varepsilon$$

$$\leq -C_D p(p-1) \int_\Omega n_\varepsilon^{m+p-3} |\nabla n_\varepsilon|^2 + p(p-1) \int_\Omega n_\varepsilon^{p-1} |\nabla n_\varepsilon| |\nabla c_\varepsilon|$$

$$\leq -\frac{C_D p(p-1)}{2} \int_\Omega n_\varepsilon^{m+p-3} |\nabla n_\varepsilon|^2 + \frac{p(p-1)}{2C_D} \int_\Omega n_\varepsilon^{p-m+1} |\nabla c_\varepsilon|^2$$

$$= -\frac{2C_D p(p-1)}{(m+p-1)^2} \int_\Omega \left| \nabla n_\varepsilon^{\frac{m+p-1}{2}} \right|^2 + \frac{p(p-1)}{2C_D} \int_\Omega n_\varepsilon^{p-m+1} |\nabla c_\varepsilon|^2$$

$$(4.5.48)$$

for each $t \in (0, T_{\max,\varepsilon})$ and all $\varepsilon \in (0, 1)$. For the rightmost integral, we deduce from (4.5.25), (4.5.46) and Lemma 4.54 that

$$\frac{p(p-1)}{2C_D} \int_\Omega n_\varepsilon^{p-m+1} |\nabla c_\varepsilon|^2$$

$$= \frac{p(p-1)}{2C_D} \int_{\{n_\varepsilon \leq 1\}} n_\varepsilon^{p-m+1} |\nabla c_\varepsilon|^2 + \frac{p(p-1)}{2C_D} \int_{\{n_\varepsilon > 1\}} n_\varepsilon^{p-m+1} |\nabla c_\varepsilon|^2$$

$$\leq \frac{p(p-1)|\Omega|}{2C_D} \|\nabla c\|_{L^\infty(\Omega)}^2 + \frac{p(p-1)}{2C_D} \|\nabla c\|_{L^\infty(\Omega)}^2 \int_\Omega n_\varepsilon^{p-m+1}$$

$$\leq \frac{p(p-1)|\Omega|}{2C_D} I_{p,\varepsilon}^{\frac{2p}{p-1} \cdot (\frac{1}{3} + \iota)}(t) + \frac{p(p-1)}{2C_D} I_{p,\varepsilon}^{\frac{2p}{p-1} \cdot (\frac{1}{3} + \iota)}(t) \cdot \int_\Omega n_\varepsilon^{p-m+1}$$

$$(4.5.49)$$

for each $t \in (0, T_{\max,\varepsilon})$ and all $\varepsilon \in (0, 1)$. Since (4.5.46) implies

$$\frac{2}{m+p-1} < \frac{2(p-m+1)}{m+p-1} < 6,$$

with $a := \frac{3(p-m)(m+p-1)}{(p-m+1)(3m+3p-4)} \in (0, 1)$ an application of the Gagliardo–Nirenberg inequality combined with (4.5.15) shows that

$$\int_\Omega n_\varepsilon^{p-m+1}$$

$$= \left\| n_\varepsilon^{\frac{m+p-1}{2}} \right\|_{L^{\frac{2(p-m+1)}{m+p-1}}(\Omega)}^{\frac{2(p-m+1)}{m+p-1}}$$

$$\leq C_1 \left\{ \left\| \nabla n_\varepsilon^{\frac{m+p-1}{2}} \right\|_{L^2(\Omega)}^a \left\| n_\varepsilon^{\frac{m+p-1}{2}} \right\|_{L^{\frac{2}{m+p-1}}(\Omega)}^{1-a} + \left\| n_\varepsilon^{\frac{m+p-1}{2}} \right\|_{L^{\frac{2}{m+p-1}}(\Omega)} \right\}^{\frac{2(p-m+1)}{m+p-1}} \quad (4.5.50)$$

$$\leq C_2 \left\| \nabla n_\varepsilon^{\frac{p+m-1}{2}} \right\|_{L^2(\Omega)}^{2 \cdot \frac{3(p-m)}{3m+3p-4}} + C_2$$

for each $t \in (0, T_{\max,\varepsilon})$ and all $\varepsilon \in (0, 1)$, where both C_1 and C_2 are positive constants. Observing that

$$\frac{3(p-m)}{3m+3p-4} - 1 = \frac{-6m+4}{3m+3p-4} < 0$$

due to $m > 1$, we again employ Young's inequality and derive from (4.5.49) and (4.5.50) that

$$
\begin{aligned}
&\frac{p(p-1)}{2C_D} \int_\Omega n_\varepsilon^{p-m+1} |\nabla c_\varepsilon|^2 \\
&\leq \frac{p(p-1)|\Omega|}{2C_D} I_{p,\varepsilon}^{\frac{2p}{p-1}\cdot(\frac{1}{3}+\iota)}(t) + \frac{p(p-1)C_2}{2C_D} I_{p,\varepsilon}^{\frac{2p}{p-1}\cdot(\frac{1}{3}+\iota)}(t) \\
&\quad + \frac{p(p-1)C_2}{2C_D} I_{p,\varepsilon}^{\frac{2p}{p-1}\cdot(\frac{1}{3}+\iota)}(t) \cdot \left\| \nabla n_\varepsilon^{\frac{p+m-1}{2}} \right\|_{L^2(\Omega)}^{2\cdot\frac{3(p-m)}{3m+3p-4}} \\
&\leq \frac{p(p-1)(|\Omega|+C_2)}{2C_D} I_{p,\varepsilon}^{\frac{2p}{p-1}\cdot(\frac{1}{3}+\iota)}(t) + \frac{C_D p(p-1)}{(m+p-1)^2} \int_\Omega \left| \nabla n_\varepsilon^{\frac{m+p-1}{2}} \right|^2 \\
&\quad + C_3 I_{p,\varepsilon}^{\frac{p}{p-1}\cdot(\frac{1}{3}+\iota)\cdot\frac{3m+3p-4}{3m-2}}(t)
\end{aligned}
$$
(4.5.51)

for each $t \in (0, T_{\max,\varepsilon})$ and all $\varepsilon \in (0, 1)$. In light of the facts that $I_{p,\varepsilon} \geq 1$ and that $I_{p,\varepsilon}$ is nondecreasing with respect to t, it follows from (4.5.48) and (4.5.51) that

$$\frac{d}{dt} \int_\Omega n_\varepsilon^p + \frac{C_D p(p-1)}{(m+p-1)^2} \int_\Omega \left| \nabla n_\varepsilon^{\frac{m+p-1}{2}} \right|^2 \leq C_4 I_{p,\varepsilon}^{\frac{p}{p-1}\cdot(\frac{1}{3}+\iota)\cdot\frac{3m+3p-4}{3m-2}}(t) \quad (4.5.52)$$

with $C_4 := \frac{p(p-1)(|\Omega|+C_2)}{2C_D} + C_3$ for each $t \in (0, T_{\max,\varepsilon})$ and all $\varepsilon \in (0, 1)$. From $m > 1$ and (4.5.46), it is clear that $p > 1$ and $p > \frac{3}{2}(1-m)$, which warrants that

$$\frac{2}{m+p-1} < \frac{2p}{m+p-1} < 6,$$

whence letting $b := \frac{3(p-1)(m+p-1)}{p(3m+3p-4)} \in (0, 1)$, we once more make use of the Gagliardo–Nirenberg inequality to have

$$
\begin{aligned}
&\left\{ \int_\Omega n_\varepsilon^p \right\}^{\frac{3m+3p-4}{3(p-1)}} \\
&= \left\| n_\varepsilon^{\frac{m+p-1}{2}} \right\|_{L^{\frac{2p}{m+p-1}}(\Omega)}^{\frac{2p}{m+p-1}\cdot\frac{3m+3p-4}{3(p-1)}} \\
&\leq C_5 \left\{ \left\| \nabla n_\varepsilon^{\frac{m+p-1}{2}} \right\|_{L^2(\Omega)}^b \left\| n_\varepsilon^{\frac{m+p-1}{2}} \right\|_{L^{\frac{2}{m+p-1}}(\Omega)}^{1-b} + \left\| n_\varepsilon^{\frac{m+p-1}{2}} \right\|_{L^{\frac{2}{m+p-1}}(\Omega)} \right\}^{\frac{2p}{m+p-1}\cdot\frac{3m+3p-4}{3(p-1)}} \\
&\leq C_6 \left\| \nabla n_\varepsilon^{\frac{p+m-1}{2}} \right\|_{L^2(\Omega)}^2 + C_6
\end{aligned}
$$

with $C_5 > 0$ and $C_6 > 0$ for each $t \in (0, T_{\max,\varepsilon})$ and all $\varepsilon \in (0, 1)$, that is

$$\int_\Omega \left| \nabla n_\varepsilon^{\frac{m+p-1}{2}} \right|^2 \geq \frac{1}{C_6} \left\{ \int_\Omega n_\varepsilon^p \right\}^{\frac{3m+3p-4}{3(p-1)}} - 1 \qquad (4.5.53)$$

for each $t \in (0, T_{\max,\varepsilon})$ and all $\varepsilon \in (0, 1)$. Also due to $I_{p,\varepsilon} \geq 1$ and its nondecreasing features, for each fixed $T \in (0, T_{\max,\varepsilon})$, a combination of (4.5.52) with (4.5.53) entails

$$\frac{d}{dt} \int_\Omega n_\varepsilon^p + C_7 \left\{ \int_\Omega n_\varepsilon^p \right\}^{\frac{3m+3p-4}{3(p-1)}} \leq C_8 I_{p,\varepsilon}^{\frac{p}{p-1} \cdot (\frac{1}{3}+\iota) \cdot \frac{3m+3p-4}{3m-2}}(T) \qquad (4.5.54)$$

with $C_7 := \frac{C_D p(p-1)}{C_6(m+p-1)^2}$ and $C_8 := C_4 + \frac{C_D p(p-1)}{(m+p-1)^2}$ for each $t \in (0, T)$ and all $\varepsilon \in (0, 1)$. By means of an ODE comparison argument, we obtain from (4.5.54) that for any fixed $T \in (0, T_{\max,\varepsilon})$

$$\int_\Omega n_\varepsilon^p \leq \max \left\{ \int_\Omega n_0^p, \left\{ \frac{C_8}{C_7} I_{p,\varepsilon}^{\frac{p}{p-1} \cdot (\frac{1}{3}+\iota) \cdot \frac{3m+3p-4}{3m-2}}(T) \right\}^{\frac{3(p-1)}{3m+3p-4}} \right\}$$

for each $t \in (0, T)$ and all $\varepsilon \in (0, 1)$, which further implies

$$\int_\Omega n_\varepsilon^p \leq C_9 \cdot I_{p,\varepsilon}^{p \cdot \frac{1+3\iota}{3m-2}}(T) = C_9 \cdot I_{p,\varepsilon}^{p\lambda}(T) \qquad (4.5.55)$$

for each $t \in (0, T)$ and all $\varepsilon \in (0, 1)$, where $C_9 := \max \left\{ \int_\Omega n_0^p, \left\{ \frac{C_8}{C_7} \right\}^{\frac{3(p-1)}{3m+3p-4}} \right\}$. Recalling (4.5.25), one can infer from (4.5.55) that

$$I_{p,\varepsilon}(T) \leq 1 + C_9^{\frac{1}{p}} I_{p,\varepsilon}^\lambda(T) \leq C_{10} I_{p,\varepsilon}^\lambda(T)$$

with $C_{10} := 1 + C_9^{\frac{1}{p}}$ for each $T \in (0, T_{\max,\varepsilon})$ and all $\varepsilon \in (0, 1)$. In view of (4.5.47), this further shows

$$I_{p,\varepsilon}(T) \leq C_{10}^{\frac{1}{1-\lambda}} \qquad (4.5.56)$$

for each $T \in (0, T_{\max,\varepsilon})$ and all $\varepsilon \in (0, 1)$, and thus (4.5.44) holds. Combining (4.5.56) with (4.5.52) and (4.5.47) entails

$$\frac{d}{dt} \int_\Omega n_\varepsilon^p + \frac{C_D p(p-1)}{(m+p-1)^2} \int_\Omega \left| \nabla n_\varepsilon^{\frac{m+p-1}{2}} \right|^2 \leq C_{11} \qquad (4.5.57)$$

with $C_{11} := C_4 \cdot C_{10}^{\frac{p}{p-1} \cdot (\frac{1}{3}+\iota) \cdot \frac{3m+3p-4}{3m-3-\iota}}$ for any $t \in (0, T_{\max,\varepsilon})$, whence upon an integration of (4.5.57) on $(0, T)$ for each $T \in (0, T_{\max,\varepsilon})$, we have

$$\int_\Omega n_\varepsilon^p(\cdot, T) + \frac{C_D p(p-1)}{(m+p-1)^2} \int_0^T \int_\Omega \left| \nabla n_\varepsilon^{\frac{m+p-1}{2}} \right|^2 \leq C_{11}T + \int_\Omega n_0^p. \quad (4.5.58)$$

Thanks to the nonnegativity of n_ε, $m > 1$ and (4.1.25), we let $p = m$ and derive from (4.5.58) that

$$\int_0^T \int_\Omega n_\varepsilon^{2m-3} |\nabla n_\varepsilon|^2 \leq C_{12}(T+1) \quad (4.5.59)$$

with $C_{12} := \frac{4}{C_D m(m-1)} \max\{C_{11}, \int_\Omega n_0^m\}$ for each $T \in (0, T_{\max,\varepsilon})$, which shows (4.5.45) by choosing $C_* = C_{12}$ and thus completes the proof.

Now, we are able to verify the uniform boundedness for the left-hand side of (4.5.14) so as to establish the global solvability of the approximated problems (4.5.7), which underlies the derivation of global boundedness and stabilization in problem (4.1.16), (4.1.22) and (4.1.23) by means of well-established arguments.

Lemma 4.56 *Let $m > 1$. Then the family of the solutions $(n_\varepsilon, c_\varepsilon, v_\varepsilon, u_\varepsilon)_{\varepsilon \in (0,1)}$ as established in Lemma 4.49 solves (4.5.7) globally and has the properties that for any $r > 3$ and all $t > 0$ there exists $C = C(r) > 0$ independent of $\varepsilon \in (0,1)$ such that*

$$\|n_\varepsilon(\cdot, t)\|_{L^\infty(\Omega)} + \|c_\varepsilon(\cdot, t)\|_{W^{1,r}(\Omega)} + \|v_\varepsilon(\cdot, t)\|_{W^{1,r}(\Omega)} + \|A^\alpha u_\varepsilon(\cdot, t)\|_{L^2(\Omega)} \leq C. \quad (4.5.60)$$

Proof At first, for any $l > 3$ and each $\alpha \in (\frac{3}{4}, 1)$, a combination of Lemma 4.55 with Lemma 4.51 provides some $C_1 > 0$ such that

$$\|u_\varepsilon(\cdot, t)\|_{L^l(\Omega)} + \|A^\alpha u_\varepsilon(\cdot, t)\|_{L^2(\Omega)} \leq C_1 \quad (4.5.61)$$

for all $t \in (0, T_{\max,\varepsilon})$ and $\varepsilon \in (0,1)$. Moreover, from (4.5.16), Lemmas 4.54 and 4.55, we can infer the existence of $C_2 > 0$ fulfilling

$$\|c_\varepsilon(\cdot, t)\|_{W^{1,\infty}(\Omega)} \leq C_2 \quad (4.5.62)$$

for all $t \in (0, T_{\max,\varepsilon})$ and $\varepsilon \in (0,1)$. In conjunction with (4.5.61) and (4.5.62), an application of a Moser-type iteration reasoning (Tao and Winkler 2012a, Lemma A.1) to n_ε-equation in (4.5.7) yields

$$\|n_\varepsilon(\cdot, t)\|_{L^\infty(\Omega)} \leq C_3 \quad (4.5.63)$$

with some $C_3 > 0$ for all $t \in (0, T_{\max,\varepsilon})$ and $\varepsilon \in (0,1)$. Apart from that, for any $r > 3$ (Liu 2020, Lemma 5.1) combined with (4.5.16) allows for a choice of $C_4 > 0$ such that

$$\|v_\varepsilon(\cdot, t)\|_{W^{1,r}(\Omega)} \leq C_4 \quad (4.5.64)$$

for all $t \in (0, T_{\max,\varepsilon})$ and $\varepsilon \in (0,1)$. As a result, a collection of (4.5.61)–(4.5.64) along with (4.5.14) shows the global solvability of (4.5.7) and the validity of (4.5.60).

4.5.4 Global Solvability

The task of this section is to construct global weak solutions of (4.1.16), (4.1.22) and (4.1.23) in the sense of Definition 4.1. As the first step toward this, some further regularity features of $(n_\varepsilon, c_\varepsilon, v_\varepsilon, u_\varepsilon)_{\varepsilon \in (0,1)}$ are essential to be provided.

Lemma 4.57 *There exists $v \in (0, 1)$ with the properties that one can find some ε-independent $C > 0$ fulfilling*

$$\|u_\varepsilon(\cdot, t)\|_{C^v(\bar{\Omega})} \le C \quad \textit{for all} \ \ t \ge 0, \tag{4.5.65}$$

$$\|c_\varepsilon\|_{C^v(\bar{\Omega} \times [t,t+1])} \le C \quad \textit{for all} \ \ t \ge 0 \tag{4.5.66}$$

and

$$\|v_\varepsilon\|_{C^v(\bar{\Omega} \times [t,t+1])} \le C \quad \textit{for all} \ \ t \ge 0, \tag{4.5.67}$$

and that for any $\tau > 0$, there exists ε-independent $C(\tau) > 0$, such that

$$\|\nabla c_\varepsilon\|_{C^v(\bar{\Omega} \times [t,t+1])} \le C(\tau) \quad \textit{for all} \ \ t \ge \tau \tag{4.5.68}$$

and

$$\|\nabla v_\varepsilon\|_{C^v(\bar{\Omega} \times [t,t+1])} \le C(\tau) \quad \textit{for all} \ \ t \ge \tau. \tag{4.5.69}$$

Proof According to the arguments of Liu (2020, Lemmas 5.4–5.6), (4.5.66)–(4.5.69) can be derived from a combination of maximal Sobolev regularity with appropriate embedding consequences, while (4.5.65) is an immediate result of (4.5.60) because of the embedding $D(A^\alpha) \hookrightarrow C^v(\bar{\Omega})$ for each $v \in \left(0, 2\alpha - \frac{3}{2}\right)$ (Giga 1981; Henry 1981), due to $\alpha \in \left(\frac{3}{4}, 1\right)$ required by (4.1.25).

In order to take limit of $(n_\varepsilon)_{\varepsilon \in (0,1)}$ by suitable extraction procedures in the sequel, it is also necessary to explore the regularity properties of time derivatives of $(n_\varepsilon)_{\varepsilon \in (0,1)}$. For expressing conveniently, throughout the sequel, we let

$$B_n := \sup_{\varepsilon \in (0,1)} \|n_\varepsilon\|_{L^\infty(\Omega \times (0,\infty))}. \tag{4.5.70}$$

Lemma 4.58 *Let $m > 1$. Then for each $T > 0$, there exists $C = C(T) > 0$ satisfying*

$$\int_0^T \|\partial_t n_\varepsilon^m(\cdot, t)\|_{(W_0^{1,\infty}(\Omega))^*} dt \le C(T) \quad \textit{for all} \ \ \varepsilon \in (0, 1). \tag{4.5.71}$$

Furthermore, one can find $C > 0$ independent of $\varepsilon \in (0, 1)$, such that

$$\|n_\varepsilon(\cdot, t) - n_\varepsilon(\cdot, s)\|_{(W_0^{2,2}(\Omega))^*} \le C|t - s| \quad \textit{for all} \ \ t \ge 0 \ \ \textit{and} \ \ s \ge 0. \tag{4.5.72}$$

Proof For any fixed $\psi \in C_0^\infty(\bar{\Omega})$ and $t \in (0, T)$, integrations by parts combined with applications of Young's inequality on the basis of the first equation in (4.5.7) entails

$$
\left| \frac{1}{m} \int_\Omega \partial_t n_\varepsilon^m (\cdot, t) \cdot \psi \right|
$$

$$
= \left| \int_\Omega n_\varepsilon^{m-1} \left\{ \nabla \cdot \left(D_\varepsilon(n_\varepsilon) \nabla n_\varepsilon - n_\varepsilon F_\varepsilon'(n_\varepsilon) \nabla c_\varepsilon - n_\varepsilon u_\varepsilon \right) - F_\varepsilon(n_\varepsilon) v_\varepsilon \right\} \cdot \psi \right|
$$

$$
= \left| -(m-1) \int_\Omega n_\varepsilon^{m-2} D_\varepsilon(n_\varepsilon) |\nabla n_\varepsilon|^2 \psi - \int_\Omega n_\varepsilon^{m-1} D_\varepsilon(n_\varepsilon) \nabla n_\varepsilon \cdot \nabla \psi \right.
$$

$$
+ (m-1) \int_\Omega n_\varepsilon^{m-1} F_\varepsilon'(n_\varepsilon) (\nabla n_\varepsilon \cdot \nabla c_\varepsilon) \psi + \int_\Omega n_\varepsilon^m F_\varepsilon'(n_\varepsilon) \nabla c_\varepsilon \cdot \nabla \psi
$$

$$
\left. + \frac{1}{m} \int_\Omega n_\varepsilon^m u_\varepsilon \cdot \nabla \psi - \int_\Omega n_\varepsilon^{m-1} F_\varepsilon(n_\varepsilon) v_\varepsilon \psi \right|
$$

$$
\leq \left\{ C_D(m-1) \int_\Omega n_\varepsilon^{2m-3} |\nabla n_\varepsilon|^2 + C_D \int_\Omega n_\varepsilon^{2m-2} |\nabla n_\varepsilon| \right.
$$

$$
+ (m-1) \int_\Omega n_\varepsilon^{m-1} |\nabla n_\varepsilon| \cdot |\nabla c_\varepsilon| + \int_\Omega n_\varepsilon^m |\nabla c_\varepsilon|
$$

$$
\left. + \frac{1}{m} \int_\Omega n_\varepsilon^m |u_\varepsilon| + \int_\Omega n_\varepsilon^m v_\varepsilon \right\} \cdot \|\psi\|_{W^{1,\infty}(\Omega)}
$$

$$
\leq \left\{ C_D(m-1) \int_\Omega n_\varepsilon^{2m-3} |\nabla n_\varepsilon|^2 + C_D \int_\Omega n_\varepsilon^{2m-3} |\nabla n_\varepsilon|^2 + C_D \int_\Omega n_\varepsilon^2 \right.
$$

$$
+ (m-1) \int_\Omega n_\varepsilon^{2m-3} |\nabla n_\varepsilon|^2 + (m-1) \int_\Omega n_\varepsilon |\nabla c_\varepsilon|^2 + \int_\Omega n_\varepsilon^m |\nabla c_\varepsilon|
$$

$$
\left. + \frac{1}{m} \int_\Omega n_\varepsilon^m |u_\varepsilon| + \int_\Omega n_\varepsilon^m v_\varepsilon \right\} \cdot \|\psi\|_{W^{1,\infty}(\Omega)}
$$

$$
\leq \left\{ (C_D m + m - 1) \int_\Omega n_\varepsilon^{2m-3} |\nabla n_\varepsilon|^2 + C_D B_n^2 |\Omega| + (m-1) B_n \int_\Omega |\nabla c_\varepsilon|^2 \right.
$$

$$
\left. + B_n^m \int_\Omega |\nabla c_\varepsilon| + \frac{B_n^m}{m} \int_\Omega |u_\varepsilon| + B_n^m M_0 |\Omega| \right\} \cdot \|\psi\|_{W^{1,\infty}(\Omega)}
$$

for all $\varepsilon \in (0, 1)$, with C_D and M_0 given by (4.1.24) and (4.5.16), respectively, which thus together with (4.5.45) and (4.5.65) yields (4.5.71). As for (4.5.72), readers can refer to Liu (2020) for its proof. $\qquad \blacksquare$

Now, we are in the position to verify global solvability of (4.1.16), (4.1.22) and (4.1.23).

Lemma 4.59 *Let $m > 1$. Then one can find $(\varepsilon_j)_{j \in \mathbb{N}} \subset (0, 1)$, a null set $\aleph \subset (0, \infty)$ and functions n, c, v and u complying with (4.5.1) and (4.5.2), such that $\varepsilon_j \searrow 0$ as $j \to \infty$, that $n \geq 0$, $c \geq 0$ and $v \geq 0$ in $\Omega \times (0, \infty)$, and that as $\varepsilon = \varepsilon_j \searrow 0$, we have*

$$n_\varepsilon \to n \quad a.e. \text{ in } \Omega \text{ for each } t \in (0, \infty)\backslash\aleph, \tag{4.5.73}$$

$$n_\varepsilon \overset{*}{\rightharpoonup} n \quad \text{in } L^\infty(\Omega \times (0, \infty)), \tag{4.5.74}$$

$$n_\varepsilon \to n \quad \text{in } C^0_{loc}\left([0, \infty); (W^{2,2}_0(\Omega))^*\right), \tag{4.5.75}$$

$$c_\varepsilon \to c \text{ in } C^0_{loc}\left(\bar{\Omega} \times [0, \infty)\right), \tag{4.5.76}$$

$$c_\varepsilon \overset{*}{\rightharpoonup} c \text{ in } L^\infty((0, \infty); W^{1,r}(\Omega)) \text{ for each } r \in (1, \infty), \tag{4.5.77}$$

$$\nabla c_\varepsilon \to \nabla c \text{ in } C^0_{loc}\left(\bar{\Omega} \times [0, \infty)\right), \tag{4.5.78}$$

$$v_\varepsilon \to v \text{ in } C^0_{loc}\left(\bar{\Omega} \times [0, \infty)\right), \tag{4.5.79}$$

$$v_\varepsilon \overset{*}{\rightharpoonup} v \text{ in } L^\infty((0, \infty); W^{1,r}(\Omega)) \text{ for each } r \in (1, \infty), \tag{4.5.80}$$

$$\nabla v_\varepsilon \to \nabla v \text{ in } C^0_{loc}\left(\bar{\Omega} \times [0, \infty)\right), \tag{4.5.81}$$

$$u_\varepsilon \to u \text{ in } C^0_{loc}\left(\bar{\Omega} \times [0, \infty)\right), \tag{4.5.82}$$

$$u_\varepsilon \overset{*}{\rightharpoonup} u \text{ in } L^\infty(\Omega \times (0, \infty)), \tag{4.5.83}$$

and

$$\nabla u_\varepsilon \rightharpoonup \nabla u \text{ in } L^2_{loc}\left(\bar{\Omega} \times [0, \infty)\right). \tag{4.5.84}$$

Furthermore, (n, c, v, u) solves (4.1.16), (4.1.22) and (4.1.23) globally in the sense of Definition 4.1.

Proof Observing that

$$\int_0^T \int_\Omega |\nabla n_\varepsilon^m|^2 = m^2 \int_0^T \int_\Omega n_\varepsilon^{2m-2}|\nabla n_\varepsilon|^2 \leq m^2 B_n \int_0^T \int_\Omega n_\varepsilon^{2m-3}|\nabla n_\varepsilon|^2$$

for each $T > 0$ and all $\varepsilon \in (0, 1)$, we thereby infer from (4.5.60) and (4.5.45) that actually $n_\varepsilon^m \in L^2_{loc}\left([0, \infty); (W^{1,2}(\Omega))\right)$. Thereupon, in line with the reasoning of Liu (2020, Lemma 7.2), the convergence claimed by (4.5.73)–(4.5.84) as well as the integral identities (4.5.3)–(4.5.6) are valid.

4.5.5 Asymptotic Behavior

Recalling (4.5.9) and (4.5.10), one can see that with some sufficiently small $\varepsilon_* \in (0, 1)$ fulfilling

$$B_n \leq \frac{1}{\varepsilon_*} \tag{4.5.85}$$

(4.5.17) can be rewritten as

$$\int_0^\infty \int_\Omega n_\varepsilon c_\varepsilon \le C \quad \text{for all } \varepsilon \in (0, \varepsilon_*),$$

where $C > 0$. This in conjunction with (4.5.18), the convergence of $(n_\varepsilon)_{\varepsilon \in (0,1)}$ and $(v_\varepsilon)_{\varepsilon \in (0,1)}$ in Lemma 4.59 as well as the uniform boundedness property of $(n_\varepsilon)_{\varepsilon \in (0,1)}$ implies the following stability of the spatial average of both n and v. The detailed reasoning thereof can be found in Liu (2020).

Lemma 4.60 *Suppose that* $\aleph \subset (0, \infty)$ *is the null set provided by Lemma 4.59. Then we have*

$$\int_\Omega n(\cdot, t) \to \left\{ \int_\Omega n_0 - \int_\Omega v_0 \right\}_+ \quad as \ (0, \infty) \backslash \aleph \ni t \to \infty \tag{4.5.86}$$

and

$$\int_\Omega v(\cdot, t) \to \left\{ \int_\Omega v_0 - \int_\Omega n_0 \right\}_+ \quad as \ (0, \infty) \backslash \aleph \ni t \to \infty. \tag{4.5.87}$$

Now, we are able to achieve the stability of both v and c as asserted by (4.1.28).

Lemma 4.61 *Both v and c have the properties that*

$$v \to v_\infty \quad in \ W^{1,\infty}(\Omega) \ as \ t \to \infty \tag{4.5.88}$$

and

$$c \to v_\infty \quad in \ W^{1,\infty}(\Omega) \ as \ t \to \infty, \tag{4.5.89}$$

respectively, where $v_\infty = \frac{1}{|\Omega|} \left\{ \int_\Omega v_0 - \int_\Omega n_0 \right\}_+$.

Proof According to the arguments of Liu (2020, Lemmas 8.3–8.4), the convergence (4.5.81) together with (4.5.18) shows the uniform boundedness features of ∇v in $L^2(\Omega \times (0, \infty))$ by Fatou's lemma, which along with the Poincaré inequality, (4.5.87) and the continuity of v implied by (4.5.79) entails the convergence $v \to v_\infty$ as $t \to \infty$ in the topology of $L^2(\Omega)$. In view of the embedding $C^{1+v}(\bar{\Omega}) \hookrightarrow W^{1,\infty}(\Omega) \hookrightarrow L^2(\Omega)$ with the first one being compact, (4.5.88) follows from an Ehrling type interpolation argument relying on the Hölder regularity property of ∇v implied by (4.5.69). With the aid of (4.5.88), the convergence $c \to v_\infty$ as $t \to \infty$ in $L^2(\Omega)$ can be derived from applications of a standard testing procedure along with the dominated convergence theorem to the second equation in (4.5.7) on the basis of (4.5.16), (4.5.76) and (4.5.79), based on which and the Hölder continuity of ∇c implied by (4.5.68), the convergence (4.5.89) is proved to be valid also from an Ehrling type lemma.

For the large time behavior of n, we intend to divide the discussion into two situations, that are $\int_\Omega n_0 \le \int_\Omega v_0$ and $\int_\Omega n_0 > \int_\Omega v_0$, where in the case when $\int_\Omega n_0 >$

$\int_{\Omega} v_0$, a quasi-energy structure which resembles that constructed in Winkler (2018c) is essential to be analyzed for detecting the corresponding stability of n.

Lemma 4.62 With $\aleph \subset (0, \infty)$ as chosen in Lemma 4.59, for $\int_{\Omega} n_0 \leq \int_{\Omega} v_0$, we have

$$n(\cdot, t) \to n_{\infty} \quad in \; L^1(\Omega) \; as \; (0, \infty) \backslash \aleph \ni t \to \infty, \tag{4.5.90}$$

while for $\int_{\Omega} n_0 > \int_{\Omega} v_0$, we have

$$n(\cdot, t) \to n_{\infty} \quad in \; L^2(\Omega) \; as \; (0, \infty) \backslash \aleph \ni t \to \infty, \tag{4.5.91}$$

where $n_{\infty} = \frac{1}{|\Omega|} \left\{ \int_{\Omega} n_0 - \int_{\Omega} v_0 \right\}_{+}$.

Proof If $\int_{\Omega} n_0 \leq \int_{\Omega} v_0$, then clearly $n_{\infty} = 0$, whence (4.5.90) is an immediate consequence of (4.5.85). Whereas, if $\int_{\Omega} n_0 > \int_{\Omega} v_0$, in line with the reasoning of Liu (2020, Lemma 8.6), it is essential to firstly establish an inequality as follows, which shows the quantity $\int_{\Omega} (n_{\varepsilon} - n_{\infty})^2$ remains small during a certain short time, that is for any fixed $t_* \geq 0$

$$\int_{\Omega} \left(n_{\varepsilon}(\cdot, t) - n_{\infty} \right)^2 \leq C_1 \cdot \left\{ \int_{\Omega} \left(n_{\varepsilon}(\cdot, t_*) - n_{\infty} \right)^2 + \int_{\Omega} |\nabla c_{\varepsilon}(\cdot, t_*)|^2 \right.$$
$$\left. + \int_{t_*}^{t} \int_{\Omega} v_{\varepsilon} n_{\varepsilon} + \sup_{s \in (t_*, t_*+1)} \int_{\Omega} |\nabla v_{\varepsilon}(\cdot, s)|^2 \right\} \tag{4.5.92}$$

for all $t \in (t_*, t_* + 1)$ and $\varepsilon \in (0, \varepsilon_*)$ with some $C_1 > 0$ and $\varepsilon_* \in (0, 1)$ satisfying (4.5.85), where $\int_{\Omega} (n_{\varepsilon}(\cdot, t_*) - n_{\infty})^2$ can be verified to be arbitrarily small whenever t_* is sufficiently large. Consequently, along with the decay properties of the last three integrals on the right-hand side of (4.5.92), as claimed by Lemma 4.50, (4.5.91) can be obtained.

Thanks to the bounds of n in $L^{\infty}(\Omega)$ and the continuity implied by (4.5.75), the topologies in which n converges to n_{∞} as $t \to \infty$ as asserted by Lemma 4.62 can be further improved.

Lemma 4.63 each $p \geq 1$,

$$n(\cdot, t) \to n_{\infty} \quad in \; L^p(\Omega) \; as \; t \to \infty \tag{4.5.93}$$

holds.

Proof As performed in the proof of Liu (2020, Corollary 8.7), the topology of the convergence claimed by (4.5.93) can be achieved by drawing on the Hölder inequality on the basis of the boundedness property of n in $L^{\infty}(\Omega)$ as well as the stability of n provided by Lemma 4.62. Moreover, in light of the continuity implied by (4.5.75), the restriction that the convergence should be valid outside null sets of times as required by Lemma 4.62 can be removed. As a result, (4.5.93) follows.

The convergence of n and v in (4.5.93) and (4.5.88), respectively, enables us to derive the large time behavior of u from employing the variation-of-constants formula along with smoothing features of analytic semigroup, as demonstrated in the arguments of Liu (2020, Lemma 8.8).

Lemma 4.64 *For u, we have*

$$u(\cdot, t) \to 0 \quad in \ L^{\infty}(\Omega) \ as \ t \to \infty. \tag{4.5.94}$$

Proof Readers can find the detailed proof in Liu (2020).

Proof of Theorem 4.6. Theorem 4.6 follows from a collection of Lemmas 4.59, 4.61, 4.63 and 4.64.

Chapter 5
Density-Suppressed Motility System

5.1 Introduction

The reaction–diffusion models can reproduce a wide variety of exquisite spatio-temporal patterns arising in embryogenesis, development and population dynamics due to the diffusion-driven (Turing) instability (Kondo and Miura 2010; Murray 2001). Many of them invoke nonlinear diffusion enhanced by the local environment condition to accounting for population pressure (cf. Méndez et al. 2012), volume exclusion (cf. Painter and Hillen 2002; Wang and Hillen 2007) or avoidance of danger (cf. Murray 2001) and so on. However, the opposite situation where the species will slow down its random diffusion rate when encountering external signals such as the predator in pursuit of the prey (Jin and Wang 2021; Kareiva and Odell 1987) and the bacterial searching food (Keller and Segel 1970, 1971b) has not been considered. Recently, a so-called "self-trapping" mechanism was introduced in Liu (2011) by a synthetic biology approach onto programmed bacterial *Escherichia coli* cells which excrete signaling molecules acyl-homoserine lactone (AHL) such that at low AHL levels, the bacteria undergo run-and-tumble random motion and are motile, while at high AHL levels, the bacteria tumble incessantly and become immotile due to the vanishing macroscopic motility. Remarkably, *Escherichia coli* cells formed the outward expanding ring (strip) patterns in the petri dish (Fig. 5.1).

To understand the underlying patterning mechanism, both two-component and three-component "density-suppressed motility" reaction–diffusion systems are proposed. In this chapter, we study the global existence, boundedness, asymptotic behavior of solutions and the existence of traveling wave solutions for density-suppressed motility models. The chapter is divided into two parts. Section 5.3 is devoted to investigate a two-component density-suppressed motility model and shows the existence of traveling wave solutions which are genuine patterns observed in the experiment of Liu (2011), whereas Sect. 5.4 shows the existence and the asymptotic behavior of global weak solution for a three-component quasilinear density-suppressed motility model.

© The Author(s) 2022   275
Y. Ke et al., *Analysis of Reaction-Diffusion Models with the Taxis Mechanism*,
Financial Mathematics and Fintech, https://doi.org/10.1007/978-981-19-3763-7_5

Fig. 5.1 Time-lapsed photographs of spatio-temporal patterns formed by the engineered *Escherichia coli* strain CL3 (see details in Liu 2011). The figure is taken from Fig. 1 in Liu (2011) for illustration

Chemotaxis plays an outstanding role in the life of many cells and microorganisms, such as the transport of embryonic cells to developing tissues and immune cells to infection sites (Isenbach 2004; Murray 2001). The celebrated mathematical model describing chemotactic migration processes at population level is the Keller–Segel system of the form

$$\begin{cases} u_t = \nabla \cdot (\gamma(u,v)\nabla u - u\phi(u,v)\nabla v), & x \in \Omega, t > 0, \\ v_t = d\Delta v - v + u, & x \in \Omega, t > 0, \end{cases} \quad (5.1.1)$$

in a bounded domain $\Omega \subset \mathbb{R}^n$ where $u = u(x,t)$ denotes the population density and $v = v(x,t)$ is the concentration of chemical substance secreted by the population itself (Keller and Segel 1970). The prominent feature of (5.1.1) is the ability of the constitutive ingredient cross-diffusion thereof to describe the collective behavior of cell populations mediated by a chemoattractant. Indeed, a rich literature has revealed that the Neumann initial-boundary value problem for the classical Keller–Segel system

$$\begin{cases} u_t = \Delta u - \nabla \cdot (u\nabla v), & x \in \Omega, t > 0, \\ v_t = d\Delta v - v + u, & x \in \Omega, t > 0 \end{cases} \quad (5.1.2)$$

possesses solutions blowing up in finite time with respect to the spatial L^∞ norm of u in two- and even higher dimensional frameworks under some condition on the mass and the moment of the initial data (Herrero and Velázquez 1996, 1997; Winkler 2013, see also the surveys Bellomo et al. 2016). Apart from that, when ϕ and γ in (5.1.1) are only smooth positive functions of u on $[0,\infty)$, a considerable literature underlines the crucial role of asymptotic beahvior of the ratio $\frac{\gamma(u)}{\phi(u)}$ at large values of u with regard to the occurrence of singularity phenomena (see recent progress in Ishida et al. 2014; Winkler 2017c, 2019e).

As a simplification of (5.1.1), the Keller–Segel system with density-dependent motility

$$\begin{cases} u_t = \nabla \cdot (\gamma(v)\nabla u - u\phi(v)\nabla v), & x \in \Omega, t > 0, \\ v_t = d\Delta v - v + u, & x \in \Omega, t > 0 \end{cases} \quad (5.1.3)$$

was proposed to describe the aggregation phase of Dictyostelium discoideum (Dd) cells in response to the chemical signal cyclic adenosine monophosphate (cAMP) secreted by Dd cells in Keller and Segel (1971b). Here, the signal-dependent diffusivity $\gamma(v)$ and chemotactic sensitivity function $\phi(v)$ are linked through

$$\phi(v) = (\alpha - 1)\gamma'(v),$$

where $\alpha \geq 0$ denotes the ratio of effective body length (i.e., distance between the signal–receptors) to the walk length (see Cai et al. 2022 for details). Notice that when $\alpha = 0$, there is only one receptor in a cell, and hence, chemotaxis is driven by the indirect effect of chemicals in the absence of the chemical gradient sensing. In this case, (5.1.3) reads as

$$\begin{cases} u_t = \Delta(\gamma(v)u), & x \in \Omega, t > 0, \\ v_t = d\Delta v - v + u, & x \in \Omega, t > 0, \end{cases} \tag{5.1.4}$$

where the considered diffusion process of the population is essentially Brownian, and the assumption $\gamma'(v) < 0$ accounts for the repressive effect of the chemical concentration on the population motility (Fu et al. 2012). In the context of acyl-homoserine lactone (AHL) density-dependent motility, the extended model of (5.1.4)

$$\begin{cases} u_t = \Delta(u\gamma(v)) + \beta \dfrac{uw^2}{w^2 + \lambda}, & x \in \Omega, \ t > 0, \\ v_t = D\Delta v + u - v, & x \in \Omega, \ t > 0, \\ w_t = \Delta w - \dfrac{uw^2}{w^2 + \lambda}, & x \in \Omega, \ t > 0 \end{cases} \tag{5.1.5}$$

was proposed in Liu (2011) to advocate that spatio-temporal pattern of Escherichia coli cells can be induced via so-called "self-trapping" mechanisms, that is, at low AHL levels, the bacteria undergo run-and-tumble random motion, while at high AHL levels, the bacteria tumble incessantly and become immotile at the macroscale.

In comparison with plenty of results on the Keller–Segel system where the diffusion depends on the density of cells, the respective knowledge seems to be much less complete when the cell dispersal explicitly depends on the chemical concentration via the motility function $\gamma(v)$, which is due to considerable challenges of the analysis caused by the degeneracy of $\gamma(v)$ as $v \to \infty$ from the mathematical point of view. Indeed, to the best of our knowledge, Yoon and Kim (2017) showed that in the case of $\gamma(v) = \frac{c_0}{v^k}$ for small c_0, problem (5.1.4) admits a global classical solutions in any dimensions. The smallness condition on c_0 is removed lately in Ahn and Yoon (2019) for the parabolic–elliptic version of (5.1.4) with $0 < k < \frac{n}{(n-2)_+}$. Furthermore, for the full parabolic system (5.1.4) in the three-dimensional setting, Tao and Winkler (2017a) showed the existence of certain global weak solutions, which become eventually smooth and bounded for suitably small initial data u_0 under the assumption

(H) $\gamma(v) \in C^3([0, \infty))$, and there exist $\gamma_1, \gamma_2, \eta > 0$ such that $0 < \gamma_1 \leq \gamma(v) \leq \gamma_2$, $|\gamma'(v)| < \eta$ for all $v \geq 0$.

It should be remarked that based on the comparison method, Fujie and Jiang (2021) obtained the uniform-in-time boundedness to (5.1.4) in two-dimensional setting for the more general motility function γ and in the three-dimensional case under a stronger growth condition on $1/\gamma$, respectively. In addition, they investigated the asymptotic behavior to the parabolic–elliptic analog of (5.1.4) under the assumption $\max\limits_{0 \leq v < +\infty} \dfrac{|\gamma'(v)|^2}{\gamma(v)} < +\infty$ or $\gamma(v) = v^{-k}$ with $0 < k < \dfrac{n}{(n-2)_+}$ in Fujie and Jiang (2020) and Jiang and Laurençot (2021).

On the considered time scales of cell migration, e.g., metastatic cells moving in semi-solid medium, often it is relevant to take into account the growth of the population. A prototypical choice to accomplish this is the addition of logistic growth terms $\kappa u - \mu u^2$ in the cell equation (Murray 2001). From the mathematical point of view, the dissipative action of logistic-like growth possibly prevents the occurrence of singularity phenomena in various chemotaxis models. For instance, for the chemotaxis-growth system (Fu et al. 2012)

$$\begin{cases} u_t = \Delta(\gamma(v)u) + u(a - bu), & x \in \Omega, t > 0, \\ v_t = \Delta v - v + u, & x \in \Omega, t > 0, \end{cases} \tag{5.1.6}$$

it is shown in Jin et al. (2018) that in two-dimensional setting, the system admits a unique global classical solution if the motility function $\gamma \in C^3([0, \infty))$ satisfies $\gamma(v) > 0$ and $\gamma'(v) < 0$ for all $v \geq 0$, $\lim_{v \to \infty} \gamma(v) = 0$ and $\lim\limits_{v \to \infty} \dfrac{\gamma'(v)}{\gamma(v)}$, and even the constant steady state $(1, 1)$ is globally asymptotically stable if $a = b > \dfrac{1}{16} \max\limits_{0 \leq v < +\infty} \dfrac{|\gamma'(v)|^2}{\gamma(v)}$. For $a = b$, the global existence thereof in the higher dimensions has been proved for large a and b (Wang and Wang 2019a), while for small a and b, the respective model can generate pattern formation (see Ma et al. 2020). The reader is referred to Lv and Wang (2020, 2021) for the other studies on the related variants involving super-quadratic degradation terms.

As recalled above, the existing results for (5.1.6) are confined to the global well-posedness and asymptotic behaviors of solutions and stationary solutions (pattern formation). However, the traveling wave solutions, which are genuinely relevant to the experiment observation of Liu (2011), are not investigated mathematically except for a special case that $\gamma(v)$ is piecewise constant. When $\gamma(v)$ is a constant, equations of (5.1.6) are decoupled each other and the first equation becomes the well-known Fisher-KPP equation—a benchmark model for the study of traveling wave solutions of reaction–diffusion equations (Murray 2001). However, once $\gamma(v)$ is non-constant, (5.1.6) becomes a coupled system with cross-diffusion, and the study of traveling wave solutions drastically becomes difficult.

The purpose of Sect. 5.3 is to make some progress in this direction and explore the existence of traveling wave solutions to (5.1.6) with allowable wave speeds. With general $\gamma(v)$, the analysis and results will be too complicated to have an elegant presentation. Noticing that the key feature of $\gamma(v)$ lies in the monotone property $\gamma'(v) < 0$, in this Section, we consider a general algebraically decreasing motility function

$$\gamma(v) = \frac{1}{(1+v)^m}, \quad m > 0. \tag{5.1.7}$$

However, our argument can be directly extended to other forms of motility function, such as the exponentially decreasing $\gamma(v) = e^{-\chi v}$ and so on.

To put things in perspective, we rewrite (5.1.6) as

$$\begin{cases} u_t = \nabla \cdot (\gamma(v)\nabla u + u\gamma'(v)\nabla v) + u(a - bu), \\ v_t = \Delta v + u - v, \end{cases} \tag{5.1.8}$$

which is a Keller–Segel-type chemotaxis model proposed in Keller and Segel (1971b) with growth. For the classical chemotaxis-growth system

$$\begin{cases} u_t = \nabla \cdot (\nabla u - \chi u \nabla v) + u(a - bu), \\ \tau v_t = \Delta v + u - v, \end{cases} \tag{5.1.9}$$

traveling wave solutions are investigated in a series of works (Nadin et al. 2008; Salako and Shen 2017a, b, 2018, 2020) for both cases $\tau = 0$ and $\tau = 1$, where $\chi > 0$ denotes the chemotactic coefficient. The existence of traveling wave solutions with minimal wave speed depending on a and χ was obtained, and the asymptotic wave speed as $\chi \to 0$ as well as the spreading speed were examined in detail in Salako and Shen (2017a, b, 2018) and Salako et al. (2019) where the major tool used therein to prove the existence of traveling wave solutions is the parabolic comparison principle. Except traveling wave solutions, the chemotaxis-growth system (5.1.9) can also drive other complex patterning dynamics (cf. Kolokolnikov et al. 2014; Ma et al. 2012; Painter and Hillen 2011). When the volume filling effect is considered in (5.1.9) (i.e., $\chi u \nabla v$ is changed to $\chi u(1 - u)\nabla v$), the traveling wave solutions with minimal wave speed were shown to exist in Ou and Yuan (2009) for small chemotactic coefficient $\chi > 0$. For the original singular Keller–Segel system generating traveling waves without cell growth, we refer to Keller and Segel (1971a), Li et al. (2014), Wang (2013) and references therein. In contrast to the classical chemotaxis-growth system (5.1.9), both diffusive and chemotactic coefficients in the system (5.1.8) are non-constant. This not only makes the analysis more complex, but also makes the parabolic comparison principle inapplicable due to the nonlinear diffusion. In this section, we shall develop some new ideas to tackle the various difficulties induced by the nonlinear motility function $\gamma(v)$ and establish the existence of traveling wave solutions to (5.1.6).

In Sect. 5.3, we shall establish the existence of traveling wave solutions and wave speed of (5.1.6) and explore how the density-suppressed motility influences traveling wave profiles and "the minimal wave speed". In the spatially homogeneous situation, the steady states are $(0, 0)$ and $(a/b, a/b)$, which are, respectively, unstable (saddle point) and stable node. This suggests that we should look for traveling wavefront solutions to (5.1.6) connecting $(a/b, a/b)$ to $(0, 0)$. Moreover, negative u and v have no physical meanings to what we have in mind in the sequel.

A nonnegative solution $(u(x, t), v(x, t))$ is called a traveling wave solution of (5.1.6) connecting $(a/b, a/b)$ to $(0, 0)$ and propagating in the direction $\xi \in S^{N-1}$ with speed c if it is of the form

$$(u(x, t), v(x, t)) = (U(x \cdot \xi - ct), V(x \cdot \xi - ct)) =: (U(z), V(z))$$

satisfying the following equations:

$$\begin{cases} (\gamma(V)U)'' + cU' + U(a - bU) = 0, \\ V'' + cV' + U - V = 0 \end{cases} \tag{5.1.10}$$

and

$$(U(-\infty), V(-\infty)) = (a/b, a/b), \quad (U(+\infty), V(+\infty)) = (0, 0), \tag{5.1.11}$$

where $' = \frac{d}{dz}$. In the first part of this chapter, we proceed to find the constraints on the parameters to exclude the spatio-temporal pattern formation and guarantee the existence of traveling wave solutions connecting the two constant steady states.

Denoting

$$b^*(m, a) = \max\left\{9m, 3m + 2\sqrt{\frac{m(m + 1)a}{1 + a}}\right\}, \tag{5.1.12}$$

we obtain the following two theorems (Li and Wang 2021c).

Theorem 5.1 *Let $\gamma(v)$ be given in (5.1.7). Then for any $c \geq 2\sqrt{a}$ and $b > b^*(m, a)$, the system (5.1.6) has a traveling wave solution $(u(x, t), v(x, t)) = (U(x \cdot \xi - ct), V(x \cdot \xi - ct))$ with speed c in the direction $\xi \in S^{N-1}$ for all $(x, t) \in \mathbb{R}^N \times [0, +\infty)$, satisfying*

$$\lim_{z \to +\infty} \frac{U(z)}{e^{-\lambda z}} = 1, \quad \lim_{z \to +\infty} \frac{V(z)}{e^{-\lambda z}} = \frac{1}{1 + a} \tag{5.1.13}$$

with $\lambda = \frac{c - \sqrt{c^2 - 4a}}{2}$ and

$$\liminf_{z \to -\infty} U(z) > 0 \quad and \quad \liminf_{z \to -\infty} V(z) > 0.$$

Moreover, if

$$\mathcal{K}(m, a) = m \sqrt{\frac{a(1+a)}{m(m+1)}} \left(\sqrt{\frac{a(1+a)}{m(m+1)}} + 1 \right)^m < 1, \tag{5.1.14}$$

we have

$$\lim_{z \to -\infty} U(z) = \lim_{z \to -\infty} V(z) = a/b$$

and

$$\lim_{z \to \pm\infty} U'(z) = \lim_{z \to \pm\infty} V'(z) = 0.$$

Theorem 5.2 *For $c < 2\sqrt{a}$, there is no traveling wave solution $(u(x, t), v(x, t)) = (U(x \cdot \xi - ct), V(x \cdot \xi - ct))$ of (5.1.6) connecting the constant solutions $(a/b, a/b)$ and $(0, 0)$ with speed c.*

Remark 5.1 Theorems 5.1 and 5.2 imply that $c = 2\sqrt{a}$ is the minimal wave speed same as the one for the classical Fisher-KPP equation and irrelevant to the decay rate of the motility function. Different from the Fisher-KPP equation, a lower bound $b^*(m, a)$ for b is induced by the density-suppressed motility. As $m \to 0$, $\gamma(v) \to 1$ and the equation for u becomes the classical Fisher-KPP equation. Noticing

$$\lim_{m \to 0} b^*(m, a) = 0 \text{ and } \lim_{m \to 0} \mathcal{K}(m, a) \to 0,$$

our result well agrees with that for the classical Fisher-KPP equation.

Proof strategies for Theorems 5.1 and 5.2. Since the model (5.1.6) is a cross-diffusion system, see also (5.1.8), many classical tools proving the existence of traveling waves such as phase plane analysis, topological methods and bifurcation analysis (cf. Volpert et al. 1994), among others, become infeasible. Motivated from excellent works of Salako and Shen (2017a, 2018, 2020) for the chemotaxis-growth model (5.1.9) by constructing super- and sub-solutions and proving the existence of traveling wave solutions as the large time limit of solutions in the moving-coordinate system based on the parabolic comparison principle, we plan to achieve our goals in a similar spirit. However, substantial differences exist between the models (5.1.6) and (5.1.9). The nonlinear motility function $\gamma(v)$ in (5.1.6) refrains us from employing the parabolic comparison principle and constructing super- and sub-solutions with the same decay rate at the far field, which are crucial ingredients used for (5.1.9) in Salako and Shen (2017a, 2018). In the first part of this chapter, we develop two innovative ideas to overcome these barriers. First, we introduce an auxiliary parabolic problem (5.3.16) with constant diffusion to which the method of super- and sub-solutions applies (see Sect. 5.3.2). This auxiliary problem subtly bypasses the barriers induced by the nonlinear diffusion but its time-asymptotic limit yields a solution to an elliptic problem (5.3.29) whose fixed points indeed correspond to solutions to (5.1.10)—namely traveling wave solutions to our concerned system (5.1.6) (see Sect. 5.3.3). Second, we construct a sequence of relaxed sub-solution $\underline{U}_n(x)$ for any $n > 1$ with a spatially inhomogeneous decay rate $\theta_1(x)$ which approaches to the constant decay rate of the super-solution $\overline{U}(x)$ as $x \to +\infty$ (see Sect. 5.2). With them, we use the method of super- and sub-solutions to construct solutions to the auxiliary parabolic problem (5.3.16) in appropriate function space and manage to show its time-asymptotic limit problem has a fixed point. This is a fresh idea substantially different from the works (Salako and Shen 2017a, 2018) where the super- and sub-solutions were directly constructed with the same decay rates by taking the advantage of constant diffusion.

We divide the proof of Theorem 5.1 into four steps. In step 1, we construct an auxiliary parabolic problem (5.3.16) with constant diffusion and prove its global boundedness uniformly in time (see Proposition 5.1) by the method of super- and sub-solutions. In step 2, we show that the limit of global solutions to (5.3.16) as $t \to \infty$ yields a semi-wavefront solution to an elliptic problem (5.3.29) with some compactness argument (see Proposition 5.2). In step 3, we show that the solution obtained in step 2 satisfies the boundary condition (5.1.11) by direct estimates under some constraints on m and a (see Proposition 5.3), which hence warrants that the semi-wavefront solution is indeed a wavefront solution in $\mathbb{R}$. Finally, in step 4, we use Schauder's fixed point theorem to prove that (5.3.29) has a fixed point which gives a solution to (5.1.10) in $\mathbb{R}$ satisfying (5.1.11) (see Sect. 5.3.3), where the trick of utilizing relaxed sub-solution $\underline{U}_n(x)$ with spatially inhomogeneous decay rate is critically used to obtain the continuity of the solution map. Theorem 5.2 is proved directly by an argument of contradiction.

Section 5.4 is devoted to the asymptotic behavior of a quasilinear Keller–Segel system with signal-suppressed motility. In the context of the diffusion of cells in a porous medium (see the discussions in Calvez and Carrillo 2006; Vázquez 2007), Winkler (2020) considered the cross-diffusion system

$$\begin{cases} u_t = \Delta(\gamma(v)u^m), & x \in \Omega, t > 0, \\ v_t = \Delta v - v + u, & x \in \Omega, t > 0 \end{cases} \tag{5.1.15}$$

in smoothly bounded convex domains $\Omega \subset \mathbb{R}^n$, where $m > 1$, γ generalizes the prototype $\gamma(v) = a + b(v+d)^{-\alpha}$ with $a \geq 0, b > 0, d \geq 0$ and $\alpha \geq 0$, and proved the boundedness of global weak solutions to the associated initial-boundary value problem under some constriction on m and α, which particularly indicates that increasing m in the cell equation goes along with a certain regularizing effect despite both the diffusion and the cross-diffusion mechanisms implicitly contained in (5.1.15) are simultaneously enhanced.

In a recent paper (Jin et al. 2020), Jin et al. considered the three-component system

$$\begin{cases} u_t = \Delta(\gamma(v)u) + \beta u f(w) - \theta u, & x \in \Omega, \ t > 0, \\ v_t = D\Delta v + u - v, & x \in \Omega, \ t > 0, \\ w_t = \Delta w - u f(w), & x \in \Omega, \ t > 0 \end{cases} \tag{5.1.16}$$

in a bounded domain $\Omega \subset \mathbb{R}^2$, where $\beta, D > 0$ and $\theta \geq 0$, the random motility function $\gamma(v)$ satisfies (H) and functional response function $f(w)$ fulfills the assumption

$$f(w) \in C^1([0, \infty)), \quad f(0) = 0, \quad f(w) > 0 \text{ in } (0, \infty) \text{ and } f'(w) > 0 \text{ on } [0, \infty). \tag{5.1.17}$$

Based on the method of energy estimates and the Moser iteration, they showed the uniform boundedness to initial-boundary value problem of (5.1.16), inter alia the asymptotic behavior thereof when parameter D is suitably large. Note that the authors of Lv and Wang (2022) showed the existence of global classical solutions to system (5.1.16) without the restriction (H) on $\gamma(v)$. In synopsis of the above results, one natural problem seems to consist in determining to which extent nonlinear diffusion of porous medium type may influence the solution behavior in chemotaxis systems involving density-suppressed motility. Accordingly, the purpose of the present work is to address this question in the context of the particular choice $\gamma(v) = v^{-\alpha}$ with $\alpha > 0$ instead of assumption (H) in (5.1.16). Specifically, we consider the asymptotic behavior to the initial-boundary value problem

$$\begin{cases} u_t = \Delta\left(\dfrac{u^m}{v^\alpha}\right) + \beta u f(w), & x \in \Omega, \ t > 0, \\ v_t = D\Delta v + u - v, & x \in \Omega, \ t > 0, \\ w_t = \Delta w - u f(w), & x \in \Omega, \ t > 0 \end{cases} \tag{5.1.18}$$

along with the initial conditions

$$u(x, 0) = u_0, \ v(x, 0) = v_0 \text{ and } w(x, 0) = w_0, \quad x \in \Omega \tag{5.1.19}$$

and under the boundary conditions

$$\frac{\partial u}{\partial v} = \frac{\partial v}{\partial v} = \frac{\partial w}{\partial v} = 0 \text{ on } \partial\Omega \tag{5.1.20}$$

in a bounded convex domain $\Omega \subset \mathbb{R}^2$ with smooth boundary $\partial\Omega$.

In what follows, for simplicity, we shall drop the differential element in the integrals without confusion, namely abbreviating $\int_\Omega f(x)dx$ as $\int_\Omega f$ and $\int_0^t \int_\Omega f(x, \tau) \, dx d\tau$ as $\int_0^t \int_\Omega f(\cdot,)d\tau$ as an important step toward a comprehensive understanding of the effect of nonlinear diffusion on the density-suppressed motility model. Our main result asserts that the weak solutions to the density-suppressed motility system (5.1.18) may approach the relevant homogeneous steady state in the large time limit if D is suitably large, which is stated as follows (Xu and Wang 2021).

Theorem 5.3 *Let $\Omega \subset \mathbb{R}^2$ be a bounded convex domain with smooth boundary, and suppose that $m > 1$, $\alpha > 0$, $\beta > 0$ and f satisfies (5.1.17). Assume that initial data $(u_0, v_0, w_0) \in (W^{1,\infty}(\Omega))^3$ with $u_0 \gneq 0$, $w_0 \gneq 0$ and $v_0 > 0$ in $\overline{\Omega}$. Then problem (5.1.18)–(5.1.20) admits at least one global weak solution (u, v, w) in the sense of Definition 2.1 below. Moreover, there exists constant $D_0 > 0$ such that if $D > D_0$,*

$$\lim_{t\to\infty} \|u(\cdot,t) - u_\star\|_{L^\infty(\Omega)} + \|v(\cdot,t) - u_\star\|_{L^\infty(\Omega)} + \|w(\cdot,t)\|_{L^\infty(\Omega)} = 0 \quad (5.1.21)$$

with $u_\star = \frac{1}{|\Omega|}\int_\Omega u_0 + \frac{\beta}{|\Omega|}\int_\Omega w_0$.

Proof strategies for Theorem 5.3. As the first step to prove the above claim, in Sect. 5.4, we give the definition of a global weak solution to problem (5.1.18)–(5.1.20) and recall that problem (5.1.18)–(5.1.20) with $m > 1$ and $\alpha > 0$ possesses a globally defined weak solution in two-dimensional setting by the approximation procedure (5.2.20). With respect to the convergence properties asserted in (5.1.21), our analysis is essentially different from that of Jin et al. (2020). In fact, thanks to $\gamma_1 \le \gamma(v) \le \gamma_2$ for all $v \ge 0$ in (H), authors of Jin et al. (2020) derived the estimate of $\|u(\cdot,t)\|_{L^2(\Omega)}$, which is the starting point of a priori estimate of $\|u(\cdot,t)\|_{L^\infty(\Omega)}$. In particular, the assumption $\gamma_1 \le \gamma(v)$ plays an essential role in constructing energy function $\mathscr{F}(u,v) := \|u(\cdot,t) - u_*\|_{L^2(\Omega)} + \|v(\cdot,t) - u_*\|_{L^2(\Omega)}$, which leads to the convergence of (u,v) if D is suitable large (see the proofs of Lemma 4.8 and Lemma 4.10 in Jin et al. 2020 for the details). Whereas our asymptotic analysis consists at its core in an analysis of the functional

$$\int_\Omega u^2 + \eta \int_\Omega |\nabla v|^2$$

for solutions of certain regularized versions of (5.1.18), provided that in dependence on the model parameter D, the positive constant η is suitably chosen when D is suitable large. This yields the finiteness of $\int_3^\infty \int_\Omega |\nabla u^{\frac{m+1}{2}}|^2$ and $\int_3^\infty \int_\Omega |\nabla v|^2$ (see Lemma 5.18) and then entails that as a consequence of these integral inequalities, all our solutions asymptotically become homogeneous in space and hence satisfy (5.1.21) (Lemmas 5.19–5.22).

Remark 5.2 (1) Note that as an apparently inherent drawback, assumption (H) in Jin et al. (2020) excludes $\gamma(v)$ decay functions such as $v^{-\alpha}$. Indeed, despite v is bounded below by δ with the help of Lemma 5.4 and thereby the upper bound for $\gamma(v)$ can be removed, an lower bound for $\gamma(v)$ in (H) is essentially required therein.

(2) Due to the results on the existence of global solutions in Winkler (2020), the asymptotic behavior of solutions herein seems to be achieved for the higher dimensional version of (5.1.18) at the cost of additional constraint on m and α.

5.2 Preliminaries

In this section, we first introduce some notations/definitions and list some basic facts which will be used in our subsequent analysis in Sect. 5.3. In particular, the construction of relaxed super and sub-solutions with spatially inhomogeneous decay rates will be presented. For $c \ge 2\sqrt{a}$, define

$$\lambda := \frac{c - \sqrt{c^2 - 4a}}{2} \quad \text{and} \quad \theta_1(x) := \frac{c - \sqrt{c^2 - 4a\left(1 + \frac{e^{-\lambda x}}{1+a}\right)^{-m}}}{2\left(1 + \frac{e^{-\lambda x}}{1+a}\right)^{-m}} \quad \forall x \in \mathbb{R},$$

(5.2.1)

for which

$$\lambda^2 - c\lambda + a = 0, \quad \left(1 + \frac{e^{-\lambda x}}{1+a}\right)^{-m} \theta_1^2(x) - c\theta_1(x) + a = 0 \quad \forall x \in \mathbb{R} \quad (5.2.2)$$

and

$$\lim_{x \to +\infty} \theta_1(x) = \lambda, \quad 0 < \theta_1(x) < \lambda \le \sqrt{a} \quad \forall x \in \mathbb{R}. \quad (5.2.3)$$

Choose

$$\theta_2(x) := \begin{cases} \theta_1(x) + \lambda/4, & c = 2\sqrt{a}, \\ \theta_1(x) + \lambda/k_0, & c > 2\sqrt{a} \end{cases} \quad \forall x \in \mathbb{R} \quad (5.2.4)$$

with $k_0 > \max\left\{\frac{2\lambda}{c-2\lambda}, 2\right\}$. Then

$$\theta_2(x) \in \left(\theta_1(x), \theta_1(x) + \frac{\lambda}{2}\right) \quad \forall x \in \mathbb{R} \quad (5.2.5)$$

and there exists $x_0 \in \mathbb{R}$ such that

$$\theta_2(x) < 2\theta_1(x) \quad \text{for } x > x_0. \quad (5.2.6)$$

Define two functions:

$$\overline{U}(x) := \min\{e^{-\lambda x}, \eta\} \quad \forall x \in \mathbb{R} \quad (5.2.7)$$

and

$$\underline{U}_n(x) := \begin{cases} \delta, & x \le x_\delta, \\ d_n e^{-\theta_1(x)x} + d_0 e^{-\theta_2(x)x}, & x > x_\delta \end{cases} \quad (5.2.8)$$

for $b > b^*(m, a)$ with $b^*(m, a)$ is defined in (5.1.12), where δ is chosen sufficiently small, $x_\delta > 0$ is the unique positive solution of the equation $d_n e^{-\theta_1(x)x} + d_0 e^{-\theta_2(x)x} = \delta$,

Fig. 5.2 A schematic of
functions $\overline{U}(x)$ and $\underline{U}_n(x)$,
where the solid black line
represents $\overline{U}(x)$ and the
dashed red line represents
$\underline{U}_n(x)$

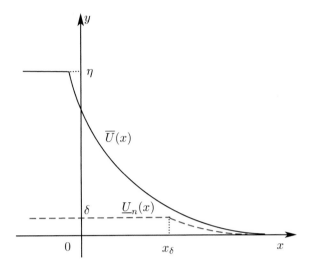

$$d_n := 1 - \frac{1}{n} \quad \text{with } 2 \le n \in \mathbb{N}, \qquad d_0 := \begin{cases} 1, c = 2\sqrt{a}, \\ -1, c > 2\sqrt{a} \end{cases}$$

and

$$\eta := \frac{2a}{b - 3m + \sqrt{(b - 3m)^2 - \frac{4m(m+1)a}{1+a}}}. \tag{5.2.9}$$

Noticing that $d_n \in (0, 1)$ and

$$\lim_{x \to +\infty} e^{(\theta_1(x) - \lambda)x} = 1,$$

which will be verified in Lemma 5.1, we can choose sufficiently small δ, with which
x_δ is large enough such that for all $x \in \mathbb{R}$,

$$0 < \underline{U}_n < \overline{U} \le \eta.$$

We note that the functions $\overline{U}(x)$ and $\underline{U}_n(x)$ will be essentially used later as the super-
and sub-solutions of an auxiliary problem we introduce in Sect. 5.3.2. A schematic
of $\overline{U}(x)$ and $\underline{U}_n(x)$ is plotted in Fig. 5.2. Note that the coefficients d_n ($n \ge 2$) and
d_0 determine the amplitude of $\underline{U}_n(x)$ and $\theta_1(x)$ and determine the decay of $\underline{U}_n(x)$
for large $x > x_\delta$. This is a new ingredient developed in the first part of Chapter 5 to
settle the difficulty of analysis caused by the nonlinear motility function $\gamma(v)$.

Denote

$$C_{\text{unif}}^b(\mathbb{R}) := \{u \in C(\mathbb{R})|\, u \text{ is uniformly continuous in } \mathbb{R} \text{ and } \sup_{z \in \mathbb{R}} |u(z)| < +\infty\},$$

which is equipped with the norm

$$\|u\| = \sup_{z \in \mathbb{R}} |u(z)|.$$

Define the function space

$$\mathscr{E}_n := \{u \in C_{\text{unif}}^b(\mathbb{R})|\underline{U}_n \le u \le \overline{U}\}, \quad \mathscr{X}_0 := \bigcap_{n>1} \mathscr{E}_n.$$

To find solutions of (5.1.10) in $\mathscr{X}_0$, we need the following Lemmas.

Lemma 5.1 *Let λ and $\theta_1(x)$ be defined in (5.2.1). Then it follows that*

$$\lim_{x \to +\infty} e^{(\theta_1(x)-\lambda)x} = 1. \tag{5.2.10}$$

Moreover, for sufficiently small $\delta > 0$, if $x > x_\delta$, then for $c = 2\sqrt{a}$,

$$0 < \theta_1'(x) \le 2K_1 e^{-\frac{\lambda}{2}x} \text{ and } -\lambda K_1 e^{-\frac{\lambda}{2}x} \le \theta_1''(x) < 0 \tag{5.2.11}$$

with $K_1 = \frac{a}{2}\sqrt{\frac{m}{1+a}}$; while for $c > 2\sqrt{a}$,

$$0 < \theta_1'(x) \le 2K_2 e^{-\lambda x} \text{ and } -2\lambda K_2 e^{-\lambda x} \le \theta_1''(x) < 0 \tag{5.2.12}$$

with $K_2 = \dfrac{4a^2 m\lambda}{\left(c+\sqrt{c^2-4a}\right)^2 \sqrt{c^2-4a}(1+a)}$.

Proof In the sequel, for notational simplicity, under $c \ge 2\sqrt{a}$, we introduce the following notations:

$$\begin{cases} \phi(x) := 1 + \dfrac{e^{-\lambda x}}{1+a}, \\ \rho(\phi(x)) := \sqrt{c^2 - 4a\phi^{-m}(x)}, \\ h(\phi(x)) := \dfrac{2a}{c + \rho(\phi(x))} = \dfrac{2a}{c + \sqrt{c^2 - 4a\phi^{-m}(x)}}. \end{cases}$$

Then from the definition of $\theta_1(x)$, we have

$$\theta_1(x) = \frac{c - \sqrt{c^2 - 4a\phi^{-m}(x)}}{2\phi^{-m}(x)} = \frac{2a}{c + \sqrt{c^2 - 4a\phi^{-m}(x)}} = h(\phi(x)).$$

With simple calculation, we find

$$h'(\phi(x)) = \frac{-4a^2m}{\rho(\phi(x))(c + \rho(\phi(x)))^2\phi^{m+1}(x)}, \quad \phi'(x) = \frac{-\lambda e^{-\lambda x}}{1 + a} \qquad (5.2.13)$$

and then

$$\theta_1'(x) = (h(\phi(x)))' = h'(\phi)\phi'(x) = \frac{4a^2m\lambda e^{-\lambda x}}{\rho(\phi(x))(c + \rho(\phi(x)))^2\phi^{m+1}(x)(1 + a)} > 0.$$
$$(5.2.14)$$

When $c = 2\sqrt{a}$, it has that $\lim\limits_{x\to+\infty} \phi(x) = 1$ and $\lim\limits_{x\to+\infty} \rho(\phi(x)) = 0$. By L'Hopital's rule, we have

$$
\begin{aligned}
\lim_{x\to+\infty} \left| \frac{e^{-\frac{\lambda}{2}x}}{\rho(\phi(x))} \right|^2 &= \lim_{x\to+\infty} \frac{e^{-\lambda x}}{c^2 - 4a\phi^{-m}(x)} \\
&= \lim_{x\to+\infty} \frac{e^{-\lambda x}}{4a\phi^m(x) - 4a} \lim_{x\to+\infty} \phi^m(x) \qquad (5.2.15) \\
&= \lim_{x\to+\infty} \frac{1 + a}{4ma\phi^{m-1}(x)} = \frac{1 + a}{4ma}.
\end{aligned}
$$

Then it can be easily verified that

$$\lim_{x\to+\infty} \frac{4a^2m\lambda e^{-\frac{\lambda}{2}x}}{\rho(\phi(x))(c + \rho(\phi(x)))^2\phi^{m+1}(x)(1 + a)} = \frac{a}{2}\sqrt{\frac{m}{1 + a}} := K_1,$$

from which and (5.2.14), by choosing sufficiently small $\delta > 0$, we can find $x_\delta > 0$ such that for $x > x_\delta$, there holds that

$$0 < \theta_1'(x) \le 2K_1 e^{-\frac{\lambda}{2}x}.$$

When $c > 2\sqrt{a}$, it has that $\lim\limits_{x\to+\infty} \phi(x) = 1$ and $\lim\limits_{x\to+\infty} \rho(\phi(x)) = \sqrt{c^2 - 4a}$. It can be directly checked that

$$
\begin{aligned}
&\lim_{x\to+\infty} \frac{4a^2m\lambda}{\rho(\phi(x))(c + \rho(\phi(x)))^2\phi^{m+1}(x)(1 + a)} \\
&= \frac{4a^2m\lambda}{\sqrt{c^2 - 4a}(c + \sqrt{c^2 - 4a})^2(1 + a)} := K_2.
\end{aligned}
$$

Then from (5.2.14), by choosing sufficiently small $\delta > 0$, for $x > x_\delta$, we obtain

$$0 < \theta_1'(x) \le 2K_2 e^{-\lambda x}.$$

The first parts of (5.2.11) and (5.2.12) are proved.

On the other hand, by L'Hôpital's rule, using (5.2.13) and (5.2.15), for $c = 2\sqrt{a}$, we obtain

$$\lim_{x \to +\infty} (h(\phi(x)) - h(1))\, x$$

$$= \lim_{x \to +\infty} h'(\phi(x)) \frac{\lambda x^2 e^{-\lambda x}}{1 + a}$$

$$= \lim_{x \to +\infty} \frac{-4a^2 m \lambda x^2 e^{-\lambda x}}{\rho(\phi(x))(c + \rho(\phi(x)))^2 \phi^{m+1}(x)(1 + a)}$$

$$= -\frac{4a^2 m \lambda}{c^2(1 + a)} \lim_{x \to +\infty} \frac{e^{-\frac{\lambda}{2}x}}{\rho(\phi(x))} \lim_{x \to +\infty} \frac{x^2}{e^{\frac{\lambda}{2}x}}$$

$$= -\frac{am\lambda}{1 + a} \sqrt{\frac{1 + a}{4ma}} \lim_{x \to +\infty} \frac{x^2}{e^{\frac{\lambda}{2}x}} = 0.$$

While for $c > 2\sqrt{a}$, we obtain

$$\lim_{x \to +\infty} (h(\phi(x)) - h(1))\, x = \lim_{x \to +\infty} h'(\phi(x)) \frac{\lambda x^2 e^{-\lambda x}}{1 + a} = \frac{\lambda h'(1)}{1 + a} \lim_{x \to +\infty} \frac{x^2}{e^{\lambda x}} = 0.$$

Summing up, for $c \geq 2\sqrt{a}$, we obtain

$$\lim_{x \to +\infty} e^{(\theta_1(x) - \lambda)x} = e^{\lim_{x \to +\infty} (h(\phi(x)) - h(1))x} = 1$$

and then (5.2.10) follows.

Now, we turn to the estimate of $\theta_1''(x)$. Noticing that $h''(\phi) = h'(\phi)(\ln(-h'(\phi)))'$ and

$$(\ln(-h'(\phi)))' = \left[\ln(4a^2 m) - 2\ln(c + \rho(\phi)) - \ln \rho(\phi) - (m+1)\ln \phi \right]'$$

$$= \frac{-4am}{\rho(\phi)(c + \rho(\phi))\phi^{m+1}} - \frac{2am}{(\rho(\phi))^2 \phi^{m+1}} - \frac{m+1}{\phi},$$

which together with (5.2.13) and the fact that $\phi'(x) = \frac{-\lambda e^{-\lambda x}}{1 + a}$ and $\phi''(x) = \frac{\lambda^2 e^{-\lambda x}}{1 + a}$ implies

$$\theta_1''(x)$$
$$= (h(\phi(x)))'' = (h'(\phi(x))\phi'(x))' = h'(\phi(x))\phi''(x) + h''(\phi(x))(\phi'(x))^2$$
$$= h'(\phi(x))[\phi''(x) + (\ln(-h'(\phi)))'(\phi'(x))^2]$$
$$= \frac{-4a^2 m \lambda^2 e^{-\lambda x}}{\rho(\phi(x))(c + \rho(\phi(x)))^2 \phi(x)^{m+1}(1 + a)}$$
$$\cdot \left\{ 1 - \frac{e^{-\lambda x}}{1 + a} \left(\frac{4am}{\rho(\phi(x))(c + \rho(\phi(x)))\phi(x)^{m+1}} + \frac{2am}{(\rho(\phi(x)))^2 \phi(x)^{m+1}} + \frac{m+1}{\phi(x)} \right) \right\}.$$

For $c = 2\sqrt{a}$, then $\lambda = \sqrt{a}$, from (5.2.15), it can be verified that

$$\lim_{x \to +\infty} \frac{-4a^2 m\lambda^2 e^{-\frac{\lambda}{2}x}}{\rho(\phi(x))(c + \rho(\phi(x)))^2 \phi(x)^{m+1}(1 + a)} = -\lambda K_1$$

and

$$\lim_{x \to +\infty} \frac{e^{-\lambda x}}{1 + a} \left(\frac{4am}{\rho(\phi(x))(c + \rho(\phi(x)))\phi(x)^{m+1}} + \frac{2am}{(\rho(\phi(x)))^2 \phi(x)^{m+1}} + \frac{m+1}{\phi(x)} \right) = \frac{1}{2}.$$

By choosing sufficiently small $\delta > 0$, we can find a $x_\delta > 0$ such that

$$0 > \theta_1''(x) \geq -\lambda K_1 e^{-\frac{\lambda}{2}x}, \text{ for } x > x_\delta.$$

While for $c > 2\sqrt{a}$, we can check that

$$\lim_{x \to +\infty} \frac{-4a^2 m\lambda^2}{\rho(\phi(x))(c + \rho(\phi(x)))^2 \phi(x)^{m+1}(1 + a)} = -\lambda K_2$$

and

$$\lim_{x \to +\infty} \frac{e^{-\lambda x}}{1 + a} \left(\frac{4am}{\rho(\phi(x))(c + \rho(\phi(x)))\phi(x)^{m+1}} + \frac{2am}{(\rho(\phi(x)))^2 \phi(x)^{m+1}} + \frac{m+1}{\phi(x)} \right) = 0.$$

Then by choosing sufficiently small $\delta > 0$ so as to generate a $x_\delta > 0$, we have

$$0 > \theta_1''(x) \geq -2\lambda K_2 e^{-\lambda x}, \text{ for } x > x_\delta.$$

Then the last parts of (5.2.11) and (5.2.12) follow. This completes the proof of Lemma 5.1.

Throughout Sect. 5.4, we shall pursue weak solutions to problem (5.1.18)–(5.1.20) specified as follows.

Definition 5.1 Let $m > 1, \alpha > 0, \beta > 0$ and f satisfies (5.1.17). Then a triple (u, v, w) of nonnegative functions

$$\begin{cases} u \in L^1_{loc}(\overline{\Omega} \times [0, \infty)) \\ v \in L^1_{loc}([0, \infty); W^{1,1}(\Omega)) \\ w \in L^1_{loc}([0, \infty); W^{1,1}(\Omega)) \end{cases}$$

will be called a global weak solution of problem (5.1.18)–(5.1.20) if

$$u^m / v^\alpha \in L^1_{loc}(\overline{\Omega} \times [0, \infty)) \tag{5.2.16}$$

and

$$-\int_0^\infty \int_\Omega u\varphi_t - \int_\Omega u_0\varphi(\cdot,0) = \int_0^\infty \int_\Omega \frac{u^m}{v^\alpha}\Delta\varphi + \beta\int_0^\infty \int_\Omega uf(w)\varphi \quad (5.2.17)$$

for all $\varphi \in C_0^\infty(\overline{\Omega} \times [0,\infty))$ such that $\frac{\partial\varphi}{\partial\nu}|_{\partial\Omega} = 0$ and

$$-\int_0^\infty \int_\Omega v\varphi_t - \int_\Omega v_0\varphi(\cdot,0) = -D\int_0^\infty \int_\Omega \nabla v \cdot \nabla\varphi - \int_0^\infty \int_\Omega v\varphi + \int_0^\infty \int_\Omega u\varphi$$
$$(5.2.18)$$

for all $\varphi \in C_0^\infty(\overline{\Omega} \times [0,\infty))$ as well as

$$\int_0^\infty \int_\Omega w\varphi_t - \int_\Omega w_0\varphi(\cdot,0) = -\int_0^\infty \int_\Omega \nabla w \cdot \nabla\varphi - \int_0^\infty \int_\Omega uf(w)\varphi \quad (5.2.19)$$

for all $\varphi \in C_0^\infty(\overline{\Omega} \times [0,\infty))$.

For $\varepsilon \in (0,1)$, we denote by $(u_\varepsilon, v_\varepsilon, w_\varepsilon)$ the solution of the regularized problem

$$\begin{cases} u_{\varepsilon t} = \varepsilon\Delta(u_\varepsilon + 1)^M + \Delta\left(u_\varepsilon(u_\varepsilon + \varepsilon)^{m-1}v_\varepsilon^{-\alpha}\right) + \beta u_\varepsilon f(w_\varepsilon), & x \in \Omega,\ t > 0, \\ v_{\varepsilon t} = D\Delta v_\varepsilon + u_\varepsilon - v_\varepsilon, & x \in \Omega,\ t > 0, \\ w_{\varepsilon t} = \Delta w_\varepsilon - u_\varepsilon f(w_\varepsilon), & x \in \Omega,\ t > 0, \\ \dfrac{\partial u_\varepsilon}{\partial\nu} = \dfrac{\partial v_\varepsilon}{\partial\nu} = \dfrac{\partial w_\varepsilon}{\partial\nu} = 0, & x \in \partial\Omega,\ t > 0, \\ u_\varepsilon(x,0) = u_0,\ v_\varepsilon(x,0) = v_0,\ w_\varepsilon(x,0) = w_0, & x \in \Omega \end{cases}$$
$$(5.2.20)$$

with $M > m$. Note that due to the a priori boundedness of w_ε, the global smooth solvability of (5.2.20) can be derived by the argument in Lemma 2.4 of Winkler (2020) with evident minor adaptations, and we may refrain from giving the details for brevity here. As for the global weak solutions of (5.1.18)–(5.1.20), we can state as follows.

Lemma 5.2 Let $m > 1, \alpha > 0, \beta > 0$ and f satisfies (5.1.17). Then there exist $(\varepsilon_j)_{j\in\mathbb{N}} \subset (0,1)$ as well as nonnegative functions

$$\begin{cases} u \in L^\infty(\overline{\Omega} \times [0,\infty)) \\ v \in C^0(\overline{\Omega} \times [0,\infty)) \bigcap L_{loc}^2([0,\infty); W^{1,2}(\Omega)) \\ w \in C^0(\overline{\Omega} \times [0,\infty)) \bigcap L_{loc}^2([0,\infty); W^{1,2}(\Omega)) \end{cases} \quad (5.2.21)$$

such that $\varepsilon_j \searrow 0$ as $j \to \infty$ and as $\varepsilon_j \searrow 0$, we have

$$u_\varepsilon \to u \quad a.e. \ in \ \Omega \times (0, \infty), \tag{5.2.22}$$

$$u_\varepsilon \to u \quad in \ \bigcap_{p \geq 1} L^p_{loc}(\overline{\Omega} \times [0, \infty)), \tag{5.2.23}$$

$$v_\varepsilon \to v \quad in \ C^0_{loc}(\overline{\Omega} \times [0, \infty)), \tag{5.2.24}$$

$$w_\varepsilon \to w \quad in \ C^0_{loc}(\overline{\Omega} \times [0, \infty)), \tag{5.2.25}$$

$$\nabla v_\varepsilon \rightharpoonup \nabla v \quad in \ L^2_{loc}(\overline{\Omega} \times [0, \infty)), \tag{5.2.26}$$

$$\nabla w_\varepsilon \rightharpoonup \nabla w \quad in \ L^2_{loc}(\overline{\Omega} \times [0, \infty)). \tag{5.2.27}$$

Moreover, $v > 0$ in $\overline{\Omega} \times (0, \infty)$ and (u, v, w) forms a global weak solution of (5.1.18)–(5.1.20) in the sense of Definition 5.1.

Proof The existence of global weak solutions of (5.1.18)–(5.1.20) can be verified on the basis of straightforward extraction procedures as in Winkler (2020). Indeed, due to the a priori boundedness of w_ε, one can derive some necessary a priori estimation for $(u_\varepsilon, v_\varepsilon, w_\varepsilon)$ such as $\int_t^{t+1} \int_\Omega u_\varepsilon^p$ with all $p < m + 1$, $(v_\varepsilon, w_\varepsilon)$ in $(W^{1,q}(\Omega))^2$ with some $q > 2$ and u_ε in $L^\infty(\Omega)$ and finally apply an Aubin–Lions lemma to obtain a weak solution of (5.1.18)–(5.1.20) with the additional information (5.2.22) (we refer the reader to the proof of Lemma 7.1 in Winkler 2020 for detail).

The following basic properties of the spatial L^1 norms of $(u_\varepsilon, v_\varepsilon, w_\varepsilon)$ as well as the L^∞ norm of w_ε are easily verified.

Lemma 5.3 *Let $(u_\varepsilon, v_\varepsilon, w_\varepsilon)$ be the classical solution of (5.2.20) in $\Omega \times (0, \infty)$. Then we have*

$$\|u_\varepsilon(\cdot, t)\|_{L^1(\Omega)} + \beta \|w_\varepsilon(\cdot, t)\|_{L^1(\Omega)} = \|u_0\|_{L^1(\Omega)} + \beta \|w_0\|_{L^1(\Omega)}, \tag{5.2.28}$$

$$\|u_\varepsilon(\cdot, t)\|_{L^1(\Omega)} \geq \|u_0\|_{L^1(\Omega)}, \tag{5.2.29}$$

$$\int_\Omega v_\varepsilon(\cdot, t) \leq \int_\Omega v_0 + \int_\Omega u_0 + \beta \int_\Omega w_0 \tag{5.2.30}$$

as well as

$$t \mapsto \|w_\varepsilon(\cdot, t)\|_{L^\infty(\Omega)} \ is \ non\text{-}increasing \ in \ [0, \infty). \tag{5.2.31}$$

Proof Multiplying w_ε-equation by β and adding the result to u_ε-equation in (5.2.20), we get

$$\beta \frac{d}{dt} \int_\Omega w_\varepsilon + \frac{d}{dt} \int_\Omega u_\varepsilon = 0, \tag{5.2.32}$$

which immediately yields (5.2.28). An integration of the first equation in (5.2.20) gives us

$$\frac{d}{dt}\int_\Omega u_\varepsilon = \int_\Omega u_\varepsilon f(w_\varepsilon) \geq 0 \qquad (5.2.33)$$

which readily entails (5.2.29). Upon the integration of the second equation in (5.2.20), we can see that

$$\frac{d}{dt}\int_\Omega v_\varepsilon + \int_\Omega v_\varepsilon \leq \int_\Omega u_\varepsilon$$

which, along with (5.2.28), leads to (5.2.30). Due to the fact that f and w_ε are non-negative, the claim in (5.2.31) results upon an application of the maximum principle to w_ε-equation in (5.2.20).

Let us first derive a positive uniform-in-time lower bound for v_ε which will alleviate the difficulties caused by the singularity of signal-dependent motility function $v^{-\alpha}$ near zero. Despite the quantitative lower estimate for solutions of the Neumann problem was established in the related literature (Hillen et al. 2013; Winkler 2020), we present a proof of our results with some necessary details to make the lower bound accessible to the sequel analysis.

Lemma 5.4 *For all $D \geq 1$ and $\varepsilon \in (0, 1)$, there exists $\delta > 0$ such that*

$$v_\varepsilon(x, t) > \delta \quad \text{for all } x \in \Omega \text{ and } t > 0. \qquad (5.2.34)$$

Proof According to the pointwise lower bound estimate for the Neumann heat semi-group $(e^{t\Delta})_{t\geq 0}$ on the convex domain Ω, one can find $C_1(\Omega) > 0$ such that

$$e^{t\Delta}\varphi \geq C_1(\Omega)\int_\Omega \varphi \quad \text{for all } t \geq 1 \text{ and each nonnegative } \varphi \in C^0(\overline{\Omega})$$

(e.g., Fujie 2016; Hillen et al. 2013).

By the time rescaling $\tilde{t} = Dt$, we can see that $\tilde{v}(x, \tilde{t}) := v_\varepsilon(x, \frac{\tilde{t}}{D})$ satisfies

$$\frac{\partial \tilde{v}}{\partial \tilde{t}} = \Delta\tilde{v} - D^{-1}\tilde{v} + D^{-1}u_\varepsilon(x, D^{-1}\tilde{t}). \qquad (5.2.35)$$

Now applying the variation-of-constants formula to (5.2.35), we have

$$\tilde{v}(\cdot, \tilde{t}) = e^{\tilde{t}(\Delta - D^{-1})}v_0(\cdot) + D^{-1}\int_0^{\tilde{t}} e^{(\tilde{t}-s)(\Delta - D^{-1})}u_\varepsilon(\cdot, D^{-1}s)ds \quad t > 0,$$

$$(5.2.36)$$

where by the comparison principle, we can see

$$e^{\tilde{t}(\Delta - D^{-1})}v_0(\cdot) \geq e^{-\tilde{t}D^{-1}}\inf_{x\in\Omega} v_0(x)$$

$$\geq e^{-2}\inf_{x\in\Omega} v_0(x) \quad \text{for all } x \in \Omega, \tilde{t} \leq 2D$$

and

$$D^{-1} \int_0^{\tilde{t}} e^{(\tilde{t}-s)(\Delta-D^{-1})} u_\varepsilon(\cdot, D^{-1}s) ds$$

$$\geq D^{-1} \int_0^{\tilde{t}-1} e^{(\tilde{t}-s)(\Delta-D^{-1})} u_\varepsilon(\cdot, D^{-1}s) ds$$

$$\geq C_1(\Omega) D^{-1} \left(\int_0^{\tilde{t}-1} e^{-D^{-1}(\tilde{t}-s)} ds \right) \inf_{s \in (0,\infty)} \int_\Omega u_\varepsilon(\cdot, s)$$

$$\geq C_1(\Omega)(e^{-D^{-1}} - e^{-D^{-1}\tilde{t}}) \int_\Omega u_0$$

$$\geq \frac{C_1(\Omega)}{2e} \int_\Omega u_0 \quad \text{for all } x \in \Omega \text{ and } \tilde{t} \geq 2D,$$

due to $D \geq 1$. Therefore, inserting above inequalities into (5.2.36), readily establish (5.2.34) with $\delta = \min\{\frac{C_1(\Omega)}{2e} \int_\Omega u_0, e^{-2} \inf_{x \in \Omega} v_0(x)\}$.

Through a straightforward semigroup argument, we formulate a favorable dependence of $\|v_\varepsilon(\cdot, t)\|_{L^p(\Omega)}$ with respect to parameter D.

Lemma 5.5 *For $p > 1$, there exists $C(p) > 0$ such that*

$$\|v_\varepsilon(\cdot, t)\|_{L^p(\Omega)} \leq C(p)(1 + D^{\frac{1}{p}-1}) \quad \text{for all } t > 0. \tag{5.2.37}$$

Proof Applying Duhamel's formula to the equation

$$\frac{\partial \tilde{v}}{\partial \tilde{t}} = \Delta \tilde{v} - D^{-1}\tilde{v} + D^{-1} u_\varepsilon(x, D^{-1}\tilde{t})$$

satisfied by $\tilde{v}(x, \tilde{t}) := v_\varepsilon(x, \frac{\tilde{t}}{D})$ and employing well-known smoothing properties of the Neumann heat semigroup $(e^{t\Delta})_{t \geq 0}$ on Ω (see Lemma 3 of Rothe 1984 or Lemma 1.3 of Winkler 2010 for example), we can find $C_p > 0$ such that for any $\tilde{t} > 0$

$$\|\tilde{v}_\varepsilon(\cdot, \tilde{t})\|_{L^p(\Omega)}$$

$$= \|e^{-D^{-1}\tilde{t}} e^{\tilde{t}\Delta} v_0(\cdot) + D^{-1} \int_0^{\tilde{t}} e^{(\tilde{t}-s)(\Delta-D^{-1})} u_\varepsilon(\cdot, D^{-1}s) ds\|_{L^p(\Omega)}$$

$$\leq e^{-D^{-1}\tilde{t}} \|v_0\|_{L^p(\Omega)} + \frac{C_p}{D} \int_0^{\tilde{t}} e^{-D^{-1}(\tilde{t}-s)}(1 + (\tilde{t}-s)^{-1+\frac{1}{p}}) \|u_\varepsilon(\cdot, D^{-1}s)\|_{L^1(\Omega)} ds$$

$$\leq \|v_0\|_{L^p(\Omega)} + \frac{C_p}{D}(\|u_0\|_{L^1(\Omega)} + \beta\|w_0\|_{L^1(\Omega)}) \int_0^{\tilde{t}} e^{-D^{-1}(\tilde{t}-s)}(1 + (\tilde{t}-s)^{-1+\frac{1}{p}}) ds$$

$$= \|v_0\|_{L^p(\Omega)} + \frac{C_p}{D}(\|u_0\|_{L^1(\Omega)} + \beta\|w_0\|_{L^1(\Omega)}) \int_0^{\tilde{t}} e^{-D^{-1}\sigma}(1 + \sigma^{-1+\frac{1}{p}}) d\sigma$$

$$\leq \|v_0\|_{L^p(\Omega)} + \frac{C_p}{D}(\|u_0\|_{L^1(\Omega)} + \beta\|w_0\|_{L^1(\Omega)})(D + D^{\frac{1}{p}} \int_0^\infty e^{-\sigma} \sigma^{-1+\frac{1}{p}} d\sigma)$$

$$\leq \|v_0\|_{L^p(\Omega)} + (1 + D^{\frac{1}{p}-1}) C_p(\|u_0\|_{L^1(\Omega)} + \beta\|w_0\|_{L^1(\Omega)})(1 + \int_0^\infty e^{-\sigma} \sigma^{-1+\frac{1}{p}} d\sigma)$$

which ends up (5.2.37) with

$$C(p) = \|v_0\|_{L^p(\Omega)} + 2C_p(\|u_0\|_{L^1(\Omega)} + \beta\|w_0\|_{L^1(\Omega)}) \left(1 + \int_0^\infty e^{-\sigma}\sigma^{-1+\frac{1}{p}}d\sigma\right).$$

5.3 Traveling Wave Solutions to a Density-Suppressed Motility Model

5.3.1 Some a Priori Estimates

Lemma 5.6 *For any $u \in \mathscr{E}_n$, denote $V(\cdot; u)$ the solution of*

$$V'' + cV' + u - V = 0. \tag{5.3.1}$$

Then for $c \geq 2\sqrt{a}$, we have

$$0 < V(x; u) \leq \min\left\{\frac{e^{-\lambda x}}{1+a}, \eta\right\}, \tag{5.3.2}$$

$$|V'(x; u)| \leq \min\left\{\frac{2\eta}{\sqrt{c^2+4}}, \frac{2e^{-\lambda x}}{\sqrt{c^2+4}}\right\} \leq \min\left\{\frac{\eta}{\sqrt{1+a}}, \frac{e^{-\lambda x}}{\sqrt{1+a}}\right\} \tag{5.3.3}$$

for all $x \in \mathbb{R}$.

Proof Denote

$$\lambda_1 = \frac{-c - \sqrt{c^2+4}}{2}, \quad \lambda_2 = \frac{-c + \sqrt{c^2+4}}{2}. \tag{5.3.4}$$

From (5.3.4) and the definition of λ in (5.2.1), we obtain

$$0 < \lambda \leq \sqrt{a}, \quad \lambda_1 < 0, \quad \lambda_2 > 0, \quad \lambda_1 + \lambda < 0, \quad \lambda_2 + \lambda > 0 \tag{5.3.5}$$

and

$$\lambda_1\lambda_2 = -1, \quad \lambda_1 + \lambda_2 = -c, \quad \lambda^2 - c\lambda - 1 = -(1+a). \tag{5.3.6}$$

By the variation of constants, the solution of (5.3.1) can be expressed as

$$V(x; u) = \frac{1}{\lambda_2 - \lambda_1}\left(\int_{-\infty}^x e^{\lambda_1(x-s)}u(s)ds + \int_x^{+\infty} e^{\lambda_2(x-s)}u(s)ds\right). \tag{5.3.7}$$

Note that $0 \leq u \leq \overline{U} = \min\{\eta, e^{-\lambda x}\}$ since $u \in \mathscr{E}_n$. Then using (5.3.5) and (5.3.6), we obtain from (5.3.7) that

$$0 \le V(x;u) \le \frac{1}{\lambda_2 - \lambda_1} \left(\int_{-\infty}^{x} e^{\lambda_1(x-s)} e^{-\lambda s} ds + \int_{x}^{+\infty} e^{\lambda_2(x-s)} e^{-\lambda s} ds \right)$$

$$= \frac{1}{\lambda_2 - \lambda_1} \left(\frac{e^{-(\lambda_1+\lambda)s} \big|_{-\infty}^{x}}{-(\lambda_1+\lambda)e^{-\lambda_1 x}} + \frac{e^{-(\lambda_2+\lambda)s} \big|_{x}^{+\infty}}{-(\lambda_2+\lambda)e^{-\lambda_2 x}} \right)$$

$$= \frac{-e^{-\lambda x}}{\lambda^2 - c\lambda - 1} = \frac{e^{-\lambda x}}{1+a}$$

and

$$0 \le V(x;u) \le \frac{1}{\lambda_2 - \lambda_1} \left(\int_{-\infty}^{x} e^{\lambda_1(x-s)} \eta \, ds + \int_{x}^{+\infty} e^{\lambda_2(x-s)} \eta \, ds \right) = \frac{-\eta}{\lambda_1 \lambda_2} = \eta.$$

Thus, the inequality in (5.3.2) follows. On the other hand, differentiating (5.3.7) with respect to x, we have

$$V'(x;u) = \frac{1}{\lambda_2 - \lambda_1} \left(\int_{-\infty}^{x} \lambda_1 e^{\lambda_1(x-s)} u(s) ds + \int_{x}^{+\infty} \lambda_2 e^{\lambda_2(x-s)} u(s) ds \right).$$

For $c \ge 2\sqrt{a}$, using (5.3.5), (5.3.6) and the fact that

$$\lambda_2 - \lambda_1 = \sqrt{c^2 + 4} \ge 2\sqrt{1+a},$$

as well as the fact $0 \le u \le \min\{\eta, e^{-\lambda x}\}$, we obtain with some simple calculations

$$|V'(x;u)| \le \frac{1}{\lambda_2 - \lambda_1} \left(\int_{-\infty}^{x} (-\lambda_1) e^{\lambda_1(x-s)} e^{-\lambda s} ds + \int_{x}^{+\infty} \lambda_2 e^{\lambda_2(x-s)} e^{-\lambda s} ds \right)$$

$$\le \frac{2e^{-\lambda x}}{\sqrt{c^2 + 4}} \le \frac{e^{-\lambda x}}{\sqrt{1+a}}$$

and

$$|V'(x;u)| \le \frac{1}{\lambda_2 - \lambda_1} \left(\int_{-\infty}^{x} (-\lambda_1) e^{\lambda_1(x-s)} \eta \, ds + \int_{x}^{+\infty} \lambda_2 e^{\lambda_2(x-s)} \eta \, ds \right)$$

$$\le \frac{2\eta}{\sqrt{c^2 + 4}} \le \frac{\eta}{\sqrt{1+a}},$$

from which the inequality in (5.3.3) follows. The Lemma is, thus, proved.

Lemma 5.7 *For any $u \in \mathcal{E}_n$, denote $V(x;u)$ the solution of*

$$V'' + cV' + u - V = 0.$$

Then for sufficiently small $\delta > 0$, if $x > x_\delta$, then

$$\gamma(V)\theta_1^2(x) - c\theta_1(x) + a \geq 0, \tag{5.3.8}$$

and

$$\gamma(V)\theta_2^2(x) - c\theta_2(x) + a \geq \frac{a}{64} \quad \text{if } c = 2\sqrt{a}, \tag{5.3.9}$$

$$\gamma(V)\theta_2^2(x) - c\theta_2(x) + a \leq -\frac{\lambda(c - 2\lambda)}{4k_0} \quad \text{if } c > 2\sqrt{a}. \tag{5.3.10}$$

Proof Noticing $V(x) \leq \frac{e^{-\lambda x}}{1+a}$ for $x > x_\delta$, we get (5.3.8) from the fact that

$$\gamma(V)\theta_1^2(x) - c\theta_1(x) + a \geq \left(1 + \frac{e^{-\lambda x}}{1+a}\right)^{-m} \theta_1^2(x) - c\theta_1(x) + a = 0.$$

With

$$\lim_{x \to +\infty} \left(1 + \frac{e^{-\lambda x}}{1+a}\right)^{-m} = 1, \quad \lim_{x \to +\infty} \theta_1(x) = \lambda,$$

by choosing sufficiently small $\delta > 0$, for all $x > x_\delta$, we have

$$\frac{15}{16}\lambda \leq \theta_1(x) < \lambda, \quad \gamma(V) = \frac{1}{(1+V)^m} \geq \left(1 + \frac{e^{-\lambda x}}{1+a}\right)^{-m} \geq \frac{33}{34}. \tag{5.3.11}$$

For the case $c = 2\sqrt{a}$, for which $\lambda = \frac{c}{2}$, noticing $\theta_2(x) = \theta_1(x) + \frac{1}{4}\lambda$, using (5.3.11), we get

$$\gamma(V)\theta_2^2(x) - c\theta_2(x) + a$$
$$= \gamma(V)\theta_1^2(x) - c\theta_1(x) + a + \gamma(V)\left(\frac{1}{16}\lambda^2 + \frac{1}{2}\lambda\theta_1(x)\right) - \frac{1}{4}c\lambda$$
$$\geq \gamma(V)\left(\frac{1}{16}\lambda^2 + \frac{1}{2}\lambda\theta_1(x)\right) - \frac{1}{4}c\lambda$$
$$\geq \frac{33}{34}\left(\frac{1}{16}\lambda^2 + \frac{15}{32}\lambda^2\right) - \frac{1}{2}\lambda^2 = \frac{a}{64}$$

for all $x > x_\delta$, from which (5.3.9) follows.

On the other hand, for the case $c > 2\sqrt{a}$, for which $\lambda < \frac{c}{2}$, noticing

$$\theta_2(x) = \theta_1(x) + \frac{1}{k_0}\lambda \quad \text{with } k_0 > \max\left\{\frac{2\lambda}{c - 2\lambda}, 2\right\},$$

we obtain

$$\frac{1}{k_0}\lambda^2 + 2\lambda\theta_1(x) - c\lambda \leq \frac{1}{2}\lambda(2\lambda - c) < 0$$

and then

$$\gamma(V)\theta_2^2(x) - c\theta_2(x) + a \le \theta_2^2(x) - c\theta_2(x) + a \tag{5.3.12}$$

$$= \theta_1(x)^2 - c\theta_1(x) + a + \frac{1}{k_0}\left(\frac{1}{k_0}\lambda^2 + 2\lambda\theta_1(x) - c\lambda\right)$$

$$\le \theta_1(x)^2 - c\theta_1(x) + a + \frac{1}{2k_0}\lambda(2\lambda - c).$$

Moreover, owing to the fact $\lim_{x\to+\infty}\theta_1(x) = \lambda$, we have

$$\lim_{x\to+\infty}(\theta_1(x)^2 - c\theta_1(x) + a) = \lambda^2 - c\lambda + a = 0.$$

Then choosing δ sufficiently small, we obtain that for all $x > x_\delta$

$$\theta_1(x)^2 - c\theta_1(x) + a \le \frac{1}{4k_0}\lambda(c - 2\lambda). \tag{5.3.13}$$

Inserting (5.3.13) into (5.3.12), we obtain $\gamma(V)\theta_2^2(x) - c\theta_2(x) + a \le \frac{1}{4k_0}\lambda(2\lambda - c) < 0$. Thus, (5.3.10) follows and Lemma 5.7 is proved.

5.3.2 Auxiliary Problems

In this section, we shall investigate some auxiliary problems which act as bridges to our concerned problem.

1. An auxiliary parabolic problem

In the sequel, for convenience, we use $\gamma'(v)$ and $\gamma''(v)$ to denote the first- and second-order derivatives of $\gamma(v)$ with respect to v, respectively. This should not be confused with U', V', U'', V'' where the prime $'$ means the differentiation with respect to x. Given $u \in \mathcal{E}_n$, we first consider the following equation:

$$V'' + cV' + u - V = 0 \tag{5.3.14}$$

which, subject to variation of constants, yields

$$V := V(x; u) = \frac{1}{\lambda_2 - \lambda_1}\left(\int_{-\infty}^{x} e^{\lambda_1(x-s)}u(s)ds + \int_{x}^{+\infty} e^{\lambda_2(x-s)}u(s)ds\right). \tag{5.3.15}$$

Now taking V in (5.3.15) as a known function, we define

$$F(U, U')$$
$$:= \frac{1}{\gamma(V)} \left\{ (2\gamma'(V)V' + c) U' + [\gamma''(V)|V'|^2 + \gamma'(V)(V - U - cV') + a] U - bU^2 \right\}.$$

By $U(x, t; u, \overline{U})$, we denote the solution of the following Cauchy problem:

$$\begin{cases} U_t = U'' + F(U, U'), & x \in \mathbb{R}, t > 0 \\ U(x, 0; u, \overline{U}) = \overline{U}(x), & x \in \mathbb{R}. \end{cases} \tag{5.3.16}$$

From Lemma 5.6 and the definition of $\gamma(\cdot)$, the boundedness of $\frac{1}{\gamma(V)}$, $\gamma'(V)$, $\gamma''(V)$, V and V' has been guaranteed. Then the comparison principle is applicable to (5.3.16). By the semigroup theory, U can be represented as

$$U(x, t; u, \overline{U}) = e^{t(\Delta - 1)}\overline{U}(x) + \int_0^t e^{-(t-s)} e^{(t-s)\Delta}(U + F(U, U'))(x, s) ds.$$

The local existence of solutions to (5.3.16) can be obtained by the well-known fixed point theorem (cf. see Salako and Shen 2017b, Theorem 1.1) along with standard parabolic estimates. We omit the details here for brevity and assume that the solution of (5.3.16) exists in an maximal interval $[0, T)$ for some $T \in (0, \infty]$ with $U(x, 0; u, \overline{U}) > 0$ for $x \in \mathbb{R}$. Then the comparison principle for (5.3.16) implies that $U(x, t; u, \overline{U}) > 0$ for all $(x, t) \in \mathbb{R} \times [0, T)$.

Proposition 5.1 *If $c \geq 2\sqrt{a}$ and $b > b^*(m, a)$ with $b^*(m, a)$ defined in (5.1.12), there exists $\delta > 0$ such that for any $u \in \mathscr{E}_n$, the solution $U(x, t; u, \overline{U})$ of (5.3.16) satisfies $U(\cdot, t; u, \overline{U}) \in \mathscr{E}_n$ for all $t \in [0, +\infty)$.*

Proof Denote

$$L(U) \tag{5.3.17}$$
$$:= \gamma(V)U'' + (2\gamma'(V)V' + c) U' + (\gamma''(V)(V')^2 + \gamma'(V)(V - U - cV') + a) U - bU^2$$

with V defined in (5.3.15). Noticing $\gamma(V) > 0$, we have

$$U'' + F(U, U') = \frac{L(U)}{\gamma(V)}.$$

Hence, a function $U(x)$ is a super-solution (resp. sub-solution) of (5.3.16) if $L(U) \leq 0$ (reps. $L(U) \geq 0$). Firstly, we need to prove that for any solution $u \in \mathscr{E}_n$, there exists $U(x, t; u, \overline{U}) \leq \overline{U}$. For any $s \geq 0$, from the definition of $\gamma(\cdot)$, we have

$$0 < \gamma(s) = \frac{1}{(1+s)^m} \leq 1, \quad -m < \gamma'(s) = -\frac{m}{(1+s)^{m+1}} < 0, \tag{5.3.18}$$

and

$$0 < \gamma''(s) = \frac{m(m+1)}{(1+s)^{m+2}} \le m(m+1). \tag{5.3.19}$$

From (5.3.17), using (5.3.2), (5.3.3), (5.3.18) and (5.3.19), by the definition of η in (5.2.9), it is easy to verify that

$$
\begin{aligned}
L(\eta) &= \left(\gamma''(V)(V')^2 + \gamma'(V)(V - \eta - cV') + a\right)\eta - b\eta^2 \\
&\le \left(\frac{m(m+1)}{1+a}\eta^2 + m\left(1 + \frac{2c}{\sqrt{c^2+4}}\right)\eta + a - b\eta\right)\eta \\
&\le \left(\frac{m(m+1)}{1+a}\eta^2 + 3m\eta + a - b\eta\right)\eta = 0.
\end{aligned}
$$

On the other hand, from (5.3.17), using (5.3.2), (5.3.3), (5.3.18) and (5.3.19), we obtain

$$
\begin{aligned}
&L(e^{-\lambda x}) \\
&= \gamma(V)\lambda^2 e^{-\lambda x} - \left(2\gamma'(V)V' + c\right)\lambda e^{-\lambda x} \\
&\quad + \left(\gamma''(V)(V')^2 e^{-\lambda x} + \gamma'(V)(V - e^{-\lambda x} - cV')\right)e^{-\lambda x} + ae^{-\lambda x} - be^{-2\lambda x} \\
&\le \lambda^2 e^{-\lambda x} + \frac{2m\lambda}{\sqrt{1+a}}e^{-2\lambda x} - c\lambda e^{-\lambda x} + \frac{m(m+1)}{1+a}e^{-4\lambda x} + \left(m + \frac{2cm}{\sqrt{4+c^2}}\right)e^{-2\lambda x} \\
&\quad + ae^{-\lambda x} - be^{-2\lambda x} \\
&\le (\lambda^2 - c\lambda + a)e^{-\lambda x} + \left(\frac{2m\sqrt{a}}{\sqrt{1+a}} + \frac{m(m+1)}{1+a}e^{-2\lambda x} + 3m - b\right)e^{-2\lambda x},
\end{aligned}
$$

where we have used the fact that $\lambda \in (0, \sqrt{a}]$. Noticing

$$b > b^*(m, a) > \frac{2m\sqrt{a}}{\sqrt{1+a}} + 3m,$$

by choosing δ sufficiently small in (5.3.26), we obtain $L(e^{-\lambda x}) \le 0$ for all $x > x_\delta$. By the comparison principle for parabolic equations, it follows that $U(x, t; u, \overline{U}) \le \overline{U}$.

Now we prove that for any $u \in \mathscr{E}_n$, we have $U(x, t; u, \overline{U}) \ge \underline{U}_n$. From (5.3.17), using (5.3.2), (5.3.3), (5.3.18) and (5.3.19), we obtain

$$
\begin{aligned}
L(\delta) &= \gamma''(V)(V')^2\delta + \gamma'(V)(V - \delta - cV')\delta + \delta(a - b\delta) \\
&\ge \gamma'(V)(V - cV')\delta + \delta(a - b\delta) \tag{5.3.20} \\
&\ge \delta\left(a - b\delta - m\eta\left(1 + \frac{2c}{\sqrt{c^2+4}}\right)\right) \\
&\ge \delta\left(a - b\delta - 3m\eta\right).
\end{aligned}
$$

Owing to the fact $b > b^*(m, a)$, we obtain

$$\eta = \frac{2a}{b - 3m + \sqrt{(b - 3m)^2 - \frac{4m(m+1)a}{1+a}}} < \frac{a}{3m}. \tag{5.3.21}$$

Substituting (5.3.21) into (5.3.20), we have $L(\delta) \geq 0$ for sufficiently small δ. On the other hand, using (5.3.17), by direct but tedious calculations, we have

$$
\begin{aligned}
& L(d_n e^{-\theta_1(x)x} + d_0 e^{-\theta_2(x)x}) \\
&= \gamma(V)(d_n e^{-\theta_1(x)x} + d_0 e^{-\theta_2(x)x})'' + (2\gamma'(V)V' + c)(d_n e^{-\theta_1(x)x} + d_0 e^{-\theta_2(x)x})' \\
&\quad + \left(\gamma''(V)(V')^2 + \gamma'(V)(V - (d_n e^{-\theta_1(x)x} + d_0 e^{-\theta_2(x)x}) - cV') + a\right) \\
&\quad \cdot (d_n e^{-\theta_1(x)x} + d_0 e^{-\theta_2(x)x}) - b(d_n e^{-\theta_1(x)x} + d_0 e^{-\theta_2(x)x}))^2 \tag{5.3.22} \\
&\geq \left(\gamma(V)\theta_1^2(x) - c\theta_1(x) + a\right) d_n e^{-\theta_1(x)x} + \left(\gamma(V)\theta_2^2(x) - c\theta_2(x) + a\right) d_0 e^{-\theta_2(x)x} \\
&\quad + d_n e^{-\theta_1(x)x} \left[\gamma(V)\left((\theta_1'(x)x)^2 + 2\theta_1'(x)\theta_1(x)x - \theta_1''(x)x - 2\theta_1'(x)\right) - c\theta_1'(x)x \right. \\
&\quad \left. -2\gamma'(V)V'(\theta_1'(x)x + \theta_1(x)) + \gamma''(V)(V')^2 + \gamma'(V)(V - cV')\right] \\
&\quad + d_0 e^{-\theta_2(x)x} \left[\gamma(V)\left((\theta_2'(x)x)^2 + 2\theta_2'(x)\theta_2(x)x - \theta_2''(x)x - 2\theta_2'(x)\right) - c\theta_2'(x)x \right. \\
&\quad \left. -2\gamma'(V)V'(\theta_2'(x)x + \theta_2(x)) + \gamma''(V)(V')^2 + \gamma'(V)(V - cV')\right] \\
&\quad - b(d_n e^{-\theta_1(x)x} + d_0 e^{-\theta_2(x)x})^2.
\end{aligned}
$$

To prove that $L(d_n e^{-\theta_1(x)x} + d_0 e^{-\theta_2(x)x}) \geq 0$, we consider the cases $c = 2\sqrt{a}$ and $c > 2\sqrt{a}$ separately.

Case 1. $c = 2\sqrt{a}$. In this case, we have $d_0 = 1$ and substitute it into (5.3.22). Using (5.3.18) and Lemmas 5.1 and 5.7, by choosing sufficiently small δ, for $x > x_\delta$, we obtain

$$\gamma'(V) < 0, \quad \gamma''(V) > 0, \quad \theta_1'(x) > 0, \quad \theta_1''(x) < 0$$

and

$$\gamma(V)\theta_1^2(x) - c\theta_1(x) + a \geq 0, \quad \gamma(V)\theta_2^2(x) - c\theta_2(x) + a \geq \frac{a}{64},$$

from which we obtain that for any $x > x_\delta$,

$$
\begin{aligned}
& L(d_n e^{-\theta_1(x)x} + e^{-\theta_2(x)x}) \tag{5.3.23} \\
&\geq \frac{a}{64} e^{-\theta_2(x)x} + d_n e^{-\theta_1(x)x} \\
&\quad \cdot \left[-2\gamma(V)\theta_1'(x) - c\theta_1'(x)x - 2\gamma'(V)V'(\theta_1'(x)x + \theta_1(x)) + \gamma'(V)(V - cV')\right] \\
&\quad + e^{-\theta_2(x)x} \left[-2\gamma(V)\theta_2'(x) - c\theta_2'(x)x - 2\gamma'(V)V'(\theta_2'(x)x + \theta_2(x)) + \gamma'(V)(V - cV')\right] \\
&\quad - b(d_n e^{-\theta_1(x)x} + e^{-\theta_2(x)x})^2.
\end{aligned}
$$

Furthermore, from (5.2.3) and Lemmas 5.1 and 5.6, we have

$$0 < \theta_1(x) < \sqrt{a}, \quad 0 < \theta_1'(x) \le 2K_1 e^{-\frac{\lambda}{2}x}$$

and

$$0 < V(x; u) \le \min\left\{\frac{e^{-\lambda x}}{1+a}, \eta\right\}, \quad |V'(x; u)| \le \min\left\{\frac{\eta}{\sqrt{1+a}}, \frac{e^{-\lambda x}}{\sqrt{1+a}}\right\}.$$

By the above estimates and (5.3.18), we arrive at the following estimates:

$$-2\gamma(V)\theta_1'(x) - c\theta_1'(x)x - 2\gamma'(V)V'(\theta_1'(x)x + \theta_1(x)) + \gamma'(V)(V - cV')$$

$$(5.3.24)$$

$$\ge -\left(4 + 2cx + \frac{4m\eta x}{\sqrt{1+a}}\right)K_1 e^{-\frac{\lambda}{2}x} - \left(\frac{2m\sqrt{a}}{\sqrt{1+a}} + \frac{m}{1+a} + \frac{cm}{\sqrt{1+a}}\right)e^{-\lambda x}$$

$$= -\left(4 + 4\sqrt{a}x + \frac{4m\eta x}{\sqrt{1+a}}\right)K_1 e^{-\frac{\lambda}{2}x} - \left(\frac{4m\sqrt{a}}{\sqrt{1+a}} + \frac{m}{1+a}\right)e^{-\lambda x}.$$

Then from the fact that $\theta_2(x) = \theta_1(x) + \frac{\lambda}{4}$, we get

$$-2\gamma(V)\theta_2'(x) - c\theta_2'(x)x - 2\gamma'(V)V'(\theta_2'(x)x + \theta_2(x)) + \gamma'(V)(V - cV')$$

$$(5.3.25)$$

$$= -2\gamma(V)\theta_1'(x) - c\theta_1'(x)x - 2\gamma'(V)V'(\theta_1'(x)x + \theta_1(x)) + \gamma'(V)(V - cV')$$

$$-\frac{1}{2}\gamma'(V)V'\lambda$$

$$\ge -\left(4 + 4\sqrt{a}x + \frac{4m\eta x}{\sqrt{1+a}}\right)K_1 e^{-\frac{\lambda}{2}x} - \left(\frac{4m\sqrt{a}}{\sqrt{1+a}} + \frac{m}{1+a} + \frac{m\sqrt{a}}{2\sqrt{1+a}}\right)e^{-\lambda x}.$$

Substituting (5.3.24) and (5.3.25) into (5.3.23), we end up with

$$L(d_n e^{-\theta_1(x)x} + e^{-\theta_2(x)x})$$

$$\ge e^{-\theta_2(x)x}\left\{\frac{a}{64} - K_1\left(4 + 4\sqrt{a}x + \frac{4m\eta x}{\sqrt{1+a}}\right)\left(e^{-\frac{\lambda}{2}x} + d_n e^{(\theta_2(x)-\theta_1(x)-\frac{\lambda}{2})x}\right)\right.$$

$$(5.3.26)$$

$$-\left(\frac{4m\sqrt{a}}{\sqrt{1+a}} + \frac{m}{1+a}\right)\left(e^{-\lambda x} + d_n e^{(\theta_2(x)-\theta_1(x)-\lambda)x}\right) - \frac{m\sqrt{a}}{2\sqrt{1+a}}e^{-\lambda x}$$

$$\left. - b\left(d_n^2 e^{(\theta_2(x)-2\theta_1(x))x} + e^{-\theta_2(x)x} + 2d_n e^{-\theta_1(x)x}\right)\right\}.$$

From (5.2.5) and (5.2.6), we have $\theta_2(x) - 2\theta_1(x) < 0$ and $\theta_2(x) - \theta_1(x) - \lambda < \theta_2(x) - \theta_1(x) - \frac{\lambda}{2} < 0$ for $x > x_\delta$, then for $c = 2\sqrt{a}$, by choosing δ sufficiently small in (5.3.26), we obtain

$$L(d_n e^{-\theta_1(x)x} + e^{-\theta_2(x)x}) \geq 0$$

for all $x > x_\delta$.

Case 2. $c > 2\sqrt{a}$. Inserting $d_0 = -1$ in (5.3.22), using Lemmas 5.1, 5.6 and 5.7, we obtain

$$0 < \theta_1(x) < \sqrt{a}, \quad 0 < \theta_1'(x) \leq 2K_2 e^{-\lambda x} \quad 0 > \theta_1''(x) \geq -2\lambda K_2 e^{-\lambda x},$$

$$0 < V(x; u) \leq \min\left\{\frac{e^{-\lambda x}}{1+a}, \eta\right\},$$

$$|V'(x; u)| \leq \min\left\{\frac{2\eta}{\sqrt{c^2+4}}, \frac{2e^{-\lambda x}}{\sqrt{c^2+4}}\right\} \leq \min\left\{\frac{\eta}{\sqrt{1+a}}, \frac{e^{-\lambda x}}{\sqrt{1+a}}\right\},$$

$$\gamma(V)\theta_1^2(x) - c\theta_1(x) + a \geq 0, \quad \gamma(V)\theta_2^2(x) - c\theta_2(x) + a \leq -\frac{\lambda(c-2\lambda)}{4k_0}.$$

By these results, (5.3.18) and (5.3.19), for any $x > x_\delta$, noticing that $\theta_2(x) = \theta_1(x) + \frac{\lambda}{k_0}$, we obtain

$$
\begin{aligned}
&L(d_n e^{-\theta_1(x)x} - e^{-\theta_2(x)x})\\
&\geq \left(\gamma(V)\theta_1^2(x) - c\theta_1(x) + a\right) d_n e^{-\theta_1(x)x} - \left(\gamma(V)\theta_2^2(x) - c\theta_2(x) + a\right) e^{-\theta_2(x)x}\\
&\quad + d_n e^{-\theta_1(x)x}\left[-2\gamma(V)\theta_1'(x) - c\theta_1'(x)x - 2\gamma'(V)V'(\theta_1'(x)x + \theta_1(x)) + \gamma'(V)(V - cV')\right]\\
&\quad - e^{-\theta_2(x)x}\left[\gamma(V)((\theta_2'(x)x)^2 + 2\theta_2'(x)\theta_2(x)x - \theta_2''(x)x) - 2\gamma'(V)V'(\theta_2'(x)x + \theta_2(x))\right.\\
&\quad \left.+\gamma''(V)(V')^2 - c\gamma'(V)V'\right] - b(d_n e^{-\theta_1(x)x} - e^{-\theta_2(x)x})^2 \qquad (5.3.27)\\
&\geq e^{-\theta_2(x)x}\left\{\frac{\lambda(c-2\lambda)}{4k_0} - b\left(d_n^2 e^{(\theta_2(x)-2\theta_1(x))x} + e^{-\theta_2(x)x}\right)\right.\\
&\quad - \left(4K_2 + 2cxK_2 + \frac{2m}{\sqrt{1+a}}(2K_2 e^{-\lambda x}x + \sqrt{a}) + m\left(\frac{1}{1+a} + \frac{2c}{\sqrt{4+c^2}}\right)\right)\\
&\quad \cdot d_n e^{(\theta_2(x)-\theta_1(x)-\lambda)x}\\
&\quad - \left((2K_2x)^2 e^{-\lambda x} + 4K_2\left(\sqrt{a} + \frac{\lambda}{k_0}\right)x + 2\lambda K_2 x + \frac{2m}{\sqrt{1+a}}\left(2K_2 x e^{-\lambda x} + \sqrt{a} + \frac{\lambda}{k_0}\right)\right.\\
&\quad \left.\left.+\frac{m(m+1)}{1+a}e^{-\lambda x} + \frac{2cm}{\sqrt{4+c^2}}\right)e^{-\lambda x}\right\}.
\end{aligned}
$$

Noticing $\theta_2(x) - 2\theta_1(x) < 0$ and $\theta_2(x) - \theta_1(x) - \lambda < 0$ for $x > x_\delta$, then for $c > 2\sqrt{a}$, by choosing δ sufficiently small in (5.3.27), we obtain

$$L(d_n e^{-\theta_1(x)x} - e^{-\theta_2(x)x}) \geq 0$$

for all $x > x_\delta$. Then by the comparison principle for parabolic equations, we obtain $U(x, t; u) \geq \underline{U}_n$ for $c \geq 2\sqrt{a}$.

Summing up, by choosing

$$\eta := \frac{2a}{b - 3m + \sqrt{(b - 3m)^2 - \frac{4m(m+1)a}{1+a}}} \qquad (5.3.28)$$

and sufficiently small δ, $(\overline{U}, \underline{U}_n)$ is a pair of super- and sub-solutions of (5.3.16) (see a schematic of super- and sub-solutions illustrated in Fig. 5.2). Denoting $U(x, t; u, \overline{U})$ the unique solution of (5.3.16), by the comparison principle for parabolic equations, we obtain $\underline{U}_n \leq U(x, t; u, \overline{U}) \leq \overline{U}$ and thus $U(x, t; u, \overline{U}) \in \mathscr{E}_n$. This completes the proof of Proposition 5.1.

2. An auxiliary elliptic problem

Now for $u \in \mathscr{X}_0 := \bigcap_{n>1} \mathscr{E}_n$, we study the following problem:

$$\begin{cases} \gamma(V)U'' + \left(2\gamma'(V)V' + c\right)U' + \left(\gamma''(V)(V')^2 + \gamma'(V)(V - U - cV') + a\right)U \\ \quad - bU^2 = 0, \\ V'' + cV' + u - V = 0, \end{cases}$$

$$(5.3.29)$$

which is equivalent to solving $L(U) = 0$.

Proposition 5.2 *For every $u \in \mathscr{X}_0$, if $c \geq 2\sqrt{a}$ and $b > b^*(m, a)$ with $b^*(m, a)$ defined in (5.1.12), denote $U(x, t; u, \overline{U})$ the solution of (5.3.16) with $U(x, 0; u, \overline{U}) = \overline{U}$, there exists a unique function $U(x; u) \in \mathscr{X}_0$ such that*

$$U(x; u) = \lim_{t \to \infty} U(x, t; u, \overline{U}) = \inf_{t > 0} U(x, t; u, \overline{U})$$

and $U(x; u)$ is the unique solution of (5.3.29) satisfying

$$\liminf_{x \to -\infty} U(x; u) > 0 \quad and \quad \lim_{x \to +\infty} \frac{U(x; u)}{e^{-\lambda x}} = 1. \qquad (5.3.30)$$

Proof From Proposition 5.1, we have

$$U(x, t; u, \overline{U}) \leq \overline{U}(x) \quad \text{for all } (x, t) \in \mathbb{R} \times [0, +\infty). \qquad (5.3.31)$$

For any $0 \leq t_1 \leq t_2$, noticing

$$U(x, t_2; u, \overline{U}) = U(x, t_1; u, U(x, t_2 - t_1; u, \overline{U})),$$

from (5.3.31), we have
$$U(x, t_2 - t_1; u, \overline{U}) \leq \overline{U}(x).$$

Then using again the comparison principle for parabolic equations, we obtain

$$U(x, t_2; u, \overline{U}) \leq U(x, t_1; u, \overline{U}),$$

which implies that $U(x, \cdot; u, \overline{U})$ is decreasing with respect to t. Noticing

$$U(x, \cdot; u, \overline{U})$$

has lower and upper bounds since $U(x, \cdot; u, \overline{U}) \in \mathscr{E}_n$ as shown in Lemma 5.1, one can conclude that there exists a unique $U(x; u)$ such that

$$U(x; u) = \lim_{t \to \infty} U(x, t; u, \overline{U}) = \inf_{t > 0} U(x, t; u, \overline{U}) \qquad (5.3.32)$$

for all $x \in \mathbb{R}$. Denote

$$U_n(x, t) = U(x, t + t_n; u, \overline{U})$$

for $(x, t) \in \mathbb{R} \times [0, \infty)$, where $\{t_n\}_{n \geq 1}$ is an increasing sequence of positive real numbers converging to $+\infty$. Then from the elliptic regularity theory for (5.3.14) and parabolic regularity theory for (5.3.16), we obtain that for all $1 < p < \infty$, $R > 0$, $T > 0$,

$$\|V\|_{W^{2,p}(-R,R)} \leq C \quad \text{and} \quad \|U_n\|_{W_p^{2,1}((-R,R) \times (0,T))} \leq C.$$

From the Sobolev embedding theorem, we obtain

$$\|V\|_{C^{1,\alpha}_{\text{loc}}(\mathbb{R})} \leq C \quad \text{and} \quad \|U_n\|_{C^{\alpha,\alpha/2}_{\text{loc}}(\mathbb{R} \times (0,+\infty))} \leq C.$$

Arzelà–Ascoli's theorem and Schauder's theory for parabolic equation (cf. Krylov 1996) imply that there is a subsequence $\{U_{n'}\}_{n' \geq 1}$ of the sequence $\{U_n\}_{n \geq 1}$ and a function $\tilde{U} \in C^{2,1}(\mathbb{R} \times (0, \infty))$, such that $\{U_{n'}\}_{n' \geq 1}$ converges to $\tilde{U}$ locally uniformly in $C^{2,1}(\mathbb{R} \times (0, \infty))$ as $n' \to \infty$. Hence, $\tilde{U}(x, t)$ solves (5.3.29) and $\tilde{U} \in \mathscr{X}_0$. On the other hand, noticing $\tilde{U}(x, t) = \lim_{t \to \infty} U(x, t; u, \overline{U})$, from (5.3.32), we have $U(x; u) = \tilde{U}(x, t)$ for every $x \in \mathbb{R}$ and $t \geq 0$, from which we obtain that $U(x; u) \in \mathscr{X}_0$ is a solution of (5.3.29). Furthermore, from (5.2.10) and the definition of $\mathscr{X}_0$, we obtain

$$\liminf_{x \to -\infty} U(x; u) > 0$$

and

$$d_n \leq \liminf_{x \to +\infty} \frac{U(x; u)}{e^{-\lambda x}} \leq \limsup_{x \to +\infty} \frac{U(x; u)}{e^{-\lambda x}} = 1 \qquad (5.3.33)$$

for any $n \geq 2$. Noticing $\lim_{n \to \infty} d_n = 1$, by taking $n \to \infty$ in (5.3.33), we obtain

$$\lim_{x \to +\infty} \frac{U(x; u)}{e^{-\lambda x}} = 1.$$

The uniqueness of $U(x; u)$ satisfying (5.3.30) follows from the same arguments as that in Lemma 3.6 in Salako and Shen (2017a). The proof is thus completed.

5.3.3 Minimal Wave Speed

In this section, we shall prove Theorems 5.1 and 5.2. To this end, we first prove the following result concerning the asymptotic behavior of solutions to (5.1.10) as $z \to \pm\infty$.

Proposition 5.3 *Assume that $a > 0$ and $m > 0$ satisfy (5.1.14). Then any solution $(U, V) \in (C^2(\mathbb{R}) \cap \mathscr{X}_0)^2$ to (5.1.10) has the property that*

$$\lim_{z \to +\infty} U(z) = \lim_{z \to +\infty} V(z) = 0, \quad \lim_{z \to -\infty} U(z) = \lim_{z \to -\infty} V(z) = a/b$$

and

$$\lim_{z \to \pm\infty} U'(z) = \lim_{z \to \pm\infty} V'(z) = 0.$$

Proof From the fact that $(U, V) \in \mathscr{X}_0^2$ and Lemma 5.6, we obtain

$$|U(z)| \leq \eta, \quad |V(z)| \leq \eta \quad \text{and} \quad |V'(z)| \leq \frac{\eta}{\sqrt{1+a}} \tag{5.3.34}$$

for all $z \in \mathbb{R}$. From the first equation of (5.1.10), by the Hölder regularity estimates for bounded solutions of elliptic equations and the Schauder theory (Gilbarg and Trudinger 2001), there exists $C > 0$ independent of z and $\alpha \in (0, 1)$ such that $\|U\|_{C^{2,\alpha}(z,z+1)} \leq C$ and $\|V\|_{C^{2,\alpha}(z,z+1)} \leq C$ for all $z \in \mathbb{R}$, from which it follows that

$$|U'(z)| \leq C, \quad |U''(z)| \leq C \quad \text{and} \quad |V''(z)| \leq C \tag{5.3.35}$$

for all $z \in \mathbb{R}$. Multiplying the first equation of (5.1.10) by $(a - bU)$, integrating over $[-R, R]$, we obtain

$$
\begin{aligned}
0 &= \int_{-R}^{R} (\gamma(V)U)''(a - bU)\,dz + c\int_{-R}^{R} U'(a - bU)\,dz + \int_{-R}^{R} U(a - bU)^2 dz \\
&= (\gamma(V)U)'(a - bU)\big|_{z=-R}^{z=R} + b\int_{-R}^{R} (\gamma'(V)V'U + \gamma(V)U')U'\,dz + caU\big|_{z=-R}^{z=R} \\
&\quad - \frac{1}{2}cbU^2\big|_{z=-R}^{z=R} + \int_{-R}^{R} U(a - bU)^2 dz.
\end{aligned}
$$

Then using (5.3.18), (5.3.34) and (5.3.35), we find a constant C_1 independent of R such that

$$\frac{b}{(1+\eta)^m} \int_{-R}^{R} |U'|^2 dz + \int_{-R}^{R} U(a-bU)^2 dz$$

$$\leq b \int_{-R}^{R} \gamma(V)|U'|^2 dz + \int_{-R}^{R} U(a-bU)^2 dz \qquad (5.3.36)$$

$$\leq b \int_{-R}^{R} |\gamma'(V)V'UU'| dz - (\gamma(V)U)'(a-bU)\big|_{z=-R}^{z=R} - caU\big|_{z=-R}^{z=R} + \frac{1}{2}cbU^2\big|_{z=-R}^{z=R}$$

$$\leq C_1 + \frac{1}{2}bm\eta \left(\int_{-R}^{R} |U'|^2 dz + \int_{-R}^{R} |V'|^2 dz \right).$$

On the other hand, multiplying the second equation of (5.1.10) by V'' and integrating the result over $[-R, R]$, we obtain

$$0 = \int_{-R}^{R} |V''|^2 dz + c \int_{-R}^{R} V'V'' dz + \int_{-R}^{R} UV'' dz - \int_{-R}^{R} VV'' dz$$

$$= \int_{-R}^{R} |V''|^2 dz + \frac{c}{2}(V')^2\big|_{z=-R}^{z=R} + UV'\big|_{z=-R}^{z=R} - \int_{-R}^{R} U'V' dz - VV'\big|_{z=-R}^{z=R} + \int_{-R}^{R} |V'|^2 dz.$$

This along with (5.3.34) and (5.3.35) yields

$$\int_{-R}^{R} |V''|^2 dz + \int_{-R}^{R} |V'|^2 dz \leq C_2 + \int_{-R}^{R} U'V' dz \leq C_2 + \frac{1}{2} \int_{-R}^{R} |U'|^2 dz + \frac{1}{2} \int_{-R}^{R} |V'|^2 dz,$$

where C_2 is a constant independent of R. Then it follows that

$$\int_{-R}^{R} |V'|^2 dz \leq 2C_2 + \int_{-R}^{R} |U'|^2 dz. \qquad (5.3.37)$$

Substituting (5.3.37) into (5.3.36), one can find a constant $C_3 = C_1 + bm\eta C_2$ independent of R such that

$$\frac{b}{(1+\eta)^m} \int_{-R}^{R} |U'|^2 dz + \int_{-R}^{R} U(a-bU)^2 dz \leq C_3 + bm\eta \int_{-R}^{R} |U'|^2 dz. \quad (5.3.38)$$

Note that (5.3.28) together with condition (5.1.14) implies

$$\frac{1}{(1+\eta)^m} - m\eta > 0.$$

Sending $R \to \infty$ in (5.3.38), we obtain

$$b \left(\frac{1}{(1+\eta)^m} - m\eta \right) \int_{\mathbb{R}} |U'|^2 dz + \int_{\mathbb{R}} U(a-bU)^2 dz \leq C_3. \qquad (5.3.39)$$

By sending $R \to \infty$ in (5.3.37), we find a constant $C_4 > 0$ such that

$$\int_{\mathbb{R}} |V'|^2 dz \le C_4. \tag{5.3.40}$$

Then (5.3.39) and (5.3.40) assert that

$$U' \in L^2(\mathbb{R}), \quad U(a - bU)^2 \in L^1(\mathbb{R}), \quad V' \in L^2(\mathbb{R}). \tag{5.3.41}$$

From (5.3.35) and (5.3.41), we obtain

$$\lim_{z \to \pm\infty} U(z) \in \{0, a/b\}, \quad \lim_{z \to \pm\infty} U'(z) = 0 \quad \text{and} \quad \lim_{z \to \pm\infty} V'(z) = 0. \tag{5.3.42}$$

Furthermore, from the definition of $\mathscr{X}_0$ and the fact that $U \in \mathscr{X}_0$, we obtain

$$\lim_{z \to +\infty} U(z) = 0 \quad \text{and} \quad \lim_{z \to -\infty} U(z) = a/b.$$

On the other hand, from the second equation of (5.1.10), we have

$$V(z) = \frac{1}{\lambda_2 - \lambda_1} \left(\int_{-\infty}^{z} e^{\lambda_1(z-s)} U(s) ds + \int_{z}^{+\infty} e^{\lambda_2(z-s)} U(s) ds \right) \tag{5.3.43}$$

with $\lambda_1 < 0$ and $\lambda_2 > 0$ defined in (5.3.4). Applying L'Hopital's rule to (5.3.43), from the fact (5.3.42), we obtain

$$
\begin{aligned}
\lim_{z \to +\infty} V(z) &= \lim_{z \to +\infty} \frac{1}{\lambda_2 - \lambda_1} \left(\frac{\int_{-\infty}^{z} e^{-\lambda_1 s} U(s) ds}{e^{-\lambda_1 z}} + \frac{\int_{z}^{+\infty} e^{-\lambda_2 s} U(s) ds}{e^{-\lambda_2 z}} \right) \\
&= \frac{1}{\lambda_2 - \lambda_1} \lim_{z \to +\infty} \left(\frac{U(z)}{-\lambda_1} + \frac{U(z)}{\lambda_2} \right) \\
&= \lim_{z \to +\infty} U(z) = 0
\end{aligned}
$$

and

$$
\begin{aligned}
\lim_{z \to -\infty} V(z) &= \lim_{z \to -\infty} \frac{1}{\lambda_2 - \lambda_1} \left(\frac{\int_{-\infty}^{z} e^{-\lambda_1 s} U(s) ds}{e^{-\lambda_1 z}} + \frac{\int_{z}^{+\infty} e^{-\lambda_2 s} U(s) ds}{e^{-\lambda_2 z}} \right) \\
&= \frac{1}{\lambda_2 - \lambda_1} \lim_{z \to -\infty} \left(\frac{U(z)}{-\lambda_1} + \frac{U(z)}{\lambda_2} \right) \\
&= \lim_{z \to -\infty} U(z) = \frac{a}{b}.
\end{aligned}
$$

This completes the proof.

1. *Proof of Theorem* 5.1. Note that a fixed point of the mapping $u \ni \mathscr{X}_0 \mapsto U(\cdot, u) \in \mathscr{X}_0$ formed in (5.3.29) is a solution to the wave equations (5.1.10). Hence, to prove the existence of traveling wave solutions to (5.1.6), it suffices to prove that

the mapping $u \ni \mathscr{X}_0 \mapsto U(\cdot, u) \in \mathscr{X}_0$ formed in (5.3.29) has a fixed point. We shall achieve this by the Schauder fixed point theorem.

First, we prove that the mapping $u \ni \mathscr{X}_0 \mapsto U(\cdot, u) \in \mathscr{X}_0$ is compact. Let $\{u_n\}_{n \geq 1}$ be a sequence in $\mathscr{X}_0$. Denote $U_n = U(\cdot, u_n)$, we have $U_n \in \mathscr{X}_0$. From the elliptic regularity theorem, we have that $\|U_n\|_{W_{\text{loc}}^{2,p}(\mathbb{R})} \leq C$ for all $p > 1$. From the Sobolev embedding theorem, we obtain $\|U_n\|_{C_{\text{loc}}^{\alpha}(\mathbb{R})} \leq C$, which along with Arzela–Ascoli's theorem implies that there is a subsequence $\{U_{n'}\}_{n' \geq 1}$ of the sequence $\{U_n\}_{n \geq 1}$ and a function $U(x) \in C(\mathbb{R})$, such that $\{U_{n'}\}_{n' \geq 1} \to U(x)$ locally uniformly in $C(\mathbb{R})$. Furthermore, we have $U(x) \in \mathscr{X}_0$. Then the mapping $u \ni \mathscr{X}_0 \mapsto U(\cdot, u) \in \mathscr{X}_0$ is compact.

Second, we prove that the mapping $u \ni \mathscr{X}_0 \mapsto U(\cdot; u) \in \mathscr{X}_0$ is continuous. To this end, denote

$$\|u\|_* = \sum_{n=1}^{\infty} \frac{1}{2^n} \|u\|_{L^{\infty}([-n,n])}.$$

Then any sequence of functions in $\mathscr{X}_0$ is convergent with respect to norm $\|\cdot\|_*$ if and only if it converges locally uniformly on $\mathbb{R}$. Let $u \in \mathscr{X}_0$ and $\{u_n\}_{n \geq 1}$ be a sequence in $\mathscr{X}_0$ such that u_n converges to u locally uniformly on $\mathbb{R}$ as $n \to \infty$. Then by the elliptic regularity theorem applied to the second equation of (5.3.29) and the Sobolev embedding theorem, we obtain

$$\|V(\cdot; u_n)\|_{C_{\text{loc}}^{1,\alpha}(\mathbb{R})} \leq C.$$

From Arzelà–Ascoli's theorem, there exists a subsequence of $\{V(\cdot; u_n)\}_{n \geq 1}$, still denoted by itself without confusion, such that

$$\lim_{n' \to \infty} V(\cdot; u_n) = V(\cdot; u) \quad \text{in } C_{\text{loc}}^1(\mathbb{R}).$$

Suppose by contradiction that the mapping $u \ni \mathscr{X}_0 \mapsto U(\cdot; u) \in \mathscr{X}_0$ is not continuous, then there exist $\delta > 0$ and a subsequence $\{u_{n'}\}_{n' \geq 1}$ such that

$$\|U(\cdot; u_{n'}) - U(\cdot; u)\|_* \geq \delta, \quad \forall n \geq 1. \tag{5.3.44}$$

By Schauder's theory (Krylov 1996) applied to the first equation of (5.3.29) and the Sobolev embedding theorem, from Arzelà–Ascoli's theorem, there is a subsequence $\{U(\cdot; u_{n''})\}_{n'' \geq 1}$ of the sequence $\{U(\cdot; u_{n'})\}_{n' \geq 1}$ and a function $U(\cdot) \in C^2(\mathbb{R})$, such that $\{U(\cdot; u_{n''})\}_{n'' \geq 1}$ converges to $U(\cdot)$ in $C_{\text{loc}}^2(\mathbb{R})$ and U is a solution of (5.3.29). Moreover, from the fact that $U(\cdot; u_{n''}) \in \mathscr{X}_0$ and

$$\lim_{n \to \infty} \|U(\cdot; u_{n''}) - U(\cdot)\|_* = 0,$$

we obtain $U(\cdot) \in \mathscr{X}_0$. Then from Proposition 5.2, we obtain $U(\cdot) = U(\cdot, u)$. By (5.3.44), then

$$\|U(\cdot; u) - U(\cdot)\|_* \geq \delta,$$

which is a contradiction. Hence, the mapping $u \ni \mathscr{X}_0 \mapsto U(\cdot; u) \in \mathscr{X}_0$ is continuous.

Now by Schauder's fixed point theorem, there is $U \in \mathscr{X}_0$ such that $U(\cdot) = U(\cdot; U)$. Denote $V(\cdot) := V(\cdot; U)$. Then (U, V) is a solution of (5.1.10). From the definition of $\mathscr{X}_0$ and (5.2.10), we obtain

$$\lim_{z \to +\infty} \frac{U(z)}{e^{-\lambda z}} = 1.$$

This along with (5.3.5)–(5.3.6) and L'Hôpital's Rule yields

$$\lim_{z \to +\infty} \frac{V(z)}{e^{-\lambda z}} = \lim_{z \to +\infty} \frac{1}{\lambda_2 - \lambda_1} \left(\frac{\int_{-\infty}^{z} e^{-\lambda_1 s} U(s) ds}{e^{-(\lambda_1 + \lambda)z}} + \frac{\int_{z}^{+\infty} e^{-\lambda_2 s} U(s) ds}{e^{-(\lambda_2 + \lambda)z}} \right)$$

$$= \frac{1}{\lambda_2 - \lambda_1} \lim_{z \to +\infty} \left(\frac{U(z)}{-(\lambda_1 + \lambda)e^{-\lambda z}} - \frac{U(z)}{-(\lambda_2 + \lambda)e^{-\lambda z}} \right) = \frac{1}{1 + a}.$$

Since $U \in \mathscr{X}_0$, it follows that $\liminf_{z \to -\infty} U(z) > 0$. On the other hand, noticing for $z < x_\delta$, $U(z) > \delta$ and then

$$V(z) = \frac{1}{\lambda_2 - \lambda_1} \left(\int_{-\infty}^{z} e^{\lambda_1(z-s)} U(s) ds + \int_{z}^{+\infty} e^{\lambda_2(z-s)} U(s) ds \right)$$

$$\geq \frac{\delta}{\lambda_2 - \lambda_1} \int_{-\infty}^{z} e^{\lambda_1(z-s)} ds = \frac{\delta}{(\lambda_2 - \lambda_1)(-\lambda_1)} > 0,$$

from which $\liminf_{z \to -\infty} V(z) > 0$ follows. Finally, by the assumption (5.1.14) and Proposition 5.3, we finish the proof of Theorem 5.1.

2. *Proof of Theorem 5.2.* Arguing by contradiction, for $c < 2\sqrt{a}$, we suppose that there is a traveling wave solution $(u(x, t), v(x, t)) = (U(x \cdot \xi - ct), V(x \cdot \xi - ct))$ of (5.1.6) connecting the constant solutions $(a/b, a/b)$ and $(0, 0)$. Take a sequence $\{z_n\}$ with $z_n \to +\infty$, then

$$\lim_{n \to +\infty} U(z_n) = \lim_{n \to +\infty} V(z_n) = \lim_{n \to +\infty} V'(z_n) = 0.$$

Now we set

$$h_n(z) = \frac{U(z + z_n)}{U(z_n)}, \quad U_n(z) = U(z + z_n), \quad V_n(z) = V(z + z_n).$$

As U is bounded and satisfies (5.1.10), the Harnack inequality implies that the shifted functions $U_n(z)$, $V_n(z)$ and $V_n'(z)$ converge to zero locally uniformly in z and the sequence h_n is locally uniformly bounded and satisfies

$$\begin{cases} \gamma''(V_n)(V_n')^2 h_n + \gamma'(V_n)(V_n - U_n - cV_n')h_n + 2\gamma'(V_n)V_n'h_n' + \gamma(V_n)h_n'' \\ \quad + ch_n' + h_n(a - bU_n) = 0, \\ V_n'' + U_n - V_n + cV_n' = 0 \end{cases}$$

in $\mathbb{R}$. Thus, up to a subsequence, the sequence $\{h_n\}_{n \geq 1}$ converges to a function h that satisfies

$$h'' + ch' + ah = 0 \quad \text{in } \mathbb{R}. \tag{5.3.45}$$

Moreover, h is nonnegative and $h(0) = 1$. Equation (5.3.45) admits such a solution if and only if $c \geq 2\sqrt{a}$, which leads to a contradiction. This denies our assumption and hence (5.1.6) admits no traveling wave solution connecting $(a/b, a/b)$ and $(0, 0)$ with speed $c < 2\sqrt{a}$.

5.3.4 Selection of Wave Profiles

By introducing some auxiliary problems and spatially inhomogeneous relaxed decay rates for super- and sub-solutions constructed, we manage to establish the existence of traveling wavefront solutions to the density-suppressed motility system (5.1.6) with decay motility function (5.1.7), where we find that there is a minimal wave speed coincident with the one for the cornerstone Fisher-KPP equation and a maximum wave speed c resulting from the nonlinear diffusion. However, we are unable to characterize further properties of wave profiles such as monotonicity, stability and so on. In this section, we shall discuss the selection of possible wave profiles motivated by some argument in Ou and Yuan (2009).

1. Trailing edge wave profiles
In the spatially homogeneous situation, the system (5.1.6) has equilibria $(0, 0)$ and $(a/b, a/b)$, which are unstable saddle and stable node, respectively. This suggests that we should look for traveling wavefront solutions to (5.1.6) connecting $(a/b, a/b)$ to $(0, 0)$ as we have done. Now we linearize the ODE system (5.1.10) at the origin $(0, 0)$ and let $U' = X$, $V' = Y$. Then we get the following linear system of (U, X, V, Y):

$$\begin{pmatrix} U' \\ X' \\ V' \\ Y' \end{pmatrix} = \begin{pmatrix} 0 & 1 & 0 & 0 \\ -\frac{a}{\gamma(0)} & -\frac{c}{\gamma(0)} & 0 & 0 \\ 0 & 0 & 0 & 1 \\ -1 & 0 & 1 & -c \end{pmatrix} \begin{pmatrix} U \\ X \\ V \\ Y \end{pmatrix}.$$

The eigenvalue λ of the above coefficient matrix is

$$\left(\lambda^2 + \frac{c}{\gamma(0)}\lambda + \frac{a}{\gamma(0)} \right)\left(\lambda^2 + c\lambda - 1 \right) = 0.$$

To ensure there is a positive trajectory connecting the equilibria $(0, 0)$ and $(a/b, a/b)$, we need to rule out the case that $(0, 0)$ is a spiral, which amounts to require

$$c \geq 2\sqrt{\gamma(0)a}. \tag{5.3.46}$$

With $\gamma(v)$ given in (5.1.7), $\gamma(0) = 1$ and (5.3.46) is equivalent to $c \geq 2\sqrt{a}$. This is well consistent with our results obtained in Theorems 5.1 and 5.2. Under the restriction (5.3.46), it can be easily checked that the origin $(0, 0)$ is either a stable node or saddle point, which indicates that the traveling wave profile around the origin $(0, 0)$ will not be oscillatory or periodic.

Next, we linearize the system (5.1.10) at $(a/b, a/b)$ and arrive at the following linearized system:

$$\begin{pmatrix} U' \\ X' \\ V' \\ Y' \end{pmatrix} = \begin{pmatrix} 0 & 1 & 0 & 0 \\ \frac{a(b+\sigma_2)}{\sigma_1 b} & -\frac{c}{\sigma_1} & -\frac{a\sigma_2}{b\sigma_1} & \frac{a\sigma_2 c}{b\sigma_1} \\ 0 & 0 & 0 & 1 \\ -1 & 0 & 1 & -c \end{pmatrix} \begin{pmatrix} U \\ X \\ V \\ Y \end{pmatrix} \tag{5.3.47}$$

where $\sigma_1 = \gamma(a/b)$ and $\sigma_2 = \gamma'(a/b)$. By some tedious computation, we find that the eigenvalue λ of the above coefficient matrix is determined by the following characteristic equation:

$$\lambda^4 + \left(c + \frac{c}{\sigma_1}\right)\lambda^3 + \left(\frac{c^2}{\sigma_1} - \frac{a(b+\sigma_2)}{\sigma_1 b} - 1\right)\lambda^2 - \frac{(a+1)c}{\sigma_1}\lambda + \frac{a}{\sigma_1} = 0. \tag{5.3.48}$$

We suppose that there are periodic solutions near the positive equilibrium $(a/b, a/b)$, namely the above characteristic equation has purely imaginary roots $\lambda = \pm\omega i$, where ω is a real number. Then the substitution of this ansatz into the equation (5.3.48) immediately yields a necessary condition $c = 0$, and consequently, we get

$$\omega^4 - \left(\frac{a(b+\sigma_2)}{\sigma_1 b} + 1\right)\omega^2 + \frac{a}{\sigma_1} = 0. \tag{5.3.49}$$

Notice that $\sigma_2 = \gamma'(a/b) < 0$. Then a necessary and sufficient condition warranting that Eq. (5.3.49) has a real root ω is

$$|\sigma_2| < \frac{b}{a}\sigma_1\left(\sqrt{\frac{a}{\sigma_1}} - 1\right)^2. \tag{5.3.50}$$

That is, the linearized system (5.3.47) at the equilibrium $(a/b, a/b)$ will have periodic solutions if the condition (5.3.50) is fulfilled. Thereof, we anticipate that the non-monotone traveling wave solutions oscillating about the critical point $(a/b, a/b)$ may exist, but whether the condition (5.3.50) is sufficient to guarantee that the nonlinear system (5.1.6) has similar oscillatory behavior around the equilibrium $(a/b, a/b)$

is very hard to determine and even to predict due to the complexity induced by the nonlinear diffusion and cross-diffusion in the system. Below we shall use numerical simulations to illustrate that indeed the condition (5.3.50) plays a critical role for the nonlinear system in determining the monotonicity of wave profiles.

We consider the motility function $\gamma(v) = \frac{1}{(1+v)^m}$ $(m > 0)$ as given in (5.3.29). With simple calculation, we find that the condition (5.3.50) amounts to

$$\sqrt{m} < \sqrt{\frac{1 + \vartheta}{\vartheta}} \left| \sqrt{a(1 + \vartheta)^m} - 1 \right|, \quad \vartheta = \frac{a}{b}. \tag{5.3.51}$$

Without loss of generality, we first choose $m = 6$ and $a = b = 0.1$. Then $\vartheta = 1$ and

$$\sqrt{\frac{1 + \vartheta}{\vartheta}} \left| \sqrt{a(1 + \vartheta)^m} - 1 \right| = 2.1635 < \sqrt{6} = 2.4495.$$

Hence, the condition (5.3.51) is violated and no oscillation around $(a/b, a/b) = (1, 1)$ is expected for the linearized system. To verify if this is the case for the nonlinear system (5.1.6), we set the initial value (u_0, v_0) as

$$u_0(x) = v_0(x) = \frac{1}{1 + e^{2(x-20)}} \tag{5.3.52}$$

and perform the numerical simulations in an interval $[0, 200]$ with Neumann boundary conditions to comply with the experiment. The numerical solution of (5.1.6) is shown in Fig. 5.3 where we observe that the solution will stabilize into monotone traveling waves although it oscillates initially. This is also well consistent with our analytical results about the existence of traveling wave solutions given in Theorem 5.1 when $\mathcal{K}(m, a) = 0.4143 < 1$ if $m = 6$ and $a = b = 0.1$. Next, we choose $m = 4$ and $a = b = 1$ such that $\sqrt{\frac{1+\vartheta}{\vartheta}} \left| \sqrt{a(1 + \vartheta)^m} - 1 \right| = 4.2426$ and hence (5.3.51) holds. But numerically we still find that the system (5.1.6) will generate monotone traveling waves qualitatively similar to the patterns shown in Fig. 5.3 (not shown here for brevity). This implies that the condition (5.3.50) is not sufficient to induce non-monotone traveling waves oscillating around $(a/b, a/b)$.

Now an important question is whether the density-suppressed motility system (5.1.6) is capable of producing persistent oscillating traveling waves to interpret (at least qualitatively) the pattern observed in the experiment (see Fig. 5.1). To explore this question numerically, we consider the following sigmoid motility function:

$$\gamma(v) = 1 - \frac{v - 1}{\sqrt{0.1 + (v - 1)^2}}$$

which decays but changes the convexity at the point $v = 1$, in contrast to the decreasing function (5.1.7) whose convexity remains unchanged. We perform the numerical simulations for (5.1.6) with $a = b = 0.2$ in an interval $[0, 200]$ with the same initial value (5.3.52). Remarkably, we find non-monotone traveling wavefronts develop

Fig. 5.3 Numerical simulations of wave propagation generated by the system (5.1.6) in [0, 200] with $\gamma(v) = \frac{1}{(1+v)^m}$ with $m = 6, a = b = 0.1, u_0 = v_0 = \frac{1}{1+e^{2(x-20)}}$

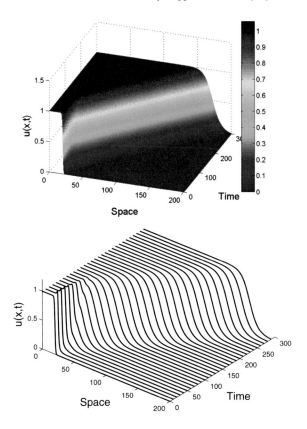

(see Fig. 5.4) and persist in time, where the wave oscillates at the trailing edge and propagates into the far field as time evolves. This is a prominent feature different from the patterns shown in Fig. 5.3 generated from the motility function (5.1.7). If we choose some other forms of decreasing function $\gamma(v)$ that changes its convexity at $v = a/b = 1$, we shall numerically find similar non-monotone traveling wavefront patterns generated by (5.1.6).

The above numerical simulations indicate, although not proved in Chapter 5, that the density-suppressed motility system (5.1.6) can generate both monotone and non-monotone traveling wavefront solutions connecting $(a/b, a/b)$ to $(0, 0)$. It numerically appears that the change of convexity of $\gamma(v)$ at $v = a/b$ is necessary to generate the non-monotone traveling wavefronts oscillating at the trailing edge around the equilibrium $(a/b, a/b)$. The underlying mechanism remains mysterious and we will leave it as an open question for future study.

Next, we are devoted to exploring the patterns in a disk to mimic the apparatus used in the experiment of Liu (2011) where the experiment was conducted in Petri dishes with bacteria initially inoculated at the center (see Fig. 5.1). In the numerical simulations, we set the domain as a disk with radius 10 and initially place the initial value $(u_0, v_0) = (4 + e^{-(x^2+y^2)}, 4 + e^{-(x^2+y^2)})$ in the center. We use the motility

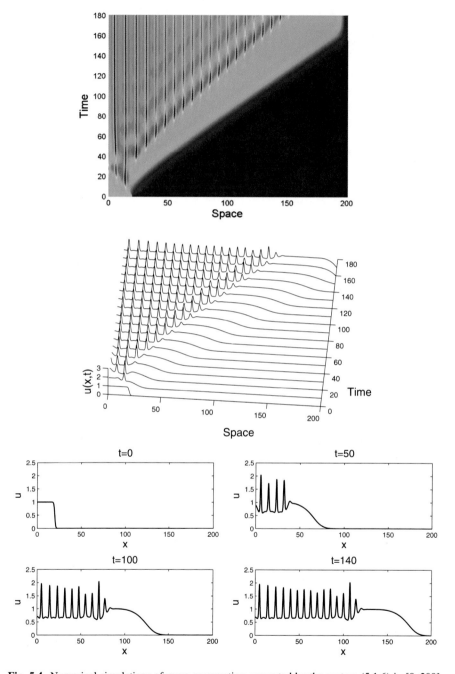

Fig. 5.4 Numerical simulations of wave propagation generated by the system (5.1.6) in $[0, 200]$ with $\gamma(v) = 1 - \frac{v-1}{\sqrt{0.1+(v-1)^2}}$, $a = b = 0.2$, $u_0 = v_0 = \frac{1}{1+e^{2(x-20)}}$

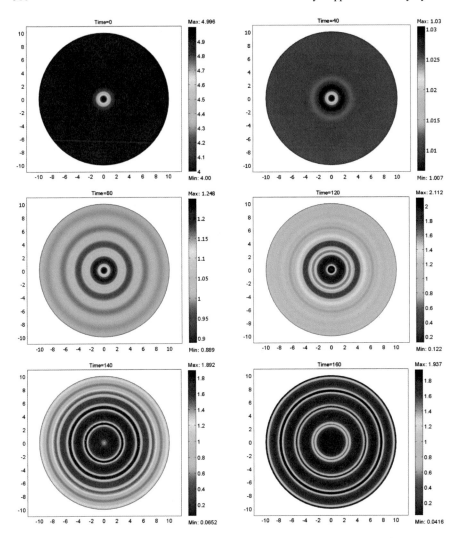

Fig. 5.5 Snapshot of numerical simulations of outward expanding ring patterns in a disk generated by the system (5.1.6) with $\gamma(v) = \frac{1}{(1+v)^6}$, $a = b = 0.1$, $u_0 = v_0 = 4 + e^{-(x^2+y^2)}$

function given in (5.1.7) with $m = 6$ and set out Neumann boundary (i.e., zero-flux) conditions aligned with the experiment reality. The snapshots of numerical patterns are recorded in Fig. 5.5, where we do observe the outward expanding ring patterns qualitatively analogous to the experiment patterns shown in Fig. 5.1. This validates the capability of model (5.1.6) to reproduce the experimental patterns. However, we should underline that it appears that the generation of oscillating patterns in two dimensions does not rely on the change of convexity of the motility function $\gamma(v)$ as shown in Fig. 5.5, which is very different from the situation in 1D as shown in

Figs. 5.3 and 5.4. This imposes another interesting question elucidating this subtle difference.

2. Leading edge wave speeds

Following the spirit of classical method as in Mollison (1977) and Murray (2001), we discuss the selection of the wave speed c from the initial conditions given at infinity. Suppose that the initial value (u_0, v_0) of the system (5.1.6) satisfies

$$\begin{cases} u_0(x) \sim Ae^{-\lambda x}, \\ v_0(x) \sim Be^{-\lambda x}, \end{cases} \quad \text{as } x \to \infty \tag{5.3.53}$$

with positive amplitudes A and B. Now we look for traveling wave solutions of (5.1.10) at the leading edge (i.e., $x \to \infty$) in the form of

$$\begin{cases} u(x, t) \sim Ae^{-\lambda(x-ct)}, \\ v(x, t) \sim Be^{-\lambda(x-ct)}. \end{cases} \tag{5.3.54}$$

We substitute (5.3.54) into the first equation of (5.1.6) and get the dispersion relation between the wave speed c and the initial decay rate λ:

$$c = \gamma(0)\lambda + \frac{a}{\lambda}. \tag{5.3.55}$$

Hence, by the standard argument as in Murray (2001), the asymptotic wave speed c of traveling wave solutions to (5.1.6) satisfies

$$c = \begin{cases} \gamma(0)\lambda + \frac{a}{\lambda}, & \text{if } 0 < \lambda < \sqrt{a}, \\ 2\sqrt{\gamma(0)a}, & \text{if } \lambda \geq \sqrt{a}. \end{cases} \tag{5.3.56}$$

Next, we plug (5.3.54) into the second equation of (5.1.6) and get the following relation on the amplitude of u and v:

$$A = [1 + a + (\gamma(0) - 1)\lambda^2]B. \tag{5.3.57}$$

Therefore, given the initial condition (5.3.53), the leading edge of traveling waves is fully determined by the ansatz (5.3.54) with wave speed (5.3.56) and amplitudes fulfilling (5.3.57).

As an example, we consider the motility function (5.1.7) chosen in the first part of this chapter, where $\gamma(0) = 1$ and hence (5.3.55) gives

$$\lambda^2 - c\lambda + a = 0$$

which is exactly the same as the equation (5.2.2). Furthermore, (5.3.57) gives $A = (1 + a)B$ which well agrees with the result (5.1.13) in Theorem 5.1.

5.4 Asymptotic Behavior of Solutions to a Signal-Suppressed Motility Model

5.4.1 Space–Time L^1-Estimates for $u_\varepsilon^{m+1} v_\varepsilon^{-\alpha}$

In this section, taking advantage of the special structure of the diffusive processes in (5.2.20) (also (5.1.16)), the classical duality arguments (cf. Cañizo et al. 2014; Tao and Winkler 2017a) are used to obtain the fundamental regularity information for a bootstrap argument. To this end, we denote by A the self-adjoint realization of $-\Delta + 1$ under homogeneous Neumann boundary condition in $L^2(\Omega)$ with its domain given by $D(A) = \left\{ \varphi \in W^{2,2}(\Omega) | \frac{\partial \varphi}{\partial \nu} = 0 \right\}$ and A is self-adjoint and possesses a family $(A^\beta)_{\beta \in \mathbb{R}}$ of corresponding densely defined self-adjoint fractional powers.

Lemma 5.8 *Assume that $m > 1$ and $D \geq 1$, then for $t > 0$*

$$\frac{d}{dt} \int_\Omega |A^{-\frac{1}{2}}(u_\varepsilon + 1)|^2 + \int_\Omega u_\varepsilon^{m+1} v_\varepsilon^{-\alpha} \leq C \int_\Omega |A^{-1}(u_\varepsilon + 1)|^{m+1} + C \quad (5.4.1)$$

with constant $C > 0$ independent of D.

Proof Due to $\partial_t(u_\varepsilon + 1) = u_{\varepsilon t}$, the first equation in (5.2.20) can be written as

$$\frac{d}{dt} A^{-1}(u_\varepsilon + 1) + \varepsilon(u_\varepsilon + 1)^M + u_\varepsilon(u_\varepsilon + \varepsilon)^{m-1} v_\varepsilon^{-\alpha}$$
$$= A^{-1} \left\{ \varepsilon(u_\varepsilon + 1)^M + u_\varepsilon(u_\varepsilon + \varepsilon)^{m-1} v_\varepsilon^{-\alpha} + \beta u_\varepsilon f(w_\varepsilon) \right\}. \quad (5.4.2)$$

Testing (5.4.2) by $u_\varepsilon + 1$, one has

$$\frac{1}{2} \frac{d}{dt} \int_\Omega |A^{-\frac{1}{2}}(u_\varepsilon + 1)|^2 + \varepsilon \int_\Omega (u_\varepsilon + 1)^{M+1} + \int_\Omega u_\varepsilon(u_\varepsilon + \varepsilon)^{m-1}(u_\varepsilon + 1) v_\varepsilon^{-\alpha}$$
$$= \varepsilon \int_\Omega (u_\varepsilon + 1)^M A^{-1}(u_\varepsilon + 1) + \int_\Omega u_\varepsilon(u_\varepsilon + \varepsilon)^{m-1} v_\varepsilon^{-\alpha} A^{-1}(u_\varepsilon + 1)$$
$$+ \beta \int_\Omega u_\varepsilon f(w_\varepsilon) A^{-1}(u_\varepsilon + 1). \quad (5.4.3)$$

Thanks to $W^{2,2}(\Omega) \hookrightarrow L^\infty(\Omega)$ in two-dimensional setting and the standard elliptic regularity in $L^2(\Omega)$, one can find $C_1 > 0$ and $C_2 > 0$ such that

$$\|\varphi\|_{L^{M+1}(\Omega)}^{M+1} \leq C_1 \|\varphi\|_{W^{2,2}(\Omega)}^{M+1} \leq C_2 \|A\varphi\|_{L^2(\Omega)}^{M+1}$$

for all $\varphi \in W^{2,2}(\Omega)$ such that $\frac{\partial \varphi}{\partial \nu}|_{\partial \Omega} = 0$. Hence, by the Young inequality, we can see that

$$\varepsilon \int_{\Omega} (u_{\varepsilon} + 1)^M A^{-1}(u_{\varepsilon} + 1) \leq \frac{\varepsilon}{2} \int_{\Omega} (u_{\varepsilon} + 1)^{M+1} + \frac{\varepsilon}{2} \int_{\Omega} |A^{-1}(u_{\varepsilon} + 1)|^{M+1}$$

$$\leq \frac{\varepsilon}{2} \|u_{\varepsilon} + 1\|_{L^{M+1}(\Omega)}^{M+1} + \frac{\varepsilon C_1}{2} \|A^{-1}(u_{\varepsilon} + 1)\|_{W^{2,2}(\Omega)}^{M+1}$$

$$= \frac{\varepsilon}{2} \int_{\Omega} (u_{\varepsilon} + 1)^{M+1} + \frac{\varepsilon C_1 C_2}{2} \|u_{\varepsilon} + 1\|_{L^2(\Omega)}^{M+1},$$

which along with the Young inequality implies that for any $\varepsilon_1 > 0$, there exists $c(\varepsilon_1) > 0$ such that $\|\varphi\|_{L^2(\Omega)} \leq \varepsilon_1 \|\varphi\|_{L^{M+1}(\Omega)} + c(\varepsilon_1) \|\varphi\|_{L^1(\Omega)}$ due to $M > 1$ and entails that

$$\varepsilon \int_{\Omega} (u_{\varepsilon} + 1)^M A^{-1}(u_{\varepsilon} + 1) \leq \frac{3\varepsilon}{4} \int_{\Omega} (u_{\varepsilon} + 1)^{M+1} + C_3 \|u_{\varepsilon} + 1\|_{L^1(\Omega)}^{M+1}.$$

Furthermore, since $\|w_{\varepsilon}(\cdot, t)\|_{L^{\infty}(\Omega)} \leq \|w_0\|_{L^{\infty}(\Omega)}$, we apply Lemma 5.4 and Young's inequality to obtain that for $t > 0$,

$$\int_{\Omega} u_{\varepsilon}(u_{\varepsilon} + \varepsilon)^{m-1} v_{\varepsilon}^{-\alpha} A^{-1}(u_{\varepsilon} + 1)$$

$$\leq \frac{1}{4} \int_{\Omega} \left\{ u_{\varepsilon}(u_{\varepsilon} + \varepsilon)^{m-1} \right\}^{\frac{m+1}{m}} v_{\varepsilon}^{-\alpha} + C_4 \int_{\Omega} |A^{-1}(u_{\varepsilon} + 1)|^{m+1} v_{\varepsilon}^{-\alpha} \qquad (5.4.4)$$

$$\leq \frac{1}{4} \int_{\Omega} u_{\varepsilon}^{\frac{m+1}{m}} (u_{\varepsilon} + \varepsilon)^{\frac{m^2-1}{m}} v_{\varepsilon}^{-\alpha} + C_4 \delta^{-\alpha} \int_{\Omega} |A^{-1}(u_{\varepsilon} + 1)|^{m+1}$$

and

$$\beta \int_{\Omega} u_{\varepsilon} f(w_{\varepsilon}) A^{-1}(u_{\varepsilon} + 1)$$

$$\leq \frac{1}{4} \int_{\Omega} u_{\varepsilon}^{m+1} v_{\varepsilon}^{-\alpha} + C_5 \int_{\Omega} v_{\varepsilon}^{\frac{\alpha}{m}} |A^{-1}(u_{\varepsilon} + 1)|^{\frac{m+1}{m}} \qquad (5.4.5)$$

$$\leq \frac{1}{4} \int_{\Omega} u_{\varepsilon}^{m+1} v_{\varepsilon}^{-\alpha} + \int_{\Omega} |A^{-1}(u_{\varepsilon} + 1)|^{m+1} + C_6 \int_{\Omega} v_{\varepsilon}^{\frac{\alpha}{m-1}}.$$

Noticing that $u_{\varepsilon} + 1 \geq \max\{u_{\varepsilon} + \varepsilon, \varepsilon\}$, we have

$$\int_{\Omega} u_{\varepsilon}(u_{\varepsilon} + \varepsilon)^{m-1}(u_{\varepsilon} + 1) v_{\varepsilon}^{-\alpha} \geq \frac{1}{4} \int_{\Omega} u_{\varepsilon}^{\frac{m+1}{m}} (u_{\varepsilon} + \varepsilon)^{\frac{m^2-1}{m}} v_{\varepsilon}^{-\alpha} + \frac{3}{4} \int_{\Omega} u_{\varepsilon}^{m+1} v_{\varepsilon}^{-\alpha},$$

and hence insert (5.4.5) and (5.4.4) into (5.4.3) to get

$$\frac{d}{dt} \int_{\Omega} |A^{-\frac{1}{2}}(u_{\varepsilon} + 1)|^2 + \int_{\Omega} u_{\varepsilon}^{m+1} v_{\varepsilon}^{-\alpha}$$

$$\leq 2(C_4 \delta^{-\alpha} + 1) \int_{\Omega} |A^{-1}(u_{\varepsilon} + 1)|^{m+1} + 2C_6 \int_{\Omega} v_{\varepsilon}^{\frac{\alpha}{m-1}},$$

which along with Lemma 5.5 and $D \geq 1$ readily arrive at (5.4.1).

By means of suitable interpolation arguments, one can appropriately estimate the integrals $\int_\Omega |A^{-1}(u_\varepsilon + 1)|^{m+1}$ and $\int_\Omega |A^{-\frac{1}{2}}(u_\varepsilon + 1)|^2$ in terms of $\int_\Omega u_\varepsilon^{m+1} v_\varepsilon^{-\alpha}$ and thereby derive estimate of the form

$$\int_t^{t+1} \int_\Omega u_\varepsilon^{m+1} v_\varepsilon^{-\alpha} \le C$$

with $C > 0$ independent of D, which can be stated as follows.

Lemma 5.9 *Let $m > 1$ and $D \ge 1$. Then there exists $C > 0$ such that for all $D \ge 1$ as well as $\varepsilon \in (0, 1)$*

$$\int_t^{t+1} \int_\Omega u_\varepsilon^{m+1} v_\varepsilon^{-\alpha} \le C \quad \text{for all } t > 0. \tag{5.4.6}$$

Proof By the standard elliptic regularity in $L^2(\Omega)$, we have

$$\int_\Omega |A^{-1}(u_\varepsilon + 1)|^{m+1} \le C_1 \|u_\varepsilon + 1\|_{L^2(\Omega)}^{m+1}.$$

Noticing that for the given $p \in (2, m + 1)$ (for example, $p := \frac{m+3}{2}$), an application of Young's inequality implies that for any $\eta > 0$, there exists $C_1(\eta) > 0$ such that

$$C_1 \|u_\varepsilon + 1\|_{L^2(\Omega)}^{m+1} \le \eta \|u_\varepsilon + 1\|_{L^p(\Omega)}^{m+1} + C_1(\eta) \|u_\varepsilon + 1\|_{L^1(\Omega)}^{m+1}.$$

On the other hand, by the Hölder inequality, we can see that

$$\int_\Omega u_\varepsilon^p = \int_\Omega \left(u_\varepsilon^{m+1} v_\varepsilon^{-\alpha}\right)^{\frac{p}{m+1}} v_\varepsilon^{\frac{p\alpha}{m+1}}$$

$$\le \left(\int_\Omega u_\varepsilon^{m+1} v_\varepsilon^{-\alpha}\right)^{\frac{p}{m+1}} \left(\int_\Omega v_\varepsilon^{\frac{p\alpha}{m+1-p}}\right)^{\frac{m+1-p}{m+1}}.$$

Hence, combining the above estimates with Lemma 5.5, we arrive at

$$\int_\Omega |A^{-1}(u_\varepsilon + 1)|^{m+1} \le \eta \|u_\varepsilon + 1\|_{L^p(\Omega)}^{m+1} + C_1(\eta) \|u_\varepsilon + 1\|_{L^1(\Omega)}^{m+1}$$

$$\le \eta \|u_\varepsilon\|_{L^p(\Omega)}^{m+1} + C_2(\eta)$$

$$\le \eta \left(\int_\Omega u_\varepsilon^{m+1} v_\varepsilon^{-\alpha}\right) \left(\int_\Omega v_\varepsilon^{\frac{p\alpha}{m+1-p}}\right)^{\frac{m+1-p}{p}} + C_2(\eta)$$

$$\le \eta C_3(\alpha, m) \left(\int_\Omega u_\varepsilon^{m+1} v_\varepsilon^{-\alpha}\right) + C_2(\eta) \quad \text{for all } t > 0. \tag{5.4.7}$$

On the other hand, by self-adjointness of $A^{-\frac{1}{2}}$ and Hölder's inequality, we get

$$\int_\Omega |A^{-\frac{1}{2}}(u_\varepsilon + 1)|^2 = \int_\Omega (u_\varepsilon + 1)A^{-1}(u_\varepsilon + 1)$$

$$\le \|u_\varepsilon + 1\|_{L^2(\Omega)}\|A^{-1}(u_\varepsilon + 1)\|_{L^2(\Omega)}$$

$$\le C_4\|u_\varepsilon + 1\|^2_{L^2(\Omega)}$$

$$\le C_5\|u_\varepsilon\|^2_{L^2(\Omega)} + C_5$$

$$\le C_6\|u_\varepsilon\|^{m+1}_{L^2(\Omega)} + C_6 \quad \text{for all } t > 0.$$

So in this position, proceeding in the same way as above, we also have

$$\int_\Omega |A^{-\frac{1}{2}}(u_\varepsilon + 1)|^2 \le C_6\eta\|u_\varepsilon + 1\|^{m+1}_{L^p(\Omega)} + C_1(\eta)C_6\|u_\varepsilon + 1\|^{m+1}_{L^1(\Omega)}$$

$$\le \eta C_3(\alpha, m)\left(\int_\Omega u_\varepsilon^{m+1}v_\varepsilon^{-\alpha}\right) + C_7(\eta) \quad \text{for all } t > 0 \quad (5.4.8)$$

Therefore, inserting (5.4.7) and (5.4.8) into (5.4.1) and taking η sufficiently small, we have

$$\frac{d}{dt}\int_\Omega |A^{-\frac{1}{2}}(u_\varepsilon + 1)|^2 + C_8\int_\Omega |A^{-\frac{1}{2}}(u_\varepsilon + 1)|^2 + C_8\int_\Omega u_\varepsilon^{m+1}v_\varepsilon^{-\alpha} \le C_9 \quad \text{for all } t > 0$$

with some $C_8 > 0, C_9 > 0$ for all $D \ge 1$. Furthermore, by Lemma 3.4 of Stinner et al. (2014), we immediately obtain (5.4.6).

As the direct consequence of Lemmas 5.9 and 5.5, we have the following.

Lemma 5.10 Let $m > 1, D \ge 1$, then for $p \in (\max\{2, \frac{m+1}{\alpha+1}\}, m + 1)$, one can find a constant $C(p) > 0$ such that

$$\int_t^{t+1}\int_\Omega u_\varepsilon^p(\cdot, s)ds \le C(p) \quad \text{for all } t > 0 \text{ and } D \ge 1. \quad (5.4.9)$$

Proof For $p \in (2, m + 1)$, we utilize Young's inequality to estimate

$$\int_t^{t+1}\int_\Omega u_\varepsilon^p = \int_t^{t+1}\int_\Omega \left(u_\varepsilon^{m+1}v_\varepsilon^{-\alpha}\right)^{\frac{p}{m+1}}v_\varepsilon^{\frac{p\alpha}{m+1}}$$

$$\le \int_t^{t+1}\int_\Omega u_\varepsilon^{m+1}v_\varepsilon^{-\alpha} + \int_t^{t+1}\int_\Omega v_\varepsilon^{\frac{p\alpha}{m+1-p}} \quad \text{for all } t > 0,$$

which leads to (5.4.9) with the help of Lemma 5.5.

5.4.2 Boundedness of Solutions (u_ε, v_ε, w_ε)

On the basis of the quite well-established arguments from parabolic regularity theory, we can turn the space–time integrability properties of u_ε^p into the integrability properties of ∇v_ε as well as ∇w_ε.

Lemma 5.11 *Let $m > 1, \alpha > 0$ and suppose that $D \geq 1$. Then for $q \in (2, \frac{2(m+1)}{(3-m)_+})$, there exists constant $C > 0$ such that for all $D \geq 1$ and $\varepsilon \in (0, 1)$*

$$\|v_\varepsilon(\cdot, t)\|_{W^{1,q}(\Omega)} \leq C \tag{5.4.10}$$

as well as

$$\|w_\varepsilon(\cdot, t)\|_{W^{1,q}(\Omega)} \leq C \tag{5.4.11}$$

for all $t > 0$.

Proof From the continuity of function $h(x) = \frac{2x}{(4-x)_+}$ for $x \in [2, 4)$, it follows that for given $q > 2$ suitably close to the number $\frac{2(m+1)}{(3-m)_+}$, one can choose $p \in (2, m+1)$ in an appropriately small neighborhood of $m + 1$ such that

$$\frac{p}{p-1} \cdot \left(\frac{1}{2} + \frac{1}{p} - \frac{1}{q}\right) < 1. \tag{5.4.12}$$

From the smoothing properties of Neumann heat semigroup $(e^{t\Delta})_{t \geq 0}$, it follows that there exist $C_i > 0 (i = 1, 2)$ such that

$$\|e^\Delta \varphi\|_{W^{1,q}(\Omega)} \leq C_1 \|\varphi\|_{L^1(\Omega)} \quad for \ \varphi \in C^0(\overline{\Omega}) \tag{5.4.13}$$

as well as

$$\|e^{t\Delta} \varphi\|_{W^{1,q}(\Omega)} \leq C_2 t^{-\frac{1}{2} - \frac{1}{2}(\frac{1}{p} - \frac{1}{q})} \|\varphi\|_{L^p(\Omega)} \quad for \ all \ t \in (0, 1) \ and \ \varphi \in C^0(\overline{\Omega}). \tag{5.4.14}$$

Therefore, by the Duhamel representation to the second equation of (5.2.35), we obtain

$$\|\tilde{v}_\varepsilon(\cdot, t)\|_{W^{1,q}(\Omega)} \tag{5.4.15}$$

$$= \left\| e^{(t-(t-1)_+)(\Delta - D^{-1})} \tilde{v}_\varepsilon(\cdot, (t-1)_+) + \frac{1}{D} \int_{(t-1)_+}^t e^{(t-s)(\Delta - D^{-1})} u_\varepsilon(\cdot, \frac{s}{D}) ds \right\|_{W^{1,q}(\Omega)}$$

$$\leq \left\| e^{(t-(t-1)_+)\Delta} \tilde{v}_\varepsilon(\cdot, (t-1)_+) \right\|_{W^{1,q}(\Omega)} + \frac{1}{D} \int_{(t-1)_+}^t \left\| e^{(t-s)\Delta} u_\varepsilon(\cdot, \frac{s}{D}) \right\|_{W^{1,q}(\Omega)} ds.$$

Due to (5.4.13) and (5.4.14), we have

$$\|e^{(t-(t-1)_+)\Delta}\tilde{v}_\varepsilon(\cdot,(t-1)_+)\|_{W^{1,q}(\Omega)} = \|e^{\Delta}\tilde{v}_\varepsilon(\cdot,t-1)\|_{W^{1,q}(\Omega)}$$
$$\leq C_1\|\tilde{v}_\varepsilon(\cdot,t-1)\|_{L^1(\Omega)} \quad \text{for } t > 1, \tag{5.4.16}$$

while for $t \leq 1$,

$$\|e^{(t-(t-1)_+)\Delta}\tilde{v}_\varepsilon(\cdot,(t-1)_+)\|_{W^{1,q}(\Omega)} = \|e^{t\Delta}v_0(\cdot)\|_{W^{1,q}(\Omega)}$$
$$\leq C_1\|v_0(\cdot)\|_{W^{1,\infty}(\Omega)}.$$

On the other hand, we can see that for $t > 0$

$$\int_{(t-1)_+}^t \|e^{-(t-s)\Delta}u_\varepsilon(\cdot,D^{-1}s)\|_{W^{1,q}(\Omega)}$$

$$\leq C_2\int_{(t-1)_+}^t (t-s)^{-\frac{1}{2}-(\frac{1}{p}-\frac{1}{q})}\|u_\varepsilon(\cdot,D^{-1}s)\|_{L^p(\Omega)}ds$$

$$\leq C_2\left\{\int_{(t-1)_+}^t (t-s)^{-\frac{p}{p-1}(\frac{1}{2}+\frac{1}{p}-\frac{1}{q})}ds\right\}^{\frac{p-1}{p}}\left\{\int_{(t-1)_+}^t \|u_\varepsilon(\cdot,D^{-1}s)\|_{L^p(\Omega)}^p ds\right\}^{\frac{1}{p}}$$

$$\leq C_2\left(\int_0^1 \sigma^{-\frac{p}{p-1}\cdot(\frac{1}{2}+\frac{1}{p}-\frac{1}{q})}d\sigma\right)^{\frac{p-1}{p}}\left\{\int_{(t-1)_+}^t \|u_\varepsilon(\cdot,D^{-1}s)\|_{L^p(\Omega)}^p ds\right\}^{\frac{1}{p}}$$

$$\leq C_2D^{\frac{1}{p}}\left(\int_0^1 \sigma^{-\frac{p}{p-1}\cdot(\frac{1}{2}+\frac{1}{p}-\frac{1}{q})}d\sigma\right)^{\frac{p-1}{p}}\left\{\int_{(D^{-1}t-D^{-1})_+}^{D^{-1}t} \|u_\varepsilon(\cdot,s)\|_{L^p(\Omega)}^p ds\right\}^{\frac{1}{p}}$$

$$\leq C_3D^{\frac{1}{p}}, \tag{5.4.17}$$

where due to $D \geq 1$ and the application of Lemma 5.10, we have

$$\int_{(D^{-1}t-D^{-1})_+}^{D^{-1}t} \|u_\varepsilon(\cdot,s)\|_{L^p(\Omega)}^p ds \leq C_4$$

and the finiteness of $\int_0^1 \sigma^{-\frac{p}{p-1}\cdot(\frac{1}{2}+\frac{1}{p}-\frac{1}{q})}d\sigma$ due to (5.4.12). Hence, combining (5.4.15) with (5.4.16) and (5.4.17) gives

$$\|v_\varepsilon(\cdot,t)\|_{W^{1,q}(\Omega)} \leq C_2\|\tilde{v}_\varepsilon(\cdot,t-1)\|_{L^1(\Omega)} + C_3D^{\frac{1}{p}-1} + C_1\|v_0(\cdot)\|_{W^{1,\infty}(\Omega)}$$
$$\leq C_2\left(\int_\Omega u_0 + \beta\int_\Omega w_0\right) + C_3 + C_1\|v_0(\cdot)\|_{W^{1,\infty}(\Omega)}$$

for all $t > 0$ and thus completes the proof of (5.4.10).

Next due to $\|w_\varepsilon(\cdot,t)\|_{L^\infty(\Omega)} \leq \|w_0\|_{L^\infty(\Omega)}$, an application of the Duhamel representation to the third equation in (5.2.20) yields

$$\|w_\varepsilon(\cdot, t)\|_{W^{1,q}(\Omega)} \leq \left\|e^\Delta w_\varepsilon\left(\cdot, (t-1)_+\right)\right\|_{W^{1,q}(\Omega)}$$
$$+ f(\|w_0\|_{L^\infty(\Omega)}) \int_{(t-1)_+}^t \left\|e^{(t-s)\Delta} u_\varepsilon(\cdot, s)\right\|_{W^{1,q}(\Omega)} ds,$$

and thereby (5.4.11) can be actually derived as above.

The following lemma will be used in the derivation of regularity features about spatial and temporal derivatives of u_ε.

Lemma 5.12 *Let $p > 0$ and $\varphi \in C^\infty(\overline{\Omega})$, then*

$$\frac{1}{p} \int_\Omega \frac{d}{dt}(u_\varepsilon + \varepsilon)^p \cdot \varphi + (p-1)M\varepsilon \int_\Omega (u_\varepsilon + \varepsilon)^{p-2}(u_\varepsilon + 1)^{M-1}|\nabla u_\varepsilon|^2 \varphi$$
$$= (1-p) \int_\Omega (mu_\varepsilon + \varepsilon)(u_\varepsilon + \varepsilon)^{m+p-4} v_\varepsilon^{-\alpha}|\nabla u_\varepsilon|^2 \varphi$$
$$+ \alpha(p-1) \int_\Omega (u_\varepsilon + \varepsilon)^{m+p-3} v_\varepsilon^{-\alpha-1} \nabla u_\varepsilon \cdot \nabla v_\varepsilon \varphi$$
$$+ (1-p) \int_\Omega (mu_\varepsilon + \varepsilon)(u_\varepsilon + \varepsilon)^{m+p-3} v_\varepsilon^{-\alpha} \nabla u_\varepsilon \cdot \nabla \varphi$$
$$- M\varepsilon \int_\Omega (u_\varepsilon + \varepsilon)^{p-1}(u_\varepsilon + 1)^{M-1} \nabla u_\varepsilon \cdot \nabla \varphi$$
$$+ \alpha \int_\Omega u_\varepsilon (u_\varepsilon + \varepsilon)^{m+p-2} v_\varepsilon^{-\alpha-1} \nabla v_\varepsilon \cdot \nabla \varphi + \beta \int_\Omega u_\varepsilon f(w_\varepsilon)(u_\varepsilon + \varepsilon)^{p-1} \varphi$$

$$(5.4.18)$$

for all $t > 0$ and $\varepsilon \in (0, 1)$.

Proof This can be verified by the straightforward computation.

Thanks to the boundedness of $\|\nabla v_\varepsilon(\cdot, t)\|_{L^q(\Omega)}$ with some $q > 2$ in Lemma 5.11, we can achieve the following D-independent L^p-estimate of u_ε with finite p.

Lemma 5.13 *Let $m > 1$. Then for all $D \geq 1$ and any $p > 1$, there exists a constant $C(p) > 0$ such that*
$$\|u_\varepsilon(\cdot, t)\|_{L^p(\Omega)} \leq C(p)$$

for all $t > 0$ and $\varepsilon \in (0, 1)$.

Proof According to Lemmas 5.4 and 5.5, one can find $C_i > 0 (i = 1, 2)$ independent of $D \geq 1$ fulfilling

$$v_\varepsilon^{-\alpha}(x, t) \geq C_1, \quad v_\varepsilon^{-\alpha-2}(x, t) \leq C_2 \quad \text{in } \Omega \times (0, \infty) \qquad (5.4.19)$$

for all $\varepsilon \in (0, 1)$.

Letting $\varphi \equiv 1$ in (5.4.18) and by Young's inequality, we have

$$\frac{d}{dt} \int_\Omega (u_\varepsilon + \varepsilon)^p + p(p-1) \int_\Omega (mu_\varepsilon + \varepsilon)(u_\varepsilon + \varepsilon)^{m+p-4} v_\varepsilon^{-\alpha} |\nabla u_\varepsilon|^2 + \int_\Omega (u_\varepsilon + \varepsilon)^p$$

$$\leq \alpha p(p-1) \int_\Omega u_\varepsilon (u_\varepsilon + \varepsilon)^{m+p-3} v_\varepsilon^{-\alpha-1} \nabla u_\varepsilon \cdot \nabla v_\varepsilon$$

$$+ \beta p \int_\Omega u_\varepsilon f(w_\varepsilon)(u_\varepsilon + \varepsilon)^{p-1} + \int_\Omega (u_\varepsilon + \varepsilon)^p$$

$$\leq \frac{p(p-1)}{2} \int_\Omega (mu_\varepsilon + \varepsilon)(u_\varepsilon + \varepsilon)^{m+p-4} v_\varepsilon^{-\alpha} |\nabla u_\varepsilon|^2$$

$$+ \frac{\alpha^2 p(p-1)}{2} \int_\Omega (u_\varepsilon + \varepsilon)^{m+p-1} v_\varepsilon^{-\alpha-2} |\nabla v_\varepsilon|^2$$

$$+ \beta p f(\|w_0\|_{L^\infty(\Omega)}) \int_\Omega u_\varepsilon (u_\varepsilon + \varepsilon)^{p-1} + \int_\Omega (u_\varepsilon + \varepsilon)^p.$$

Furthermore, recalling (5.4.19), we can find $C_3 > 0$ and $C_4 > 0$ independent of p such that

$$\frac{d}{dt} \int_\Omega (u_\varepsilon + \varepsilon)^p + C_3 \int_\Omega |\nabla (u_\varepsilon + \varepsilon)^{\frac{m+p-1}{2}}|^2 + \int_\Omega (u_\varepsilon + \varepsilon)^p$$

$$\leq C_4 p^2 \int_\Omega (u_\varepsilon + \varepsilon)^{m+p-1} |\nabla v_\varepsilon|^2 + C_4 p \int_\Omega (u_\varepsilon + \varepsilon)^p. \tag{5.4.20}$$

According to (5.4.10), $\|\nabla v_\varepsilon\|^2_{L^q(\Omega)} \leq C_5$ for any fixed $q \in (2, \frac{2(m+1)}{(3-m)_+})$, and hence, the Hölder inequality yields

$$C_4 p^2 \int_\Omega (u_\varepsilon + \varepsilon)^{m+p-1} |\nabla v_\varepsilon|^2$$

$$\leq C_4 p^2 \left\{ \int_\Omega (u_\varepsilon + \varepsilon)^{\frac{(m+p-1)q}{q-2}} \right\}^{1-\frac{2}{q}} \|\nabla v_\varepsilon\|^2_{L^q(\Omega)}$$

$$\leq C_4 C_5 p^2 \|(u_\varepsilon + \varepsilon)^{\frac{m+p-1}{2}}\|^2_{L^{\frac{2q}{q-2}}(\Omega)} \tag{5.4.21}$$

$$\leq \frac{C_3}{4} \int_\Omega |\nabla (u_\varepsilon + \varepsilon)^{\frac{m+p-1}{2}}|^2 + C_6(p),$$

where we have used an Ehrling-type inequality due to $W^{1,2}(\Omega) \hookrightarrow L^{\frac{2q}{q-2}}(\Omega)$ in two-dimensional setting and (5.2.28).

On the other hand, since

$$C_4 p \int_\Omega (u_\varepsilon + \varepsilon)^p$$

$$\leq \eta \int_\Omega (u_\varepsilon + \varepsilon)^{m+p-1} + \frac{(C_4 p)^{\frac{m-1+p}{m-1}} |\Omega|}{\eta^{\frac{p}{m-1}}} = \eta \|(u_\varepsilon + \varepsilon)^{\frac{m+p-1}{2}}\|^2_{L^2(\Omega)} + \frac{(C_4 p)^{\frac{m-1+p}{m-1}} |\Omega|}{\eta^{\frac{p}{m-1}}}$$

for any $\eta > 0$, we also have

$$C_4 p \int_\Omega (u_\varepsilon + \varepsilon)^p \le \frac{C_3}{4} \int_\Omega |\nabla(u_\varepsilon + \varepsilon)^{\frac{m+p-1}{2}}|^2 + C_7(p) \tag{5.4.22}$$

with some $C_7(p) > 0$. Now inserting (5.4.22) and (5.4.21) into (5.4.20), we infer that for all $t > 0$

$$\frac{d}{dt} \int_\Omega (u_\varepsilon + \varepsilon)^p + \int_\Omega (u_\varepsilon + \varepsilon)^p \le C_8(p)$$

with $C_8(p) > 0$ independent of $D \ge 1$, which along with a standard comparison argument implies that

$$\int_\Omega u_\varepsilon^p(\cdot, t) \le \max\{C_8(p), \|u_0\|_{L^p(\Omega)}^p + 1\}$$

for all $t \ge 0$ and thus yields the claimed conclusion.

With the L^p-estimate of u_ε at hand, the standard Moser-type iteration can be immediately applied in our approaches to obtain further regularity concerning L^∞-norm of u_ε (see Lemma A.1 of Tao and Winkler 2012a for example) and so we refrain from giving the details here.

Lemma 5.14 *Assume that $m > 1, \alpha > 0$ and $D \ge 1$, then there exists $C > 0$ such that*

$$\|u(\cdot, t)\|_{L^\infty(\Omega)} \le C \text{ for all } t \ge 0. \tag{5.4.23}$$

Remark 5.3 It should be mentioned that when $m > 1, \alpha > 0$ and $D > 0$, one can obtain the boundedness of L^∞-norm of u_ε for all $t > 0$ by the above argument (also see Winkler 2020 for reference). However, the explicit dependence of $\|u_\varepsilon(\cdot, t)\|_{L^p(\Omega)}$ on D is required to investigate the large time behavior of solutions in the sequel. Hence, $D \ge 1$ is imposed specially for the convenience of our discussion below.

At the end of this section, based on the above results, we derive a regularity property for v which goes beyond those in Lemma 5.11.

Lemma 5.15 *Let $m > 1, \alpha > 0$. Then there exists $C > 0$ such that for all $D \ge 1$ and $\varepsilon \in (0, 1)$ such that*

$$\|\nabla v_\varepsilon(\cdot, t)\|_{L^\infty(\Omega)} \le C \tag{5.4.24}$$

as well as

$$\|\nabla w_\varepsilon(\cdot, t)\|_{L^\infty(\Omega)} \le C \tag{5.4.25}$$

for all $t > D$.

Proof Due to $\|\nabla e^{\Delta}\tilde{v}(\cdot, \tilde{t} - 1)\|_{L^\infty(\Omega)} \le C_1 \|\tilde{v}(\cdot, \tilde{t} - 1)\|_{L^1(\Omega)}$, as the proof of Lemma 5.11, we use the Duhamel formula of (5.2.35) in the following way:

$$\|\nabla \tilde{v}(\cdot, \tilde{t})\|_{L^\infty(\Omega)}$$

$$= \left\| \nabla e^{(\Delta - D^{-1})} \tilde{v}(\cdot, \tilde{t} - 1) + D^{-1} \int_{\tilde{t}-1}^{\tilde{t}} \nabla e^{(t-s)(\Delta - D^{-1})} u(\cdot, D^{-1}s) ds \right\|_{L^\infty(\Omega)}$$

$$\leq \|\nabla e^\Delta \tilde{v}(\cdot, \tilde{t} - 1)\|_{L^\infty(\Omega)} + D^{-1} \int_{\tilde{t}-1}^{\tilde{t}} \|\nabla e^{(\tilde{t}-s)\Delta} u(\cdot, D^{-1}s)\|_{L^\infty(\Omega)} ds$$

$$\leq C_1 \|\tilde{v}(\cdot, \tilde{t} - 1)\|_{L^1(\Omega)} + C_2 D^{-1} \int_{\tilde{t}-1}^{\tilde{t}} (1 + (\tilde{t} - s)^{-\frac{3}{4}}) ds \max_{\tilde{t}-1 \leq s \leq \tilde{t}} \|u(\cdot, D^{-1}s)\|_{L^4(\Omega)}.$$

for all $\tilde{t} > 1$, which along with (5.4.23) readily leads to (5.4.24). It is obvious that (5.4.25) can be proved similarly.

5.4.3 Asymptotic Behavior

1. Weak decay information
The standard parabolic regularity property becomes applicable to improve the regularity of u, v and w as follows.

Lemma 5.16 *Let (u, v, w) be the nonnegative global solution of (5.1.18)–(5.1.20) obtained in Lemma 5.2. Then there exist $\kappa \in (0, 1)$ and $C > 0$ such that for all $t > D$*

$$\|u\|_{C^{\kappa, \frac{\kappa}{2}}(\overline{\Omega} \times [t, t+1])} \leq C \tag{5.4.26}$$

as well as

$$\|v\|_{C^{2+\kappa, 1+\frac{\kappa}{2}}(\overline{\Omega} \times [t, t+1])} + \|w\|_{C^{2+\kappa, 1+\frac{\kappa}{2}}(\overline{\Omega} \times [t, t+1])} \leq C. \tag{5.4.27}$$

Proof We rewrite the first equation of (5.2.20) in the form

$$u_{\varepsilon t} = \nabla \cdot a(x, t, u_\varepsilon, \nabla u_\varepsilon) + b(x, t, u_\varepsilon, \nabla u_\varepsilon)$$

where

$$a(x, t, u_\varepsilon, \nabla u_\varepsilon) = (\varepsilon M(u_\varepsilon + 1)^{M-1} + m u_\varepsilon^{m-1} v_\varepsilon^{-\alpha}) \nabla u_\varepsilon - \alpha u_\varepsilon^m v_\varepsilon^{-\alpha-1} \nabla v_\varepsilon$$

and

$$b(x, t, u_\varepsilon, \nabla u_\varepsilon) = \beta u_\varepsilon f(w_\varepsilon).$$

According to Lemmas 5.4, 5.5, 5.14 and 5.15, there exist two constants $C_1 > 0$ and $C_2 > 0$ independent of $D \geq 1$ satisfying

$$C_1 \leq v_\varepsilon^{-\alpha}(x, t) \leq C_2 \text{ in } \Omega \times (D, \infty)$$

and

$$\|v_\varepsilon^{-\alpha-1}(\cdot,t)\|_{L^\infty(\Omega)} + \|\nabla v_\varepsilon(\cdot,t)\|_{L^\infty(\Omega)} + \|u_\varepsilon(\cdot,t)\|_{L^\infty(\Omega)} + \|w_\varepsilon(\cdot,t)\|_{L^\infty(\Omega)} \leq C_2$$

for $t \geq D$. This guarantees that for all $(x,t) \in \Omega \times (D,\infty)$

$$a(x,t,u_\varepsilon,\nabla u_\varepsilon) \cdot \nabla u_\varepsilon \geq \frac{C_1 m}{2} u_\varepsilon^{m-1} |\nabla u_\varepsilon|^2 - C_3,$$

$$|a(x,t,u_\varepsilon,\nabla u_\varepsilon)| \leq mC_4 u_\varepsilon^{m-1} |\nabla u_\varepsilon| + C_4 |u_\varepsilon|^{\frac{m-1}{2}}$$

and

$$|b(x,t,u_\varepsilon,\nabla u_\varepsilon)| \leq C_5$$

with some constants $C_i > 0$ ($i = 3,4,5$) for all $t > D$ and $\varepsilon \in (0,1)$. Therefore, as an application of the known result on the Hölder regularity in scalar parabolic equations (Porzio and Vespri 1993), there exist $\kappa_1 \in (0,1)$ and $C > 0$ such that for all $t > D$ and $\varepsilon \in (0,1)$,

$$\|u_\varepsilon\|_{C^{\kappa_1,\frac{\kappa_1}{2}}(\overline{\Omega}\times[t,t+1])} \leq C,$$

which along with (5.2.22) readily entails (5.4.26) with $\kappa = \kappa_1$. Similarly, one can also conclude that there exist $\kappa_2 \in (0,1)$ and $C > 0$ such that

$$\|v\|_{C^{\kappa_2,\frac{\kappa_2}{2}}(\overline{\Omega}\times[t,t+1])} + \|w\|_{C^{\kappa_2,\frac{\kappa_2}{2}}(\overline{\Omega}\times[t,t+1])} \leq C \quad \text{for all } t > D.$$

Moreover, since $f \in C^1[0,\infty)$, we have

$$\|uf(w)\|_{C^{\kappa_3,\frac{\kappa_3}{2}}(\overline{\Omega}\times[t,t+1])} \leq C \quad \text{for all } t > D$$

with $\kappa_3 = \min\{\kappa_1,\kappa_2\}$. Thereupon (5.4.27) with $\kappa = \kappa_3$ follows from the parabolic regularity estimates (Ladyzenskaja et al. 1968, Chap. IV, Theorem 5.3).

The first step toward establishing the stabilization result in Theorem 5.3 consists in the following observation.

Lemma 5.17 *Assuming that $m > 1$ and $D \geq 1$, we have*

$$\int_0^\infty \int_\Omega uf(w) < \infty \tag{5.4.28}$$

and

$$\int_0^\infty \int_\Omega |\nabla w|^2 < \infty. \tag{5.4.29}$$

Proof An integration of the third equation in (5.1.18) yields

$$\int_\Omega w_\varepsilon(\cdot,t) + \int_0^t \int_\Omega u_\varepsilon f(w_\varepsilon) = \int_\Omega w_0 \quad \text{for all } t > 0.$$

Since $w_\varepsilon \geq 0$, this entails

$$\int_0^\infty \int_\Omega u_\varepsilon f(w_\varepsilon) \leq \int_\Omega w_0 \tag{5.4.30}$$

which implies (5.4.28) on an application of Fatou's lemma, because $u_\varepsilon f(w_\varepsilon) \to uf(w)$ a.e. in $\Omega \times (0, \infty)$.

We test the same equation by w_ε to see that

$$\frac{1}{2}\int_\Omega w_\varepsilon^2(\cdot, t) + \int_0^t \int_\Omega |\nabla w_\varepsilon|^2 = \frac{1}{2}\int_\Omega w_0^2 - \int_0^t \int_\Omega u_\varepsilon f(w_\varepsilon)w_\varepsilon \leq \frac{1}{2}\int_\Omega w_0^2$$

and thereby verifies (5.4.29) via (5.2.32).

The above decay information of w_ε seems to be weak for the derivation of the large time behavior of u_ε and v_ε. Indeed, under additional constraint on D, we obtain the decay information concerning the gradient of u_ε and v_ε which makes our latter analysis possible.

Lemma 5.18 *Let $m > 1$ and $\alpha > 0$. There exists $D_0 \geq 1$ such that whenever $D > D_0$, the solution of (5.1.18)–(5.1.20) constructed in Lemma 5.2 satisfies*

$$\int_3^\infty \int_\Omega |\nabla u^{\frac{m+1}{2}}|^2 < \infty \tag{5.4.31}$$

as well as

$$\int_3^\infty \int_\Omega |\nabla v|^2 < \infty. \tag{5.4.32}$$

Proof Testing the first equation of (5.2.20) by $(u_\varepsilon + \varepsilon)$ and applying Young's inequality, we obtain that

$$\frac{1}{2}\frac{d}{dt}\int_\Omega (u_\varepsilon + \varepsilon)^2 + \int_\Omega (u_\varepsilon + \varepsilon)^{m-1} v_\varepsilon^{-\alpha} |\nabla u_\varepsilon|^2$$

$$\leq \alpha \int_\Omega (u_\varepsilon + \varepsilon)^m v_\varepsilon^{-\alpha-1} \nabla u_\varepsilon \cdot \nabla v_\varepsilon + \beta \int_\Omega u_\varepsilon (u_\varepsilon + \varepsilon) f(w_\varepsilon)$$

$$\leq \frac{1}{2}\int_\Omega (u_\varepsilon + \varepsilon)^{m-1} v_\varepsilon^{-\alpha} |\nabla u_\varepsilon|^2 + \frac{\alpha^2}{2}\int_\Omega v_\varepsilon^{-\alpha-2}(u_\varepsilon + \varepsilon)^{m+1}|\nabla v_\varepsilon|^2$$

$$+ \beta \int_\Omega (u_\varepsilon + \varepsilon)u_\varepsilon f(w_\varepsilon),$$

and hence

$$\frac{d}{dt} \int_{\Omega} (u_{\varepsilon} + \varepsilon)^2 + \int_{\Omega} (u_{\varepsilon} + \varepsilon)^{m-1} v_{\varepsilon}^{-\alpha} |\nabla u_{\varepsilon}|^2$$

$$\leq \alpha^2 \int_{\Omega} v_{\varepsilon}^{-\alpha-2} (u_{\varepsilon} + \varepsilon)^{m+1} |\nabla v_{\varepsilon}|^2 + 2\beta \int_{\Omega} (u_{\varepsilon} + \varepsilon) u_{\varepsilon} f(w_{\varepsilon}).$$

(5.4.33)

On the other hand, let $\mu_{\varepsilon}(t) = \left(\frac{1}{|\Omega|} \int_{\Omega} u_{\varepsilon}^{\frac{m+1}{2}} (\cdot, t) \right)^{\frac{2}{m+1}}$, then testing the second equation of (5.2.20) by $-\Delta v_{\varepsilon}$ shows

$$\frac{d}{dt} \int_{\Omega} |\nabla v_{\varepsilon}|^2 + 2D \int_{\Omega} (\Delta v_{\varepsilon})^2 + 2 \int_{\Omega} |\nabla v_{\varepsilon}|^2$$

$$= 2 \int_{\Omega} (u_{\varepsilon}(\cdot, t) - \mu_{\varepsilon}(t)) \Delta v_{\varepsilon}$$

$$\leq \frac{1}{D} \int_{\Omega} |u_{\varepsilon}(\cdot, t) - \mu_{\varepsilon}(t)|^2 + D \int_{\Omega} (\Delta v_{\varepsilon})^2,$$

and thus

$$\frac{d}{dt} \int_{\Omega} |\nabla v_{\varepsilon}|^2 + D \int_{\Omega} (\Delta v_{\varepsilon})^2 + 2 \int_{\Omega} |\nabla v_{\varepsilon}|^2 \leq \frac{1}{D} \int_{\Omega} |u_{\varepsilon}(\cdot, t) - \mu_{\varepsilon}(t)|^2. \quad (5.4.34)$$

Hence, combining (5.4.33) and (5.4.34), we have

$$\frac{d}{dt} \left(\int_{\Omega} (u_{\varepsilon} + \varepsilon)^2 + \eta \int_{\Omega} |\nabla v_{\varepsilon}|^2 \right) + \eta D \int_{\Omega} |\Delta v_{\varepsilon}|^2 + 2\eta \int_{\Omega} |\nabla v_{\varepsilon}|^2$$

$$+ \int_{\Omega} v_{\varepsilon}^{-\alpha} (u_{\varepsilon} + \varepsilon)^{m-1} |\nabla u_{\varepsilon}|^2$$

$$\leq \frac{\eta}{D} \int_{\Omega} |u_{\varepsilon}(\cdot, t) - \mu_{\varepsilon}(t)|^2 + \alpha^2 \int_{\Omega} v_{\varepsilon}^{-\alpha-2} (u_{\varepsilon} + \varepsilon)^{m+1} |\nabla v_{\varepsilon}|^2$$

$$+ 2 \int_{\Omega} (u_{\varepsilon} + \varepsilon) u_{\varepsilon} f(w_{\varepsilon})$$

(5.4.35)

for parameter $\eta > 0$ which will be determined later.

In view of Lemmas 5.4 and 5.5, there exist $C_i > 0 (i = 1, 2)$ independent of $D \geq 1$ satisfying

$$v_{\varepsilon}^{-\alpha}(x, t) \geq C_1, \quad v_{\varepsilon}^{-\alpha-2}(x, t) \leq C_2 \text{ in } \Omega \times (2, \infty)$$

for all $\varepsilon \in (0, 1)$. Therefore, from (5.4.35), it follows that

$$\frac{d}{dt}\left(\int_\Omega (u_\varepsilon + \varepsilon)^2 + \eta \int_\Omega |\nabla v_\varepsilon|^2\right) + \eta D \int_\Omega |\Delta v_\varepsilon|^2 + 2\eta \int_\Omega |\nabla v_\varepsilon|^2$$

$$+ C_1 \int_\Omega (u_\varepsilon + \varepsilon)^{m-1} |\nabla u_\varepsilon|^2$$

$$\leq \frac{\eta}{D} \int_\Omega |u_\varepsilon(\cdot, t) - \mu_\varepsilon(t)|^2 + \alpha^2 C_2 \int_\Omega (u_\varepsilon + \varepsilon)^{m+1} |\nabla v_\varepsilon|^2 + 2 \int_\Omega (u_\varepsilon + \varepsilon) u_\varepsilon f(w_\varepsilon).$$

$$(5.4.36)$$

According to Lemma 5.13 with $p = 2(m + 1)$, we have

$$\left(\int_\Omega (u_\varepsilon + \varepsilon)^{2(m+1)}\right)^{\frac{1}{2}} \leq C_3,$$

and then use the Gagliardo–Nirenberg inequality and the Hölder inequality to arrive at

$$\int_\Omega (u_\varepsilon + \varepsilon)^{m+1} |\nabla v_\varepsilon|^2$$

$$\leq \left(\int_\Omega (u_\varepsilon + \varepsilon)^{2(m+1)}\right)^{\frac{1}{2}} \left(\int_\Omega |\nabla v_\varepsilon|^4\right)^{\frac{1}{2}}$$

$$\leq C_4 \left(\int_\Omega (u_\varepsilon + \varepsilon)^{2(m+1)}\right)^{\frac{1}{2}} \left(\|\Delta v_\varepsilon\|^2 + \|\nabla v_\varepsilon\|^2_{L^2(\Omega)}\right)$$

$$\leq C_3 C_4 (\|\Delta v_\varepsilon\|^2 + \|\nabla v_\varepsilon\|^2_{L^2(\Omega)}).$$

$$(5.4.37)$$

Therefore, inserting (5.4.37) into (5.4.36) yields

$$\frac{d}{dt}\left(\int_\Omega (u_\varepsilon + \varepsilon)^2 + \eta \int_\Omega |\nabla v_\varepsilon|^2\right) + \eta D \int_\Omega |\Delta v_\varepsilon|^2 + 2\eta \int_\Omega |\nabla v_\varepsilon|^2$$

$$+ \frac{4C_1}{(m+1)^2} \int_\Omega |\nabla (u_\varepsilon + \varepsilon)^{\frac{m+1}{2}}|^2$$

$$\leq \frac{\eta}{D} \int_\Omega |u_\varepsilon(\cdot, t) - \mu_\varepsilon(t)|^2 + \alpha^2 C_2 C_3 C_4 (\|\Delta v_\varepsilon\|^2_{L^2(\Omega)} + \|\nabla v_\varepsilon\|^2_{L^2(\Omega)})$$

$$+ 2 \int_\Omega (u_\varepsilon + \varepsilon) u_\varepsilon f(w_\varepsilon).$$

$$(5.4.38)$$

By the elementary inequality:

$$\frac{\xi^\mu - \delta^\mu}{\xi - \delta} \geq \delta^{\mu-1} \quad \text{for } \mu \geq 1, \xi \geq 0, \delta \geq 0 \text{ and } \xi \neq \delta,$$

we have

$$|u_\varepsilon^{\frac{m+1}{2}}(\cdot, t) - \mu_\varepsilon^{\frac{m+1}{2}}| \geq \mu_\varepsilon(\cdot, t)^{\frac{m-1}{2}} |u_\varepsilon(\cdot, t) - \mu_\varepsilon(t)|$$

and thus

$$\mu_\varepsilon^{m-1}(t) \int_\Omega |u_\varepsilon(\cdot,t) - \mu_\varepsilon(t)|^2 \leq \int_\Omega |u_\varepsilon^{\frac{m+1}{2}}(\cdot,t) - \mu_\varepsilon^{\frac{m+1}{2}}(t)|^2.$$

Furthermore, by the Hölder inequality and the noncreasing property of $t \mapsto \int_\Omega u_\varepsilon$ $(\cdot,t)$,

$$\mu_\varepsilon(t) \geq \frac{1}{|\Omega|} \int_\Omega u_\varepsilon(\cdot,t) \geq \frac{1}{|\Omega|} \int_\Omega u_0$$

and thereby the Poincaré inequality entails that for some $C_5 > 0$

$$
\begin{aligned}
\overline{u_0}^{m-1} &\int_\Omega |u_\varepsilon(\cdot,t) - \mu_\varepsilon(t)|^2 \\
&\leq \int_\Omega |u_\varepsilon^{\frac{m+1}{2}}(\cdot,t) - \mu_\varepsilon^{\frac{m+1}{2}}(t)|^2 \\
&\leq C_5 \int_\Omega |\nabla u_\varepsilon^{\frac{m+1}{2}}|^2 \\
&\leq C_5 \int_\Omega |\nabla (u_\varepsilon + \varepsilon)^{\frac{m+1}{2}}|^2.
\end{aligned}
\tag{5.4.39}
$$

Hence, substituting (5.4.39) into (5.4.38) shows that

$$
\begin{aligned}
\frac{d}{dt}&\left(\int_\Omega (u_\varepsilon + \varepsilon)^2 + \eta \int_\Omega |\nabla v_\varepsilon|^2\right) + \left(\frac{4C_1}{(m+1)^2} - \frac{\eta C_5}{D\overline{u_0}^{m-1}}\right) \int_\Omega |\nabla (u_\varepsilon + \varepsilon)^{\frac{m+1}{2}}|^2 \\
&\leq (\alpha^2 C_2 C_3 C_4 - \eta D)\|\Delta v_\varepsilon\|_{L^2(\Omega)}^2 + (\alpha^2 C_2 C_3 C_4 - 2\eta)\|\nabla v_\varepsilon\|_{L^2(\Omega)}^2 \\
&\quad + 2 \int_\Omega (u_\varepsilon + \varepsilon) u_\varepsilon f(w_\varepsilon) \\
&\leq (\alpha^2 C_2 C_3 C_4 - \eta)\|\Delta v_\varepsilon\|_{L^2(\Omega)}^2 \\
&\quad + (\alpha^2 C_2 C_3 C_4 - 2\eta)\|\nabla v_\varepsilon\|_{L^2(\Omega)}^2 + 2\|u_\varepsilon(\cdot,t)\|_{L^\infty(\Omega)} + 1) \int_\Omega u_\varepsilon f(w_\varepsilon)
\end{aligned}
$$

and hence completes the proof upon the choice of $D_0 := \max\{1, \frac{\alpha^2 C_2 C_3 C_4 C_5 (m+1)^2}{3 C_1 \overline{u_0}^{m-1}}\}$. Indeed, for any $D > D_0$, it is possible to find $\eta > 0$ such that

$$\frac{3C_1}{(m+1)^2} \geq \frac{\eta C_5}{D\overline{u_0}^{m-1}}, \quad \alpha^2 C_2 C_3 C_4 \leq \eta$$

and thereby

$$
\begin{aligned}
\frac{d}{dt}&\left(\int_\Omega |u_\varepsilon + \varepsilon|^2 + \eta \int_\Omega |\nabla v_\varepsilon|^2\right) + \frac{C_1}{(m+1)^2} \int_\Omega |\nabla (u_\varepsilon + \varepsilon)^{\frac{m+1}{2}}|^2 + \eta \int_\Omega |\nabla v_\varepsilon|^2 \\
&\leq 2(\|u_\varepsilon(\cdot,t)\|_{L^\infty(\Omega)} + 1) \int_\Omega u_\varepsilon f(w_\varepsilon).
\end{aligned}
$$

Therefore, in view of (5.4.30), (5.4.23) and (5.4.24), we see that for any $t > 3$,

$$\int_3^t \int_\Omega |\nabla (u_\varepsilon + \varepsilon)^{\frac{m+1}{2}}|^2 + \int_3^t \int_\Omega |\nabla v_\varepsilon|^2 \leq C_6 + C_6 \int_3^\infty \int_\Omega u_\varepsilon f(w_\varepsilon) \leq C_6 + C_6 \int_\Omega w_0$$
$$(5.4.40)$$

with constant $C_6 > 0$ independent of ε and time t, which implies that (5.4.31) and (5.4.32) are valid due to the lower semi-continuity of norms.

2. Decay of w

The integrability statement in Lemma 5.17 can be turned into the decay property of w with respect to the norm in $L^\infty(\Omega)$, thanks to the fact that $\|u(\cdot, t)\|_{L^1(\Omega)}$ is increasing with time, while $\|w(\cdot, t)\|_{L^\infty(\Omega)}$ is non-increasing.

Lemma 5.19 *The third component of the weak solution of (5.1.18)–(5.1.20) constructed in Lemma 5.2 fulfills*

$$\|w(\cdot, t)\|_{L^\infty(\Omega)} \to 0 \quad as \; t \to \infty. \tag{5.4.41}$$

Proof Writing $\overline{u_0} := \frac{1}{|\Omega|} \int_\Omega u_0$ and $\overline{f(w)} := \frac{1}{|\Omega|} \int_\Omega f(w)$, we use the Cauchy–Schwarz inequality and the Poincaré inequality to see that for all $t > 0$

$$\overline{u_0} \cdot \int_\Omega f(w) = \int_\Omega u \overline{f(w)}$$
$$= \int_\Omega u f(w) - \int_\Omega u(f(w) - \overline{f(w)})$$
$$\leq \int_\Omega u f(w) + C_1 \|u\|_{L^\infty(\Omega)} \|f'(w)\|_{L^\infty(\Omega)} \left\{ \int_\Omega |\nabla w|^2 \right\}^{\frac{1}{2}}.$$

Thanks to the boundedness of u and w, we have

$$\overline{u_0}^2 \cdot \left\{ \int_\Omega f(w) \right\}^2 \leq 2 \left\{ \int_\Omega u f(w) \right\}^2 + C_2 \int_\Omega |\nabla w|^2$$
$$\leq C_3 \int_\Omega u f(w) + C_2 \int_\Omega |\nabla w|^2.$$

Hence, from Lemma 5.17, it follows that

$$\int_1^\infty \|f(w(\cdot, t))\|_{L^1(\Omega)}^2 dt < \infty,$$

which, along with the uniform Hölder estimate from Lemma 5.16, implies that

$$f(w(\cdot, t)) \to 0 \quad \text{in } L^1(\Omega) \quad \text{as } t \to \infty$$

and thereby we may extract a subsequence $(t_j)_{j \in \mathbb{N}} \subset \mathbb{N}$ such that as $t_j \to \infty$, $f(w(\cdot, t_j)) \to 0$ almost everywhere in Ω. Recalling function f is positive on $(0, \infty)$

and $f(0) = 0$, this necessarily requires that $w(\cdot, t_j) \to 0$ almost everywhere in Ω as $t_j \to \infty$. Furthermore, the dominated convergence theorem ensures that

$$w(\cdot, t_j) \to 0 \quad \text{in } L^1(\Omega) \quad \text{as } t_j \to \infty.$$

Now invoking the Gagliardo–Nirenberg inequality in two dimensional setting, we have

$$\|w(\cdot, t_j)\|_{L^\infty(\Omega)} \leq C_4 \|\nabla w(\cdot, t_j)\|_{L^4(\Omega)}^{\frac{4}{5}} \|w(\cdot, t_j)\|_{L^1(\Omega)}^{\frac{1}{5}} + C_4 \|w(\cdot, t_j)\|_{L^1(\Omega)}$$

and thus

$$\|w(\cdot, t_j)\|_{L^\infty(\Omega)} \to 0 \quad \text{as } t_j \to \infty. \tag{5.4.42}$$

Since $t \mapsto \|w(\cdot, t)\|_{L^\infty(\Omega)}$ is noncreasing by Lemma 5.4, (5.4.41) indeed results from (5.4.42).

3. Convergence of u

Now we will show that u stabilizes toward the constant $\overline{u_0} + \beta \overline{w_0}$ as $t \to \infty$. Note that a first step in this direction is provided by the finiteness of $\int_3^\infty \int_\Omega |\nabla u^{\frac{m+1}{2}}|^2$ in Lemma 5.18, which implies that $\|\nabla u^{\frac{m+1}{2}}(\cdot, t_k)\|_{L^2(\Omega)}$ along a suitable sequence of numbers $t_k \to \infty$. However, in order to make sure convergence along the entire net $t \to \infty$, a certain decay property of u_t seems to be required.

Lemma 5.20 *We have*

$$\int_3^\infty \|u_t(\cdot, t)\|_{(W_0^{1,2}(\Omega))^*}^2 dt < \infty. \tag{5.4.43}$$

Proof For any $\varphi \in C_0^\infty(\Omega)$, multiplying the first equation in (5.2.20) by φ and integrating by parts over Ω yield

$$
\begin{aligned}
&\left| \int_\Omega u_{\varepsilon t} \varphi \right| \\
&= \left| \int_\Omega \varepsilon \nabla (u_\varepsilon + 1)^M \cdot \nabla \varphi + \nabla (u_\varepsilon (u_\varepsilon + \varepsilon)^{m-1} v_\varepsilon^{-\alpha}) \cdot \nabla \varphi + \beta u_\varepsilon f(w_\varepsilon) \varphi \right| \\
&\leq \int_\Omega (M(u_\varepsilon + 1)^{M-1} |\nabla u_\varepsilon| + m v_\varepsilon^{-\alpha} (u_\varepsilon + \varepsilon)^{m-1} |\nabla u_\varepsilon| + \alpha (u_\varepsilon + 1)^m v_\varepsilon^{-\alpha-1} |\nabla v_\varepsilon|) |\nabla \varphi| \\
&\quad + \beta \int_\Omega |u_\varepsilon f(w_\varepsilon)| \|\varphi\|_{L^\infty(\Omega)} \\
&\leq C_1 \left(\left\{ \int_\Omega |\nabla (u_\varepsilon + \varepsilon)^{\frac{m+1}{2}}|^2 \right\}^{\frac{1}{2}} + \left\{ \int_\Omega |\nabla v_\varepsilon|^2 \right\}^{\frac{1}{2}} \right) \|\varphi\|_{W^{1,2}(\Omega)} \\
&\quad + \beta \int_\Omega u_\varepsilon f(w_\varepsilon) \|\varphi\|_{L^\infty(\Omega)}
\end{aligned}
$$

with $C_1 > 0$ independent of φ and ε, where we have used the boundedness of u_ε and v_ε.

As in the considered two-dimensional setting we have $W^{1,2}(\Omega) \hookrightarrow L^\infty(\Omega)$, the above inequality implies that

$$\|u_{\varepsilon t}(\cdot, t)\|_{(W_0^{1,2}(\Omega))^*} \le C_1 \left(\left\{ \int_\Omega |\nabla (u_\varepsilon + \varepsilon)^{\frac{m+1}{2}}|^2 \right\}^{\frac{1}{2}} + \left\{ \int_\Omega |\nabla v_\varepsilon|^2 \right\}^{\frac{1}{2}} \right) + \beta \int_\Omega u_\varepsilon f(w_\varepsilon)$$

for all $t > 3$ and hence for all $T > 4$,

$$\int_3^T \|u_{\varepsilon t}(\cdot, t)\|_{(W^{1,2}(\Omega))^*}^2 \, dt$$

$$\le C_2 \left(\int_3^T \int_\Omega |\nabla(u_\varepsilon + \varepsilon)^{\frac{m+1}{2}}|^2 + \int_3^T \int_\Omega |\nabla v_\varepsilon|^2 + \int_3^T \int_\Omega u_\varepsilon f(w_\varepsilon) \right)$$

which together with (5.4.40) leads to

$$\int_3^\infty \|u_{\varepsilon t}(\cdot, t)\|_{(W_0^{1,2}(\Omega))^*}^2 \, dt \le C_3$$

with $C_3 > 0$ independent of ε. Hence, (5.4.43) results from lower semi-continuity of the norm in the Hilbert space $L^2((3, \infty); (W_0^{1,2}(\Omega))^*)$ with respect to weak convergence.

Thanks to the above estimates, we adapt the argument in Winkler (2015b) to show that u actually stabilizes toward $\overline{u_0} + \beta \overline{w_0}$ in the claimed sense beyond in the weak-$*$ sense in $L^\infty(\Omega)$.

Lemma 5.21 *Let $m > 1$, $\alpha > 0$ and suppose that $D \ge D_0$ with D_0 as in Lemma 5.18. Then we have*

$$\|u(\cdot, t) - u_\star\|_{L^\infty(\Omega)} \to 0 \quad \text{as } t \to \infty, \tag{5.4.44}$$

where $u_\star = \frac{1}{|\Omega|} \int_\Omega u_0 + \frac{\beta}{|\Omega|} \int_\Omega w_0$.

Proof According to Lemmas 5.18 and 5.20, one can conclude that

$$u(\cdot, t) \stackrel{w^*}{\rightharpoonup} u_\star \quad \text{in } L^\infty(\Omega) \quad \text{as } t \to \infty. \tag{5.4.45}$$

In fact, if this conclusion does not hold, then one can find a sequence $(t_k)_{k \in \mathbb{N}} \subset (0, \infty)$ such that $t_k \to \infty$ as $k \to \infty$, and some $\tilde{\psi} \in L^1(\Omega)$ such that

$$\int_\Omega u(x, t_k) \tilde{\psi} \, dx - \int_\Omega u_\star \tilde{\psi} \, dx \ge C_1 \quad \text{for all } k \in \mathbb{N}$$

with some $C_1 > 0$. Furthermore, by the boundedness of u and the density of $C_0^\infty(\Omega)$ in $L^1(\Omega)$, we can choose $\psi \in C_0^\infty(\Omega)$ closing $\tilde{\psi}$ in $L^1(\Omega)$ enough that

$$\int_\Omega u(x, t_k) \psi \, dx - \int_\Omega u_\star \psi \, dx \ge \frac{3C_1}{4} \quad \text{for all } k \in \mathbb{N}$$

and then

$$\int_{t_k}^{t_k+1} \int_\Omega u(x,t)\psi\, dx\, dt - \int_{t_k}^{t_k+1} \int_\Omega u_\star \psi\, dx\, dt \geq \frac{C_1}{2} \quad \text{for all sufficiently large } k \in \mathbb{N},$$

(5.4.46)

where we have used the fact that

$$\left| \int_{t_k}^{t_k+1} \int_\Omega (u(x,t) - u(x,t_k))\psi\, dx \right|$$

$$= \left| \int_{t_k}^{t_k+1} \int_{t_k}^{t} \langle u_t(\cdot,s), \psi(\cdot)\rangle ds\, dt \right|$$

$$\leq \int_{t_k}^{t_k+1} \int_{t_k}^{t} \|u_t(\cdot,s)\|_{(W_0^{1,2}(\Omega))^*} ds\, dt \cdot \|\psi\|_{W_0^{1,2}(\Omega)}$$

$$\leq \int_{t_k}^{t_k+1} \left\{ \int_{t_k}^{t} \|u_t(\cdot,s)\|^2_{(W_0^{1,2}(\Omega))^*} ds \right\}^{\frac{1}{2}} |t - t_k|^{\frac{1}{2}} dt \cdot \|\psi\|_{W_0^{1,2}(\Omega)}$$

$$\leq \left\{ \int_{t_k}^{t_k+1} \int_{t_k}^{t} \|u_t(\cdot,s)\|^2_{(W_0^{1,2}(\Omega))^*} ds\, dt \right\}^{\frac{1}{2}} \cdot \|\psi\|_{W_0^{1,2}(\Omega)}$$

$$\leq \left\{ \int_{t_k}^{\infty} \|u_t(\cdot,s)\|^2_{(W_0^{1,2}(\Omega))^*} ds \right\}^{\frac{1}{2}} \cdot \|\psi\|_{W_0^{1,2}(\Omega)}$$

$$\longrightarrow 0 \quad \text{as } k \to \infty,$$

due to Lemma 5.20.

Let $\mu(t) = \left(\frac{1}{|\Omega|} \int_\Omega u^{\frac{m+1}{2}}(\cdot,t) \right)^{\frac{2}{m+1}}$. Then as in (5.4.39), we have

$$\overline{u_0}^{m-1} \int_\Omega |u(\cdot,t) - \mu(t)|^2 \leq \int_\Omega |u^{\frac{m+1}{2}}(\cdot,t) - \mu^{\frac{m+1}{2}}(t)|^2 \leq C_5 \int_\Omega |\nabla u^{\frac{m+1}{2}}|^2$$

and thus

$$\overline{u_0}^{m-1} \int_{t_k}^{t_k+1} \int_\Omega |u(\cdot,t) - \mu(t)|^2 \leq C_5 \int_{t_k}^{t_k+1} \int_\Omega |\nabla u^{\frac{m+1}{2}}(\cdot,t)|^2.$$

(5.4.47)

We now introduce

$$u_k(x,s) := u(x, t_k + s), \ (x,s) \in \Omega \times (0,1)$$

and

$$\mu_k(x,s) := \mu(x, t_k + s), \ (x,s) \in \Omega \times (0,1)$$

for $k \in \mathbb{N}$. Then (5.4.47) implies that

$$\overline{u_0}^{m-1} \int_0^1 \int_\Omega |u_k(\cdot, s) - \mu_k(s)|^2 ds \leq C_5 \int_{t_k}^{t_k+1} \int_\Omega |\nabla u^{\frac{m+1}{2}}(\cdot, t)|^2$$
$$\to 0 \text{ as } k \to \infty,$$

due to (5.4.31) in Lemma 5.18. This means that

$$u_k(x, s) - \mu_k(s) \to 0 \text{ in } L^2(\Omega \times (0, 1)) \text{ as } k \to \infty,$$

which in particular allows us to get

$$\int_0^1 \int_\Omega (u_k(\cdot, s) - \mu_k(s))\psi ds \to 0 \text{ as } k \to \infty \qquad (5.4.48)$$

as well as

$$\int_0^1 \int_\Omega (u_k(\cdot, s) - \mu_k(s)) ds \to 0 \text{ as } k \to \infty. \qquad (5.4.49)$$

Moreover, by Lemma 5.19, we have

$$\int_{t_k}^{t_k+1} \int_\Omega w(\cdot, t) dt \leq |\Omega| \|w(\cdot, t_k)\|_{L^\infty(\Omega)} \to 0 \text{ as } k \to \infty$$

and thereby

$$|\Omega| \int_0^1 \mu_k(s) ds = \int_0^1 \int_\Omega u_k(\cdot, s) ds - \int_0^1 \int_\Omega (u_k(\cdot, s) - \mu_k(s)) ds$$
$$= |\Omega| u_* - \beta \int_{t_k}^{t_k+1} \int_\Omega w(\cdot, t) dt - \int_0^1 \int_\Omega (u_k(\cdot, s) - \mu_k(s)) ds$$
$$\to |\Omega| u_* \text{ as } k \to \infty \qquad (5.4.50)$$

due to (5.4.49) and (5.4.41).

Therefore, from (5.4.46), (5.4.48) and (5.4.50), it follows that

$$\frac{C_1}{2} \leq \int_{t_k}^{t_k+1} \int_\Omega u(\cdot, t)\psi dt - \int_{t_k}^{t_k+1} \int_\Omega u_* \psi dt$$
$$= \int_0^1 \int_\Omega (u_k(\cdot, s) - \mu_k(s))\psi ds + \int_0^1 \int_\Omega \mu_k(s)\psi ds - u_* \int_\Omega \psi$$
$$= \int_0^1 \int_\Omega (u_k(\cdot, s) - \mu_k(s))\psi ds + \int_0^1 \mu_k(s) ds \int_\Omega \psi - u_* \int_\Omega \psi$$
$$\to 0 \text{ as } k \to \infty,$$

which is absurd and hence proves that actually (5.4.45) is valid.

Let us suppose on the contrary that (5.4.44) be false. Then without loss of generality, there exist sequence $\{x_k\}_{k\in\mathbb{N}}$ and $\{t_k\}_{k\in\mathbb{N}} \in (0, \infty)$ with $t_k \to \infty$ as $k \to \infty$ such that for some $C_1 > 0$

$$u(x_k, t_k) - u_* = \max_{x\in\Omega} |u(x, t_k) - u_*| \geq C_1 \text{ for all } k \in \mathbb{N}.$$

In view of the compactness of $\overline{\Omega}$, where passing to subsequences we can find $x_0 \in \overline{\Omega}$ such that $x_k \to x_0$ as $k \to \infty$. Furthermore, because u is uniformly continuous in $\bigcup_{k\in\mathbb{N}}(\overline{\Omega} \times t_k)$, this entails that one can extract a further subsequence if necessary such that

$$u(x, t_k) - u_* \geq \frac{C_1}{2} \text{ for all } x \in B := B_\delta(x_0) \cap \Omega \text{ and } k \in \mathbb{N}$$

for some $\delta > 0$. Noticing that if $x_0 \in \partial\Omega$, the smoothness of $\partial\Omega$ ensures the existence of $\hat{x}_0 \in \Omega$ and a smaller $\hat{\delta} > 0$ such that $B_{\hat{\delta}}(\hat{x}_0) \subset B$. Now taking the nonnegative function $\psi \in C_0^\infty(B_{\hat{\delta}}(\hat{x}_0)))$ such as a smooth truncated function in $B_{\hat{\delta}}(\hat{x}_0))$, we then have

$$\int_\Omega (u(x, t_k) - u_\star)\psi dx = \int_{B_{\hat{\delta}}(\hat{x}_0)} (u(x, t_k) - u_\star)\psi dx \geq \frac{C_1}{2} \cdot \int_\Omega \psi dx,$$

which contradicts (5.4.45) and hence proves the lemma.

4. Stabilization of v

In what follows, based on the uniform Hölder bounds of v and decay of ∇v implied by (5.4.27) and (5.4.32), respectively, we shall show the corresponding stabilization result for v by a contradiction argument.

Lemma 5.22 *Let $m > 1$ and (u, v, w) be the solution of* (5.1.18)–(5.1.20) *obtained in Lemma 5.2. Then we have*

$$\|v(\cdot, t) - u_\star\|_{L^\infty(\Omega)} \to 0 \quad as\ t \to \infty. \tag{5.4.51}$$

Proof According to the uniform Hölder bounds of v and decay of ∇v implied by (5.4.27) and (5.4.32), respectively, (5.4.51) may be derived by a contradiction argument. Indeed, assume that (5.4.51) was false, then we can find a sequence $(t_k)_{k\in\mathbb{N}}$ with $t_k \to \infty$ as $k \to \infty$, and constant $C_1 > 0$ such that

$$\|v(\cdot, t_k) - u_\star\|_{L^\infty} \geq C_1.$$

Furthermore, the uniform Hölder continuity of v in $\Omega \times [t, t + 1]$ warrants the existence of $(x_k)_{k\in\mathbb{N}}$ and $r > 0$ such that

$$|v(x, t) - u_\star| > \frac{C_1}{2}$$

for every $x \in B_r(x_k)$ and $t \in (t_k, t_k + \tau)$ and hence

$$\int_{t_k}^{t_k+\tau} \int_\Omega |v(\cdot, t) - u_\star|^2 > \frac{|\Omega| \tau c_1^2}{4}.$$ (5.4.52)

On the other hand, the Poincaré inequality indicates

$$\int_{t_k}^{t_k+\tau} \int_\Omega |v(\cdot, t) - u_\star|^2 \leq C \int_{t_k}^{t_k+\tau} \int_\Omega |\nabla v|^2 + C \int_{t_k}^{t_k+\tau} \int_\Omega \overline{|v(\cdot, t) - u_\star|^2}.$$
$$(5.4.53)$$

Therefore, (5.4.53) yields a contradiction to (5.4.52) thanks to

$$\int_{t_k}^{t_k+\tau} \int_\Omega |\nabla v|^2 \to 0 \quad \text{as } t_k \to \infty.$$

Now the convergence result in the flavor of Theorem 5.3 has actually been proved already.

Proof of Theorem 5.3. The claimed assertion in Theorem 5.3 is the consequence of Lemmas 5.19, 5.21 and 5.22.

Chapter 6
Multi-taxis Cross-Diffusion System

6.1 Introduction

Multi-taxis appears in society interactions and cancer treatment. Society interactions can lead to the complex dynamical behavior in biology and even in criminology (Eftimie et al. 2007; Guttal and Couzin 2010; Short et al. 2008). A particular example in this direction is mixed-species foraging flocks, such as the formation of Alaska's shearwater flocks through attraction to kittiwake foragers (Hoffman et al. 1981). Oncolytic viruses (OV) are a kind of viruses that preferentially infect and destroy cancer cells. Oncolytic viruses can be engineered by some of the less virulent viruses in nature and be readily combined with other agents. A diverse range of viruses has been investigated as potential cancer therapeutics, such as herpesvirus, adenovirus, vaccinia virus measles virus and polio virus, and oncolytic virotherapy offers a novel promising cancer treatment modality (Breitbach and Parato 2015; Goldsmith et al. 1998; Msaouel et al. 2013).

This chapter is concerned with the multi-taxis diffusion systems modeling foraging–scrounging interplay or oncolytic virotherapy. Section 6.3 is concerned with the asymptotic behavior in a doubly tactic resource consumption model with proliferation. Toward better understanding of the effect of foraging–scrounging interplay on spatio-temporal dynamics, the authors of Tania et al. (2012) proposed the forager–scrounger system given by

$$\begin{cases} u_t = \Delta u - \chi_1 \nabla \cdot (u \nabla w), \\ v_t = \Delta v - \chi_2 \nabla \cdot (v \nabla u), \\ w_t = d \Delta w - \lambda(u + v)w - \mu w + r \end{cases} \qquad (6.1.1)$$

with positive parameters d, χ_1, χ_2, λ and nonnegative parameters μ and r, for the unknown population densities $u = u(x, t)$ and $v = v(x, t)$ of foragers and scroungers and nutrient concentration $w = w(x, t)$, respectively. The term $-\nabla \cdot (u \nabla w)$ accounts for the tendency of foragers moving toward the increasing resource concentration,

© The Author(s) 2022
Y. Ke et al., *Analysis of Reaction-Diffusion Models with the Taxis Mechanism*,
Financial Mathematics and Fintech, https://doi.org/10.1007/978-981-19-3763-7_6

and $-\nabla \cdot (v\nabla u)$ models the movement of scroungers following the actively searching foragers rather the resource. Due to the sequential taxis-type cross-diffusion
mechanisms in (6.1.1), the considerable extra difficulties seem to be expected when
compared to the corresponding scrounger-free system

$$\begin{cases} u_t = \Delta u - \chi_1 \nabla \cdot (u\nabla w), \\ w_t = d\Delta w - \lambda uw - \mu w + r. \end{cases} \tag{6.1.2}$$

Indeed, in the prototypical case $\mu = r = 0$, the two-dimensional version of (6.1.2)
exhibits a substantially stronger tendency toward spatial homogeneous equilibria,
which is also valid for its 3D analog at least after some waiting times (Tao and Winkler 2012c). This result implies that any destabilization of the taxis mechanism in
(6.1.2) can be suppressed by the relaxation of the diffusion process together with
nutrient consumption and thereby allows for a certain entropy-like structure. The
feature of (6.1.1) is the sequential taxis, that is, the nutrient-taxis mechanism from
(6.1.2) coupled with forager-taxis mechanism. In this situation, the mild relaxation
of foragers may not suppress the potential of destabilization driven by the forager-
taxis mechanism and thus limits the accessibility of energy-like techniques from the
mathematical point of view. Accordingly, to the best of our knowledge, the analytical results in the literature are available only for the low dimensions or certain
generalized solutions, and thereby, the comprehensive understanding of (6.1.1) is
still far from complete (Black 2020; Cao 2020; Cao and Tao 2021; Liu 2019; Liu
and Zhuang 2020; Tao and Winkler 2019b; Wang and Wang 2020; Winkler 2019c).
For example, Tao and Winkler (2019b) established the existence of global classical
solutions to the corresponding Neumann initial-boundary value problem of (6.1.1) in
the one-dimensional setting for suitably regular initial data, as well as an exponential
stabilization provided that the initial masses of either u or v are suitably small. As
for the higher dimensional model (6.1.1), only generalized solutions are considered
in Winkler (2019c) under an explicit condition on the initial datum for w and r, and
moreover, they can approach spatially homogeneous equilibria in the large time limit
if r decays sufficiently fast. For more related works on smooth properties of solutions to the variants of (6.1.1), inter alia accounting for the superlinear degradation
mechanisms of two populations, we refer the readers to Black (2020) and Wang and
Wang (2020).

On the other hand, (6.1.2) may be viewed as a kind of the predator–prey system
with prey-taxis:

$$\begin{cases} u_t = \Delta u - \chi_1 \nabla \cdot (u\nabla w) + uf(w, u) + h(u), \\ w_t = d\Delta w - \lambda uf(w, u) + g(w), \end{cases} \tag{6.1.3}$$

where $u(x, t)$ and $w(x, t)$ are predator density and prey density, respectively; $\chi_1 \nabla w$
is the velocity of predators pursuing preys (i.e., prey-taxis); $h(u)$ and $g(w)$ represent
the intra-specific interaction of predators and preys, while $f(w, u)$ is the functional
response, and its typical form in the literature is $f(w, u) = w$ (Lotka–Volterra type)

and $\frac{1}{\lambda}$ is the biomass conversion rate from the prey loss to predator gain. In contrast to the attractive Keller–Segel model, prey-taxis in most cases of (6.1.3) tends to stabilize the predator–prey interactions and may actually lead to the lack of pattern formation, which contradicts intuitive assumptions (Chakraborty et al. 2007; Lee et al. 2008, 2009; Lewis 1994). It also has been recognized that the possibility of spatial pattern formation in (6.1.3) crucially depends on the death rate of predators, the prey growth kinetics $g(w)$ and inter alia functional forms of functional response $f(w, u)$ (Cai et al. 2022; Lee et al. 2009; Wang et al. 2015). In addition to the pattern formation in (6.1.3), the question of which extent the intrinsic predator–prey interaction may preclude the population overcrowding has received considerable attention (see Jin and Wang 2017; Wang and Wang 2019b; Wu et al. 2018; Xiang 2018 and references therein).

In synopsis of the above results, it is natural to consider the dynamical behavior of (6.1.1) when the proliferation of foragers and scroungers is taken into account, which thus indicates that the population proliferation essentially relies on the availability of nutrient resources. Specially, this work will be concerned with the initial-boundary value problem

$$
\begin{cases}
u_t = \Delta u - \chi_u \nabla \cdot (u \nabla w) + uw, & x \in \Omega, t > 0, \\
v_t = \Delta v - \chi_v \nabla \cdot (v \nabla u) + vw, & x \in \Omega, t > 0, \\
w_t = \Delta w - \lambda(u + v)w - \mu w, & x \in \Omega, t > 0, \\
\nabla u \cdot \nu = \nabla v \cdot \nu = \nabla w \cdot \nu = 0, & x \in \partial\Omega, t > 0, \\
u(x, 0) = u_0(x), \ v(x, 0) = v_0(x), \ w(x, 0) = w_0(x), & x \in \Omega
\end{cases}
\tag{6.1.4}
$$

in a smoothly bounded domain $\Omega \subset \mathbb{R}^N$, $N \geq 1$, where ν denotes the outward normal vector field on $\partial\Omega$.

It is worthwhile to mention that (6.1.4) can be regarded as a relative of

$$
\begin{cases}
u_t = \Delta u - \chi \nabla \cdot (u \nabla w) + uw, & x \in \Omega, t > 0, \\
v_t = \alpha vw, & x \in \Omega, t > 0, \\
w_t = \Delta w - \beta uw - \gamma vw, & x \in \Omega, t > 0, \\
\nabla u \cdot \nu = \nabla w \cdot \nu = 0, & x \in \partial\Omega, t > 0, \\
u(x, 0) = u_0(x), \ v(x, 0) = v_0(x), \ w(x, 0) = w_0(x), & x \in \Omega,
\end{cases}
\tag{6.1.5}
$$

which describes the competition between the populations u and v feeding on a common single non-renewable resource w. The authors of Krzyżanowski et al. (2019) asserted global solvability of problem (6.1.5) within a natural weak solution concept and moreover provided an analytical evidence which indicates that under suitably small initial nutrient distributions, in the long time perspective, the motility ability of population u will turn out to be a competitive advantage irrespectively of the competitive kinetics thereof. It should be remarked that the structure of (6.1.5) is comparatively simple enough to allow for the quasi-dissipative property, which seems to be lost due to the taxis-type cross-diffusive term in the second equation of (6.1.4). Inspired by Cao and Lankeit (2016), Myowin et al. (2020) and Li et al. (2019b), we shall consider the asymptotic behavior of (6.1.4) under suitably small initial data. Our standing assumptions on the initial data herein will be that

$$
\begin{cases}
v_0 \in W^{1,\infty}(\Omega) \text{ is nonnegative with } v_0 \not\equiv 0 \text{ and that} \\
(u_0, w_0) \in (W^{2,\infty}(\Omega))^2 \text{ is nonnegative with } \dfrac{\partial u_0}{\partial v} = 0, \ \dfrac{\partial w_0}{\partial v} = 0.
\end{cases}
\tag{6.1.6}
$$

In this setting, all of the solutions of (6.1.4) approach spatially homogeneous profiles in the large time limit when suitably regular initial data satisfy a certain small condition, which reads as follows (Li and Wang 2021a). It is remarked that in comparison with the relative results of Wang and Wang (2020), the small restriction on initial data does not involve v_0 herein.

Theorem 6.1 *Let* $\Omega \subset \mathbb{R}^N$ ($N \geq 1$) *be a bounded domain with smooth boundary and* $m_\infty = \frac{1}{|\Omega|} \int_\Omega (\lambda u_0 + \lambda v_0 + w_0)$. *Then there exists* $\varepsilon_0 > 0$ *such that for all* $\varepsilon < \varepsilon_0$ *and*

$$
\|(\lambda u_0 + \lambda v_0 + w_0)(\cdot) - m_\infty\|_{L^\infty(\Omega)} \leq \varepsilon, \quad \|\nabla u_0\|_{L^{2p_0}(\Omega)} \leq \varepsilon,
$$
$$
\|w_0\|_{L^\infty(\Omega)} \leq \varepsilon, \quad \|\nabla w_0\|_{L^{2p_0}(\Omega)} \leq \varepsilon, \quad \|\Delta w_0\|_{L^{p_0}(\Omega)} \leq \varepsilon
$$

with some $p_0 \in \mathbb{N}$ *satisfying* $p_0 > 1 + \frac{N}{2}$, *the problem* (6.1.4) *admits a unique non-negative global classical solution* $(u, v, w) \in (C(\overline{\Omega} \times [0, \infty)) \cap C^{2,1}(\overline{\Omega} \times (0, \infty)))^3$. *Moreover, there exist constants* $u^* \in (0, \frac{m_\infty}{\lambda})$, $v^* \in (0, \frac{m_\infty}{\lambda})$ *and* $K_i > 0$ ($i = 1, 2, 3$) *such that for all* $t \in (0, \infty)$, *we have*

$$
\|u(\cdot, t) - u^*\|_{L^\infty(\Omega)} \leq K_1 \varepsilon e^{-\alpha t},
$$
$$
\|v(\cdot, t) - v^*\|_{L^\infty(\Omega)} \leq K_2 \varepsilon e^{-\alpha t},
$$
$$
\|w(\cdot, t)\|_{L^\infty(\Omega)} \leq K_3 \varepsilon e^{-\alpha t},
$$

where $\alpha = \min\{\lambda_1, \mu\}$ *for* $\mu > 0$ *and* $\alpha = \min\{\lambda_1, m_\infty\}$ *for* $\mu = 0$ *with* $\lambda_1 > 0$ *the first nonzero eigenvalue of* $-\Delta$ *in* Ω *under the Neumann boundary condition.*

It is noted that the $L^p - L^q$ estimate for the Neumann heat semigroup $e^{t\Delta}$: there exists $C > 0$ such for all $\omega \in L^q(\Omega)$ with $\int_\Omega \omega = 0$,

$$
\|e^{t\Delta}\omega\|_{L^p(\Omega)} \leq C \left(1 + t^{-\frac{N}{2}\left(\frac{1}{q} - \frac{1}{p}\right)}\right) e^{-\lambda_1 t} \|\omega\|_{L^q(\Omega)}
$$

plays an important role in the derivation of decay estimations in Cao and Lankeit (2016), Myowin et al. (2020) and Li et al. (2019b). However, for the doubly tactic model (6.1.4) with $\mu > 0$, despite its dissipative feature, a more subtle effort is required in rigorous analysis due to the invalid of the mass conservation of $\lambda u(\cdot, t) + \lambda v(\cdot, t) + w(\cdot, t)$. Indeed, the core of our argument is to verify that the interval $(0, T)$ on which solutions enjoy some exponential decay properties can be extended to $(0, \infty)$ in which the application of above $L^p - L^q$ estimate seems to be necessary. To this end, a nonnegative auxiliary quantity $z(\cdot, t) = \mu \int_0^t e^{(t-s)\Delta} w(\cdot, s) ds$ is introduced and accordingly allows us to apply $L^p - L^q$ estimate in our argument since the mass of $\lambda u(\cdot, t) + \lambda v(\cdot, t) + w(\cdot, t) + z(\cdot, t)$ is conserved now. It should be remarked that our approach is also valid when system (6.1.4) takes into account nutrient renewal $r(x, t)$ with a certain temporal decay.

Oncolytic virotherapy offers a novel promising cancer treatment modality and currently has some limitations in the oncolytic efficacy, which might be the result of virus clearance and the physical barriers inside tumors such as the interstitial fluid pressure, extracellular matrix (ECM) deposits and tight inter-cellular junctions. The next two sections of this chapter focus on the boundedness and asymptotic behavior for solutions to oncolytic virotherapy models involving triply haptotactic terms. To better understand the physical barriers that limit virus spread, the authors of Alzahrani et al. (2019) recently proposed the PDE-ODE system of the form

$$
\begin{cases}
u_t = D_u \Delta u - \xi_u \nabla \cdot (u \nabla v) + \mu_u u (1 - u) - \rho_u u z, \\
w_t = D_w \Delta w - \xi_w \nabla \cdot (w \nabla v) - \delta_w w + \rho_w u z, \\
v_t = -(\alpha_u u + \alpha_w w) v + \mu_v v (1 - v), \\
z_t = D_z \Delta z - \xi_z \nabla \cdot (z \nabla v) - \delta_z z - \rho_z u z + \beta w
\end{cases}
\tag{6.1.7}
$$

to describe the coupled dynamics of uninfected cancer cells u, OV-infected cancer cells w, ECM v and oncolytic viruses (OV) z. Herein, the underlying modeling hypotheses are that in addition to random diffusion with the respective motility coefficient D_u and D_w, cancer cells can direct their movement toward regions of higher ECM densities with the haptotactic coefficient ξ_u, ξ_w, respectively, and that uninfected cells, apart from proliferating logistically at rate μ_u, are converted into an infected state upon contact with virus particles, whereas infected cells die owing to lysis at a rate δ_w. It is assumed that the static ECM is degraded by both types of cancer cells, possibly remodeled with rate μ_v in the sense of spontaneous renewal of healthy tissue. Finally, it is also supposed that besides the random motion with D_z the random motility coefficient, virus particles move up the gradient of ECM with the ECM-OV-taxis rate ξ_z, increase at a rate β due to the release of free virus particles through infected cells and undergo decay at the rate δ_z accounting for the natural virions' death as well as the trapping of these virus particles into the cancer cells.

From a mathematical perspective, model (6.1.7) on the one hand involves three simultaneous haptotaxis processes, but on the other hand contains the production term $\rho u z$ in w−equation which distinguishes (6.1.7) from the most of the previous haptotaxis (Fontelos et al. 2002; Marciniak-Czochra and Ptashnyk 2010; Liţcanu and Morales-Rodrigo 2010b; Tao 2011; Winkler 2018b) and chemotaxis–haptotaxis models (Pang and Wang 2018; Stinner et al. 2014; Tao and Winkler 2019a). In fact, the haptotactic migration of u, z toward higher densities v simultaneously, in which no smoothing action on the spatial regularity of v can be expected, renders us unable to apply smoothing estimates for the Neumann heat semigroup to gain a priori boundedness information on u and z beyond the norm in $L^1(\Omega)$. Accordingly, this superlinear production term $\rho u z$ in (6.1.7) seems likely to increase the destabilizing potential in the sense of enhancing the tendency toward blow-up of solutions and thus becomes the key contributor to mathematical challenges already given in the derivation of global solvability theory of (6.1.7), which is also indicated in the qualitative analysis of chemotaxis-May–Nowak model (Bellomo and Tao 2020; Bellomo et al. 2019; Hu and Lankeit 2018; Winkler 2019d).

Though the methodological limitations seem to widely restrict the theoretical understanding of the full model (6.1.7), some analytical works on simplifications of the latter have recently been achieved in Tao and Winkler (2020a, b, 2021). Indeed, upon neglecting haptotactic migration processes of infected tumor cells and oncolytic viroses, renewal of ECM as well as proliferation of infected tumor cells, Tao and Winkler considered the corresponding Neumann initial-boundary value problem for

$$
\begin{cases}
u_t = \Delta u - \nabla \cdot (u \nabla v) - \rho u z, \\
v_t = -(u + w)\, v, \\
w_t = D_w \Delta w - w + u z, \\
z_t = D_z \Delta z - z - u z + \beta w
\end{cases}
\tag{6.1.8}
$$

in a bounded domain $\Omega \subset \mathbb{R}^2$ and obtained that the globally defined classical solution is bounded if $0 < \beta < 1$, $\rho \geq 0$ (Tao and Winkler 2020a), whereas for $\beta > 1$ and $\int_\Omega u(\cdot, 0) > |\Omega|/(\beta - 1)$, infinite-time blow-up occurs at least in the particular case when $\rho = 0$ (Tao and Winkler 2021). In order to provide an complement to this, the study in Tao and Winkler (2022) reveals that for any $\rho \geq 0$ and arbitrary $\beta > 0$, at each prescribed level $\gamma \in (0, 1/(\beta - 1)_+)$, one can identify an L^∞-neighborhood of the homogeneous distribution $(u, v, w, z) \equiv (\gamma, 0, 0, 0)$ within which all initial data lead to globally bounded solutions that stabilize toward the constant equilibrium $(u_\infty, 0, 0, 0)$ with some $u_\infty > 0$. On the other hand, in Tao and Winkler (2020c), it is proved that if $\beta \in (0, 1)$, for any choice of $M > 0$, one can find initial data such that the globally defined classical solution satisfies $u \geq M$ in $\Omega \times (0, \infty)$.

Moreover, for the doubly haptotactic version of (6.1.7) with $\xi_z = 0$, the global classical solvability to the corresponding initial-boundary value problem for more comprehensive systems of the form

$$
\begin{cases}
u_t = D_u \Delta u - \xi_u \nabla \cdot (u \nabla v) + \mu_u u (1 - u) - \rho_u u z, \\
w_t = D_w \Delta w - \xi_w \nabla \cdot (w \nabla v) - \delta_w w + \rho_w u z, \\
v_t = -(\alpha_u u + \alpha_w w) v + \mu_v v (1 - v), \\
z_t = D_z \Delta z - \delta_z z - \rho_z u z + \beta w.
\end{cases}
\tag{6.1.9}
$$

is proved in Tao and Winkler (2020b). This is achieved by discovering a quasi-Lyapunov functional structure that allows to appropriately cope with the presence of nonlinear zero-order interaction terms which apparently form the most significant additional mathematical challenge of the considered system in comparison to previously studied haptotaxis models.

The purpose of Sect. 6.4 is to a more comprehensive understanding of model (6.1.7) in the biologically most relevant constellation in which the haptotactic motion of virus particles is taken into account particularly, and either the production term uz or proliferating term $\mu_u u(1 - u)$ is adjusted to $\dfrac{uz}{k_u + \theta u}$ of the Beddington–deAngelis type with positive parameters k_u, θ (Bellomo and Tao 2020) or $\mu_u u(1 - u^r)$ of superquadratic type, respectively. We are concerned with the PDE-ODE system given by

$$
\begin{cases}
u_t = D_u \Delta u - \xi_u \nabla \cdot (u \nabla v) + \mu_u u(1 - u^r) - \dfrac{\rho u z}{k_u + \theta u}, & x \in \Omega, t > 0, \\[2mm]
w_t = D_w \Delta w - \xi_w \nabla \cdot (w \nabla v) - \delta_w w + \dfrac{\rho u z}{k_u + \theta u}, & x \in \Omega, t > 0, \\[2mm]
v_t = -(\alpha_u u + \alpha_w w)v + \mu_v v(1 - v), & x \in \Omega, t > 0, \\[2mm]
z_t = D_z \Delta z - \xi_z \nabla \cdot (z \nabla v) - \delta_z z - \dfrac{\rho u z}{k_u + \theta u} + \beta w, & x \in \Omega, t > 0, \\[2mm]
(D_u \nabla u - \xi_u u \nabla v) \cdot \nu = 0, & x \in \partial\Omega, t > 0, \\[2mm]
(D_w \nabla w - \xi_w w \nabla v) \cdot \nu = 0, & x \in \partial\Omega, t > 0, \\[2mm]
(D_w \nabla z - \xi_z z \nabla v) \cdot \nu = 0, & x \in \partial\Omega, t > 0, \\[2mm]
u(x, 0) = u_0(x), w(x, 0) = w_0(x), v(x, 0) = v_0(x), z(x, 0) = z_0(x), & x \in \Omega
\end{cases}
$$
$$(6.1.10)$$

in a bounded domain $\Omega \subset \mathbb{R}^2$ with smooth boundary, where for the initial data (u_0, w_0, v_0, z_0), we suppose throughout Sect. 6.4 that

$$
\begin{cases}
u_0, w_0, z_0 \text{ and } v_0 \text{ are nonnegative functions from } C^{2+\vartheta}(\bar{\Omega}) \text{ for some } \vartheta \in (0, 1), \\[2mm]
\text{with } u_0 \not\equiv 0, \ w_0 \not\equiv 0, \ z_0 \not\equiv 0, \ v_0 \not\equiv 0 \text{ and } \dfrac{\partial w_0}{\partial \nu} = 0 \text{ on } \partial\Omega.
\end{cases}
$$
$$(6.1.11)$$

Beyond the global classical solvability, in Sect. 6.4, we focus on the global boundedness of classical solutions to (6.1.10)–(6.1.11) stated as follows, which can be regarded as a first step toward the qualitative comprehension of (6.1.10) (Li and Wang 2021b).

Theorem 6.2 *Let $\Omega \subset \mathbb{R}^2$ be a bounded domain with smooth boundary, D_u, D_w, D_z, $\xi_u, \xi_w, \xi_z, \mu_u, \mu_v, \rho, k_u, \alpha_u, \alpha_w, \beta, \delta_w$ and δ_z are positive parameters. Suppose that $r = 1, \theta > 0$ or $r > 1, \theta \geq 0$. Then for any choice of (u_0, w_0, v_0, z_0) fulfilling (6.1.11), there exists $C > 0$ such that if $\xi_w \alpha_w < C$, (6.1.10) admits a unique global classical solution (u, w, v, z), where $\|u(\cdot, t)\|_{L^\infty(\Omega)}, \|w(\cdot, t)\|_{L^\infty(\Omega)}$ and $\|z(\cdot, t)\|_{L^\infty(\Omega)}$ are uniformly bounded for $t \in (0, \infty)$.*

Remark 6.1 In line with the above discussion, the boundedness result on (6.1.10) with $r = 1, \theta = 0$ is also valid when $\xi_z = 0$.

Remark 6.2 When $\mu_v = 0$, one can see that the restriction on $\xi_w \alpha_w$ in Theorem 6.2 can be replaced by a certain small condition on v_0.

A cornerstone of our analysis is to show that for the suitably small $\xi_w \alpha_w$, the functional

$$
\begin{aligned}
\mathscr{F}(t) \\
:= A \int_\Omega e^{\chi_w v(\cdot,t)} b(\cdot,t) \ln b(\cdot,t) + \int_\Omega e^{\chi_u v(\cdot,t)} a(\cdot,t) \ln a(\cdot,t) + \int_\Omega e^{\chi_z v(\cdot,t)} c(\cdot,t) \ln c(\cdot,t)
\end{aligned}
$$

with $a = ue^{-\chi_u v}$, $b = we^{-\chi_w v}$ and $c = ze^{-\chi_z v}$ enjoys a certain quasi-dissipative property under appropriate choice of the positive constant A (of (6.4.32)). As the first step in this direction, we perform the variable change used in several precedents, by which the crucial haptotactic contribution to the equations in (6.1.10) is reduced to zero-order terms $\chi_u a(\alpha_u u + \alpha_w w)v - \chi_u \mu_v av(1 - v)$, $\chi_w b(\alpha_u u + \alpha_w w)v - \chi_w \mu_v bv(1 - v)$ and $\chi_z c(\alpha_u u + \alpha_w w)v - \chi_z \mu_v cv(1 - v)$, respectively (see (6.4.1) below). Thanks to a variant of the Gagliardo–Nirenberg inequality involving certain $L \log L$-type norms, the latter offers the sufficient regularity so as to allow for the L^∞-bounds of solutions in the present two-dimensional setting.

Section 6.5 is devoted to understand the dynamics behavior of (6.1.7) to a considerable extent in higher dimensional settings in light of the above-mentioned results and a recent consideration of global classical solutions to the one-dimensional (6.1.7) in Tao (2021). Specially, taking into account the linear degradation instead of the renewal of ECM and neglecting the proliferation of uninfected tumor cells in (6.1.7), we are concerned with the following Neumann initial-boundary problem in $\Omega \subset \mathbb{R}^N (N \geq 1)$:

$$
\begin{cases}
u_t = \Delta u - \xi_u \nabla \cdot (u \nabla v) - \rho_u uz, & x \in \Omega, t > 0, \\
w_t = \Delta w - \xi_w \nabla \cdot (w \nabla v) - \delta_w w + \rho_w uz, & x \in \Omega, t > 0, \\
v_t = -(\alpha_u u + \alpha_w w)v - \delta_v v, & x \in \Omega, t > 0, \\
z_t = \Delta z - \xi_z \nabla \cdot (z \nabla v) - \delta_z z - \rho_z uz + \beta w, & x \in \Omega, t > 0, \\
(\nabla u - \xi_u u \nabla v) \cdot \nu = (\nabla w - \xi_w w \nabla v) \cdot \nu = (\nabla z - \xi_z z \nabla v) \cdot \nu = 0, & x \in \partial \Omega, t > 0, \\
u(x, 0) = u_0(x), w(x, 0) = w_0(x), v(x, 0) = v_0(x), z(x, 0) = z_0(x), & x \in \Omega,
\end{cases}
$$
$$(6.1.12)$$

where $\xi_u, \xi_w, \xi_z, \rho_u, \rho_w, \rho_z, \delta_w, \delta_v, \delta_z, \alpha_u, \alpha_w$ and β are positive parameters, for the initial data (u_0, w_0, v_0, z_0), we suppose throughout the third part of Chap. 6 that

$$
\begin{cases}
u_0, w_0, v_0, z_0 \text{ are nonnegative functions from } C^{2+\nu}(\overline{\Omega}) \text{ for some } \nu \in (0, 1), \\
u_0 \not\equiv 0, \ w_0 \not\equiv 0, \ v_0 \not\equiv 0, \ z_0 \not\equiv 0 \text{ and } \dfrac{\partial u_0}{\partial \nu} = \dfrac{\partial w_0}{\partial \nu} = \dfrac{\partial v_0}{\partial \nu} = \dfrac{\partial z_0}{\partial \nu} = 0 \text{ on } \partial \Omega.
\end{cases}
$$
$$(6.1.13)$$

Our main result makes sure that for suitably small initial data, these solutions will be globally bounded and approach some constant profiles asymptotically (Wei et al. 2022).

Theorem 6.3 *Let $\Omega \subset \mathbb{R}^N$ ($N \geq 1$) be a bounded domain with smooth boundary. Assume (6.1.13) holds and $\beta \leq \frac{\rho_u + \rho_z}{\rho_w} \delta_w$. Then if for some $p_0 > \max\{1, \frac{N}{2}\}$, there exists $\varepsilon > 0$ which depends on $\xi_u, \xi_w, \xi_z, \rho_u, \rho_w, \rho_z, \alpha_y, \alpha_w$ such that*

$$\left\| u_0 + \frac{\rho_u + \rho_z}{\rho_w} w_0 + z_0 \right\|_{L^\infty(\Omega)} \le \varepsilon,$$

$$\|\nabla u_0(\cdot)\|_{L^{2p_0}(\Omega)} \le \varepsilon, \quad \|\nabla w_0(\cdot)\|_{L^{2p_0}(\Omega)} \le \varepsilon,$$

$$\|\nabla v_0(\cdot)\|_{L^{2p_0}(\Omega)} \le \varepsilon, \quad \|\nabla z_0(\cdot)\|_{L^{2p_0}(\Omega)} \le \varepsilon,$$

$$\|\Delta u_0(\cdot)\|_{L^{p_0}(\Omega)} \le \varepsilon, \quad \|\Delta w_0(\cdot)\|_{L^{p_0}(\Omega)} \le \varepsilon, \quad \|\Delta v_0(\cdot)\|_{L^{p_0}(\Omega)} \le \varepsilon,$$

the problem (6.1.12) has a unique nonnegative global classical solution

$$(u, v, w, z) \in \left(C\left(\bar{\Omega} \times [0, \infty) \right) \cap C^{2,1}\left(\bar{\Omega} \times (0, \infty) \right) \right)^4.$$

Moreover, there exists a nonnegative constant u^* such that

$$\|u(\cdot, t) - u^*\|_{L^\infty(\Omega)} \to 0,$$

$$\|w(\cdot, t)\|_{L^\infty(\Omega)} \to 0,$$

$$\|z(\cdot, t)\|_{L^\infty(\Omega)} \to 0,$$

$$\|v(\cdot, t)\|_{L^\infty(\Omega)} \to 0$$

as $t \to \infty$.

Remark 6.3 Our result indicates that the infected cancer cells and virus particle population can become extinct asymptotically and the density of uninfected cancer cells tends to a nonnegative constant u^* which is less than $\bar{u}_0$. This result implies that the oncolytic virotherapy is effective. Unfortunately, the condition under which u^* equals to zero is left as an open problem.

Same to the analysis in Sect. 6.3, a more subtle effort seems to be required for our analysis in Sect. 6.5 due to the decreasing of $\int_\Omega \left(u + \frac{\rho_u + \rho_z}{\rho_w} w + z \right)(\cdot, s)ds$. To this end, a nonnegative auxiliary quantity

$$Q(\cdot, t) = \int_0^t e^{(t-s)\Delta} \left[-\left(\beta - \frac{\rho_u + \rho_z}{\rho_w} \delta_w \right) w + \delta_z z \right](\cdot, s)ds$$

is introduced and accordingly allows us to apply $L^p - L^q$ estimate in our argument since the mass of $u + \frac{\rho_u + \rho_z}{\rho_w} w + z + Q$ is conserved now.

6.2 Preliminaries

In this section, we provide some preliminary results that will be used in the subsequent sections.

By applying the maximal Sobolev regularity (Theorem 3.1 of Hieber and Prüss 1997), we can obtain the following lemma, which together with Lemmas 1.1, 3.2 and 4.3 will play an important role in the proof of our main results in Sects. 6.3 and 6.5.

Lemma 6.1 (Ishida et al. 2014, Lemma 2.1; Yang et al. 2015, Lemma 2.2) *Let* $r \in (1, \infty)$ *and consider the following evolution equation:*

$$
\begin{cases}
h_t = \Delta h + f, & (x, t) \in \Omega \times (0, T), \\
\nabla h \cdot \nu = 0, & (x, t) \in \partial\Omega \times (0, T), \\
h(x, 0) = h_0(x), & x \in \Omega.
\end{cases}
\tag{6.2.1}
$$

Then for each $h_0 \in W^{2,r}(\Omega)$ *with* $\nabla h_0 \cdot \nu = 0$ *on* $\partial\Omega$ *and any* $f \in L^r((0, T), L^r(\Omega))$, *(6.2.1) admits a unique mild solution* $h \in W^{1,r}((0, T); L^r(\Omega)) \cap L^r((0, T); W^{2,r}(\Omega))$. *Moreover, there exists* $C_r > 0$, *such that*

$$
\int_0^T \int_\Omega |\Delta h|^r \leq C_r \int_0^T \int_\Omega |f|^r + C_r(\|h_0\|_{L^r(\Omega)}^r + \|\Delta h_0\|_{L^r(\Omega)}^r).
\tag{6.2.2}
$$

The following lemma is a special case of Lemma A.5 in Tao and Winkler (2014b) and can be regarded as a variant of a Gagliardo–Nirenberg inequality originally derived in Biler et al. (1994), which will be of importance in the later analysis in Sect. 6.4.

Lemma 6.2 *Let* $\Omega \subset \mathbb{R}^2$ *be a bounded domain with smooth boundary, and let* $p \in (1, \infty)$ *and* $\varepsilon > 0$. *Then there exists* $K(p, \varepsilon) > 0$ *such that*

$$
\|\varphi\|_{L^{\frac{2(p+1)}{p}}(\Omega)}^{\frac{2(p+1)}{p}} \leq \varepsilon \|\nabla\varphi\|_{L^2(\Omega)}^2 \cdot \int_\Omega |\varphi|^{\frac{2}{p}} |\ln|\varphi|| + K(p, \varepsilon) \left(\|\varphi\|_{L^{\frac{2}{p}}(\Omega)}^{\frac{2(p+1)}{p}} + 1 \right)
$$

holds for all $\varphi \in W^{1,2}(\Omega)$.

The following Lemma is based on simple calculations on the maximal value of the function $f(t) = t^a e^{-bt}$ for $t \geq 0$, which is used in the analysis in Sect. 6.5.

Lemma 6.3 *For all* $a > 0$, $b > 0$, *we have*

$$
t^a e^{-bt} \leq \left(\frac{a}{be} \right)^a
$$

holds for all $t \geq 0$.

6.3 Asymptotic Behavior of Solutions to a Doubly Tactic Resource Consumption Model

At the beginning of this subsection, we provide a basic state on local existence and extensibility of solutions to (6.1.4), which can be readily proved by the Amann theory. Similar proof thereof can be found in Tao and Winkler (2019b, Lemma 2.1) and hence we omit the detail here.

Lemma 6.4 *There exist a maximal existence time $T_{max} \in (0, \infty]$ and $(u, v, w) \in C(\overline{\Omega} \times [0, T_{max})) \cap C^{2,1}(\overline{\Omega} \times (0, T_{max}))$ such that (u, v, w) is the unique nonnegative solution of (6.1.4) in $[0, T_{max})$. Furthermore, if $T_{max} < +\infty$, then*

$$\lim_{t \to T_{max}} \left(\|u(\cdot, t)\|_{W^{1,q}(\Omega)} + \|v(\cdot, t)\|_{W^{1,q}(\Omega)} + \|w(\cdot, t)\|_{W^{1,q}(\Omega)} \right) = \infty$$

for all $q > N$.

Theorem 6.1 is the consequence of the following lemmas. In the proof of these lemmas, the constants $c_i > 0, i = 0, 1, \ldots, 4$, refer to those in Lemmas 1.1 and 4.3, respectively.

We first collect some easily verifiable observations in the following lemma:

Lemma 6.5 *Under the assumptions of Theorem 6.1, there exist $M_i > 1 (i = 1, 2, 3)$, and $\varepsilon > 0$ such that*

$$2c_{10}c_4(1 + m_\infty)(\chi_u M_4 + \chi_v M_3) \leq \frac{M_1}{2}, \tag{6.3.1}$$

$$e^{\frac{2}{\alpha}} \leq \frac{M_2}{2}, \tag{6.3.2}$$

$$2c_3 + 2c_{10}c_2 \left((\chi_u M_6 + M_2|\Omega|^{\frac{1}{2p_0}}) \frac{1 + m_\infty}{\lambda} + 1 \right) \leq \frac{M_3}{2}, \tag{6.3.3}$$

$$M_2\varepsilon \leq 1, \quad M_5\varepsilon \leq 1, \quad M_3\varepsilon \leq 1, \quad \chi_u M_4 M_3\varepsilon \leq 1, \quad 4\chi_v c_2 M_3\varepsilon \leq 1, \tag{6.3.4}$$

where

$$M_4 := 2c_3 + 2|\Omega|^{\frac{1}{2p_0}}(1 + m_\infty + \mu)c_{10}c_2 M_2, \quad M_5 := M_1 + 2c_1,$$

$$M_6 := 2c_1 + 2c_4 M_2 c_{10}\lambda \left(|\Omega|^{\frac{1}{2p_0}} + C_5|\Omega|^{\frac{p_0-2}{2p_0(p_0-1)}} \right) + 2c_4 M_4 c_{10}(1 + m_\infty + \mu)|\Omega|^{\frac{1}{2p_0}}$$

and constant $C_5 > 0$ is given in Lemma 6.10 below which is independent of $M_i (i = 1, 2, 3)$.

To obtain a conservation law of mass, we introduce an nonnegative variable z satisfying

$$\begin{cases} z_t = \Delta z + \mu w, & (x, t) \in \Omega \times (0, T), \\ \nabla z \cdot \nu = 0, & (x, t) \in \partial\Omega \times (0, T), \\ z(x, 0) = z_0(x) \equiv 0, & x \in \Omega, \end{cases} \tag{6.3.5}$$

then it is easy to see that for any $t \in [0, T)$,

$$\int_\Omega (\lambda u + \lambda v + w + z)(\cdot, t) = \int_\Omega (\lambda u_0 + \lambda v_0 + w_0). \tag{6.3.6}$$

Let

$$T \triangleq \sup \left\{ \widetilde{T} \in (0, T_{max}) \left| \begin{array}{l} \|(\lambda u + \lambda v + w + z)(\cdot, t) - e^{t\Delta}(\lambda u_0 + \lambda v_0 + w_0)(\cdot)\|_{L^\infty(\Omega)} \\ \quad \leq M_1 \varepsilon e^{-\alpha t} \text{ for all } t \in [0, \widetilde{T}); \\ \|w(\cdot, t)\|_{L^\infty(\Omega)} \leq M_2 \varepsilon e^{-\alpha t} \text{ for all } t \in [0, \widetilde{T}); \\ \|\nabla u(\cdot, t)\|_{L^{2p_0}(\Omega)} \leq M_3 \varepsilon e^{-\alpha t} \text{ for all } t \in [0, \widetilde{T}). \end{array} \right. \right\} \tag{6.3.7}$$

By Lemma 6.4 and the smallness condition on initial data in Theorem 6.1, $T > 0$ is well-defined. We first show $T = T_{max}$. To this end, we will show that all of the estimates mentioned in (6.3.7) are valid with even smaller coefficients on the right-hand side. The derivation of these estimates will mainly rely on $L^p - L^q$ estimates for the Neumann heat semigroup and the fact that the classical solutions on $(0, T_{max})$ can be represented as

$$(\lambda u + \lambda v + w + z)(\cdot, t) = e^{t\Delta}(\lambda u_0 + \lambda v_0 + w_0)(\cdot) \tag{6.3.8}$$

$$- \lambda \int_0^t e^{(t-s)\Delta}(\chi_u \nabla \cdot (u\nabla w) + \chi_v \nabla \cdot (v\nabla u))(\cdot, s)ds,$$

$$u(\cdot, t) = e^{t\Delta}u_0(\cdot) - \int_0^t e^{(t-s)\Delta}(\chi_u \nabla \cdot (u\nabla w) - uw)(\cdot, s)ds, \tag{6.3.9}$$

$$v(\cdot, t) = e^{t\Delta}v_0(\cdot) - \int_0^t e^{(t-s)\Delta}(\chi_v \nabla \cdot (v\nabla u) - vw)(\cdot, s)ds, \tag{6.3.10}$$

$$w(\cdot, t) = e^{t\Delta}w_0(\cdot) - \int_0^t e^{(t-s)\Delta}(\lambda(u + v)w + \mu w)(\cdot, s)ds, \tag{6.3.11}$$

$$z(\cdot, t) = \mu \int_0^t e^{(t-s)\Delta}w(\cdot, s)ds \tag{6.3.12}$$

for all $t \in (0, T_{max})$ as per the variation-of-constants formula.

Lemma 6.6 *Suppose that the assumptions from Theorem 6.1 hold. Then for all $t \in (0, T)$, we have*

$$\|(\lambda u + \lambda v + w + z)(\cdot, t) - m_\infty\|_{L^\infty(\Omega)} \leq M_5 \varepsilon e^{-\alpha t} \tag{6.3.13}$$

with $M_5 := M_1 + 2c_1$.

Proof Due to

$$e^{t\Delta} m_\infty = m_\infty, \quad \int_\Omega [(\lambda u_0 + \lambda v_0 + w_0)(\cdot) - m_\infty] = 0,$$

$$\|(\lambda u_0 + \lambda v_0 + w_0)(\cdot) - m_\infty\|_{L^\infty(\Omega)} \leq \varepsilon,$$

and the assumption of Theorem 6.1, the definition of T along with Lemma 1.1(*i*) implies that for all $t \in (0, T)$,

$$\begin{aligned}
&\|(\lambda u + \lambda v + w + z)(\cdot, t) - m_\infty\|_{L^\infty(\Omega)} \\
&\leq \|(\lambda u + \lambda v + w + z)(\cdot, t) - e^{t\Delta}(\lambda u_0 + \lambda v_0 + w_0)\|_{L^\infty(\Omega)} \\
&\quad + \|e^{t\Delta}[(\lambda u_0 + \lambda v_0 + w_0)(\cdot) - m_\infty]\|_{L^\infty(\Omega)} \\
&\leq M_1 \varepsilon e^{-\alpha t} + 2c_1 \|(\lambda u_0 + \lambda v_0 + w_0)(\cdot) - m_\infty\|_{L^\infty(\Omega)} e^{-\lambda_1 t} \\
&\leq (M_1 + 2c_1)\varepsilon e^{-\alpha t} \\
&= M_5 \varepsilon e^{-\alpha t}.
\end{aligned}$$

Lemma 6.7 *Under the assumptions of Theorem 6.1, we have*

$$\|w(\cdot, t)\|_{L^\infty(\Omega)} \leq \frac{M_2}{2} \varepsilon e^{-\alpha t} \quad \text{for all } t \in (0, T). \tag{6.3.14}$$

Proof Multiplying the third equation in (6.1.4) by kw^{k-1} and integrating the result over Ω, we get $\frac{d}{dt} \int_\Omega w^k \leq -k \int_\Omega (\lambda u + \lambda v + \mu) w^k$ on $(0, T)$.

In what follows, we shall show (6.3.14) in two cases: $\mu > 0$ and $\mu = 0$.

(I) The case $\mu > 0$. Since $-(\lambda u + \lambda v + \mu) \leq -\mu$, we have

$$\frac{d}{dt} \int_\Omega w^k(\cdot, t) \leq -\mu k \int_\Omega w^k(\cdot, t),$$

and thus

$$\int_\Omega w^k(\cdot, t) \leq \int_\Omega w_0^k(\cdot) e^{-\mu k t},$$

which implies that

$$\|w(\cdot, t)\|_{L^\infty(\Omega)} \leq \varepsilon e^{-\mu t} \leq \frac{1}{2} M_2 \varepsilon e^{-\mu t}$$

for any $t \in (0, T)$, where we have used (6.3.2).

(II) The case $\mu = 0$. Note that $z \equiv 0$ for all $(x, t) \in \Omega \times [0, T)$, and thus

$$-(\lambda u + \lambda v) \le |\lambda u + \lambda v + w + z - m_\infty| + w - m_\infty.$$

From the definition of T and Lemma 6.6, it follows that for any $t \in (0, T)$,

$$\frac{d}{dt} \int_\Omega w^k(\cdot, t)$$

$$\le k \int_\Omega w^k(\cdot, t) |(\lambda u + \lambda v + w + z)(\cdot, t) - m_\infty| + k \int_\Omega w^{k+1}(\cdot, t) - km_\infty \int_\Omega w^k(\cdot, t)$$

$$\le k \|(\lambda u + \lambda v + w + z)(\cdot, t) - m_\infty\|_{L^\infty(\Omega)} \int_\Omega w^k(\cdot, t) + k \|w(\cdot, t)\|_{L^\infty(\Omega)} \int_\Omega w^k(\cdot, t)$$

$$- km_\infty \int_\Omega w^k(\cdot, t)$$

$$\le k((M_5 + M_2)\varepsilon e^{-\alpha t} - m_\infty) \int_\Omega w^k(\cdot, t)$$

and hence

$$\int_\Omega w^k(\cdot, t) \le \int_\Omega w_0^k(\cdot) e^{k\left((M_5+M_2)\varepsilon \int_0^t e^{-\alpha s} ds - m_\infty t\right)}$$

$$\le \|w_0(\cdot)\|_{L^k(\Omega)}^k e^{k\left(\frac{M_5+M_2}{\alpha}\varepsilon - m_\infty t\right)},$$

which implies that for any $t \in (0, T)$,

$$\|w(\cdot, t)\|_{L^k(\Omega)} \le \|w_0(\cdot)\|_{L^k(\Omega)} e^{\frac{(M_5+M_2)\varepsilon}{\alpha} - m_\infty t}. \tag{6.3.15}$$

Thanks to $\|w_0\|_{L^\infty(\Omega)} \le \varepsilon$ and (6.3.2), (6.3.4), we obtain that for any $t \in (0, T)$,

$$\|w(\cdot, t)\|_{L^\infty(\Omega)} \le \frac{1}{2} M_2 \varepsilon e^{-m_\infty t}$$

by letting $k \to \infty$ in (6.3.15).

Lemma 6.8 *Let the conditions from Theorem 6.1 be fulfilled. Then for all $t \in (0, T)$ and $p_0 > \frac{N}{2}$, we have*

$$\|\nabla w(\cdot, t)\|_{L^{2p_0}(\Omega)} \le M_4 \varepsilon e^{-\alpha t}$$

with $M_4 := 2c_3 + 2|\Omega|^{\frac{1}{2p_0}} (1 + m_\infty + \mu) c_{10}c_2 M_2$.

Proof By (6.3.11) and Lemma 1.1(*iii*), noticing $\|\nabla w_0(\cdot)\|_{L^{2p_0}(\Omega)} \le \varepsilon$, we have

$$\|\nabla w(\cdot, t)\|_{L^{2p_0}(\Omega)} \tag{6.3.16}$$

$$\leq \|\nabla e^{t\Delta} w_0(\cdot)\|_{L^{2p_0}(\Omega)} + \int_0^t \|\nabla e^{(t-s)\Delta}((\lambda u + \lambda v + \mu)w)(\cdot, s)\|_{L^{2p_0}(\Omega)} ds$$

$$\leq 2c_3 \varepsilon e^{-\lambda_1 t} + \int_0^t \|\nabla e^{(t-s)\Delta}((\lambda u + \lambda v + \mu)w)(\cdot, s)\|_{L^{2p_0}(\Omega)} ds.$$

Now, we estimate the last two integrals on the right-hand side of the above inequality. From the definition of T, Lemmas 1.1(*ii*), 4.3 and 6.6, it follows that

$$\int_0^t \|\nabla e^{(t-s)\Delta}((\lambda u + \lambda v + \mu)w)(\cdot, s)\|_{L^{2p_0}(\Omega)} ds$$

$$\leq c_2 \int_0^t (1 + (t-s)^{-\frac{1}{2}}) e^{-\lambda_1(t-s)} \|(\lambda u + \lambda v + \mu)w(\cdot, s)\|_{L^{2p_0}(\Omega)} \tag{6.3.17}$$

$$\leq c_2 |\Omega|^{\frac{1}{2p_0}} \int_0^t (1 + (t-s)^{-\frac{1}{2}}) e^{-\lambda_1(t-s)} \|w(\cdot, s)\|_{L^\infty(\Omega)} \|(\lambda u + \lambda v + \mu)(\cdot, s)\|_{L^\infty(\Omega)} ds$$

$$\leq c_2 |\Omega|^{\frac{1}{2p_0}} M_2 \varepsilon \int_0^t (1 + (t-s)^{-\frac{1}{2}}) e^{-\lambda_1(t-s)} e^{-\alpha s} (1 + m_\infty + \mu) ds$$

$$\leq 2|\Omega|^{\frac{1}{2p_0}} (1 + m_\infty + \mu) c_{10} c_2 M_2 \varepsilon e^{-\alpha t}.$$

Inserting (6.3.17) into (6.3.16), we get

$$\|\nabla w\|_{L^{2p_0}(\Omega)} \leq \left(2c_3 + 2|\Omega|^{\frac{1}{2p_0}} (1 + m_\infty + \mu) c_{10} c_2 M_2\right) \varepsilon e^{-\alpha t}$$

$$= M_4 \varepsilon e^{-\alpha t},$$

and thereby complete the proof.

Lemma 6.9 *Under the assumptions of Theorem 6.1, for all $p_0 > \frac{N}{2}$, there exists a constant $C > 0$ independent of T such that*

$$\int_0^T \int_\Omega |\Delta w(x, s)|^{p_0} dx ds \leq C, \tag{6.3.18}$$

$$\int_0^T \int_\Omega |\Delta u(x, s)|^{p_0} dx ds \leq C. \tag{6.3.19}$$

Proof Noticing that w satisfies

$$\begin{cases} w_t = \Delta w + F(x, t), & (x, t) \in \Omega \times (0, T), \\ \nabla w \cdot v = 0, & (x, t) \in \partial\Omega \times (0, T), \\ w(x, 0) = w_0, & x \in \Omega \end{cases} \tag{6.3.20}$$

and u satisfies

$$\begin{cases} u_t = \Delta u + G(x, t), & (x, t) \in \Omega \times (0, T), \\ \nabla u \cdot \nu = 0, & (x, t) \in \partial\Omega \times (0, T), \\ u(x, 0) = u_0, & x \in \Omega, \end{cases} \qquad (6.3.21)$$

with $F(x, t) = -\lambda(u + v)w - \mu w$ and $G(x, t) = -\chi_u(\nabla u \nabla w + u \Delta w) + uw$, respectively. By Lemmas 6.6 and 6.7, we can see that

$$\int_0^T \int_\Omega |F(x, s)|^{p_0} dx ds \leq \frac{M_2^{p_0}((1 + m_\infty)^{p_0} + \mu^{p_0})}{\alpha p_0} |\Omega|.$$

Hence, thanks to Lemma 6.1, we can find $C_1 > 0$ independent of T such that

$$\int_0^T \int_\Omega |\Delta w(x, s)|^{p_0} dx ds \leq C_1.$$

Similarly, by the definition of T, (6.3.18) and Lemma 6.8, there exists $C_2 > 0$ independent of T fulfilling

$$\int_0^T \int_\Omega |G(x, s)|^{p_0} dx ds \leq C_2.$$

Applying Lemma 6.1 to (6.3.21) once more, we have

$$\int_0^T \int_\Omega |\Delta u(x, s)|^{p_0} dx ds \leq C_3$$

for some $C_3 > 0$ independent of T and thereby complete the proof. $\qquad \blacksquare$

Thanks to the decay property of w, v, ∇u and space–time L^{p_0}-estimate for Δu, we can establish an $L^{2(p_0-1)}$ bound for ∇v based on Lemmas 1.1 and 4.3.

Lemma 6.10 *Suppose that the requirements from Theorem 6.1 are met. Then for all* $p_0 > 1 + \frac{N}{2}$, *there exists* $C_5 > 0$ *independent of* T *and* $M_i (i = 1, 2, 3)$ *such that*

$$\|\nabla v(\cdot, t)\|_{L^{2(p_0-1)}(\Omega)} \leq C_5 \quad \text{for all } t \in (0, T). \qquad (6.3.22)$$

Proof By (6.3.10), we have

$$\|\nabla v(\cdot, t)\|_{L^{2(p_0-1)}(\Omega)} \qquad (6.3.23)$$

$$\leq \|\nabla e^{t\Delta} v_0(\cdot)\|_{L^{2(p_0-1)}(\Omega)} + \chi_v \int_0^t \|\nabla e^{(t-s)\Delta} \nabla \cdot (v \nabla u)(\cdot, s)\|_{L^{2(p_0-1)}(\Omega)} ds$$

$$+ \int_0^t \|\nabla e^{(t-s)\Delta}(vw)(\cdot, s)\|_{L^{2(p_0-1)}(\Omega)} ds.$$

From Lemma 1.1(iii), we obtain

$$\|\nabla e^{t\Delta} v_0(\cdot)\|_{L^{2(p_0-1)}(\Omega)} \leq 2c_3 \|\nabla v_0(\cdot)\|_{L^{2(p_0-1)}(\Omega)} e^{-\lambda_1 t}. \tag{6.3.24}$$

Now, we estimate the last two integrals on the right-hand side of (6.3.23). From the definition of T, Lemmas $1.1(ii)$, 4.3, 6.6 and 6.9, it follows that

$$\chi_v \int_0^t \|\nabla e^{(t-s)\Delta} \nabla \cdot (v\nabla u)(\cdot, s)\|_{L^{2(p_0-1)}(\Omega)} ds$$

$$\leq c_2 \chi_v \int_0^t e^{-(t-s)\lambda_1} \left(1 + (t-s)^{-\frac{1}{2} - \frac{N}{2p_0} + \frac{N}{4(p_0-1)}}\right) \|v(\cdot, s)\Delta u(\cdot, s)\|_{L^{p_0}(\Omega)} ds$$

$$+ c_2 \chi_v \int_0^t e^{-(t-s)\lambda_1} \left(1 + (t-s)^{-\frac{1}{2} - \frac{N(2p_0-1)}{4p_0(p_0-1)} + \frac{N}{4(p_0-1)}}\right)$$

$$\cdot \|\nabla v(\cdot, s)\nabla u(\cdot, s)\|_{L^{\frac{2p_0(p_0-1)}{2p_0-1}}(\Omega)} ds$$

$$\leq c_2 \chi_v \frac{1 + m_\infty}{\lambda} \int_0^t \|\Delta u(\cdot, s)\|_{L^{p_0}(\Omega)}^{p_0} ds \tag{6.3.25}$$

$$+ c_2 \chi_v \frac{1 + m_\infty}{\lambda} \int_0^t e^{-(t-s)\frac{\lambda_1 p_0}{p_0-1}} \left(1 + (t-s)^{-\frac{1}{2} - \frac{N}{2p_0} + \frac{N}{4(p_0-1)}}\right)^{\frac{p_0}{p_0-1}} ds$$

$$+ c_2 \chi_v \sup_{t \in (0,T)} \|\nabla v(\cdot, s)\|_{L^{2(p_0-1)}(\Omega)}$$

$$\cdot \int_0^t e^{-(t-s)\lambda_1} \left(1 + (t-s)^{-\frac{1}{2} - \frac{N(2p_0-1)}{4p_0(p_0-1)} + \frac{N}{4(p_0-1)}}\right) M_3 \varepsilon e^{-\alpha s} ds$$

$$\leq c_2 \chi_v C_1 \frac{1 + m_\infty}{\lambda} + c_2 \chi_v C_2 \frac{1 + m_\infty}{\lambda} + 2c_2 c_{10} \chi_v M_3 \varepsilon e^{-\alpha t} \sup_{t \in (0,T)} \|\nabla v(\cdot, s)\|_{L^{2(p_0-1)}(\Omega)}$$

where $C_1 := \int_0^t \|\Delta u(\cdot, s)\|_{L^{p_0}(\Omega)}^{p_0} ds$ is bounded by Lemma 6.9 and

$$C_2 := \int_0^t e^{-(t-s)\frac{\lambda_1 p_0}{p_0-1}} \left(1 + (t-s)^{-\frac{1}{2} - \frac{N}{2p_0} + \frac{N}{4(p_0-1)}}\right)^{\frac{p_0}{p_0-1}} ds < +\infty$$

for $p_0 > 1 + \frac{N}{2}$.

Next, by Lemmas $1.1(ii)$, 4.3, 6.6 and 6.7, we get

$$\int_0^t \|\nabla e^{(t-s)\Delta} vw\|_{L^{2(p_0-1)}(\Omega)} ds$$

$$\leq c_2 |\Omega|^{\frac{1}{2(p_0-1)}} \int_0^t (1 + (t-s)^{-\frac{1}{2}}) e^{-\lambda_1(t-s)} \|w\|_{L^\infty(\Omega)} \|v\|_{L^\infty(\Omega)} ds \tag{6.3.26}$$

$$\leq c_2 |\Omega|^{\frac{1}{2(p_0-1)}} M_2 \varepsilon \int_0^t (1 + (t-s)^{-\frac{1}{2}}) e^{-\lambda_1(t-s)} e^{-\alpha s} \frac{1 + m_\infty}{\lambda} ds$$

$$\leq 2|\Omega|^{\frac{1}{2(p_0-1)}} c_{10} c_2 M_2 \frac{1 + m_\infty}{\lambda} \varepsilon e^{-\alpha t}.$$

Inserting (6.3.24)–(6.3.26) into (6.3.23) and using (6.3.4), we readily get

$$\sup_{t\in(0,T)} \|\nabla v\|_{L^{2(p_0-1)}(\Omega)} \leq 2c_3\|\nabla v_0(\cdot)\|_{L^{2(p_0-1)}(\Omega)} + 2c_2\chi_v C_1\frac{1+m_\infty}{\lambda} + 2c_2\chi_v C_2\frac{1+m_\infty}{\lambda}$$

$$+ 4|\Omega|^{\frac{1}{2(p_0-1)}}c_{10}c_2\frac{1+m_\infty}{\lambda},$$

and thereby from Lemma 6.5, we arrive at

$$\int_\Omega |\nabla v(x,t)|^{2(p_0-1)}dx \leq C_5, \quad t\in(0,T) \tag{6.3.27}$$

with some $C_5 > 0$ independent of T, $M_i (i = 1, 2, 3)$ and hence complete the proof.

Beyond the weak information of $\triangle w$ in Lemma 6.9, we now turn the boundedness of ∇v into a statement on decay of $\triangle w$.

Lemma 6.11 *Under the assumptions of Theorem 6.1, for all $p_0 > \frac{N}{2}$,*

$$\|\triangle w(\cdot,t)\|_{L^{p_0}(\Omega)} \leq M_6\varepsilon e^{-\alpha t} \quad \text{for all } t\in(0,T) \tag{6.3.28}$$

with

$$M_6 := 2c_1 + 2c_4 M_2 c_{10}\lambda\left(|\Omega|^{\frac{1}{2p_0}} + C_5|\Omega|^{\frac{p_0-2}{2p_0(p_0-1)}}\right) + 2c_4 M_4 c_{10}(1+m_\infty+\mu)|\Omega|^{\frac{1}{2p_0}}.$$

Proof From (6.3.8), we have

$$\|\triangle w(\cdot,t)\|_{L^{p_0}(\Omega)}$$

$$\leq \|e^{t\triangle}\triangle w_0\|_{L^{p_0}(\Omega)} + \int_0^t \|e^{(t-s)\triangle}\triangle((\lambda(u+v)+\mu)w)(\cdot,s)\|_{L^{p_0}(\Omega)}ds \tag{6.3.29}$$

$$\leq 2c_1 e^{-\lambda_1 t}\|\triangle w_0\|_{L^{p_0}(\Omega)}$$

$$+ \int_0^t \|e^{(t-s)\triangle}\nabla\cdot(\lambda(\nabla u+\nabla v)w+(\lambda(u+v)+\mu)\nabla w)(\cdot,s)\|_{L^{p_0}(\Omega)}ds$$

$$\leq 2c_1 e^{-\lambda_1 t}\|\triangle w_0\|_{L^{p_0}(\Omega)} + \lambda\int_0^t \|e^{(t-s)\triangle}\nabla\cdot((\nabla u+\nabla v)w)(\cdot,s)\|_{L^{p_0}(\Omega)}ds$$

$$+ \int_0^t \|e^{(t-s)\triangle}\nabla\cdot((\lambda(u+v)+\mu)\nabla w)(\cdot,s)\|_{L^{p_0}(\Omega)}ds.$$

From Lemma 1.1(i) and the fact that $\|\triangle w_0\|_{L^{p_0}(\Omega)} \leq \varepsilon$, we obtain that

$$\|e^{t\triangle}\triangle w_0\|_{L^{p_0}(\Omega)} \leq 2c_1\varepsilon e^{-\lambda_1 t}. \tag{6.3.30}$$

Now, we estimate the last two integrals on the right-hand side of the above inequality. From the definition of T, Lemmas 1.1(iv), 4.3, 6.6, 6.8, 6.10 and (6.3.4), it follows that

$$\lambda \int_0^t \|e^{(t-s)\Delta} \nabla \cdot ((\nabla u(\cdot, s) + \nabla v(\cdot, s))w(\cdot, s))\|_{L^{p_0}(\Omega)} ds$$

$$\leq c_4 \lambda \int_0^t (1 + (t-s)^{-\frac{1}{2}})e^{-\lambda_1(t-s)} \|w(\cdot, s)\|_{L^\infty(\Omega)} \|\nabla u(\cdot, s)\|_{L^{p_0}(\Omega)} ds \quad (6.3.31)$$

$$+ c_4 \lambda \int_0^t (1 + (t-s)^{-\frac{1}{2}})e^{-\lambda_1(t-s)} \|w(\cdot, s)\|_{L^\infty(\Omega)} \|\nabla v(\cdot, s)\|_{L^{p_0}(\Omega)} ds$$

$$\leq c_4 M_2 \varepsilon \lambda |\Omega|^{\frac{1}{2p_0}} \int_0^t (1 + (t-s)^{-\frac{1}{2}})e^{-\lambda_1(t-s)} e^{-2\alpha s} M_3 \varepsilon ds$$

$$+ c_4 M_2 \varepsilon \lambda |\Omega|^{\frac{p_0-2}{2p_0(p_0-1)}} \int_0^t (1 + (t-s)^{-\frac{1}{2}})e^{-\lambda_1(t-s)} e^{-\alpha s} C_5 ds$$

$$\leq 2c_4 M_2 c_{10} \lambda \left(|\Omega|^{\frac{1}{2p_0}} + C_5 |\Omega|^{\frac{p_0-2}{2p_0(p_0-1)}} \right) \varepsilon e^{-\alpha t}$$

and

$$\int_0^t \|e^{(t-s)\Delta} \nabla \cdot ((\lambda(u+v) + \mu)(\cdot, s)\nabla w(\cdot, s))\|_{L^{p_0}(\Omega)} ds$$

$$\leq c_4 \int_0^t (1 + (t-s)^{-\frac{1}{2}})e^{-\lambda_1(t-s)} \|\nabla w(\cdot, s)\|_{L^{p_0}(\Omega)} \|(\lambda(u+v) + \mu)(\cdot, s)\|_{L^\infty(\Omega)} ds$$

$$\leq c_4 M_4 \varepsilon (1 + m_\infty + \mu) |\Omega|^{\frac{1}{2p_0}} \int_0^t (1 + (t-s)^{-\frac{1}{2}})e^{-\lambda_1(t-s)} e^{-\alpha s} ds \quad (6.3.32)$$

$$\leq 2c_4 M_4 c_{10} (1 + m_\infty + \mu) |\Omega|^{\frac{1}{2p_0}} \varepsilon e^{-\alpha t}.$$

Inserting (6.3.30), (6.3.31) and (6.3.32) into (6.3.29), we obtain

$$\|\Delta w(\cdot, t)\|_{L^{p_0}(\Omega)} \leq M_6 \varepsilon e^{-\alpha t}$$

and thereby complete the proof.

Lemma 6.12 *Under the assumptions of Theorem 6.1, for all $p_0 > \frac{N}{2}$,*

$$\|\nabla u(\cdot, t)\|_{L^{2p_0}(\Omega)} \leq \frac{M_3}{2} \varepsilon e^{-\alpha t} \quad \text{for all } t \in (0, T). \quad (6.3.33)$$

Proof By (6.3.9), we have

$$\|\nabla u(\cdot, t)\|_{L^{2p_0}(\Omega)} \leq \|\nabla e^{t\Delta} u_0(\cdot)\|_{L^{2p_0}(\Omega)} + \chi_u \int_0^t \|\nabla e^{(t-s)\Delta} \nabla \cdot (u \nabla w)(\cdot, s)\|_{L^{2p_0}(\Omega)} ds$$

$$\quad (6.3.34)$$

$$+ \int_0^t \|\nabla e^{(t-s)\Delta} (uw)(\cdot, s)\|_{L^{2p_0}(\Omega)} ds.$$

From Lemma 1.1(iii) and the fact that $\|\nabla u_0(\cdot)\|_{L^{2p_0}(\Omega)} \leq \varepsilon$, we obtain

$$\|\nabla e^{t\Delta} u_0(\cdot)\|_{L^{2p_0}(\Omega)} \le 2c_3 \varepsilon e^{-\lambda_1 t}. \tag{6.3.35}$$

Now, we estimate the last two integrals on the right-hand side of (6.3.34). From the definition of T, Lemmas 1.1(ii), 4.3, 6.6, 6.8 and 6.11, it follows that

$$\chi_u \int_0^t \|\nabla e^{(t-s)\Delta} \nabla \cdot (u\nabla w)(\cdot, s)\|_{L^{2p_0}(\Omega)} ds$$

$$\le c_2 \chi_u \int_0^t e^{-(t-s)\lambda_1} \left(1 + (t-s)^{-\frac{1}{2} - \frac{N}{4p_0}}\right) \|u(\cdot, s)\Delta w(\cdot, s)\|_{L^{p_0}(\Omega)} ds \tag{6.3.36}$$

$$+ c_2 \chi_u \int_0^t e^{-(t-s)\lambda_1} \left(1 + (t-s)^{-\frac{1}{2} - \frac{N}{4p_0}}\right) \|\nabla u(\cdot, s)\nabla w(\cdot, s)\|_{L^{p_0}(\Omega)} ds$$

$$\le c_2 \chi_u \int_0^t e^{-(t-s)\lambda_1} \left(1 + (t-s)^{-\frac{1}{2} - \frac{N}{4p_0}}\right) M_6 \varepsilon e^{-\alpha s} \frac{1 + m_\infty}{\lambda} ds$$

$$+ c_2 \chi_u \int_0^t e^{-(t-s)\lambda_1} \left(1 + (t-s)^{-\frac{1}{2} - \frac{N}{4p_0}}\right) M_4 M_3 \varepsilon^2 e^{-2\alpha s} ds$$

$$\le 2c_{10} c_2 \chi_u \left(M_6 \frac{1 + m_\infty}{\lambda} + M_4 M_3 \varepsilon\right) \varepsilon e^{-\alpha t}$$

and

$$\int_0^t \|\nabla e^{(t-s)\Delta} uw\|_{L^{2p_0}(\Omega)} ds$$

$$\le c_2 |\Omega|^{\frac{1}{2p_0}} \int_0^t \left(1 + (t-s)^{-\frac{1}{2}}\right) e^{-\lambda_1(t-s)} \|w\|_{L^\infty(\Omega)} \|u\|_{L^\infty(\Omega)} ds \tag{6.3.37}$$

$$\le c_2 M_2 \varepsilon |\Omega|^{\frac{1}{2p_0}} \int_0^t \left(1 + (t-s)^{-\frac{1}{2}}\right) e^{-\lambda_1(t-s)} e^{-\alpha s} \frac{1 + m_\infty}{\lambda} ds$$

$$\le 2c_{10} c_2 M_2 \frac{1 + m_\infty}{\lambda} \varepsilon |\Omega|^{\frac{1}{2p_0}} e^{-\alpha t}.$$

Inserting (6.3.35)–(6.3.37) into (6.3.18) and using (6.3.3), we readily get

$$\|\nabla u\|_{L^{2p_0}(\Omega)} \le 2c_3 \varepsilon e^{-\lambda_1 t} + 2c_{10} c_2 \left(\left(\chi_u M_6 + M_2 |\Omega|^{\frac{1}{2p_0}}\right) \frac{1 + m_\infty}{\lambda} + \chi_u M_4 M_3 \varepsilon\right) \varepsilon e^{-\alpha t}$$

$$\le \frac{M_3}{2} \varepsilon e^{-\alpha t}$$

and thereby complete the proof.

Lemma 6.13 *Under the assumptions of Theorem 6.1, for all $t \in (0, T)$,*

$$\|(\lambda(u + v) + w + z)(\cdot, t) - e^{t\Delta}(\lambda(u_0 + v_0) + w_0)\|_{L^\infty(\Omega)} \le \frac{M_1}{2} \varepsilon e^{-\alpha t}.$$

Proof According to (6.3.7) and Lemma 1.1(iv), we have

$$\|(\lambda(u+v)+w+z)(\cdot,t)-e^{t\Delta}(\lambda(u_0+v_0)+w_0)\|_{L^\infty(\Omega)}$$

$$\leq \lambda\int_0^t \|e^{(t-s)\Delta}(\nabla\cdot(\chi_u u\nabla w+\chi_v v\nabla u))(\cdot,s)\|_{L^\infty(\Omega)}ds \qquad (6.3.38)$$

$$\leq \lambda\chi_u\int_0^t \|e^{(t-s)\Delta}\nabla\cdot(u\nabla w)(\cdot,s)\|_{L^\infty(\Omega)}ds + \lambda\chi_v\int_0^t \|e^{(t-s)\Delta}\nabla\cdot(v\nabla u)(\cdot,s)\|_{L^\infty(\Omega)}ds$$

$$\leq c_4\lambda\chi_u\int_0^t \left(1+(t-s)^{-\frac{1}{2}-\frac{N}{4p_0}}\right)e^{-\lambda_1(t-s)}\|u(\cdot,s)\|_{L^\infty(\Omega)}\|\nabla w(\cdot,s)\|_{L^{2p_0}(\Omega)}ds$$

$$+ c_4\lambda\chi_v\int_0^t \left(1+(t-s)^{-\frac{1}{2}-\frac{N}{4p_0}}\right)e^{-\lambda_1(t-s)}\|v(\cdot,s)\|_{L^\infty(\Omega)}\|\nabla u(\cdot,s)\|_{L^{2p_0}(\Omega)}ds$$

$$=:I_1+I_2.$$

Now, we need to estimate I_1 and I_2. Firstly, from the definition of T, Lemmas 4.3, 6.6 and 6.8, we obtain

$$I_1 \leq c_4\chi_u(1+m_\infty)M_4\varepsilon\int_0^t \left(1+(t-s)^{-\frac{1}{2}-\frac{N}{4p_0}}\right)e^{-\lambda_1(t-s)}e^{-\alpha s}ds \qquad (6.3.39)$$

$$\leq 2c_{10}c_4\chi_u(1+m_\infty)M_4\varepsilon e^{-\alpha t}$$

and

$$I_2 \leq c_4\lambda\chi_v(1+m_\infty)M_3\varepsilon\int_0^t \left(1+(t-s)^{-\frac{1}{2}-\frac{N}{4p_0}}\right)e^{-\lambda_1(t-s)}e^{-\alpha s}ds$$

$$\leq 2c_{10}c_4\chi_v(1+m_\infty)M_3\varepsilon e^{-\alpha t}. \qquad (6.3.40)$$

Combining (6.3.38)–(6.3.40) along with (6.3.1) leads to

$$\|(\lambda(u+v)+w+z)(\cdot,t)-e^{t\Delta}(\lambda(u_0+v_0)+w_0)\|_{L^\infty(\Omega)}$$

$$\leq 2c_{10}c_4(1+m_\infty)(\chi_u M_4+\chi_v M_3)\varepsilon e^{-\alpha t}$$

$$\leq \frac{1}{2}M_1\varepsilon e^{-\alpha t}$$

and hence ends the proof.

Now, we have prepared the major parts of the proof of Theorem 6.1 and thus can verify asymptotic properties stated there.

Proof of the Theorem 6.1.   First we claim that $T=T_{max}$. In fact, if $T<T_{max}$, then by Lemmas 6.7, 6.12 and 6.13, we have

$$\|w(\cdot, t)\|_{L^\infty(\Omega)} \le \frac{M_2}{2}\varepsilon e^{-\alpha t},$$

$$\|\nabla u(\cdot, t)\|_{L^\infty(\Omega)} \le \frac{M_3}{2}\varepsilon e^{-\alpha t},$$

$$\|(\lambda(u + v) + w + z)(\cdot, t) - e^{t\Delta}(\lambda(u_0 + v_0) + w_0)\|_{L^\infty(\Omega)} \le \frac{M_1}{2}\varepsilon e^{-\alpha t}$$

for all $t \in (0, T)$, which contradicts the definition of T in (6.3.7).

Next, we show that $T_{max} = \infty$. In fact, if $T_{max} < \infty$, then in view of the definition of T, Lemmas 6.6, 6.8 and 6.10, we obtain that for any $p_0 > 1 + \frac{N}{2}$,

$$\lim_{t \to T_{max}} \left(\|u(\cdot, t)\|_{W^{1,2p_0}(\Omega)} + \|v(\cdot, t)\|_{W^{1,2(p_0-1)}(\Omega)} + \|w(\cdot, t)\|_{W^{1,2p_0}(\Omega)} \right) < \infty,$$

which contradicts with Lemma 6.4. Therefore, we have $T_{max} = \infty$.

Integrating the first equation in (6.1.4) over Ω, we have

$$\int_\Omega u(x, t)dx = \int_\Omega u_0(x)dx + \int_0^t \int_\Omega (uw)(x, s)dxds,$$

which, along with the nonnegative property of u, w and the fact that

$$\|u(\cdot, t)\|_{L^\infty(\Omega)} < \frac{1 + m_\infty}{\lambda}, \quad \|w(\cdot, t)\|_{L^\infty(\Omega)} < M_2\varepsilon e^{-\alpha t},$$

warrants

$$\lim_{t \to \infty} \int_\Omega u(x, t)dx = \int_\Omega u_0(x)dx + \int_0^{+\infty} \int_\Omega (uw)(x, s)dxds, \qquad (6.3.41)$$

as well as

$$\lim_{t \to \infty} \bar{u}(t) = \frac{1}{|\Omega|} \left(\int_\Omega u_0(x)dx + \int_0^{+\infty} \int_\Omega (uw)(x, s)dxds \right) := u^*. \qquad (6.3.42)$$

As a consequence of the latter, we immediately have

$$\begin{aligned}
0 \le u^* - \bar{u}(t) &= \frac{1}{|\Omega|} \int_t^{+\infty} \int_\Omega (uw)(x, s)dxds \\
&\le \frac{1 + m_\infty}{\lambda|\Omega|} \int_t^{+\infty} \int_\Omega w(x, s)dxds \qquad (6.3.43) \\
&\le \frac{1 + m_\infty}{\alpha\lambda} M_2\varepsilon e^{-\alpha t}.
\end{aligned}$$

On the other hand, by Poincare's inequality,

$$\|u - \bar{u}\|_{L^{2p_0}(\Omega)} \leq C_1 \|\nabla u\|_{L^{2p_0}(\Omega)},$$

and thanks to $W^{1,2p_0}(\Omega) \hookrightarrow L^\infty(\Omega)$ for $p_0 > \frac{N}{2}$, we can find $C_2 > 0$ such that

$$\|u - \bar{u}\|_{L^\infty(\Omega)} \leq C_2 \|u - \bar{u}\|_{W^{1,2p_0}(\Omega)} \leq C_2(1 + C_1)\|\nabla u\|_{L^{2p_0}(\Omega)}. \tag{6.3.44}$$

Therefore, by (6.3.43) and the fact that $\|\nabla u\|_{L^{2p_0}(\Omega)} \leq M_3 \varepsilon e^{-\alpha t}$, we can pick $K_1 > 0$ such that

$$\begin{aligned}
\|u - u^*\|_{L^\infty(\Omega)} &\leq \|u - \bar{u}\|_{L^\infty(\Omega)} + \|\bar{u} - u^*\|_{L^\infty(\Omega)} \\
&\leq C_2(1 + C_1)\|\nabla u\|_{L^{2p_0}(\Omega)} + |\bar{u}(t) - u^*| \\
&\leq K_1 \varepsilon e^{-\alpha t}.
\end{aligned} \tag{6.3.45}$$

On the other hand, from (6.3.12) and Lemma 1.1(ii), we infer that

$$\begin{aligned}
\|\nabla z(\cdot, t)\|_{L^{2p_0}} &= \mu \int_0^t \|\nabla e^{(t-s)\Delta} w(\cdot, s)\|_{L^{2p_0}(\Omega)} ds \\
&\leq \mu c_2 |\Omega|^{\frac{1}{2p_0}} \int_0^t e^{-(t-s)\lambda_1}(1 + t^{-\frac{1}{2}})\|w(\cdot, s)\|_{L^\infty(\Omega)} ds \\
&\leq 2\mu c_2 c_{10} M_2 |\Omega|^{\frac{1}{2p_0}} \varepsilon e^{-\alpha t}.
\end{aligned} \tag{6.3.46}$$

By similar procedure as that in the derivation of (6.3.45), there exists constant $K_2 > 0$ such that

$$\|z - z^*\|_{L^\infty(\Omega)} \leq K_2 \varepsilon e^{-\alpha t} \tag{6.3.47}$$

with

$$z^* := \frac{\mu}{|\Omega|} \int_0^{+\infty} \int_\Omega w(x, s) dx ds. \tag{6.3.48}$$

Then from the fact that

$$\begin{aligned}
\lambda \|v - v^*\|_{L^\infty(\Omega)} &\leq \|(\lambda u + \lambda v + w + z) - m_\infty\|_{L^\infty(\Omega)} + \lambda \|u - u^*\|_{L^\infty(\Omega)} \\
&\quad + \|w\|_{L^\infty(\Omega)} + \|z - z^*\|_{L^\infty(\Omega)}
\end{aligned}$$

with

$$v^* := \frac{1}{\lambda}(m_\infty - z^*) - u^*, \tag{6.3.49}$$

using (6.3.13), (6.3.14), (6.3.45) and (6.3.47), there exists $K_3 > 0$ such that

$$\|v - v^*\|_{L^\infty(\Omega)} \le K_3 \varepsilon e^{-\alpha t}.$$

The decay estimates claimed in Theorem 6.1 readily follow and the proof of this theorem is thus completed.

6.4 Boundedness of Solutions to an Oncolytic Virotherapy Model

6.4.1 Some Basic a Prior Estimates

For the convenience in our subsequent estimation procedure, we let

$$\chi_u := \frac{\xi_u}{D_u}, \quad \chi_w := \frac{\xi_w}{D_w} \quad \text{and} \quad \chi_z := \frac{\xi_z}{D_z}$$

and introduce the variable change used in several precedents (Fontelos et al. 2002; Pang and Wang 2018; Tao and Winkler 2014b)

$$a = ue^{-\chi_u v} \quad b = we^{-\chi_w v} \quad \text{and} \quad c = ze^{-\chi_z v}$$

upon which (6.1.10) takes the following form:

$$\begin{cases}
a_t = D_u e^{-\chi_u v} \nabla \cdot (e^{\chi_u v} \nabla a) + f(a, b, v, c), & x \in \Omega, t > 0, \\
b_t = D_w e^{-\chi_w v} \nabla \cdot (e^{\chi_w v} \nabla b) + g(a, b, v, c), & x \in \Omega, t > 0, \\
v_t = -(\alpha_u a e^{\chi_u v} + \alpha_w b e^{\chi_w v})v + \mu_v v(1 - v), & x \in \Omega, t > 0, \\
c_t = D_z e^{-\chi_z v} \nabla \cdot (e^{\chi_z v} \nabla c) + h(a, b, v, c), & x \in \Omega, t > 0, \\
\dfrac{\partial a}{\partial v} = \dfrac{\partial b}{\partial v} = \dfrac{\partial c}{\partial v} = 0, & x \in \partial\Omega, t > 0, \\
a(x, 0) = u_0(x)e^{-\chi_u v_0(x)}, \ b(x, 0) = w_0(x)e^{-\chi_w v_0(x)}, & x \in \Omega, \\
v_0(x, 0) = v_0(x), \ c(x, 0) = z_0(x)e^{-\chi_z v_0(x)}, & x \in \Omega
\end{cases} \tag{6.4.1}$$

with

$$f(a, b, v, c) := \mu_u a(1 - a^r e^{r\chi_u v}) - \frac{\rho a c e^{\chi_z v}}{k_u + \theta a e^{\chi_u v}} + \chi_u a(\alpha_u a e^{\chi_u v} + \alpha_w b e^{\chi_w v})v$$

$$\qquad - \chi_u \mu_v a v(1 - v)$$

$$g(a, b, v, c) := -\delta_w b + \frac{\rho a c e^{(\chi_u + \chi_z - \chi_w)v}}{k_u + \theta a e^{\chi_u v}} + \chi_w b(\alpha_u a e^{\chi_u v} + \alpha_w b e^{\chi_w v})v$$

$$\qquad - \chi_w \mu_v b v(1 - v)$$

$$h(a, b, v, c) := -\delta_z c - \frac{\rho a c e^{\chi_u v}}{k_u + \theta a e^{\chi_u v}} + \beta b e^{(\chi_w - \chi_z)v} + \chi_z c(\alpha_u a e^{\chi_u v} + \alpha_w b e^{\chi_w v})v$$

$$\qquad - \chi_z \mu_v c v(1 - v).$$

It is noted that (6.1.10) and (6.4.1) are equivalent in this framework of classical solutions. The following basic statement on the local existence and extensibility criterion of classical solutions to (6.4.1) can be proved by a straightforward adaptation of the reasoning in Pang and Wang (2018) and Tao and Winkler (2020b).

Lemma 6.14 *Let D_u, D_w, D_z, ξ_u, ξ_w, ξ_z, μ_u, μ_v, ρ, k_u, α_u, α_w, β, δ_w and δ_z are positive parameters, and assume that $r \geq 1$, $\theta \geq 0$. Then there exist $T_{max} \in (0, \infty]$ and a uniquely determined quadruple $(a, b, v, c) \in (C^{2,1}(\overline{\Omega} \times [0, T_{max})))^4$ which solves (6.4.1) in the classical sense and $a > 0$, $b > 0$, $c > 0$ and $v > 0$ in $\Omega \times (0, T_{max})$, and that if $T_{max} < +\infty$, then*

$$\|a(\cdot, t)\|_{L^\infty(\Omega)} + \|b(\cdot, t)\|_{L^\infty(\Omega)} + \|\nabla v(\cdot, t)\|_{L^4(\Omega)} + \|c(\cdot, t)\|_{L^\infty(\Omega)} \to \infty \ \text{as} \ t \nearrow T_{max}. \tag{6.4.2}$$

Proof Invoking well-established fixed point arguments and applying the standard parabolic regularity theory, one can readily verify the local existence and uniqueness of classical solutions, as well as the extensibility criterion (6.4.2) (cf. Pang and Wang 2018; Tao and Winkler 2014b for instance). With the help of the maximum principle, we can also verify the asserted positivity of the solutions.

From now on without any further explicit mentioning, we shall suppose that the assumptions of Theorem 6.2 are satisfied, and let (a, b, v, c) and $T_{max} \in (0, \infty]$ be as provided by Lemma 6.14. Moreover, we may tacitly switch between these variables and the quadruple (u, w, v, z) if necessary.

The following important properties of solutions of (6.1.10) can be easily checked.

Lemma 6.15 *Let $T > 0$. Then solution (u, w, v, z) of (6.1.10) satisfies*

$$\|u(\cdot, t)\|_{L^1(\Omega)} \leq u^* := \max\{|\Omega|, \|u_0\|_{L^1(\Omega)}\} \quad \text{for all } t \in (0, \hat{T}), \tag{6.4.3}$$

$$\|v(\cdot, t)\|_{L^\infty(\Omega)} \leq v^* := \max\{1, \|v_0\|_{L^\infty(\Omega)}\} \quad \text{for all } t \in (0, \hat{T}) \tag{6.4.4}$$

and

$$\|w(\cdot, t)\|_{L^1(\Omega)} \leq w^* := \max\left\{\|u_0\|_{L^1(\Omega)} + \|w_0\|_{L^1(\Omega)}, \frac{4\mu_u|\Omega|}{\min\{\mu_u, \delta_w\}}\right\} \tag{6.4.5}$$

for all $t \in (0, \hat{T})$ as well as

$$\|z(\cdot, t)\|_{L^1(\Omega)} \leq z^* := \max\left\{\|z_0\|_{L^1(\Omega)}, \frac{\beta w^*}{\delta_z}\right\} \quad \text{for all } t \in (0, \hat{T}) \tag{6.4.6}$$

where $\hat{T} := \min\{T, T_{max}\}$.

Proof Integrating the first equation in (6.1.10) over Ω yields

$$\frac{d}{dt} \int_\Omega u \le \mu_u \int_\Omega u - \mu_u \int_\Omega u^{r+1} \tag{6.4.7}$$

due to $z \ge 0$. Since $(\int_\Omega u)^{r+1} \le |\Omega|^r \int_\Omega u^{r+1}$ by the Cauchy–Schwartz inequality, (6.4.7) implies that $y(t) := \int_\Omega u(\cdot, t)$ satisfies

$$y'(t) \le \mu_u y(t) - \frac{\mu_u}{|\Omega|^r} y^{1+r}(t) \quad \text{for all } (0, \hat{T}),$$

from which (6.4.3) follows by the Bernoulli inequality. On the other hand, due to the nonnegativity of u, w and v in $\overline{\Omega} \times (0, \hat{T})$, the comparison principle entails that $v_t \le \mu_v v(1 - v)$ and thus the estimate in (6.4.4) follows similarly.

Once more integrating the equations in (6.1.10) over Ω and using the fact that $2u \le u^{r+1} + 4$, we can see that

$$\frac{d}{dt} \int_\Omega u + \mu_u \int_\Omega u \le 4\mu_u |\Omega| - \rho \int_\Omega \frac{uz}{k_u + \theta u} \tag{6.4.8}$$

and

$$\frac{d}{dt} \int_\Omega w + \delta_w \int_\Omega w \le \rho \int_\Omega \frac{uz}{k_u + \theta u} \tag{6.4.9}$$

as well as

$$\frac{d}{dt} \int_\Omega z + \delta_z \int_\Omega z \le -\rho \int_\Omega \frac{uz}{k_u + \theta u} + \beta \int_\Omega w. \tag{6.4.10}$$

Combining (6.4.8)–(6.4.9), we obtain that

$$\frac{d}{dt} \left(\int_\Omega u + \int_\Omega w \right) + \mu_u \int_\Omega u + \delta_w \int_\Omega w \le 4\mu_u |\Omega|,$$

which entails that $y(t) := \int_\Omega u + \int_\Omega w$ satisfies

$$y'(t) + \min\{\mu_u, \delta_w\} y(t) \le 4\mu_u |\Omega|.$$

Hence, using the Bernoulli inequality to the above inequality, we get the estimate in (6.4.5). Further, it follows from (6.4.10) that

$$\frac{d}{dt} \int_\Omega z + \delta_z \int_\Omega z \le \beta w^*$$

and thereby derive (6.4.6) by an ODE comparison argument.

6.4.2 Bounds for a, b and c in $L \log L$

This section aims to construct an Lyapunov-like functional involving the logarithmic entropy of a, b and c, rather than that of u, w and z, which provides some regularity information of solutions that forms the crucial step in establishing L^∞ bounds for u, w and z in the present spatially two-dimensional setting. It should be mentioned that upon the special structure of (6.1.9), inter alia neglecting haptotactic migration processes of oncolytic viruses z, the energy-like functional $\mathscr{F}$ in Tao and Winkler (2020b) can be achieved by appropriately combining the logarithmic entropy of u, w, Dirichlet integral of $\sqrt{v}$ and integral of z^2 in line with some precedent studies (see Tao and Winkler 2014b; Winkler 2018b).

The first step of our approaches consists in testing the first equation of (6.4.1) against $\ln a$.

Lemma 6.16 For any $\varepsilon \in (0, 1)$, there exists $K_1(\varepsilon) > 0$ such that

$$\frac{d}{dt} \int_\Omega e^{\chi_u v} a \ln a + D_u \int_\Omega e^{\chi_u v} \frac{|\nabla a|^2}{a} + \frac{\mu_u}{2} \int_\Omega (\ln a + 1) u^{r+1} \le \varepsilon \int_\Omega w^2 + K_1(\varepsilon). \tag{6.4.11}$$

Proof From the first equation in (6.4.1), it follows

$$(a e^{\chi_u v})_t = D_u \nabla \cdot (e^{\chi_u v} \nabla a) + \mu_u u (1 - u^r) - \frac{\rho u z}{k_u + \theta u}.$$

By the positivity of a in $\overline{\Omega} \times (0, \infty)$, testing the first equation in (6.4.1) by $\ln a$ then shows that

$$\frac{d}{dt} \int_\Omega e^{\chi_u v} a \ln a$$

$$= \int_\Omega (e^{\chi_u v} a)_t \ln a + \int_\Omega e^{\chi_u v} a_t$$

$$= -D_u \int_\Omega e^{\chi_u v} \frac{|\nabla a|^2}{a} + \int_\Omega \left(\mu_u u (1 - u^r) - \frac{\rho u z}{k_u + \theta u} \right) \ln a + \int_\Omega f(a, b, v, z) e^{\chi_u v}$$

$$= -D_u \int_\Omega e^{\chi_u v} \frac{|\nabla a|^2}{a} + \int_\Omega (\ln a + 1) \left(\mu_u u (1 - u^r) - \frac{\rho u z}{k_u + \theta u} \right) \tag{6.4.12}$$

$$+ \chi_u \int_\Omega u (\alpha_u u + \alpha_w w) v - \chi_u \mu_v \int_\Omega u v (1 - v)$$

$$\le -D_u \int_\Omega e^{\chi_u v} \frac{|\nabla a|^2}{a} - \mu_u \int_\Omega (\ln a + 1)(u^{r+1} - u) - \rho \int_\Omega \frac{uz}{k_u + \theta u} \ln a$$

$$+ \chi_u \int_\Omega u (\alpha_u u + \alpha_w w) v + \chi_u \mu_v \int_\Omega u v^2.$$

By (6.4.6), we see that

$$-\rho \int_\Omega \frac{uz}{k_u + \theta u} \ln a = -\rho \int_\Omega \frac{z e^{\chi_u v}}{k_u + \theta u} a \ln a \le \frac{\rho e^{\chi_u v^*}}{k_u e} \int_\Omega z \le \frac{\rho e^{\chi_u v^*}}{k_u e} z^*,$$

since $a \ln a \ge -e^{-1}$ for all $a > 0$ and $v(x, t) \le v^*$ for all $x \in \Omega, t > 0$ by (6.4.4). Apart from that, for any $\varepsilon \in (0, 1)$, there exists $C_1(\varepsilon) > 0$ such that

$$\mu_u \int_\Omega (\ln a + 1)u + \chi_u \int_\Omega u(\alpha_u u + \alpha_w w)v + \chi_u \mu_v \int_\Omega uv^2$$
$$\le \frac{\mu_u}{2} \int_\Omega (\ln a + 1)u^2 + \varepsilon \int_\Omega w^2 + C_1(\varepsilon)$$

due to $a^2 \le \varepsilon_1 a^2 \ln a + e^{\frac{2}{\varepsilon_1}}$, $a \ln a \le \varepsilon_1 a^2 \ln a - \varepsilon_1^{-1} \ln \varepsilon_1$ and $a \le \varepsilon_1 a^2 \ln a + 2e^{\frac{2}{\varepsilon_1}}$ for any $\varepsilon_1 \in (0, 1)$. Therefore, inserting above two inequalities into (6.4.12), we arrive at (6.4.11).

Lemma 6.17 *There exists $c^* > 0$ with the property that if $\xi_w \alpha_w < c^*$, then one can find $\varepsilon_0 \in (0, 1)$ and $K_2 > 0$ such that for all $\varepsilon \in (0, \varepsilon_0)$,*

$$\frac{d}{dt} \int_\Omega e^{\chi_w v} b \ln b + \frac{D_w}{4} \int_\Omega e^{\chi_w v} \frac{|\nabla b|^2}{b} + \delta_w \int_\Omega e^{\chi_w v} b \ln b$$
$$\le \varepsilon \|c\|^2_{L^2(\Omega)} + \frac{K_2}{\varepsilon} \|a\|^2_{L^{r+1}(\Omega)} + \frac{K_2}{\varepsilon},$$
$$(6.4.13)$$

where $r = 1$ if $\theta > 0$ and $r > 1$ if $\theta \ge 0$.

Proof From the second equation in (6.4.1), it follows that

$$(be^{\chi_w v})_t = D_w \nabla \cdot (e^{\chi_w v} \nabla b) - \delta_w w + \frac{\rho u z}{k_u + \theta u}.$$

By straightforward calculation relying on $0 \le v \le 1$ in $\overline{\Omega} \times (0, \infty)$ and the Young inequality, we then see that for any $\varepsilon > 0$,

$$\frac{d}{dt} \int_\Omega e^{\chi_w v} b \ln b + \delta_w \int_\Omega e^{\chi_w v} b \ln b + D_w \int_\Omega e^{\chi_w v} \frac{|\nabla b|^2}{b} \qquad (6.4.14)$$
$$= \int_\Omega (\ln b + 1) \left(\frac{\rho u z}{k_u + \theta u} - \delta_w w \right) + \chi_w \int_\Omega w(\alpha_u u + \alpha_w w)v - \chi_w \mu_v \int_\Omega wv(1 - v)$$
$$+ \delta_w \int_\Omega w \ln b$$
$$\le \rho \int_\Omega \frac{uz}{k_u + \theta u} \ln b + \rho \int_\Omega \frac{uz}{k_u + \theta u} + \chi_w \mu_v v^* \int_\Omega w + \chi_w \alpha_u v^* \int_\Omega wu + \chi_w \alpha_w v^* \int_\Omega w^2.$$

The first summand on the right-hand side of (6.4.14) will be estimated in the case $r = 1, \theta > 0$ and $r > 1, \theta \ge 0$, respectively.

For $r = 1$ and $\theta > 0$, we have

$$\frac{d}{dt}\int_\Omega e^{\chi_w v} b \ln b + \delta_w \int_\Omega e^{\chi_w v} b \ln b + D_w \int_\Omega e^{\chi_w v} \frac{|\nabla b|^2}{b} \tag{6.4.15}$$

$$\leq \frac{\rho}{\theta} \int_{b>1} z \ln b + (\varepsilon + \chi_w \alpha_w v^*) \int_\Omega w^2 + \frac{\rho}{\theta} \int_\Omega z + \frac{\chi_w^2 \alpha_u^2 v^{*2}}{\varepsilon} \int_\Omega u^2 + \chi_w \mu_v v^* \int_\Omega w.$$

Since $\ln^2 s \leq \frac{4}{e^2} s$ for all $s > 1$, an application of the Hölder inequality leads to

$$\frac{\rho}{\theta} \int_{b>1} z \ln b \leq \varepsilon \|z\|_{L^2(\Omega)}^2 + \frac{\rho^2}{\varepsilon \theta^2} \int_\Omega b.$$

In conjunction with (6.4.15), we can see that

$$\frac{d}{dt}\int_\Omega e^{\chi_w v} b \ln b + D_w \int_\Omega e^{\chi_w v} \frac{|\nabla b|^2}{b} + \delta_w \int_\Omega e^{\chi_w v} b \ln b \tag{6.4.16}$$

$$\leq (\varepsilon + \chi_w \alpha_w v^*) \|w\|_{L^2(\Omega)}^2 + \varepsilon \|z\|_{L^2(\Omega)}^2 + \frac{\chi_w^2 \alpha_u^2 v^{*2}}{\varepsilon} \|u\|_{L^2(\Omega)}^2$$

$$+ (\chi_w \mu_v v^* + \frac{\rho^2}{\varepsilon \theta^2}) \|w\|_{L^1(\Omega)} + \frac{\rho}{\theta} \|z\|_{L^1(\Omega)}.$$

To estimate the first term on the right-hand side of (6.4.16), by means of the two-dimensional Gagliardo–Nirenberg inequalities, we can find $K_g > 0$ such that

$$\|\varphi\|_{L^4(\Omega)}^4 \leq K_g \|\nabla \varphi\|_{L^2(\Omega)}^2 \|\varphi\|_{L^2(\Omega)}^2 + K_g \|\varphi\|_{L^2(\Omega)}^4 \quad \text{for all } \varphi \in W^{1,2}(\Omega). \tag{6.4.17}$$

Thereby thanks to (6.4.4) and (6.4.5), there exists $C_1 > 0$ such that

$$(\varepsilon + \chi_w \alpha_w v^*) \|w\|_{L^2(\Omega)}^2 \leq e^{2\chi_w v^*} (\varepsilon + \chi_w \alpha_w v^*) \|\sqrt{b}\|_{L^4(\Omega)}^4$$

$$\leq \frac{e^{2v^* \chi_w} K_g (\varepsilon + \chi_w \alpha_w v^*)}{4} \int_\Omega b \int_\Omega e^{\chi_w v} \frac{|\nabla b|^2}{b} + C_1$$

$$\leq \frac{e^{2\chi_w v^*} K_g (\varepsilon + \chi_w \alpha_w v^*)}{4} w^* \cdot \int_\Omega e^{\chi_w v} \frac{|\nabla b|^2}{b} + C_1$$

$$\tag{6.4.18}$$

$$\leq \frac{3D_w}{4} \int_\Omega e^{\chi_w v} \frac{|\nabla b|^2}{b} + C_1,$$

provided that $\xi_w \alpha_w < c^* := \frac{2D_w}{e^{2\chi_w v^*} K_g w^* v^*}$ and any $0 < \varepsilon < \varepsilon_0 := \min\{1, \frac{D_w}{e^{2\chi_w v^*} K_g w^* v^*}\}$. Therefore, along with Lemma 6.15 and the Hölder inequality, we insert (6.4.18) into (6.4.16) to arrive at (6.4.13).

While for $r > 1, \theta \geq 0$, we have

$$\frac{d}{dt} \int_\Omega e^{\chi_w v} b \ln b + \delta_w \int_\Omega e^{\chi_w v} b \ln b + D_w \int_\Omega e^{\chi_w v} \frac{|\nabla b|^2}{b} \qquad (6.4.19)$$

$$\leq \frac{\rho}{k_u} \int_\Omega zu \ln b + (\varepsilon + \chi_w \alpha_w v^*) \int_\Omega w^2 + \frac{\rho}{k_u} \int_\Omega uz + \frac{\chi_w^2 \alpha_u^2 v^{*2}}{\varepsilon} \int_\Omega u^2 + \chi_w \mu_v v^* \int_\Omega w.$$

Here, an apparently challenging issue is to estimate $\int_\Omega zu \ln b$ appropriately in terms of expression which can be controlled by the dissipation terms in (6.4.11). Since there exists $C_2 > 0$ such that $\ln^{\frac{2(r+1)}{r-1}} s \leq s + C_2$ for all $s > 1$, we can infer from (6.4.6) and the Hölder inequality that

$$\frac{\rho}{k_u} \int_\Omega zu \ln b \leq \frac{\rho}{k_u} \int_{\{b>1\}} zu \ln b$$

$$\leq \frac{\rho}{k_u} \|u\|_{L^{r+1}(\Omega)} \|z\|_{L^2(\Omega)} \left\{ \int_{\{b>1\}} (\ln b)^{\frac{2(r+1)}{r-1}} \right\}^{\frac{r-1}{2(r+1)}}$$

$$\leq \frac{\rho}{k_u} \|u\|_{L^{r+1}(\Omega)} \|z\|_{L^2(\Omega)} (\|b\|_{L^1(\Omega)} + C_2 |\Omega|)^{\frac{r-1}{2(r+1)}}$$

$$\leq \frac{\varepsilon}{2} \|z\|_{L^2(\Omega)}^2 + \frac{C_3}{\varepsilon} \|u\|_{L^{r+1}(\Omega)}^2$$

with $C_3 = \frac{\rho^2}{k_u^2} (\|b\|_{L^1(\Omega)} + C_2 |\Omega|)^{\frac{r-1}{r+1}}$.

Apart from that, by the Hölder inequality and the Young inequality, it is easy to see that

$$\frac{\rho}{k_u} \int_\Omega zu \leq \frac{\rho}{k_u} \|u\|_{L^2(\Omega)} \|z\|_{L^2(\Omega)}$$

$$\leq \frac{\varepsilon}{2} \|z\|_{L^2(\Omega)}^2 + \frac{\rho^2}{2k_u^2 \varepsilon} \|u\|_{L^2(\Omega)}^2.$$

In conjunction with (6.4.19), we get

$$\frac{d}{dt} \int_\Omega e^{\chi_w v} b \ln b + D_w \int_\Omega e^{\chi_w v} \frac{|\nabla b|^2}{b} + \delta_w \int_\Omega e^{\chi_w v} b \ln b$$

$$\leq (\varepsilon + \chi_w \alpha_w v^*) \|w\|_{L^2(\Omega)}^2 + \varepsilon \|z\|_{L^2(\Omega)}^2 + \frac{C_3}{\varepsilon} \|u\|_{L^{r+1}(\Omega)}^2$$

$$+ \left(\frac{\chi_w^2 \alpha_u^2 v^{*2}}{\varepsilon} + \frac{\rho^2}{2k_u^2 \varepsilon} \right) \|u\|_{L^2(\Omega)}^2 + \chi_w \mu_v v^* \|w\|_{L^1(\Omega)}$$

$$\leq (\varepsilon + \chi_w \alpha_w v^*) \|w\|_{L^2(\Omega)}^2 + \varepsilon \|z\|_{L^2(\Omega)}^2$$

$$+ \left(\frac{C_3 + \chi_w^2 \alpha_u^2 v^{*2}}{\varepsilon} + \frac{\rho^2}{2k_u^2 \varepsilon} \right) \|u\|_{L^{r+1}(\Omega)}^2 + \chi_w \mu_v v^* \|w\|_{L^1(\Omega)} + C_4$$

for some $C_4 > 0$, which together with (6.4.18) and (6.4.5) implies that (6.4.13) holds.

Lemma 6.18 *There exists $K_3 > 0$ such that*

$$\frac{d}{dt}\int_\Omega e^{\chi_z v}c\ln c + \frac{3D_z}{4}\int_\Omega e^{\chi_z v}\frac{|\nabla c|^2}{c} + \delta_z\int_\Omega e^{\chi_z v}c\ln c$$
$$\leq K_3\|b\|_{L^2(\Omega)}^2 + K_3\|a\|_{L^2(\Omega)}^2 + K_3. \tag{6.4.20}$$

Proof By the fourth equation in (6.4.1), we can see that

$$(ce^{\chi_z v})_t = D_z\nabla\cdot(e^{\chi_z v}\nabla c) - \delta_z z - \frac{\rho u z}{k_u + \theta u} + \beta w. \tag{6.4.21}$$

Proceeding as above, we test (6.4.21) by $\ln c$ and integrate by parts to see that for $\varepsilon > 0$,

$$\frac{d}{dt}\int_\Omega e^{\chi_z v}c\ln c + \delta_z\int_\Omega e^{\chi_z v}c\ln c + D_z\int_\Omega e^{\chi_z v}\frac{|\nabla c|^2}{c} \tag{6.4.22}$$

$$\leq \int_\Omega(\ln c + 1)\left(\beta w - \frac{\rho u z}{k_u + \theta u}\right) + \chi_z\int_\Omega z(\alpha_u u + \alpha_w w)v - \chi_z\mu_v\int_\Omega cv(1-v)$$

$$\leq -\rho\int_\Omega\frac{uz}{k_u+\theta u}\ln c + \beta\int_\Omega w + \beta\int_\Omega w\ln c + \chi_z\alpha_u v^*\int_\Omega cu$$

$$+ \chi_z\alpha_w v^*\int_\Omega cw + \chi_z\mu_v\int_\Omega c$$

$$\leq \frac{\rho e^{\chi_z v^*}}{k_u}u^* + \varepsilon\int_\Omega c^2 + \frac{\chi_z^2\alpha_u^2 v^{*2}}{\varepsilon}\int_\Omega u^2 + \left(\frac{\chi_z^2\alpha_w^2 v^{*2}}{\varepsilon}+1\right)\int_\Omega w^2$$

$$+ (\chi_z\mu_v + \beta^2)\int_\Omega c + \beta\int_\Omega w$$

due to

$$-\rho\int_\Omega\frac{uz}{k_u+\theta u}\ln c = -\rho\int_\Omega\frac{ue^{\chi_z v}}{k_u+\theta u}c\ln c$$
$$\leq -\rho\int_{\{c<1\}}\frac{ue^{\chi_z v}}{k_u+\theta u}c\ln c$$
$$\leq \frac{\rho e^{\chi_z v^*}}{k_u}u^*$$

and

$$\beta\int_{\{c>1\}}w\ln c \leq \|w\|_{L^2(\Omega)}^2 + \frac{\beta^2}{4}\int_{\{c>1\}}\ln^2 c$$
$$\leq \|w\|_{L^2(\Omega)}^2 + \beta^2\int_\Omega c.$$

Now according to the two-dimensional Gagliardo–Nirenberg inequality (6.4.17) and Lemma 6.15, we pick $\varepsilon = \frac{D_z}{8K_g z^*}$ and thereafter obtain some $C_1 > 0$ such that

$$
\begin{aligned}
\varepsilon \|c\|_{L^2(\Omega)}^2 &= \varepsilon \|\sqrt{c}\|_{L^4(\Omega)}^4 \\
&\leq \left(K_g \varepsilon \int_\Omega c\right) \int_\Omega \frac{|\nabla c|^2}{c} + \varepsilon K_g \left(\int_\Omega c\right)^2 \\
&\leq \frac{D_z}{4} \int_\Omega e^{\chi_z v} \frac{|\nabla c|^2}{c} + C_1.
\end{aligned}
\tag{6.4.23}
$$

Therefore, along with Lemma 6.15, in conjunction with (6.4.22) and (6.4.23), we readily arrive at (6.4.20).

We are now ready to obtain the bounds for a, b and c in $L \log L$ by taking suitable linear combinations of the inequalities provided by Lemmata 6.16–6.18, stated as follows.

Lemma 6.19 *Let $T > 0$. Then there exists $K_4 > 0$ such that*

$$
\int_\Omega a(\cdot, t) |\ln a(\cdot, t)| \leq K_4,
\tag{6.4.24}
$$

$$
\int_\Omega b(\cdot, t) |\ln b(\cdot, t)| \leq K_4
\tag{6.4.25}
$$

and

$$
\int_\Omega c(\cdot, t) |\ln c(\cdot, t)| \leq K_4
\tag{6.4.26}
$$

for all $t \in (0, \hat{T})$ with $\hat{T} := \min\{T, T_{max}\}$.

Proof From (6.4.18) and (6.4.23), it follows that there exists $C_1 > 0$ such that

$$
\|b\|_{L^2(\Omega)}^2 \leq C_1 \int_\Omega \frac{|\nabla b|^2}{b} + C_1
\tag{6.4.27}
$$

$$
\|c\|_{L^2(\Omega)}^2 \leq C_1 \int_\Omega \frac{|\nabla c|^2}{c} + C_1.
\tag{6.4.28}
$$

Multiplying (6.4.13) by $A := \frac{8K_3 C_1}{D_w}$ and adding the resulting inequality to (6.4.20), using (6.4.27) and (6.4.28), we have

$$
\begin{aligned}
&\frac{d}{dt} \int_\Omega \left(A e^{\chi_w v} b \ln b + \int_\Omega e^{\chi_z v} c \ln c\right) + \frac{D_w}{8} \int_\Omega e^{\chi_w v} \frac{|\nabla b|^2}{b} + A \delta_w \int_\Omega e^{\chi_w v} b \ln b \\
&+ \frac{3D_z}{4} \int_\Omega e^{\chi_z v} \frac{|\nabla c|^2}{c} + \delta_z \int_\Omega e^{\chi_z v} c \ln c
\end{aligned}
\tag{6.4.29}
$$

$$\leq A\varepsilon \|c\|_{L^2(\Omega)}^2 + \frac{AK_2}{\varepsilon}\|a\|_{L^{r+1}(\Omega)}^2 + \frac{AK_2}{\varepsilon} + K_3\|a\|_{L^2(\Omega)}^2 + K_3$$

$$\leq A\varepsilon C_1 \int_\Omega e^{\chi_z v}\frac{|\nabla c|^2}{c} + \frac{AK_2}{\varepsilon}\|a\|_{L^{r+1}(\Omega)}^2 + \frac{AK_2}{\varepsilon} + K_3\|a\|_{L^2(\Omega)}^2 + A\varepsilon C_1 + K_3.$$

Taking $\varepsilon = \frac{3D_z}{8AC_1}$ in (6.4.29), we can find $C_2 > 0$ such that

$$\frac{d}{dt}\int_\Omega \left(Ae^{\chi_w v}b\ln b + \int_\Omega e^{\chi_z v}c\ln c\right) + \frac{D_w}{8}\int_\Omega e^{\chi_w v}\frac{|\nabla b|^2}{b} + A\delta_w \int_\Omega e^{\chi_w v}b\ln b$$

$$+ \frac{3D_z}{8}\int_\Omega e^{\chi_z v}\frac{|\nabla c|^2}{c} + \delta_z \int_\Omega e^{\chi_z v}c\ln c$$

$$\leq C_2\|a\|_{L^{r+1}(\Omega)}^2 + C_2.$$

$$(6.4.30)$$

Combining (6.4.11) with (6.4.30) and using (6.4.18), we can pick $\varepsilon > 0$ in (6.4.11) appropriately small to derive that for some $C_3 > 0$

$$\frac{d}{dt}\int_\Omega \left(Ae^{\chi_w v}b\ln b + \int_\Omega e^{\chi_u v}a\ln a + \int_\Omega e^{\chi_z v}c\ln c\right) + \frac{D_w}{9}\int_\Omega e^{\chi_w v}\frac{|\nabla b|^2}{b}$$

$$+ A\delta_w \int_\Omega e^{\chi_w v}b\ln b$$

$$(6.4.31)$$

$$+ \frac{3D_z}{8}\int_\Omega e^{\chi_z v}\frac{|\nabla c|^2}{c} + \delta_z \int_\Omega e^{\chi_z v}c\ln c + D_u \int_\Omega e^{\chi_u v}\frac{|\nabla a|^2}{a} + \frac{\mu_u}{2}\int_\Omega (\ln a + 1)u^{r+1}$$

$$\leq C_3\|a\|_{L^{r+1}(\Omega)}^2 + C_3,$$

which, along with $a^{r+1} \leq \varepsilon a^{r+1}\ln a + c(\varepsilon)$ for some $c(\varepsilon) > 0$, implies that there exist $C_4 > 0$ and $C_5 > 0$ fulfilling

$$\frac{d}{dt}\mathscr{F}(t) + C_4\mathscr{F}(t) \leq C_5$$

$$(6.4.32)$$

with

$$\mathscr{F}(t) := A\int_\Omega e^{\chi_w v(\cdot,t)}b(\cdot,t)\ln b(\cdot,t) + \int_\Omega e^{\chi_u v(\cdot,t)}a(\cdot,t)\ln a(\cdot,t)$$

$$+ \int_\Omega e^{\chi_z v(\cdot,t)}c(\cdot,t)\ln c(\cdot,t),$$

and thereby

$$\mathscr{F}(t) \leq C_6$$

$$(6.4.33)$$

is valid for some $C_6 > 0$.

Now, by the inequality $a \ln a \geq -e^{-1}$ for all $a > 0$,

$$\int_\Omega a(\cdot, t) |\ln a(\cdot, t)| = \int_\Omega a(\cdot, t) \ln a(\cdot, t) - 2 \int_{a<1} a(\cdot, t) \ln a(\cdot, t)$$

$$\leq \int_\Omega a(\cdot, t) \ln a(\cdot, t) + 2|\Omega|,$$

and similarly,

$$\int_\Omega b(\cdot, t) |\ln b(\cdot, t)| \leq \int_\Omega b(\cdot, t) \ln b(\cdot, t) + 2|\Omega|$$

as well as

$$\int_\Omega c(\cdot, t) |\ln c(\cdot, t)| \leq \int_\Omega c(\cdot, t) \ln c(\cdot, t) + 2|\Omega|.$$

Hence, (6.4.24)–(6.4.26) result readily from (6.4.33).

6.4.3 L^∞-Bounds for a, b and c

By means of some quite straightforward L^p testing procedures, combining Lemma 2.1 with appropriate interpolation, we can now proceed to turn the outcome of Lemma 6.19 into the L^∞-bounds for a, b and c.

Lemma 6.20 *Let (a, b, v, c) be the classical solution of (6.4.1) in $\Omega \times [0, T_{max})$. Then one can find $C > 0$ fulfilling*

$$\|a(\cdot, t)\|_{L^\infty(\Omega)} \leq C \qquad\qquad (6.4.34)$$

and

$$\|b(\cdot, t)\|_{L^\infty(\Omega)} \leq C \qquad\qquad (6.4.35)$$

as well as

$$\|c(\cdot, t)\|_{L^\infty(\Omega)} \leq C \qquad\qquad (6.4.36)$$

for all $t \in (0, T_{max})$.

Proof Testing the first equation in (6.4.1) by $e^{\xi_u v} a^{p-1}$ with $p > 4$, integrating by parts and using the Young inequality, we can find $C_1 > 0$ and $C_2 := C_2(p) > 0$ such that

$$\frac{d}{dt} \int_{\Omega} e^{\chi_u v} a^p + \int_{\Omega} e^{\chi_u v} a^p$$

$$= p \int_{\Omega} e^{\chi_u v} a^{p-1} a_t + \chi_u \int_{\Omega} e^{\chi_u v} a^p v_t + \int_{\Omega} e^{\chi_u v} a^p$$

$$= p \int_{\Omega} e^{\chi_u v} a^{p-1} \{D_u e^{-\chi_u v} \nabla \cdot (e^{\chi_u v} \nabla a) + f(a, b, v, c)\} + \chi_u \int_{\Omega} e^{\chi_u v} a^p v_t + \int_{\Omega} e^{\chi_u v} a^p$$

$$\leq -\frac{4D_u(p-1)}{p} \int_{\Omega} |\nabla a^{\frac{p}{2}}|^2 + p \chi_u \alpha_u \int_{\Omega} a^{p+1} e^{2\chi_u v} + C_1 p \int_{\Omega} a^p + C_1 p \int_{\Omega} a^p b$$

$$\leq -\frac{4D_u(p-1)}{p} \int_{\Omega} |\nabla a^{\frac{p}{2}}|^2 + C_2 \int_{\Omega} a^{p+1} + C_2 \int_{\Omega} b^{p+1} + C_2.$$

$$(6.4.37)$$

Similarly, based on the other equations in (6.4.1), we infer the existence of $C_3 > 0$ such that

$$\frac{d}{dt} \int_{\Omega} e^{\xi_w v} b^p + \int_{\Omega} e^{\xi_w v} b^p$$

$$\leq -\frac{4D_w(p-1)}{p} \int_{\Omega} |\nabla b^{\frac{p}{2}}|^2 + C_3 \int_{\Omega} a^{p+1} + C_3 \int_{\Omega} b^{p+1} + C_3 \int_{\Omega} c^{p+1} + C_3$$

$$(6.4.38)$$

as well as

$$\frac{d}{dt} \int_{\Omega} e^{\xi_z v} c^p + \int_{\Omega} e^{\xi_z v} c^p$$

$$\leq -\frac{4D_z(p-1)}{p} \int_{\Omega} |\nabla c^{\frac{p}{2}}|^2 + C_3 \int_{\Omega} a^{p+1} + C_3 \int_{\Omega} b^{p+1} + C_3 \int_{\Omega} c^{p+1} + C_3.$$

$$(6.4.39)$$

Collecting (6.4.37)–(6.4.39), we then have

$$\frac{d}{dt} \left\{ \int_{\Omega} e^{\xi_u v} a^p + \int_{\Omega} e^{\xi_w v} b^p + \int_{\Omega} e^{\xi_z v} c^p \right\} + \int_{\Omega} e^{\xi_u v} a^p + \int_{\Omega} e^{\xi_w v} b^p + \int_{\Omega} e^{\xi_z v} c^p$$

$$\leq -\frac{4(p-1)}{p} \left(D_u \int_{\Omega} |\nabla a^{\frac{p}{2}}|^2 + D_w \int_{\Omega} |\nabla b^{\frac{p}{2}}|^2 + D_z \int_{\Omega} |\nabla c^{\frac{p}{2}}|^2 \right) + (C_2 + 2C_3) \int_{\Omega} a^{p+1}$$

$$+ (C_2 + 2C_3) \int_{\Omega} b^{p+1} + (C_2 + 2C_3) \int_{\Omega} c^{p+1} + C_2 + 2C_3.$$

$$(6.4.40)$$

Now on the basis of Lemma 6.19, we employ Lemma 2.1 to estimate $\int_{\Omega} a^{p+1}, \int_{\Omega} b^{p+1}$ and $\int_{\Omega} c^{p+1}$ in term of $\int_{\Omega} |\nabla a^{\frac{p}{2}}|^2, \int_{\Omega} |\nabla b^{\frac{p}{2}}|^2$ and $\int_{\Omega} |\nabla c^{\frac{p}{2}}|^2$, respectively.

Indeed, applying Lemma 2.1 to $\varphi = a^{\frac{p}{2}}$, we have

$$(C_2 + 2C_3) \int_{\Omega} a^{p+1}$$

$$= (C_2 + 2C_3)\|a^{\frac{p}{2}}\|_{L^{\frac{2(p+1)}{p}}(\Omega)}^{\frac{2(p+1)}{p}} \tag{6.4.41}$$

$$\leq (C_2 + 2C_3)\varepsilon\|\nabla a^{\frac{p}{2}}\|_{L^2(\Omega)}^2 \cdot \int_{\Omega} a|\ln a^{\frac{p}{2}}| + (C_2 + 2C_3)K(p,\varepsilon)\left(\|a^{\frac{p}{2}}\|_{L^{\frac{2}{p}}(\Omega)}^{\frac{2(p+1)}{p}} + 1\right)$$

$$= \frac{p(C_2 + 2C_3)\varepsilon}{2}\|\nabla a^{\frac{p}{2}}\|_{L^2(\Omega)}^2 \cdot \int_{\Omega} a|\ln a| + (C_2 + 2C_3)K(p,\varepsilon)\left(\left(\int_{\Omega} a\right)^{p+1} + 1\right)$$

which along with (6.4.24) and the appropriate choice of ε readily shows that for $C_4(p) > 0$

$$(C_2 + 2C_3) \int_{\Omega} a^{p+1} \leq \frac{4(p-1)D_u}{p} \int_{\Omega} |\nabla a^{\frac{p}{2}}|^2 + C_4(p).$$

Similarly,

$$(C_2 + 2C_3) \int_{\Omega} b^{p+1} \leq \frac{4(p-1)D_w}{p} \int_{\Omega} |\nabla b^{\frac{p}{2}}|^2 + C_5(p)$$

as well as

$$(C_2 + 2C_3) \int_{\Omega} c^{p+1} \leq \frac{4(p-1)D_z}{p} \int_{\Omega} |\nabla c^{\frac{p}{2}}|^2 + C_6(p).$$

Therefore, (6.4.40) shows that

$$\frac{d}{dt}\left\{\int_{\Omega} e^{\xi_u v} a^p + \int_{\Omega} e^{\xi_w v} b^p + \int_{\Omega} e^{\xi_z v} c^p\right\} + \int_{\Omega} e^{\xi_u v} a^p + \int_{\Omega} e^{\xi_w v} b^p + \int_{\Omega} e^{\xi_z v} c^p$$

$$\leq C_7(p),$$

which entails that for all $p \geq 2$ there exists $C_8(p) > 0$ such that

$$\int_{\Omega} a^p(\cdot, t) + \int_{\Omega} b^p(\cdot, t) + \int_{\Omega} c^p(\cdot, t) \leq C_8(p) \tag{6.4.42}$$

for all $t \in (0, T_{max})$.

Furthermore, by adapting a well-established Moser-type iteration, one can readily turn the latter into the L^∞ bounds for a, b, c. However, since the procedure is rather standard (see Tao and Winkler 2014b, 2020b for example), we give the details only in places which are characteristic of the present setting.

By a straightforward calculation and three integrations by parts, we get

$$
\frac{d}{dt}\left\{\int_{\Omega} e^{\xi_u v}a^p + e^{\xi_w v}b^p + e^{\xi_z v}c^p\right\} + \int_{\Omega}\left\{e^{\xi_u v}a^p + e^{\xi_w v}b^p + e^{\xi_z v}c^p\right\}
$$

$$
\leq -2\min\{D_u, D_w, D_z\}\int_{\Omega}\left\{|\nabla a^{\frac{p}{2}}|^2 + |\nabla b^{\frac{p}{2}}|^2 + |\nabla c^{\frac{p}{2}}|^2\right\} \tag{6.4.43}
$$

$$
+ C_9 p\int_{\Omega}\left\{a^{p+1} + b^{p+1} + c^{p+1}\right\} + C_9
$$

where $C_9 > 0$ as all subsequently appearing constants $C_{10}, C_{11}, \ldots$ is independent of $p \geq 4$.

It is observed that by the Gagliardo–Nirenberg inequality, due to $2 \leq \frac{2(p+1)}{p} \leq 2.5$ for $p \geq 4$, one can pick $C_{10} > 1$ such that for all $p \geq 4$,

$$
\|\varphi\|_{L^{\frac{2(p+1)}{p}}(\Omega)} \leq C_{10}\|\nabla\varphi\|_{L^2(\Omega)}^{\frac{p+2}{2(p+1)}}\|\varphi\|_{L^1(\Omega)}^{\frac{p}{2(p+1)}} + C_{10}\|\varphi\|_{L^1(\Omega)} \quad \text{for all } \varphi \in W^{1,2}(\Omega).
$$

Applying this together with the Young inequality, we obtain that for some $C_{11} > 0$,

$$
C_9 p\int_{\Omega}a^{p+1} = C_9 p\|a^{\frac{p}{2}}\|_{L^{\frac{2(p+1)}{p}}(\Omega)}^{\frac{2(p+1)}{p}}
$$

$$
\leq C_9 p\left\{C_{10}\|\nabla a^{\frac{p}{2}}\|_{L^2(\Omega)}^{\frac{p+2}{2(p+1)}} \cdot \|a^{\frac{p}{2}}\|_{L^1(\Omega)}^{\frac{p}{2(p+1)}} + C_{10}\|a^{\frac{p}{2}}\|_{L^1(\Omega)}\right\}^{\frac{2(p+1)}{p}}
$$

$$
\leq 8C_9 C_{10}^3 p\|\nabla a^{\frac{p}{2}}\|_{L^2(\Omega)}^{\frac{p+2}{p+1}} \cdot \|a^{\frac{p}{2}}\|_{L^1(\Omega)} + 8C_9 C_{10}^3 p\|a^{\frac{p}{2}}\|_{L^1(\Omega)}^{\frac{2(p+1)}{p}}
$$

$$
\leq \min\{D_u, D_w, D_z\}\|\nabla a^{\frac{p}{2}}\|_{L^2(\Omega)}^2 + C_{11}p^4\max\{1, \|a^{\frac{p}{2}}\|_{L^1(\Omega)}\}^{\frac{2p}{p-2}},
$$

where the fact that $\frac{2(p+1)}{p} \leq \frac{2p}{p-2} \leq 4$ for any $p \geq 4$ is used.

Similarly, we have

$$
C_9 p\int_{\Omega}b^{p+1} \leq \min\{D_u, D_w, D_z\}\|\nabla b^{\frac{p}{2}}\|_{L^2(\Omega)}^2 + C_{11}p^4\max\{1, \|b^{\frac{p}{2}}\|_{L^1(\Omega)}\}^{\frac{2p}{p-2}}
$$

as well as

$$
C_9 p\int_{\Omega}c^{p+1} \leq \min\{D_u, D_w, D_z\}\|\nabla c^{\frac{p}{2}}\|_{L^2(\Omega)}^2 + C_{11}p^4\max\{1, \|c^{\frac{p}{2}}\|_{L^1(\Omega)}\}^{\frac{2p}{p-2}}.
$$

Consequently, inserting the above inequalities into (6.4.43) yields the existence of $C_{12} > 0$ such that

$$
\frac{d}{dt}\left\{\int_{\Omega} e^{\xi_u v}a^p + e^{\xi_w v}b^p + e^{\xi_z v}c^p\right\} + \int_{\Omega}\left\{e^{\xi_u v}a^p + e^{\xi_w v}b^p + e^{\xi_z v}c^p\right\} \tag{6.4.44}
$$

$$
\leq C_{12}p^4\max\{1, \|a^{\frac{p}{2}}\|_{L^1(\Omega)} + \|b^{\frac{p}{2}}\|_{L^1(\Omega)} + \|c^{\frac{p}{2}}\|_{L^1(\Omega)}\}^{\frac{2p}{p-2}}.
$$

Now let $p_k = 4 \cdot 2^k$ and $M_k = \max\{1, \sup_{t \in (0, T_{max})} \int_{\Omega} a^{p_k}(\cdot, t) + b^{p_k}(\cdot, t) + c^{p_k}(\cdot, t)\}$
for $k = 0, 1, 2, \ldots$. Then (6.4.44) implies that for $k = 1, 2, \ldots$

$$\frac{d}{dt}\left\{\int_{\Omega} e^{\xi_u v} a^{p_k} + e^{\xi_w v} b^{p_k} + e^{\xi_z v} c^{p_k}\right\} + \int_{\Omega}\left\{e^{\xi_u v} a^{p_k} + e^{\xi_w v} b^{p_k} + e^{\xi_z v} c^{p_k}\right\}$$
$$\leq C_{12} p_k{}^4 M_{k-1}^4,$$

which entails the existence of $L > 1$ independent of k such that

$$M_k \leq \max\{L^k M_{k-1}^4, |\Omega|(\|u_0\|_{L^\infty(\Omega)}^{p_k} + \|w_0\|_{L^\infty(\Omega)}^{p_k} + \|z_0\|_{L^\infty(\Omega)}^{p_k})\} \text{ for all } k \geq 1.$$

Therefore, by means of a standard recursive argument (see Pang and Wang 2018; Tao and Winkler 2014b for example), both when $L^k M_{k-1}^4 \leq |\Omega|(\|u_0\|_{L^\infty(\Omega)}^{p_k} + \|w_0\|_{L^\infty(\Omega)}^{p_k} + \|z_0\|_{L^\infty(\Omega)}^{p_k})$ for infinitely many $k \geq 1$, and as well in the opposite case, we can obtain some $C_{13} > 0$ such that for all $k \geq 1$

$$M_k^{\frac{1}{p_k}} \leq C_{13},$$

from which, after taking $k \to \infty$, the claims (6.4.34)–(6.4.36) readily follow.

According to Lemma 6.14, it remains for us to establish a priori estimates for $\|\nabla v(\cdot, t)\|_{L^4(\Omega)}$.

Lemma 6.21 *Let $T > 0$. Then there exists $C(\hat{T}) > 0$ such that $\|\nabla v(\cdot, t)\|_{L^4(\Omega)} \leq C(\hat{T})$ for all $t < \hat{T}$, where $\hat{T} := \min\{T, T_{max}\}$.*

Proof This can be achieved through an appropriate combination of three further testing processes, essentially relying on the L^∞-estimates for a, b and c just asserted. We refrain from giving the proof and refer to Tao and Winkler (2020b) or Tao and Winkler (2014b) for details in a closely related setting.

We are now in the position to prove Theorem 6.2.
Proof of Theorem 6.2. Thanks to the equivalence of (6.1.10) and (6.4.1) in the considered framework of classical solutions and in particular the extensibility criterion provided by Lemma 6.14, the proof is an evident consequence of Lemmas 6.20 and 6.21.

6.5 Asymptotic Behavior of Solutions to an Oncolytic Virotherapy Model

At the beginning of this subsection, in light of the method used in Horstmann and Winkler (2005) and Pang and Wang (2017), we provide the following statement on the local existence and extensibility of solutions to (6.1.12) as below.

Lemma 6.22 *Suppose that $\Omega \subset \mathbb{R}^N$ ($N \geq 1$) be a bounded domain with smooth boundary. Then one can find $T_{max} \in (0, \infty]$ and a unique quadruple of nonnegative functions $(u, v, w, z) \in \left(C\left(\bar{\Omega} \times [0, T_{max})\right) \cap C^{2,1}\left(\bar{\Omega} \times (0, T_{max})\right)\right)^4$ which solves (6.1.12) classically in $\Omega \times (0, T_{max})$. Moreover, if $T_{max} < +\infty$, then*

$$\lim_{t \nearrow T_{max}} \left(\|u\,(\cdot, t)\|_{W^{1,2q}(\Omega)} + \|v\,(\cdot, t)\|_{W^{2,q}(\Omega)} + \|w\,(\cdot, t)\|_{W^{1,2q}(\Omega)} + \|z\,(\cdot, t)\|_{W^{1,2q}(\Omega)}\right) = \infty$$

(6.5.1)

for all $q > \frac{N}{2}$.

To make the system mass-conserved, we introduce a nonnegative variable Q satisfying

$$\begin{cases} Q_t = \Delta Q + \left(\dfrac{\rho_u + \rho_z}{\rho_w}\delta_w - \beta\right)w + \delta_z z, & (x, t) \in \Omega \times (0, T), \\ \nabla Q \cdot \nu = 0, & (x, t) \in \partial\Omega \times (0, T), \\ Q(x, 0) = 0, & x \in \Omega, \end{cases}$$

(6.5.2)

where $\frac{\rho_u + \rho_z}{\rho_w}\delta_w - \beta \geq 0$. Then it is easy to see that for all $t \in (0, T)$,

$$\int_\Omega \left[\left(u + \frac{\rho_u + \rho_z}{\rho_w}w + z + Q\right)(\cdot, t)\right]_t = 0,$$

which means

$$\int_\Omega \left(u + \frac{\rho_u + \rho_z}{\rho_w}w + z + Q\right)(\cdot, t) = \int_\Omega \left(u_0 + \frac{\rho_u + \rho_z}{\rho_w}w_0 + z_0\right). \quad (6.5.3)$$

We first collect some easily verifiable observations in the following lemma. The constants c_i, ($i = 1, 2, 3, 4$), c_{10} and C_{p_0} refer to Lemmas 1.1, 4.3 and 3.2 respectively.

Lemma 6.23 *Under the assumptions of Theorem 6.3, there exist $M_i > 1$ ($i = 0, 1, \cdots, 6$) and $\varepsilon(\xi_u, \xi_w, \xi_z, \rho_u, \rho_w, \rho_z, \alpha_y, \alpha_w) > 0$ such that*

$$2c_3 + 2c_{10}c_2\xi_u\varepsilon(M_1M_4 + M_0M_5) + 2c_{10}c_3\rho_u M_0\varepsilon(M_1 + M_3) \leq \frac{M_1}{2}, \quad (6.5.4)$$

$$2c_3 + 2c_{10}c_2\xi_w\varepsilon(M_2M_4 + M_0M_5) + 2c_{10}c_3\rho_w M_0\varepsilon(M_1 + M_3) \leq \frac{M_2}{2}, \quad (6.5.5)$$

$$2c_3 + 2c_{10}c_2\xi_z\varepsilon (M_3M_4 + M_0M_5) + 2c_{10}c_3\rho_z M_0\varepsilon(M_1 + M_3) + 2c_{10}c_3\beta M_2 \leq \frac{M_3}{2},$$
$$(6.5.6)$$

$$1 + \frac{(\alpha_u M_1 + \alpha_w M_2)\, \|v_0\,(\cdot)\|_{L^\infty(\Omega)} \left(\frac{2p_0-1}{2p_0 e(\delta_v - \alpha)}\right)^{\frac{2p_0-1}{2p_0}}}{(2\alpha p_0)^{\frac{1}{2p_0}}} \leq \frac{M_4}{2}, \qquad (6.5.7)$$

$$1 + \varepsilon\|v_0(\cdot)\|_{L^\infty(\Omega)} \frac{(\alpha_u M_1 + \alpha_w M_2)^2}{(2\alpha p_0)^{\frac{1}{p_0}}} \left(\frac{2p_0 - 1}{p_0 e(\delta_v - \alpha)}\right)^{\frac{2p_0-1}{p_0}}$$

$$+ 2\varepsilon \frac{\alpha_u M_1 + \alpha_w M_2}{(2\alpha p_0)^{\frac{1}{2p_0}}} \left(\frac{2p_0 - 1}{2p_0 e(\delta_v - \alpha)}\right)^{\frac{2p_0-1}{2p_0}}$$

$$+ \|v_0(\cdot)\|_{L^\infty(\Omega)} C_{p_0}\alpha_u \left(\frac{p_0 - 1}{p_0 e(\delta_v - \alpha)}\right)^{\frac{p_0-1}{p_0}}$$

$$\cdot \left(\frac{\varepsilon\xi_u M_1 M_4}{((2\alpha - k)p_0)^{\frac{1}{p_0}}} + \frac{\varepsilon\xi_u M_0 M_5}{((\alpha - k)p_0)^{\frac{1}{p_0}}} + 1 + |\Omega|^{\frac{1}{p_0}}\right) \qquad (6.5.8)$$

$$+ \|v_0(\cdot)\|_{L^\infty(\Omega)} C_{p_0}\alpha_w \left(\frac{p_0 - 1}{p_0 e(\delta_v - \alpha)}\right)^{\frac{p_0-1}{p_0}}$$

$$\cdot \left(\frac{\varepsilon\xi_w M_2 M_4}{((2\alpha - k)p_0)^{\frac{1}{p_0}}} + \frac{\varepsilon\xi_w M_0 M_5}{((\alpha - k)p_0)^{\frac{1}{p_0}}} + 1 + |\Omega|^{\frac{1}{p_0}}\right)$$

$$+ \frac{\|v_0(\cdot)\|_{L^\infty(\Omega)} C_{p_0} M_0 |\Omega|^{\frac{1}{p_0}}}{(kp_0)^{\frac{1}{p_0}}} \left(\frac{p_0 - 1}{p_0 e(\delta_v - \alpha - k)}\right)^{\frac{p_0-1}{p_0}}$$

$$\cdot \left(\varepsilon\alpha_u\rho_u M_0 + \frac{1}{2}\alpha_u(\delta_v - \alpha) + \varepsilon\alpha_w\rho_w M_0\right) \leq \frac{M_5}{2},$$

$$2c_{10}c_4 M_0 M_4\varepsilon \left(\xi_u + \frac{\rho_u + \rho_z}{\rho_w}\xi_w + \xi_z\right) \leq \frac{M_6}{2}, \qquad (6.5.9)$$

where

$$M_0 = 1 + M_6 + 2c_1, \quad k = \min\left\{\frac{1}{2}(\delta_v - \alpha), \delta_w\right\} \in (0, \alpha),$$

with $\alpha \in (0, \min\{\lambda_1, \delta_v\})$ and $\lambda_1 > 0$ the first nonzero eigenvalue of $-\Delta$ in Ω under the Neumann condition.

For constants $\alpha \in (0, \min\{\lambda_1, \delta_v\})$ and $M_i > 1 (i = 0, 1, \ldots, 6)$ referring to Lemma 6.23, let

$$
T \triangleq \sup \left\{ \tilde{T} \in (0, T_{max}) \left|
\begin{array}{l}
\|\nabla u\,(\cdot, t)\|_{L^{2p_0}(\Omega)} \leq M_1 \varepsilon e^{-\alpha t} \; for \; all \; t \in [0, \tilde{T}), \\
\|\nabla w\,(\cdot, t)\|_{L^{2p_0}(\Omega)} \leq M_2 \varepsilon e^{-\alpha t} \; for \; all \; t \in [0, \tilde{T}), \\
\|\nabla z\,(\cdot, t)\|_{L^{2p_0}(\Omega)} \leq M_3 \varepsilon e^{-\alpha t} \; for \; all \; t \in [0, \tilde{T}), \\
\|\nabla v\,(\cdot, t)\|_{L^{2p_0}(\Omega)} \leq M_4 \varepsilon e^{-\alpha t} \; for \; all \; t \in [0, \tilde{T}), \\
\|\Delta v\,(\cdot, t)\|_{L^{p_0}(\Omega)} \leq M_5 \varepsilon e^{-\alpha t} \; for \; all \; t \in [0, \tilde{T}), \\
\|(u + \frac{\rho_u + \rho_z}{\rho_w} w + z + Q)\,(\cdot, t) - e^{t\Delta}(u_0 + \frac{\rho_u + \rho_z}{\rho_w} w_0 \\
+ z_0)\,(\cdot)\|_{L^{\infty}(\Omega)} \leq M_6 \varepsilon e^{-\alpha t} \; for \; all \; t \in [0, \tilde{T}).
\end{array}
\right. \right\}
$$

$$(6.5.10)$$

By Lemma 6.22 and the smallness condition on the initial data in Theorem 6.3, $T > 0$ is well-defined. We first show $T = T_{max}$. To this end, we will show that all of the estimates mentioned in (6.5.10) are valid with even smaller coefficients on the right-hand side. The derivation of these estimates will mainly rely on $L^p - L^q$ estimates for the Neumann heat semigroup and the fact that the classical solutions on $(0, T_{max})$ can be represented as

$$
\left(u + \frac{\rho_u + \rho_z}{\rho_w} w + z + Q \right) (\cdot, t)
$$
$$
= e^{t\Delta} \left(u_0 + \frac{\rho_u + \rho_z}{\rho_w} w_0 + z_0 \right) (\cdot)
$$
$$
- \int_0^t e^{(t-s)\Delta} \left[\xi_u \nabla \cdot (u \nabla v) + \frac{\rho_u + \rho_z}{\rho_w} \xi_w \nabla \cdot (w \nabla v) + \xi_z \nabla \cdot (z \nabla v) \right] (\cdot, s) ds,
$$

$$(6.5.11)$$

$$
u(\cdot, t) = e^{t\Delta} u_0 (\cdot) + \int_0^t e^{(t-s)\Delta} \left[-\xi_u \nabla \cdot (u \nabla v) - \rho_u u z \right] (\cdot, s) ds, \qquad (6.5.12)
$$

$$
w(\cdot, t) = e^{t(\Delta - \delta_w)} w_0 (\cdot) + \int_0^t e^{(t-s)(\Delta - \delta_w)} \left[-\xi_w \nabla \cdot (w \nabla v) + \rho_w u z \right] (\cdot, s) ds,
$$

$$(6.5.13)$$

$$
z(\cdot, t) = e^{t(\Delta - \delta_z)} z_0 (\cdot) + \int_0^t e^{(t-s)(\Delta - \delta_z)} \left[-\xi_z \nabla \cdot (z \nabla v) - \rho_z u z + \beta w \right] (\cdot, s) ds
$$

$$(6.5.14)$$

for all $t \in (0, T_{max})$ as per the variation-of-constants formula.

The global boundedness for solutions of (6.1.12) can be obtained directly from the following lemmas.

Lemma 6.24 *Under the assumptions of Theorem 6.3, for all $t \in (0, T)$, we have*

$$\left\| \left(u + \frac{\rho_u + \rho_z}{\rho_w} w + z + Q \right) (\cdot, t) \right\|_{L^\infty(\Omega)} \leq M_0 \varepsilon, \qquad (6.5.15)$$

where $M_0 = 1 + M_6 + 2c_1$ with c_1 defined in Lemma 1.1.

Proof Set $m_\infty = \frac{1}{|\Omega|} \int_\Omega \left(u_0 + \frac{\rho_u + \rho_z}{\rho_w} w_0 + z_0 \right)$, then $m_\infty \leq \varepsilon$. It is obvious that

$$e^{t\Delta} m_\infty = m_\infty, \quad \int_\Omega \left[\left(u_0 + \frac{\rho_u + \rho_z}{\rho_w} w_0 + z_0 \right) (\cdot) - m_\infty \right] = 0,$$

$$\left\| \left(u_0 + \frac{\rho_u + \rho_z}{\rho_w} w_0 + z_0 \right) (\cdot) - m_\infty \right\|_{L^\infty(\Omega)} \leq \varepsilon.$$

According to the Lemma 1.1(i), we know for all $t \in (0, T)$,

$$\left\| e^{t\Delta} \left[\left(u_0 + \frac{\rho_u + \rho_z}{\rho_w} w_0 + z_0 \right) (\cdot) - m_\infty \right] \right\|_{L^\infty(\Omega)} \leq 2c_1 \varepsilon e^{-\lambda_1 t}.$$

Then due to the definition of T and Lemma 1.1(i), we have

$$\left\| \left(u + \frac{\rho_u + \rho_z}{\rho_w} w + z + Q - m_\infty \right) (\cdot, t) \right\|_{L^\infty(\Omega)}$$

$$= \left\| \left(u + \frac{\rho_u + \rho_z}{\rho_w} w + z + Q \right) (\cdot, t) - e^{t\Delta} \left(u_0 + \frac{\rho_u + \rho_z}{\rho_w} w_0 + z_0 \right) (\cdot) \right\|_{L^\infty(\Omega)}$$

$$+ \left\| e^{t\Delta} \left[\left(u_0 + \frac{\rho_u + \rho_z}{\rho_w} w_0 + z_0 \right) (\cdot) - m_\infty \right] \right\|_{L^\infty(\Omega)}$$

$$\leq M_6 \varepsilon e^{-\alpha t} + 2c_1 \varepsilon e^{-\lambda_1 t}$$

$$\leq (M_6 + 2c_1) \varepsilon e^{-\alpha t}$$

and hence end the proof.

Lemma 6.25 *Under the assumptions of Theorem 6.3, for all $t \in (0, T)$, we have*

$$\| \nabla u (\cdot, t) \|_{L^{2p_0}(\Omega)} \leq \frac{M_1}{2} \varepsilon e^{-\alpha t}.$$

Proof Applying (6.5.12), Lemma 1.1 (iii), we have

$$\|\nabla u\left(\cdot,t\right)\|_{L^{2p_0}(\Omega)}$$

$$\leq \left\|\nabla e^{t\Delta}u_0\left(\cdot\right)\right\|_{L^{2p_0}(\Omega)} + \int_0^t \left\|\nabla e^{(t-s)\Delta}\left[-\xi_u \nabla \cdot (u\nabla v) - \rho_u uz\right](\cdot,s)\right\|_{L^{2p_0}(\Omega)} ds$$

$$\leq 2c_3 e^{-\lambda_1 t}\|\nabla u_0\left(\cdot\right)\|_{L^{2p_0}(\Omega)} + \xi_u \int_0^t \left\|\nabla e^{(t-s)\Delta}\nabla \cdot (u\nabla v)\left(\cdot,s\right)\right\|_{L^{2p_0}(\Omega)} ds$$

$$\tag{6.5.16}$$

$$+ \rho_u \int_0^t \left\|\nabla e^{(t-s)\Delta}(uz)\left(\cdot,s\right)\right\|_{L^{2p_0}(\Omega)} ds.$$

From Lemmas 1.1(ii) and 4.3, we obtain

$$\xi_u \int_0^t \left\|\nabla e^{(t-s)\Delta}\nabla \cdot (u\nabla v)\left(\cdot,s\right)\right\|_{L^{2p_0}(\Omega)} ds$$

$$\leq \xi_u \int_0^t \left\|\nabla e^{(t-s)\Delta}(\nabla u \nabla v)\left(\cdot,s\right)\right\|_{L^{2p_0}(\Omega)} ds + \xi_u \int_0^t \left\|\nabla e^{(t-s)\Delta}(u\Delta v)\left(\cdot,s\right)\right\|_{L^{2p_0}(\Omega)} ds$$

$$\leq \xi_u c_2 \int_0^t \left[1 + (t-s)^{-\frac{1}{2}-\frac{N}{4p_0}}\right]e^{-\lambda_1(t-s)}\|\nabla u \nabla v\left(\cdot,s\right)\|_{L^{p_0}(\Omega)} ds$$

$$+ \xi_u c_2 \int_0^t \left[1 + (t-s)^{-\frac{1}{2}-\frac{N}{4p_0}}\right]e^{-\lambda_1(t-s)}\|u\Delta v\left(\cdot,s\right)\|_{L^{p_0}(\Omega)} ds$$

$$\leq \xi_u c_2 \int_0^t \left[1 + (t-s)^{-\frac{1}{2}-\frac{N}{4p_0}}\right]e^{-\lambda_1(t-s)}\left(\|\nabla u\left(\cdot,s\right)\|_{L^{2p_0}(\Omega)} \cdot \|\nabla v\left(\cdot,s\right)\|_{L^{2p_0}(\Omega)}\right) ds$$

$$+ \xi_u c_2 \int_0^t \left[1 + (t-s)^{-\frac{1}{2}-\frac{N}{4p_0}}\right]e^{-\lambda_1(t-s)}\left(\|u\left(\cdot,s\right)\|_{L^\infty(\Omega)} \cdot \|\Delta v\left(\cdot,s\right)\|_{L^{p_0}(\Omega)}\right) ds$$

$$\leq 2c_{10}c_2\xi_u(M_1 M_4 \varepsilon^2 e^{-\min\{2\alpha,\lambda_1\}t} + M_0 M_5 \varepsilon^2 e^{-\alpha t})$$

$$\leq 2c_{10}c_2\xi_u \varepsilon^2 e^{-\alpha t}(M_1 M_4 + M_0 M_5).$$

From Hölder's inequality, using Lemmas 1.1(iii) and 4.3, we get

$$\rho_u \int_0^t \left\|\nabla e^{(t-s)\Delta}(uz)\left(\cdot,s\right)\right\|_{L^{2p_0}(\Omega)} ds$$

$$\leq 2\rho_u c_3 \int_0^t e^{-\lambda_1(t-s)}\left(\|z\nabla u(\cdot,s)\|_{L^{2p_0}(\Omega)} + \|u\nabla z(\cdot,s)\|_{L^{2p_0}(\Omega)}\right) ds$$

$$\leq 2\rho_u c_3 \int_0^t e^{-\lambda_1(t-s)}$$

$$\cdot \left(\|\nabla u\left(\cdot,s\right)\|_{L^{2p_0}(\Omega)} \cdot \|z\left(\cdot,s\right)\|_{L^\infty(\Omega)} + \|u\left(\cdot,s\right)\|_{L^\infty(\Omega)} \cdot \|\nabla z\left(\cdot,s\right)\|_{L^{2p_0}(\Omega)}\right) ds$$

$$\leq 2c_{10}c_3\rho_u(M_0 M_1 \varepsilon^2 e^{-\alpha t} + M_0 M_3 \varepsilon^2 e^{-\alpha t})$$

$$= 2c_{10}c_3\rho_u M_0 \varepsilon^2 e^{-\alpha t}(M_1 + M_3).$$

Therefore, inserting the above two results into (6.5.16), we arrive at

$$\|\nabla u\,(\cdot, t)\|_{L^{2p_0}(\Omega)}$$
$$\leq 2c_3\varepsilon e^{-\alpha t} + 2c_{10}c_2\xi_u\varepsilon^2 e^{-\alpha t}(M_1 M_4 + M_0 M_5) + 2c_{10}c_3\rho_u M_0\varepsilon^2 e^{-\alpha t}(M_1 + M_3)$$
$$\leq [2c_3 + 2c_{10}c_2\xi_u\varepsilon(M_1 M_4 + M_0 M_5) + 2c_{10}c_3\rho_u M_0\varepsilon(M_1 + M_3)]\,\varepsilon e^{-\alpha t}.$$

According to (6.5.4), we thereby complete the proof.

Lemma 6.26 *Under the assumptions of Theorem 6.3, for all $t \in (0, T)$, we have*

$$\|\nabla w\,(\cdot, t)\|_{L^{2p_0}(\Omega)} \leq \frac{M_2}{2}\varepsilon e^{-\alpha t}. \tag{6.5.17}$$

Proof From (6.5.13), using Lemmas 1.1(iii) and 4.3, we have

$$\|\nabla w\,(\cdot, t)\|_{L^{2p_0}(\Omega)}$$
$$\leq \left\|\nabla e^{t(\Delta-\delta_w)}w_0(\cdot)\right\|_{L^{2p_0}(\Omega)} + \xi_w \int_0^t \left\|\nabla e^{(t-s)(\Delta-\delta_w)}\nabla\cdot(w\nabla v)\,(\cdot, s)\right\|_{L^{2p_0}(\Omega)} ds$$
$$+ \rho_w \int_0^t \left\|\nabla e^{(t-s)(\Delta-\delta_w)}(uz)\,(\cdot, s)\right\|_{L^{2p_0}(\Omega)} ds$$
$$\leq 2c_3\varepsilon e^{-(\lambda_1+\delta_w)t} + \xi_w c_2 \int_0^t [1 + (t-s)^{-\frac{1}{2}-\frac{N}{4p_0}}]e^{-(\lambda_1+\delta_w)(t-s)}$$
$$\cdot\left(\|\nabla w(\cdot, s)\|_{L^{2p_0}(\Omega)}\|\nabla v(\cdot, s)\|_{L^{2p_0}(\Omega)}\right) ds$$
$$+ \xi_w c_2 \int_0^t [1 + (t-s)^{-\frac{1}{2}-\frac{N}{4p_0}}]e^{-(\lambda_1+\delta_w)(t-s)}\left(\|w(\cdot, s)\|_{L^{\infty}(\Omega)}\|\Delta v(\cdot, s)\|_{L^{p_0}(\Omega)}\right) ds$$
$$+ 2\rho_w c_3 \int_0^t e^{-(\lambda_1+\delta_w)(t-s)}\left(\|\nabla u(\cdot, s)\|_{L^{2p_0}(\Omega)}\|z(\cdot, s)\|_{L^{\infty}(\Omega)}\right) ds$$
$$+ 2\rho_w c_3 \int_0^t e^{-(\lambda_1+\delta_w)(t-s)}\left(\|u(\cdot, s)\|_{L^{\infty}(\Omega)}\|\nabla z(\cdot, s)\|_{L^{2p_0}(\Omega)}\right) ds$$
$$\leq [2c_3 + 2c_{10}c_2\xi_w\varepsilon\,(M_2 M_4 + M_0 M_5) + 2c_{10}c_3\rho_w M_0\varepsilon(M_1 + M_3)]\,\varepsilon e^{-\alpha t},$$

which along with (6.5.5) implies that (6.5.17) is valid.

Similar as done in Lemma 6.26, we also have the following.

Lemma 6.27 *Under the assumptions of Theorem 6.3, for all $t \in (0, T)$, we have*

$$\|\nabla z\,(\cdot, t)\|_{L^{2p_0}(\Omega)} \leq \frac{M_3}{2}\varepsilon e^{-\alpha t}. \tag{6.5.18}$$

Proof From (6.5.14), using Lemmas 1.1(iii) and 4.3, we have

$$\|\nabla z\,(\cdot,t)\|_{L^{2p_0}(\Omega)}$$

$$\leq \left\|\nabla e^{t(\Delta-\delta_z)}z_0(\cdot)\right\|_{L^{2p_0}(\Omega)} + \xi_z \int_0^t \left\|\nabla e^{(t-s)(\Delta-\delta_z)}\nabla\cdot(z\nabla v)\,(\cdot,s)\right\|_{L^{2p_0}(\Omega)}ds$$

$$+ \int_0^t \left\|\nabla e^{(t-s)(\Delta-\delta_z)}(-\rho_z uz + \beta w)\,(\cdot,s)\right\|_{L^{2p_0}(\Omega)}ds$$

$$\leq 2c_3\varepsilon e^{-(\lambda_1+\delta_z)t} + \xi_z c_2 \int_0^t [1 + (t-s)^{-\frac{1}{2}-\frac{N}{4p_0}}]e^{-(\lambda_1+\delta_z)(t-s)}$$

$$\cdot \|\nabla z(\cdot,s)\|_{L^{2p_0}(\Omega)} \|\nabla v(\cdot,s)\|_{L^{2p_0}(\Omega)}\,ds$$

$$+ \xi_z c_2 \int_0^t [1 + (t-s)^{-\frac{1}{2}-\frac{N}{4p_0}}]e^{-(\lambda_1+\delta_z)(t-s)} \|z(\cdot,s)\|_{L^\infty(\Omega)} \|\Delta v(\cdot,s)\|_{L^{p_0}(\Omega)}\,ds$$

$$+ \rho_z c_3 \int_0^t 2e^{-(\lambda_1+\delta_z)(t-s)}$$

$$\cdot \left(\|\nabla u\,(\cdot,s)\|_{L^{2p_0}(\Omega)} \cdot \|z\,(\cdot,s)\|_{L^\infty(\Omega)} + \|u\,(\cdot,s)\|_{L^\infty(\Omega)} \cdot \|\nabla z\,(\cdot,s)\|_{L^{2p_0}(\Omega)}\right)ds$$

$$+ 2\beta c_3 \int_0^t e^{-(\lambda_1+\delta_z)(t-s)} \|\nabla w(\cdot,s)\|_{L^{2p_0}(\Omega)}\,ds$$

$$\leq \left[2c_3 + 2c_{10}c_2\xi_z\varepsilon\,(M_3M_4 + M_0M_5) + 2c_{10}c_3\rho_z M_0\varepsilon(M_1 + M_3) + 2c_{10}c_3\beta M_2\right]\varepsilon e^{-\alpha t},$$

which together with (6.5.6) already implies that (6.5.18) holds.

Lemma 6.28 *Under the assumptions of Theorem 6.3, for all* $t \in (0, T)$, *we have*

$$\|\nabla v\,(\cdot,t)\|_{L^{2p_0}(\Omega)} \leq \frac{M_4}{2}\varepsilon e^{-\alpha t}. \tag{6.5.19}$$

Proof We know that

$$v_t + (\alpha_u u + \alpha_w w + \delta_v)v = 0,$$

from which we obtain

$$v\,(\cdot,t) = v_0\,(\cdot)\,e^{-\int_0^t (\alpha_u u + \alpha_w w + \delta_v)ds}. \tag{6.5.20}$$

It follows that

$$\|\nabla v\,(\cdot,t)\|_{L^{2p_0}(\Omega)}$$

$$= \left\|\nabla\left[v_0(\cdot)e^{-\int_0^t(\alpha_u u + \alpha_w w + \delta_v)ds}\right]\right\|_{L^{2p_0}(\Omega)} \tag{6.5.21}$$

$$\leq \left\|\nabla v_0\,(\cdot)\,e^{-\int_0^t(\alpha_u u + \alpha_w w + \delta_v)ds}\right\|_{L^{2p_0}(\Omega)} + \left\|v_0\,(\cdot)\,\nabla e^{-\int_0^t(\alpha_u u + \alpha_w w + \delta_v)ds}\right\|_{L^{2p_0}(\Omega)}$$

$$\leq \left\|\nabla v_0\,(\cdot)\,e^{-\delta_v t}\right\|_{L^{2p_0}(\Omega)} + \|v_0\,(\cdot)\|_{L^\infty(\Omega)}\left\|e^{-\delta_v t}\int_0^t (\alpha_u\nabla u + \alpha_w\nabla w)\,(\cdot,s)\,ds\right\|_{L^{2p_0}(\Omega)}.$$

Noticing that

$$\left(\int_0^t |\nabla u\,(\cdot,s)|\,ds\right)^{2p_0} \leq \left[\left(\int_0^t |\nabla u\,(\cdot,s)|^{2p_0}\,ds\right)^{\frac{1}{2p_0}}\left(\int_0^t 1ds\right)^{\frac{2p_0-1}{2p_0}}\right]^{2p_0}$$

$$\leq t^{2p_0-1}\int_0^t |\nabla u\,(\cdot,s)|^{2p_0}\,ds,$$

then

$$\left\|\int_0^t \nabla u(\cdot,s)ds\right\|_{L^{2p_0}(\Omega)}$$

$$=\left[\int_\Omega \left|\int_0^t \nabla u\,(\cdot,s)\,ds\right|^{2p_0}\,dx\right]^{\frac{1}{2p_0}}$$

$$\leq \left[t^{2p_0-1}\int_0^t\int_\Omega |\nabla u\,(\cdot,s)|^{2p_0}\,dxds\right]^{\frac{1}{2p_0}} \tag{6.5.22}$$

$$\leq \left(\int_0^t\int_\Omega |\nabla u\,(\cdot,s)|^{2p_0}\,dxds\right)^{\frac{1}{2p_0}} t^{\frac{2p_0-1}{2p_0}}$$

$$\leq \left(\int_0^t (M_1\varepsilon e^{-\alpha s})^{2p_0}ds\right)^{\frac{1}{2p_0}} t^{\frac{2p_0-1}{2p_0}}$$

$$\leq \frac{M_1\varepsilon t^{\frac{2p_0-1}{2p_0}}}{(2\alpha p_0)^{\frac{1}{2p_0}}}.$$

Similarly, we have

$$\left\|\int_0^t \nabla w(\cdot,s)ds\right\|_{L^{2p_0}(\Omega)} \leq \frac{M_2\varepsilon t^{\frac{2p_0-1}{2p_0}}}{(2\alpha p_0)^{\frac{1}{2p_0}}}. \tag{6.5.23}$$

From Lemma 6.3, noticing $p_0 > 1$, $\alpha - \delta_v < 0$, for all $t \geq 0$, we obtain

$$t^{\frac{2p_0-1}{2p_0}} e^{(\alpha-\delta_v)t} \leq \left(\frac{2p_0-1}{2ep_0(\delta_v-\alpha)}\right)^{\frac{2p_0-1}{2p_0}}. \tag{6.5.24}$$

Inserting (6.5.22) and (6.5.23) into (6.5.21) and using (6.5.24), we obtain

$$\|\nabla v\,(\cdot,t)\|_{L^{2p_0}(\Omega)}$$

$$\leq \left[1 + \frac{(\alpha_u M_1 + \alpha_w M_2)\,\|v_0\,(\cdot)\|_{L^\infty(\Omega)}}{(2\alpha p_0)^{\frac{1}{2p_0}}}\left(\frac{2p_0-1}{2ep_0(\delta_v-\alpha)}\right)^{\frac{2p_0-1}{2p_0}}\right]\varepsilon e^{-\alpha t}.$$

Therefore, (6.5.19) results from (6.5.7).

To obtain the estimate of $\|\Delta v(\cdot, t)\|_{L^{p_0}(\Omega)}$ for $t \geq 0$, we need the following lemma.

Lemma 6.29 *Under the assumptions of Theorem 6.3, for all $t \in (0, T)$, we have*

$$\left(\int_0^t \int_\Omega |\Delta u(x, s)|^{p_0} \, dx ds \right)^{\frac{1}{p_0}} \leq K_1(t) \tag{6.5.25}$$

and

$$\left(\int_0^t \int_\Omega |\Delta w(x, s)|^{p_0} \, dx ds \right)^{\frac{1}{p_0}} \leq K_2(t), \tag{6.5.26}$$

where

$$K_1(t) = C_{p_0} \left(\frac{\varepsilon^2 \xi_u M_1 M_4}{((2\alpha - k) p_0)^{\frac{1}{p_0}}} + \frac{\varepsilon^2 \xi_u M_0 M_5}{((\alpha - k) p_0)^{\frac{1}{p_0}}} + \frac{\varepsilon M_0 |\Omega|^{\frac{1}{p_0}} e^{kt} \left(\varepsilon \rho_u M_0 + \frac{1}{2} (\delta_v - \alpha) \right)}{(k p_0)^{\frac{1}{p_0}}} \right.$$

$$\left. + \varepsilon \left(1 + |\Omega|^{\frac{1}{p_0}} \right) \right)$$

and

$$K_2(t) = C_{p_0} \left(\frac{\varepsilon^2 \xi_w M_2 M_4}{((2\alpha - k) p_0)^{\frac{1}{p_0}}} + \frac{\varepsilon^2 \xi_w M_0 M_5}{((\alpha - k) p_0)^{\frac{1}{p_0}}} + \frac{\varepsilon^2 \rho_w M_0^2 |\Omega|^{\frac{1}{p_0}} e^{kt}}{(k p_0)^{\frac{1}{p_0}}} + \varepsilon \left(1 + |\Omega|^{\frac{1}{p_0}} \right) \right)$$

with $k = \min \left\{ \frac{1}{2} (\delta_v - \alpha), \delta_w \right\} \in (0, \alpha)$.

Proof Denote $G(x, t) = -\xi_u \nabla \cdot (u \nabla v) - \rho_u u z + \frac{1}{2} (\delta_v - \alpha) u$, then $u_t = \Delta u - \frac{1}{2} (\delta_v - \alpha) u + G(x, t)$ and

$$\left(\int_0^t e^{k p_0 s} \int_\Omega |G(x, s)|^{p_0} \, dx ds \right)^{\frac{1}{p_0}}$$

$$\leq \left(\int_0^t e^{k p_0 s} \left\| \left(-\xi_u \nabla u \nabla v - \xi_u u \Delta v - \rho_u u z + \frac{1}{2} (\delta_v - \alpha) u \right)(\cdot, s) \right\|_{L^{p_0}(\Omega)}^{p_0} \, ds \right)^{\frac{1}{p_0}}$$

$$\leq \left(\int_0^t e^{k p_0 s} \left[\xi_u M_1 M_4 \varepsilon^2 e^{-2\alpha s} + \xi_u M_0 M_5 \varepsilon^2 e^{-\alpha s} + \rho_u M_0^2 \varepsilon^2 |\Omega|^{\frac{1}{p_0}} \right. \right. \tag{6.5.27}$$

$$\left. \left. + \frac{1}{2} (\delta_v - \alpha) M_0 \varepsilon |\Omega|^{\frac{1}{p_0}} \right]^{p_0} \, ds \right)^{\frac{1}{p_0}}$$

$$\leq \frac{\varepsilon^2 \xi_u M_1 M_4}{((2\alpha - k) p_0)^{\frac{1}{p_0}}} + \frac{\varepsilon^2 \xi_u M_0 M_5}{((\alpha - k) p_0)^{\frac{1}{p_0}}} + \frac{\varepsilon M_0 |\Omega|^{\frac{1}{p_0}} e^{kt} \left(\varepsilon \rho_u M_0 + \frac{1}{2} (\delta_v - \alpha) \right)}{(k p_0)^{\frac{1}{p_0}}}.$$

According to Lemma 3.2, we obtain

$$\left(\int_0^t \int_\Omega |\Delta u\,(x,s)|^{p_0}\, dxds \right)^{\frac{1}{p_0}} \le K_1(t).$$

Similarly, denote $F\,(x,t) = -\xi_w \nabla \cdot (w \nabla v) + \rho_w uz$, then $w_t = \Delta w - \delta_w w + F(x,t)$ and

$$\left(\int_0^t e^{kp_0 s} \int_\Omega |F\,(x,s)|^{p_0}\, dxds \right)^{\frac{1}{p_0}}$$

$$\le \left(\int_0^t e^{kp_0 s} \|(-\xi_w \nabla w \nabla v - \xi_w w \Delta v + \rho_w uz)(\cdot,s)\|^{p_0}_{L^{p_0}(\Omega)}\, ds \right)^{\frac{1}{p_0}}$$

$$\le \left(\int_0^t e^{kp_0 s} \left[\xi_w M_2 M_4 \varepsilon^2 e^{-2\alpha s} + \xi_w M_0 M_5 \varepsilon^2 e^{-\alpha s} + \rho_w M_0^2 \varepsilon^2 |\Omega|^{\frac{1}{p_0}} \right]^{p_0} ds \right)^{\frac{1}{p_0}}$$

$$\le \frac{\varepsilon^2 \xi_w M_2 M_4}{((2\alpha - k)p_0)^{\frac{1}{p_0}}} + \frac{\varepsilon^2 \xi_w M_0 M_5}{((\alpha - k)p_0)^{\frac{1}{p_0}}} + \frac{\varepsilon^2 \rho_w M_0^2 |\Omega|^{\frac{1}{p_0}} e^{kt}}{(kp_0)^{\frac{1}{p_0}}},$$

thus from Lemma 3.2, we obtain

$$\left(\int_0^t \int_\Omega |\Delta w\,(x,s)|^{p_0}\, dxds \right)^{\frac{1}{p_0}} \le K_2(t)$$

and thereby complete the proof.

Lemma 6.30 *Under the assumptions of Theorem 6.3, for all $t \in (0, T)$,*

$$\|\Delta v\,(\cdot,t)\|_{L^{p_0}(\Omega)} \le \frac{M_5}{2} \varepsilon e^{-\alpha t}. \tag{6.5.28}$$

Proof By $v\,(\cdot,t) = v_0\,(\cdot)\, e^{-\int_0^t (\alpha_u u + \alpha_w w + \delta_v)ds}$, we get

$$\|\Delta v\,(\cdot,s)\|_{L^{p_0}(\Omega)}$$

$$\le \left\| div \left(\nabla v_0(\cdot) e^{-\int_0^t (\alpha_u u + \alpha_w w + \delta_v)(\cdot,s)ds} \right) \right\|_{L^{p_0}(\Omega)}$$

$$+ \left\| div \left(v_0(\cdot) \nabla e^{-\int_0^t (\alpha_u u + \alpha_w w + \delta_v)(\cdot,s)ds} \right) \right\|_{L^{p_0}(\Omega)}$$

$$\le \left\| \Delta v_0(\cdot) e^{-\int_0^t (\alpha_u u + \alpha_w w + \delta_v)(\cdot,s)ds} \right\|_{L^{p_0}(\Omega)} + \left\| v_0(\cdot) \Delta e^{-\int_0^t (\alpha_u u + \alpha_w w + \delta_v)(\cdot,s)ds} \right\|_{L^{p_0}(\Omega)}$$

$$+ 2 \left\| \nabla v_0(\cdot) e^{-\int_0^t (\alpha_u u + \alpha_w w + \delta_v)(\cdot,s)ds} \int_0^t (\alpha_u \nabla u + \alpha_w \nabla w)\,(\cdot,s)ds \right\|_{L^{p_0}(\Omega)} \tag{6.5.29}$$

$$\le \left\| \Delta v_0(\cdot) e^{-\delta_v t} \right\|_{L^{p_0}(\Omega)}$$

$$
+ \left\| v_0(\cdot) e^{-\int_0^t (\alpha_u u + \alpha_w w + \delta_v)(\cdot, s)ds} \left[\int_0^t (\alpha_u \nabla u + \alpha_w \nabla w)(\cdot, s)ds \right]^2 \right\|_{L^{p_0}(\Omega)}
$$

$$
+ \left\| v_0(\cdot) e^{-\int_0^t (\alpha_u u + \alpha_w w + \delta_v)(\cdot, s)ds} \int_0^t (\alpha_u \Delta u + \alpha_w \Delta w)(\cdot, s)\, ds \right\|_{L^{p_0}(\Omega)}
$$

$$
+ 2 \left\| e^{-\delta_v t} \nabla v_0(\cdot) \int_0^t (\alpha_u \nabla u + \alpha_w \nabla w)(\cdot, s)\, ds \right\|_{L^{p_0}(\Omega)}
$$

$$
\leq \left\| \Delta v_0(\cdot) e^{-\delta_v t} \right\|_{L^{p_0}(\Omega)} + \left\| v_0(\cdot) \right\|_{L^\infty(\Omega)} e^{-\delta_v t} \left\| \left[\int_0^t (\alpha_u \nabla u + \alpha_w \nabla w)(\cdot, s) \right]^2 \right\|_{L^{p_0}(\Omega)}
$$

$$
+ 2 \left\| \nabla v_0(\cdot) \right\|_{L^{2p_0}(\Omega)} e^{-\delta_v t} \left\| \int_0^t (\alpha_u \nabla u + \alpha_w \nabla w)(\cdot, s)ds \right\|_{L^{2p_0}(\Omega)}
$$

$$
+ \left\| v_0(\cdot) \right\|_{L^\infty(\Omega)} e^{-\delta_v t} \left\| \int_0^t (\alpha_u \Delta u + \alpha_w \Delta w)(\cdot, s)\, ds \right\|_{L^{p_0}(\Omega)}.
$$

From (6.5.22) and (6.5.23), we obtain

$$
e^{-(\delta_v - \alpha)t} \left\| \left[\int_0^t (\alpha_u \nabla u + \alpha_w \nabla w)(\cdot, s)\, ds \right]^2 \right\|_{L^{p_0}(\Omega)}
$$

$$
\leq \alpha_u^2 e^{-(\delta_v - \alpha)t} \left\| \left(\int_0^t \nabla u(\cdot, s)\, ds \right)^2 \right\|_{L^{p_0}(\Omega)} + \alpha_w^2 e^{-(\delta_v - \alpha)t} \left\| \left(\int_0^t \nabla w(\cdot, s)\, ds \right)^2 \right\|_{L^{p_0}(\Omega)}
$$

$$
+ 2\alpha_u \alpha_w e^{-(\delta_v - \alpha)t} \left\| \int_0^t \nabla u(\cdot, s)\, ds \int_0^t \nabla w(\cdot, s)\, ds \right\|_{L^{p_0}(\Omega)}
$$

$$
\leq \alpha_u^2 e^{-(\delta_v - \alpha)t} \left\| \int_0^t \nabla u(\cdot, s)\, ds \right\|_{L^{2p_0}(\Omega)}^2 + \alpha_w^2 e^{-(\delta_v - \alpha)t} \left\| \int_0^t \nabla w(\cdot, s)\, ds \right\|_{L^{2p_0}(\Omega)}^2
$$

$$
+ 2\alpha_u \alpha_w e^{-(\delta_v - \alpha)t} \left\| \int_0^t \nabla u(\cdot, s)\, ds \right\|_{L^{2p_0}(\Omega)} \left\| \int_0^t \nabla w(\cdot, s)\, ds \right\|_{L^{2p_0}(\Omega)} \tag{6.5.30}
$$

$$
\leq \frac{(\alpha_u M_1 \varepsilon)^2 t^{\frac{2p_0 - 1}{p_0}} e^{-(\delta_v - \alpha)t}}{(2\alpha p_0)^{\frac{1}{p_0}}} + \frac{2\alpha_u \alpha_w M_1 M_2 \varepsilon^2 t^{\frac{2p_0 - 1}{p_0}} e^{-(\delta_v - \alpha)t}}{(2\alpha p_0)^{\frac{1}{p_0}}}
$$

$$
+ \frac{(\alpha_w M_2 \varepsilon)^2 t^{\frac{2p_0 - 1}{p_0}} e^{-(\delta_v - \alpha)t}}{(2\alpha p_0)^{\frac{1}{p_0}}}
$$

$$
\leq \frac{(\alpha_u M_1 + \alpha_w M_2)^2 \varepsilon^2}{(2\alpha p_0)^{\frac{1}{p_0}}} \left(\frac{2p_0 - 1}{e p_0 (\delta_v - \alpha)} \right)^{\frac{2p_0 - 1}{p_0}}
$$

and

$$2 \left\| \nabla v_0(\cdot) \right\|_{L^{2p_0}(\Omega)} e^{-\delta_v t} \left\| \int_0^t (\alpha_u \nabla u + \alpha_w \nabla w)(\cdot, s) ds \right\|_{L^{2p_0}(\Omega)}$$

$$\leq 2\varepsilon^2 e^{-\alpha t} \frac{\alpha_u M_1 + \alpha_w M_2}{(2\alpha p_0)^{\frac{1}{2p_0}}} \left(\frac{2p_0 - 1}{2p_0 e(\delta_v - \alpha)} \right)^{\frac{2p_0 - 1}{2p_0}}. \tag{6.5.31}$$

From Lemma 6.3, noticing $0 < k < \delta_v - \alpha$, $p_0 > 1$, we obtain

$$t^{\frac{p_0 - 1}{p_0}} e^{-(\delta_v - \alpha)t} \leq \left(\frac{p_0 - 1}{p_0 e(\delta_v - \alpha)} \right)^{\frac{p_0 - 1}{p_0}}$$

and

$$t^{\frac{p_0 - 1}{p_0}} e^{-(\delta_v - \alpha - k)t} \leq \left(\frac{p_0 - 1}{p_0 e(\delta_v - \alpha - k)} \right)^{\frac{p_0 - 1}{p_0}}$$

for all $t \geq 0$, which together with Lemma 6.28 implies

$$e^{-(\delta_v - \alpha)t} \left\| \int_0^t (\alpha_u \Delta u + \alpha_w \Delta w)(\cdot) d\tau \right\|_{L^{p_0}(\Omega)}$$

$$\leq \left(\int_0^t \int_\Omega |\alpha_u \Delta u + \alpha_w \Delta w|^{p_0}(x, s) dx ds \right)^{\frac{1}{p_0}} t^{\frac{p_0 - 1}{p_0}} e^{-(\delta_v - \alpha)t}$$

$$\leq (\alpha_u K_1(t) + \alpha_w K_2(t)) t^{\frac{p_0 - 1}{p_0}} e^{-(\delta_v - \alpha)t} \tag{6.5.32}$$

$$= \varepsilon C_{p_0} \alpha_u t^{\frac{p_0 - 1}{p_0}} e^{-(\delta_v - \alpha)t} \left(\frac{\varepsilon \xi_u M_1 M_4}{((2\alpha - k)p_0)^{\frac{1}{p_0}}} + \frac{\varepsilon \xi_u M_0 M_5}{((\alpha - k)p_0)^{\frac{1}{p_0}}} + 1 + |\Omega|^{\frac{1}{p_0}} \right)$$

$$+ \varepsilon C_{p_0} \alpha_w t^{\frac{p_0 - 1}{p_0}} e^{-(\delta_v - \alpha)t} \left(\frac{\varepsilon \xi_w M_2 M_4}{((2\alpha - k)p_0)^{\frac{1}{p_0}}} + \frac{\varepsilon \xi_w M_0 M_5}{((\alpha - k)p_0)^{\frac{1}{p_0}}} + 1 + |\Omega|^{\frac{1}{p_0}} \right)$$

$$+ \varepsilon \frac{C_{p_0} M_0 |\Omega|^{\frac{1}{p_0}}}{(kp_0)^{\frac{1}{p_0}}} t^{\frac{p_0 - 1}{p_0}} e^{-(\delta_v - \alpha - k)t} \left(\varepsilon \alpha_u \rho_u M_0 + \frac{1}{2}\alpha_u(\delta_v - \alpha) + \varepsilon \rho_w \alpha_w M_0 \right)$$

$$\leq \varepsilon C_{p_0} \alpha_u \left(\frac{p_0 - 1}{p_0 e(\delta_v - \alpha)} \right)^{\frac{p_0 - 1}{p_0}} \left(\frac{\varepsilon \xi_u M_1 M_4}{((2\alpha - k)p_0)^{\frac{1}{p_0}}} + \frac{\varepsilon \xi_u M_0 M_5}{((\alpha - k)p_0)^{\frac{1}{p_0}}} + 1 + |\Omega|^{\frac{1}{p_0}} \right)$$

$$+ \varepsilon C_{p_0} \alpha_w \left(\frac{p_0 - 1}{p_0 e(\delta_v - \alpha)} \right)^{\frac{p_0 - 1}{p_0}} \left(\frac{\varepsilon \xi_w M_2 M_4}{((2\alpha - k)p_0)^{\frac{1}{p_0}}} + \frac{\varepsilon \xi_w M_0 M_5}{((\alpha - k)p_0)^{\frac{1}{p_0}}} + 1 + |\Omega|^{\frac{1}{p_0}} \right)$$

$$+ \varepsilon \frac{C_{p_0} M_0 |\Omega|^{\frac{1}{p_0}}}{(kp_0)^{\frac{1}{p_0}}} \left(\frac{p_0 - 1}{p_0 e(\delta_v - \alpha - k)} \right)^{\frac{p_0 - 1}{p_0}} \left(\varepsilon \alpha_u \rho_u M_0 + \frac{1}{2}\alpha_u(\delta_v - \alpha) + \varepsilon \rho_w \alpha_w M_0 \right).$$

Inserting (6.5.30), (6.5.31) and (6.5.32) into (6.5.29), we get

$$\|\Delta v\,(\cdot, t)\|_{L^{p_0}(\Omega)}$$

$$\leq \varepsilon e^{-\alpha t} + \varepsilon^2 \|v_0(\cdot)\|_{L^{\infty}(\Omega)} e^{-\alpha t} \frac{(\alpha_u M_1 + \alpha_w M_2)^2}{(2\alpha p_0)^{\frac{1}{p_0}}} \left(\frac{2p_0 - 1}{p_0 e(\delta_v - \alpha)}\right)^{\frac{2p_0 - 1}{p_0}}$$

$$+ 2\varepsilon^2 e^{-\alpha t} \frac{\alpha_u M_1 + \alpha_w M_2}{(2\alpha p_0)^{\frac{1}{2p_0}}} \left(\frac{2p_0 - 1}{2p_0 e(\delta_v - \alpha)}\right)^{\frac{2p_0 - 1}{2p_0}} \tag{6.5.33}$$

$$+ \varepsilon \|v_0(\cdot)\|_{L^{\infty}(\Omega)} e^{-\alpha t} C_{p_0} \alpha_u \left(\frac{p_0 - 1}{p_0 e(\delta_v - \alpha)}\right)^{\frac{p_0 - 1}{p_0}}$$

$$\cdot \left(\frac{\varepsilon \xi_u M_1 M_4}{((2\alpha - k)p_0)^{\frac{1}{p_0}}} + \frac{\varepsilon \xi_u M_0 M_5}{((\alpha - k)p_0)^{\frac{1}{p_0}}} + 1 + |\Omega|^{\frac{1}{p_0}}\right)$$

$$+ \varepsilon \|v_0(\cdot)\|_{L^{\infty}(\Omega)} e^{-\alpha t} C_{p_0} \alpha_w \left(\frac{p_0 - 1}{p_0 e(\delta_v - \alpha)}\right)^{\frac{p_0 - 1}{p_0}}$$

$$\cdot \left(\frac{\varepsilon \xi_w M_2 M_4}{((2\alpha - k)p_0)^{\frac{1}{p_0}}} + \frac{\varepsilon \xi_w M_0 M_5}{((\alpha - k)p_0)^{\frac{1}{p_0}}} + 1 + |\Omega|^{\frac{1}{p_0}}\right)$$

$$+ \frac{\varepsilon \|v_0(\cdot)\|_{L^{\infty}(\Omega)} e^{-\alpha t} C_{p_0} M_0 |\Omega|^{\frac{1}{p_0}}}{(kp_0)^{\frac{1}{p_0}}} \left(\frac{p_0 - 1}{p_0 e(\delta_v - \alpha - k)}\right)^{\frac{p_0 - 1}{p_0}}$$

$$\cdot \left(\varepsilon \alpha_u \rho_u M_0 + \frac{1}{2}\alpha_u(\delta_v - \alpha) + \varepsilon \rho_w \alpha_w M_0\right).$$

Therefore, (6.5.28) follows from (6.5.8).

Lemma 6.31 *Under the assumptions of Theorem 6.3, for all $t \in (0, T)$,*

$$\left\|\left(u + \frac{\rho_u + \rho_z}{\rho_w} w + z + Q\right)(\cdot, t) - e^{t\Delta}\left(u_0 + \frac{\rho_u + \rho_z}{\rho_w} w_0 + z_0\right)(\cdot)\right\|_{L^{\infty}(\Omega)}$$

$$\leq \frac{M_6}{2} \varepsilon e^{-\alpha t}. \tag{6.5.34}$$

Proof From Lemma 1.1(iv) and (6.5.11), using Lemma 4.3, it follows that

$$\left\|\left(u + \frac{\rho_u + \rho_z}{\rho_w} w + z + Q\right)(\cdot, t) - e^{t\Delta}\left(u_0 + \frac{\rho_u + \rho_z}{\rho_w} w_0 + z_0\right)(\cdot)\right\|_{L^{\infty}(\Omega)}$$

$$\leq \int_0^t \left\|e^{(t-s)\Delta}\left[\xi_u \nabla \cdot (u\nabla v) + \frac{\rho_u + \rho_z}{\rho_w} \xi_w \nabla \cdot (w\nabla v) + \xi_z \nabla \cdot (z\nabla v)\right](\cdot, s)\right\|_{L^{\infty}(\Omega)} ds$$

$$\leq \xi_u c_4 \int_0^t \left[1 + (t - s)^{-\frac{1}{2} - \frac{N}{4p_0}}\right] e^{-\lambda_1(t-s)} \|(u\nabla v)\,(\cdot, s)\|_{L^{2p_0}(\Omega)} ds$$

$$+ \frac{\rho_u + \rho_z}{\rho_w} \xi_w c_4 \int_0^t \left[1 + (t - s)^{-\frac{1}{2} - \frac{N}{4p_0}}\right] e^{-\lambda_1(t-s)} \|(w\nabla v)\,(\cdot, s)\|_{L^{2p_0}(\Omega)} ds$$

$$+ \xi_z c_4 \int_0^t \left[1 + (t-s)^{-\frac{1}{2} - \frac{N}{4p_0}} \right] e^{-\lambda_1(t-s)} \|(z\nabla v)(\cdot, s)\|_{L^{2p_0}(\Omega)} \, ds$$

$$\leq 2c_{10}c_4 M_0 M_4 \varepsilon^2 e^{-\alpha t} \left(\xi_u + \frac{\rho_u + \rho_z}{\rho_w} \xi_w + \xi_z \right),$$

and in view of (6.5.9), we already arrive at (6.5.34) and complete the proof.

Proof of Theorem 6.3. First let us verify $T = T_{max}$ by contraction. In fact, suppose that $T < T_{max}$, then from Lemmas 6.25–6.31, it follows

$$\|\nabla u(\cdot, t)\|_{L^{2p_0}(\Omega)} \leq \frac{M_1}{2} \varepsilon e^{-\alpha t},$$

$$\|\nabla w(\cdot, t)\|_{L^{2p_0}(\Omega)} \leq \frac{M_2}{2} \varepsilon e^{-\alpha t},$$

$$\|\nabla z(\cdot, t)\|_{L^{2p_0}(\Omega)} \leq \frac{M_3}{2} \varepsilon e^{-\alpha t},$$

$$\|\nabla v(\cdot, t)\|_{L^{2p_0}(\Omega)} \leq \frac{M_4}{2} \varepsilon e^{-\alpha t},$$

$$\|\Delta v(\cdot, t)\|_{L^{p_0}(\Omega)} \leq \frac{M_5}{2} \varepsilon e^{-\alpha t},$$

$$\left\| \left(u + \frac{\rho_u + \rho_z}{\rho_w} w + z + Q \right)(\cdot, t) - e^{t\Delta} \left(u_0 + \frac{\rho_u + \rho_z}{\rho_w} w_0 + z_0 \right)(\cdot) \right\|_{L^\infty(\Omega)} \leq \frac{M_6}{2} \varepsilon e^{-\alpha t},$$

for all $t \in (0, T)$, which contradicts the definition of T.

Next, we show that $T_{max} = \infty$. In fact, if $T_{max} < \infty$, then in view of the definition of T, we obtain

$$\lim_{t \nearrow T_{max}} \left(\|u(\cdot, t)\|_{W^{1,2p_0}(\Omega)} + \|v(\cdot, t)\|_{W^{2,p_0}(\Omega)} + \|w(\cdot, t)\|_{W^{1,2p_0}(\Omega)} + \|z(\cdot, t)\|_{W^{1,2p_0}(\Omega)} \right)$$

$$< \infty,$$

which contradicts with (6.5.1) in Lemma 6.22. Therefore, we have $T_{max} = \infty$.

Integrating the equation of u in (6.1.12) over Ω, we have

$$\int_\Omega u(x, t) \, dx = \int_\Omega u_0(x) dx - \rho_u \int_0^t \int_\Omega (uz)(x, s) \, dx ds,$$

which along with the nonnegative property of u, z and the fact that $\|u(\cdot, t)\|_{L^\infty(\Omega)} \leq M_0 \varepsilon$ warrants that $\bar{u}(t) := \frac{1}{|\Omega|} \int_\Omega u(x, t) \, dx$ is noncreasing with respect to time t and and its limit $t \to \infty$ exists, that is,

$$\lim_{t \to \infty} \int_{\Omega} u(x,t)\,dx = \int_{\Omega} u_0(x)\,dx - \rho_u \int_0^{+\infty} \int_{\Omega} (uz)(x,s)\,dxds$$

as well as

$$\lim_{t \to \infty} \bar{u}(t) = \frac{1}{|\Omega|} \left(\int_{\Omega} u_0(x)\,dx - \rho_u \int_0^{+\infty} \int_{\Omega} (uz)(x,s)\,dxds \right) := u^*,$$

which implies

$$0 \le \frac{\rho_u}{|\Omega|} \int_t^{\infty} \int_{\Omega} (uz)(x,s)\,dxds = \bar{u}(t) - u^* \to 0$$

as $t \to \infty$. On the other hand, by Poincare's inequality, there exists $k_1 > 0$ such that

$$\|u(\cdot,t) - \bar{u}(t)\|_{L^{2p_0}(\Omega)} \le k_1 \|\nabla u(\cdot,t)\|_{L^{2p_0}(\Omega)}.$$

By Embedding theorem, we know $W^{1,2p_0}(\Omega) \hookrightarrow C^{1-\frac{N}{p_0}}(\overline{\Omega})$, for $p_0 > \max\{1, \frac{N}{2}\}$. There exists $k_2 > 0$, such that

$$\|u(\cdot,t) - \bar{u}(t)\|_{L^{\infty}(\Omega)} \le k_2 \|u(\cdot,t) - \bar{u}(t)\|_{W^{1,2p_0}(\Omega)}$$
$$\le k_2(1+k_1) \|\nabla u(\cdot,t)\|_{L^{2p_0}(\Omega)} \to 0$$

as $t \to \infty$. Thus,

$$\left\| u(\cdot,t) - u^* \right\|_{L^{\infty}(\Omega)} \le \|u(\cdot,t) - \bar{u}(t)\|_{L^{\infty}(\Omega)} + \left\| \bar{u}(t) - u^* \right\|_{L^{\infty}(\Omega)} \to 0.$$

Now, we consider a linear combination of u and w

$$H := \rho_w u + \rho_u w,$$

then

$$H_t = \rho_w u_t + \rho_u w_t = \Delta H - \rho_w \xi_u \nabla \cdot (u\nabla v) - \rho_u \xi_w \nabla \cdot (w\nabla v) - \rho_u \delta_w w.$$

Accordingly,

$$\int_{\Omega} H(x,t)\,dx = \int_{\Omega} H(x,0)dx - \rho_u \delta_w \int_0^t \int_{\Omega} w(x,s)\,dxds.$$

Similarly, we obtain

$$\lim_{t \to \infty} \overline{H}(t) = \frac{1}{|\Omega|} \left(\int_{\Omega} H(x,0)\,dx - \rho_u \delta_w \int_0^{+\infty} \int_{\Omega} w(x,s)\,dxds \right) := H^*,$$

and

$$\left\|H(\cdot,t)-\bar{H}(t)\right\|_{L^{\infty}(\Omega)} \le k_2 \left\|H(\cdot,t)-\bar{H}(t)\right\|_{W^{1,2p_0}(\Omega)}$$
$$\le k_2 \left(1+k_1\right) \|\nabla H(\cdot,t)\|_{L^{2p_0}(\Omega)} \to 0.$$

Then

$$\|H(\cdot,t)-H^*\|_{L^{\infty}(\Omega)} \le \|H(\cdot,t)-\overline{H}(t)\|_{L^{\infty}(\Omega)} + \|\overline{H}(t)-H^*\|_{L^{\infty}(\Omega)} \to 0$$

as $t \to \infty$. Then denote $w^* := \frac{1}{\rho_u}(H^*-\rho_w u^*)$, as $t \to \infty$, we obtain

$$\rho_u \left\|w(\cdot,t)-w^*\right\|_{L^{\infty}(\Omega)}$$
$$= \left\|\rho_u w(\cdot,t)-(H^*-\rho_w u^*)\right\|_{L^{\infty}(\Omega)}$$
$$\le \left\|H(\cdot,t)-H^*\right\|_{L^{\infty}(\Omega)} + \rho_w \left\|u(\cdot,t)-u^*\right\|_{L^{\infty}(\Omega)} \to 0.$$

Next, we consider a linear combination of u, w and z. Let

$$I = \rho_z(u+w) + (\rho_u+\rho_w)z.$$

Then

$$I_t = \Delta I - \rho_z \xi_u \nabla \cdot (u\nabla v) - \rho_z \xi_w \nabla \cdot (w\nabla v) - (\rho_u+\rho_w)\xi_z \nabla \cdot (z\nabla v)$$
$$- (\rho_z \delta_w - (\rho_u+\rho_w)\beta)\, w - (\rho_u+\rho_w)\delta_z z.$$

Accordingly,

$$\int_{\Omega} I(x,t)\, dx$$
$$= \int_{\Omega} I(x,0)dx - \int_0^t \int_{\Omega} [(\rho_z \delta_w - (\rho_u+\rho_w)\beta)\, w(x,s) + (\rho_u+\rho_w)\delta_z z(x,s)]dxds,$$

where $\rho_z \delta_w - (\rho_u+\rho_w)\beta > 0$.
  Similarly, we obtain

$$\lim_{t\to\infty} \overline{I}(t) = \frac{1}{|\Omega|} \left(\int_{\Omega} I(x,0)\, dx \right.$$
$$\left. - \int_0^{+\infty} \int_{\Omega} [(\rho_z \delta_w - (\rho_u+\rho_w)\beta)\, w(x,s) + (\rho_u+\rho_w)\delta_z z(x,s)]dxds \right)$$
$$:= I^*,$$

and

$$\left\| I(\cdot,t) - \bar{I}(t) \right\|_{L^\infty(\Omega)} \le k_2 \left\| I(\cdot,t) - \bar{I}(t) \right\|_{W^{1,2p_0}(\Omega)}$$
$$\le k_2 \left(1 + k_1 \right) \left\| \nabla I(\cdot,t) \right\|_{L^{2p_0}(\Omega)} \to 0.$$

Then

$$\left\| I(\cdot,t) - I^* \right\|_{L^\infty(\Omega)} \le \left\| I(\cdot,t) - \bar{I}(t) \right\|_{L^\infty(\Omega)} + \left\| \bar{I}(t) - I^* \right\|_{L^\infty(\Omega)} \to 0$$

as $t \to \infty$. Then denote $z^* := \frac{1}{\rho_u + \rho_w}(I^* - \rho_z(u^* + w^*))$, as $t \to \infty$, we obtain

$$(\rho_u + \rho_w) \left\| z(\cdot,t) - z^* \right\|_{L^\infty(\Omega)}$$
$$= \left\| (\rho_u + \rho_w) z(\cdot,t) - (I^* - \rho_z(u^* + w^*)) \right\|_{L^\infty(\Omega)}$$
$$\le \left\| I(\cdot,t) - I^* \right\|_{L^\infty(\Omega)} + \rho_z \left\| u(\cdot,t) - u^* \right\|_{L^\infty(\Omega)} + \rho_z \left\| w(\cdot,t) - w^* \right\|_{L^\infty(\Omega)} \to 0.$$

By contradiction, if $w^* > 0$ or $z^* > 0$, then there exists $t^* > 0$ such that for all $t > t^*$,

$$\frac{d}{dt} \int_\Omega Q(x,t)dx = \int_\Omega \left(\left(\frac{\rho_u + \rho_z}{\rho_w} \delta_w - \beta \right) w(x,t) + \delta_z z(x,t) \right) dx$$
$$\ge \frac{|\Omega|}{2} \left(\left(\frac{\rho_u + \rho_z}{\rho_w} \delta_w - \beta \right) w^* + \delta_z z^* \right) > 0,$$

which implies that $\int_\Omega Q(x,t)dx \to \infty$ as $t \to \infty$ and thus contradicts with that $\|Q(\cdot,t)\|_{L^\infty(\Omega)} \le M_0 \varepsilon$. Hence, we have $w^* = z^* = 0$. On the other hand, by (6.5.20), it is easy to see that

$$\|v(\cdot,t)\|_{L^\infty(\Omega)} \to 0.$$

So the proof of Theorem 6.3 is complete.

References

Adler, J. (1966). Chemotaxis in bacteria. *Science, 153*, 708–716.

Ahn, J., Kangy, K., Kang, K., Kim, J., & Lee, J. (2017). Lower bound of mass in a chemotactic model with advection and absorbing reaction. *SIAM Journal on Mathematical Analysis, 49*, 723–755.

Ahn, J., & Yoon, C. (2019). Global well-posedness and stability of constant equilibria in parabolic-elliptic chemotaxis systems without gradient sensing. *Nonlinearity, 32*, 1327–1351.

Alikakos, N. D. (1979). An application of the invariance principle to reaction-diffusion equations. *Journal of Differential Equations, 33*, 201–225.

Alzahrani, T., Raluca Eftimie, R., & Dumitru Trucu, D. (2019). Multiscale modelling of cancer response to oncolytic viral therapy. *Mathematical Biosciences, 310*, 76–95.

Anderson, A. R. A., & Chaplain, M. A. J. (1998). A mathematical model for capillary network formation in the absence of endothelial cell proliferation. *Applied Mathematics Letters, 11*, 109–116.

Anderson, A. R. A., & Chaplain, M. A. J. (1998). Continuous and discrete mathematical models of tumor-induced angiogenesis. *Bulletin of Mathematical Biology, 60*, 857–899.

Aznavoorian, S., Stracke, M. L., Krutzsch, H., Schiffmann, E., & Liotta, L. A. (1990). Signal transduction for chemotaxis and haptotaxis by matrix molecules in tumor cells. *Journal of Cell Biology, 110*, 1427–1438.

Bellomo, N., Bellouquid, A., Tao, Y., & Winkler, M. (2015). Toward a mathematical theory of Keller-Segel models of pattern formation in biological tissues. *Mathematical Models and Methods in Applied Sciences, 25*, 1663–1763.

Bellomo, N., Bellouquid, A., & Chouhad, N. (2016). From a multiscale derivation of nonlinear cross-diffusion models to Keller-Segel models in a Navier-Stokes fluid. *Mathematical Models and Methods in Applied Sciences, 26*, 2041–2069.

Bellomo, N., Painter, K. J., Tao, Y., & Winkler, M. (2019). Occurrence versus absence of taxis-driven instabilities in a May-Nowak model for virus infection. *SIAM Journal on Applied Mathematics, 79*, 1990–2010.

Bellomo, N., & Tao, Y. (2020). Stabilization in a chemotaxis model for virus infection. *Discrete and Continuous Dynamical Systems-Series S, 13*, 105–117.

Biler, P., Hebisch, W., & Nadzieja, T. (1994). The Debye system: Existence and large time behavior of solutions. *Nonlinear Analysis, 23*, 1189–1209.

Black, T. (2018). Eventual smoothness of generalized solutions to a singular chemotaxis-Stokes system in 2D. *Journal of Differential Equations, 265*, 2296–2339.

Black, T., Lankeit, J., & Mizukami, M. (2018). Singular sensitivity in a Keller-Segel-fluid system. *Journal of Evolution Equations, 18*, 561–581.

Black, T., Lankeit, J., & Mizukami, M. (2019). A Keller-Segel-fluid system with singular sensitivity: Generalized solutions. *Mathematicsl Methods in the Applied Sciences, 42*, 3002–3020.

Black, T. (2020). Global generalized solutions to a forager-exploiter model with superlinear degradation and their eventual regularity properties. *Mathematical Models and Methods in Applied Sciences, 30*, 1075–1117.

Breitbach, C. J., Parato, K., et al. (2015). Pexa-Vec double agent engineered vaccinia: Oncolytic and active immunotherapeutic. *Current Opinion in Virology, 13*, 49–54.

Cai, Y., Cao, Q., & Wang, Z. A. Asymptotic dynamics and spatial patterns of a ratio-dependent predator-prey system with prey-taxis. *Applicable Analysis*. https://doi.org/10.1080/00036811.2020.1728259.

Calvez, V., & Carrillo, J. A. (2006). Volume effects in the Keller-Segel model: Energy estimates preventing blow-up. *Journal de Mathématiques Pures et Appliquées, 9*(86), 155–175.

Cañizo, J. A., Desvillettes, L., & Fellner, K. (2014). Improved duality estimates and applications to reaction-diffusion equations. *Communications in Partial Differential Equations, 39*, 1185–1284.

Cao, X. (2015). Global bounded solutions of the higher-dimensional Keller-Segel system under smallness conditions in optimal spaces. *Discrete and Continuous Dynamical Systems, 35*, 1891–1904.

Cao, X. (2016). Boundedness in a three-dimensional chemotaxis-haptotaxis model. *Zeitschrift fur Angewandte Mathematik und Physik, 67*, 1–13.

Cao, X., & Lankeit, J. (2016). Global classical small-data solutions for a 3D chemotaxis Navier-Stokes system involving matrix-valued sensitivities. *Calculus of Variations and Partial Differential Equations, 55*, 55–107.

Cao, X., & Winkler, M. (2018). Sharp decay estimates in a bioconvection model with quadratic degradation in bounded domains. *Proceedings of the Royal Society of Edinburgh Section A, 148*, 939–955.

Cao, X. (2020). Global radial renormalized solution to a producer-scrounger model with singular sensitivities. *Mathematical Models and Methods in Applied Sciences, 30*, 1119–1165.

Cao, X., & Tao, Y. (2021). Boundedness and stabilization enforced by mild saturation of taxis in a producer-scrounger model. *Nonlinear Analysis: Real World Applications, 57*, 103189.

Chakraborty, A., Singh, M., Lucy, D., & Ridland, P. (2007). Predator-prey model with prey-taxis and diffusion. *Mathematical and Computer Modelling, 46*, 482–498.

Chaplain, M. A. J., & Stuart, A. M. (1993). A model mechanism for the chemotactic response of endothelial cells to tumor angiogenesis factor. *IMA Journal of Mathematics Applied in Medicine and Biology, 10*, 149–168.

Chaplain, M. A. J., & Lolas, G. (2005). Mathematical modelling of cancer invasion of tissue: The role of the urokinase plasminogen activation system. *Mathematical Models and Methods in Applied Sciences, 15*, 1685–1734.

Chaplain, M. A. J., & Lolas, G. (2006). Mathematical modelling of cancer invasion of tissue: dynamic heterogeneity. *Networks and Heterogeneous Media, 1*, 399–439.

Chertock, A., Fellner, K., Kurganov, A., Lorz, A., & Markowich, P. A. (2012). Sinking, merging and stationary plumes in a coupled chemotaxis-fluid model: A high-resolution numerical approach. *Journal of Fluid Mechanics, 694*, 155–190.

Cieślak, T., & Winkler, M. (2008). Finite-time blow-up in a quasilinear system of chemotaxis. *Nonlinearity, 21*, 1057–1076.

Cieślak, T., & Stinner, C. (2012). Finite-time blowup and global-in-time unbounded solutions to a parabolic-parabolic quasilinear Keller-Segel system in higher dimensions. *Journal of Differential Equations, 252*, 5832–5851.

Coll, J. C., et al. (1994). Chemical aspects of mass spawning in corals. I. Sperm-atractant molecules in the eggs of the scleractinian coral Montipora digitata. *Marine Biology, 118*, 177–182.

Coll, J. C., et al. (1995). Chemical aspects of mass spawning in corals. II. (-)-Epi-thunbergol, the sperm attractant in the eggs of the soft coral Lobophytum crassum (Cnidaria: Octocorallia). *Marine Biology, 123*, 137–143.

Corrias, L., Perthame, B., & Zaag, H. (2003). A chemotaxis model motivated by angiogenesis. *Comptes Rendus de l'Académie des Sciences-Series I-Mathematics, 336*, 141–146.

Corrias, L., Perthame, B., & Zaag, H. (2004). Global solutions of some chemotaxis and angiogenesis systems in high space dimensions. *Milan Journal of Mathematics, 72*, 1–28.

Difrancesco, M., Lorz, A., & Markowich, P. A. (2010). Chemotaxis-fluid coupled model for swimming bacteria with nonlinear diffusion: Global existence and asymptotic behavior. *Discrete and Continuous Dynamical Systems, 28*, 1437–1453.

Dillon, R., Maini, P. K., & Othmer, H. G. (1994). Pattern formation in generalised turing systems I. Steady-state patterns in systems with mixed boundary conditions. *Journal of Mathematical Biology, 32*, 345–393.

Ding, M., & Zhao, X. (2018). Global existence, boundedness and asymptotic behavior to a logistic chemotaxis model with density-signal governed sensitivity and signal absorption. arXiv:1806.09914.

Duan, R., & Xiang, Z. (2014). A note on global existence for the chemotaxis-Stokes model with nonlinear diffusion. *International Mathematics Research Notices IMRN, 7*, 1833–1852.

Eftimie, R., De Vries, G., & Lewis, M. A. (2007). Complex spatial group patterns result from different animal communication mechanisms. *Proceedings of the National academy of Sciences of the United States of America, 104*, 6974–6980.

Engwer, C., Stinner, C., & Surulescu, C. (2017). On a structured multiscale model for acid-mediated tumor invasion: The effects of adhesion and proliferation. *Mathematical Models and Methods in Applied Sciences, 27*, 1355–1390.

Espejo, E. E., & Suzuki, T. (2015). Reaction terms avoiding aggregation in slow fluids. *Nonlinear Analysis: Real World Applications, 21*, 110–126.

Espejo, E. E., & Suzuki, T. (2017). Reaction enhancement by chemotaxis. *Nonlinear Analysis: Real World Applications, 35*, 102–131.

Espejo, E. E., & Winkler, M. (2018). Global classical solvability and stabilization in a two-dimensional chemotaxis-Navier-Stokes system modeling coral fertilization. *Nonlinearity, 31*, 1227–1259.

Evans, L. C. (2010). Partial differential equations. In: *Graduate Studies in Mathematics* (2nd ed., vol. 19, pp. 211–223). Providence, RI: American Mathematical Society.

Fontelos, M. A., Friedman, A., & Hu, B. (2002). Mathematical analysis of a model for the initiation of angiogenesis. *SIAM Journal on Mathematical Analysis, 33*, 1330–1355.

Francesco, M. D., Lorz, A., & Markowich, P. (2010). Chemotaxis-fluid coupled model for swimming bacteria with nonlinear diffusion: Global existence and asymptotic behavior. *Discrete and Continuous Dynamical Systems, 28*, 1437–1453.

Friedman, A. (1969). *Partial differential equations*. Rinehart & Winston, New York: Holt.

Friedman, A., & Lolas, G. (2005). Analysis of a mathematical model of tumor lymphangiogenesis. *Mathematical Models and Methods in Applied Sciences, 15*, 95–107.

Fu, X., Tang, L. H., Liu, C., Huang, J. D., Hwa, T., & Lenz, P. (2012). Stripe formation in bacterial system with density-suppressed motility. *Physical Review Letters, 108*, 198102.

Fujie, K. (2015). Boundedness in a fully parabolic chemotaxis system with singular sensitivity. *Journal of Mathematical Analysis and Applications, 424*, 675–684.

Fujie, K. (2016). Study of reaction-diffusion systems modeling chemotaxis. Doctoral thesis.

Fujie, K., & Senba, T. (2017). Application of an Adams type inequality to a two-chemical substances chemotaxis system. *Journal of Differential Equations, 263*, 88–148.

Fujie, K., & Jiang, J. (2020). Global existence for a kinetic model of pattern formation with density-suppressed motilities. *Journal of Differential Equations, 269*, 5338–5378.

Fujie, K., & Jiang, J. (2021). Comparison methods for a Keller-Segel-type model of pattern formations with density-suppressed motilities. *Calculus of Variations and Partial Differential Equations, 60*, 1–37.

Fujiwara, D., & Morimoto, H. (1977). An L^r-theorem of the Helmholtz decomposition of vector fields. Journal of the Faculty of Science of the University of Tokyo, 24, 685–700.

Gatenby, R. A., & Gawlinski, E. T. (1996). A reaction-diffusion model of cancer invasion. *Cancer Research, 56*, 5745–5753.

Giga, Y. (1981). The Stokes operator in L_r spaces. Proceedings of the Japan Academy, Series A, Mathematical Sciences, 57, 85–89.

Giga, Y. (1986). Solutions for semilinear parabolic equations in L^p and regularity of weak solutions of the Navier-Stokes system. Journal of Differential Equations, 61, 186–212.

Gilbarg, D., & Trudinger, N. S. (2001). *Elliptic partial differential equations of second order*. New York: Springer.

Goldsmith, K., Chen, W., Johnson, D. C., & Hendricks, R. L. (1998). Infected cell protein(ICP) 47 enhances herpes simplex virus neurovirulence by blocking the $CD8^+T$ cell response. Journal of Experimental Medicine, 187, 341–348.

Guttal, V., & Couzin, I. D. (2010). Social interactions, information use, and the evolution of collective migration. *Proceedings of the National academy of Sciences of the United States of America, 107*, 16172–16177.

Haroske, D. D., & Triebel, H. (2008). *Distributions, sobolev spaces*. Elliptic Equations: EMS Textbooks in Mathematics, European Mathematical Society (EMS), Zürich.

Henry, D. (1981). *Geometric theory of semilinear parabolic equations*. Berlin: Springer.

Herrero, M. A., & Velázquez, J. J. L. (1996). Singularity patterns in a chemotaxis model. *Mathematische Annalen, 306*, 583–623.

Herrero, M. A., & Velázquez, J. J. L. (1997). A blow-up mechanism for a chemotaxis model. *Annali della Scuola Normale Superiore di Pisa - Classe di Scienze, 24*(4), 633–683.

Hieber, M., & Prüss, J. (1997). Heat kernels and maximal $L^p - L^q$ estimate for parabolic evolution equations. Communications in Partial Differential Equations, 22, 1647–1669.

Hillen, T., & Painter, K. J. (2009). A user's guide to PDE models for chemotaxis. *Journal of Mathematical Biology, 58*, 183–217.

Hillen, T., Painter, K. J., & Winkler, M. (2013). Convergence of a cancer invasion model to a logistic chemotaxis model. *Mathematical Models and Methods in Applied Sciences, 23*, 165–198.

Hoffman, W., Heinemann, D., & Wiens, J. A. (1981). The ecology of seabird feeding flocks in Alaska. *The Auk: Ornithological Advances, 98*, 437–456.

Horstmann, D. (2003). From 1970 until present: The Keller-Segel model in chemotaxis and its consequences. *I. Jahresber. Deutsche Mathematiker-Vereinigung, 105*, 103–165.

Horstmann, D., Winkler, M.: Boundedness vs. blow-up in a chemotaxis system. J. Differential Equations, **215**, 52–107 (2005)

Htwe, M., & Wang, Y. (2019). Decay profile for the chemotactic model with advection and quadratic degradation in bounded domains. *Applied Mathematics Letters, 98*, 36–40.

Htwe, M., Pang, P. Y. H., & Wang, Y. (2020). Asymptotic behavior of classical solutions of a three-dimensional Keller-Segel-Navier-Stokes system modeling coral fertilization. *Zeitschrift für Angewandte Mathematik und Physik, 71*(90).

Hu, B., & Tao, Y. (2016). To the exclusion of blow-up in a three-dimensional chemotaxis-growth model with indirect attractant production. *Mathematical Models and Methods in Applied Sciences, 26*, 2111–2128.

Hu, B., & Lankeit, J. (2018). Boundedness of solutions to a virus infection model with chemotaxis. *Journal of Mathematical Analysis and Applications, 468*, 344–358.

Isenbach, M. (2004). *Chemotaxis*. London: Imperial College Press.

Ishida, S., Seki, K., & Yokota, T. (2014). Boundedness in quasilinear Keller-Segel systems of parabolic-parabolic type on non-convex bounded domains. *Journal of Differential Equations, 256*, 2993–3010.

Jäger, W., & Luckhaus, S. (1992). On explosions of solutions to a system of partial differential equations modelling chemotaxis. *Transactions of the American Mathematical Society, 329*, 819–824.

Jia, Z., & Yang, Z. (2019). Global existence to a chemotaxis-consumption model with nonlinear diffusion and singular sensitivity. *Applicable Analysis, 98*, 2916–2929.

Jiang, J., & Laurençot, P. (2021). Global existence and uniform boundedness in a chemotaxis model with signal-dependent motility. *Journal of Differential Equations, 299*, 513–541.

Jin, C. (2018). Global classical solution and boundedness to a chemotaxis-haptotaxis model with re-establishment mechanisms. *The Bulletin of the London Mathematical Society, 50*, 598–618.

Jin, H. Y., & Wang, Z. A. (2017). Global stability of prey-taxis systems. *Journal of Differential Equations, 262*, 1257–1290.

Jin, H. Y., Kim, Y. J., & Wang, Z. A. (2018). Boundedness, stabilization, and pattern formation driven by density-suppressed motility. *SIAM Journal on Applied Mathematics, 78*, 1632–1657.

Jin, H. Y., Shi, S., & Wang, Z. A. (2020). Boundedness and asymptotics of a reaction-diffusion system with density-dependent motility. *Journal of Differential Equations, 269*, 6758–6793.

Jin, H. Y., & Wang, Z. A. (2021). Global dynamics and spatio-temporal patterns of predator-prey systems with density-dependent motion. *European Journal of Applied Mathematics, 32*, 652–682.

Kareiva, P., & Odell, G. (1987). Swarms of predators exhibit "preytaxis" if individual predators use area-restricted search. *American Naturalist, 130*, 233–270.

Ke, Y., & Zheng, J. (2018). A note for global existence of a two-dimensional chemotaxis haptotaxis model with remodeling of non-diffusible attractant. *Nonlinearity, 31*, 4602–4620.

Ke, Y., & Zheng, J. (2019). An optimal result for global existence in a three-dimensional Keller-Segel-Navier-Stokes system involving tensor-valued sensitivity with saturation. *Calculus of Variations and Partial Differential Equations, 58*, 58–109.

Keller, E. F., & Segel, L. A. (1970). Initiation of slime mold aggregation viewed as an instability. *Journal of Theoretical Biology, 26*, 399–415.

Keller, E. F., & Segel, L. A. (1971). Traveling bands of chemotactic bacteria: A theoretical analysis. *Journal of Theoretical Biology, 30*, 235–248.

Keller, E. F., & Segel, L. A. (1971). Model for chemotaxis. *Journal of Theoretical Biology, 30*, 225–234.

Kinderlehrer, D., & Stampacchia, G. (1980). *An introduction to variational inequalities and their applications* (Vol. 88). New York-London: Pure and Applied Mathematics, Academic Press Inc.

Kiselev, A., & Ryzhik, L. (2012). Biomixing by chemotaxis and enhancement of biological reactions. *Communications in Partial Differential Equations, 37*, 298–318.

Kiselev, A., & Ryzhik, L. (2012). Biomixing by chemotaxis and efficiency of biological reactions: The critical reaction case. *Journal of Mathematical Physics, 53*, 115609.

Kiselev, A., & Xu, X. (2016). Suppression of chemotactic explosion by mixing. *Archive for Rational Mechanics and Analysis, 222*, 1077–1112.

Kolmogorov, A. N., Petrovskii, I. G., & Piskunov, N. S. (1937). A study of the equation of diffusion with increase in the quantity of matter, and its application to a biological problem. *Moscow University Bulletin of Mathematics, 1*, 1–25.

Kolokolnikov, T., Wei, J., & Alcolado, A. (2014). Basic mechanisms driving complex spike dynamics in a chemotaxis model with logistic growth. *SIAM Journal on Applied Mathematics, 74*, 1375–1396.

Kondo, S., & Miura, T. (2010). Reaction-diffusion model as a framework for understanding biological pattern formation. *Science, 329*, 1616–1620.

Kowalczyk, R., & Szymańska, Z. (2008). On the global existence of solutions to an aggregation model. *Journal of Mathematical Analysis and Applications, 343*, 379–398.

Krylov, N. V. (1996). Lectures on elliptic and parabolic equations in Hölder spaces. In *Graduate studies in mathematics* (Vol. 12). Providence, RI: American Mathematical Society

Krzyżanowski, P., Winkler, M., & Wrzosek, D. (2019). Migration-driven benefit in a two-species nutrient taxis system. *Nonlinear Analysis: Real World Applications, 48*, 94–116.

Ladyzenskaja, O. A., Solonnikov, V. A., & Ural'eva, N. N. (1968). Linear and quasilinear equations of parabolic type. (Russian) Translated from the Russian by S. Smith translations of mathematical monographs (Vol. 23). Providence, R.I.: American Mathematical Society

Lankeit, E., & Lankeit, J. (2019). Classical solutions to a logistic chemotaxis model with signal sensitivity and signal absorption. *Nonlinear Analysis: Real World Applications, 46*, 421–445.

Lankeit, E., & Lankeit, J. (2019). On the global generalized solvability of a chemotaxis model with signal absorption and logistic growth terms. *Nonlinearity, 32*, 1569–1596.

Lankeit, J. (2015). Eventual smoothness and asymptotics in a three-dimensional chemotaxis system with logistic source. *Journal of Differential Equations, 258*, 1158–1191.

Lankeit, J. (2016). Long-term behaviour in a chemotaxis-fluid system with logistic source. *Mathematical Models and Methods in Applied Sciences, 26*, 2071–2109.

Lankeit, J. (2016). A new approach toward boundedness in a two-dimensional parabolic chemotaxis system with singular sensitivity. *Mathematicsl Methods in the Applied Sciences, 39*, 394–404.

Lankeit, J. (2017). Locally bounded global solutions to a chemotaxis consumption model with singular sensitivity and nonlinear diffusion. *Journal of Differential Equations, 262*, 4052–4084.

Lankeit, J., & Wang, Y. (2017). Global existence, boundedness and stabilization in a high-dimensional chemotaxis system with consumption. *Discrete and Continuous Dynamical Systems-Series B, 37*, 6099–6121.

Lankeit, J., & Winkler, M. (2017). A generalized solution concept for the Keller-Segel system with logarithmic sensitivity: Global solvability for large nonradial data. *NoDEA Nonlinear Differential Equations Applications, 24*(49).

Lankeit, J., & Viglialoro, G. (2020). Global existence and boundedness of solutions to a chemotaxis-consumption model with singular sensitivity. *Acta Applicandae Mathematicae, 167*, 75–97.

Lee, J. M., Hillen, T., & Lewis, M. A. (2008). Continuous traveling waves for prey-taxis. *Bulletin of Mathematical Biology, 70*, 654–676.

Lee, J. M., Hillen, T., & Lewis, M. A. (2009). Pattern formation in prey-taxis systems. *Journal of Biological Dynamics, 3*, 551–573.

Levine, H. A., Sleeman, B. D., & Nilsen-Hamilton, M. (2000). A mathematical model for the roles of pericytes and macrophages in the initiation of angiogenesis. I. The role of protease inhibitors in preventing angiogenesis. *Mathematical Biosciences, 168*, 71–115.

Levine, H. A., Sleeman, B. D., & Nilsen-Hamilton, M. (2001). Mathematical modeling of the onset of capillary formation initiating angiogenesis. *Journal of Mathematical Biology, 42*, 195–238.

Lewis, M. A. (1994). Spatial coupling of plant and herbivore dynamics: The contribution of herbivore dispersal to transient and persistent waves of damage. *Theoretocal Population Biology, 45*, 277–312.

Li, D., Mu, C., Zheng, P., & Ke, K. (2019). Boundedness in a three-dimensional Keller-Segel-Stokes system involving tensor-valued sensitivity with saturation. *Discrete and Continuous Dynamical Systems-Series B, 24*, 831–849.

Li, J., Pang, P. Y. H., & Wang, Y. (2019). Global boundedness and decay property of a three-dimensional Keller-Segel-Stokes system modeling coral fertilization. *Nonlinearity, 32*, 2815–2847.

Li, J., & Wang, Y. (2018). Repulsion effects on boundedness in the higher dimensional fully parabolic attraction-repulsion chemotaxis system. *Journal of Mathematical Analysis and Applications, 467*, 1066–1079.

Li, J., & Wang, Y. (2021). Asymptotic behavior in a doubly tactic resource consumption model with proliferation. *Zeitschrift für Angewandte Mathematik und Physik, 72*(21).

Li, J., & Wang, Y. (2021). Boundedness in a haptotactic cross-diffusion system modeling oncolytic virotherapy. *Journal of Differential Equations, 270*, 94–113.

Li, J., & Wang, Z. A. (2021). Traveling wave solutions to the density-suppressed motility model. *Journal of Differential Equations, 301*, 1–36.

Li, J. Y., Li, T., & Wang, Z. A. (2014). Stability of traveling waves of the Keller-Segel system with logarithmic sensitivity. *Mathematical Models and Methods in Applied Sciences, 24*, 2819–2849.

Li, T., Suen, A., Xue, C., & Winkler, M. (2015). Global small-data solutions of a two-dimensional chemotaxis system with rotational flux term. *Mathematical Models and Methods in Applied Sciences, 25*, 721–746.

Li, X., Wang, Y., & Xiang, Z. (2016). Global existence and boundedness in a 2D Keller-Segel-Stokes system with nonlinear diffusion and rotational flux. *Communications in Mathematical Sciences, 14*, 1889–1910.

Li, X. (2019). Global classical solutions in a Keller-Segal(-Navier)-Stokes system modeling coral fertilization. *Journal of Differential Equations, 11*, 6290–6315.

Li, Y., Lin, K., & Mu, C. (2015). Boundedness and asymptotic behavior of solutions to a chemotaxis haptotaxis model in high dimensions. *Applied Mathematics Letters, 50*, 91–97.

Li, Y., & Lankeit, J. (2016). Boundedness in a chemotaxis-haptotaxis model with nonlinear diffusion. *Nonlinearity, 29*, 1564–1595.

Liţcanu, G., & Morales-Rodrigo, C. (2010). Global solutions and asymptotic behavior for a parabolic degenerate coupled system arising from biology. *Nonlinear Analysis, 72*, 77–98.

Liţcanu, G., & Morales-Rodrigo, C. (2010). Asymptotic behavior of global solutions to a model of cell invasion. *Mathematical Models and Methods in Applied Sciences, 20*, 1721–1758.

Liu, C., et al. (2011). Sequential establishment of stripe patterns in an expanding cell population. *Science, 334*, 238–241.

Liu, D. (2018). Global classical solution to a chemotaxis consumption model with singular sensitivity. *Nonlinear Analysis: Real World Applications, 41*, 497–508.

Liu, J., & Wang, Y. (2016). Boundedness and decay property in a three-dimensional Keller-Segel-Stokes system involving tensor-valued sensitivity with saturation. *Journal of Differential Equations, 261*, 967–999.

Liu, J., & Wang, Y. (2017). Global weak solutions in a three-dimensional Keller-Segel-Navier-Stokes system involving a tensor-valued sensitivity with saturation. *Journal of Differential Equations, 262*, 5271–5305.

Liu, J. (2020). Boundedness in a chemotaxis-(Navier-)Stokes system modeling coral fertilization with slow p-Laplacian diffusion. *The Journal of Mathematical Fluid Mechanics, 22*(10).

Liu, J. (2020). Large time behavior in a three-dimensional degenerate chemotaxis-Stokes system modeling coral fertilization. *Journal of Differential Equations, 269*, 1–55.

Liu, L., Zheng, J., & Bao, G. (2020). Global weak solutions in a three-dimensional Keller-Segel-Navier-Stokes system modeling coral fertilization. *Discrete and Continuous Dynamical Systems-Series B, 25*, 3437–3460.

Liu, Y. (2019). Global existence and boundedness of classical solutions to a forager-exploiter model with volume-filling effects. *Nonlinear Analysis: Real World Applications, 50*, 519–531.

Liu, Y., & Zhuang, Y. (2020). Boundedness in a high-dimensional forager-exploiter model with nonlinear resource consumption by two species. *Zeitschrift für Angewandte Mathematik und Physik, 71*(151).

Lorz, A. (2010). Coupled chemotaxis fluid equations. *Mathematical Models and Methods in Applied Sciences, 20*, 987–1004.

Lorz, A. (2012). Coupled Keller-Segel-Stokes model: Global existence for small initial data and blow-up delay. *Communications in Mathematical Sciences, 10*, 374–555.

Lv, W., & Wang, Q. (2020). Global existence for a class of chemotaxis systems with signal-dependent motility, indirect signal production and generalized logistic source. *Zeitschrift für Angewandte Mathematik und Physik, 71*(53).

Lv, W., & Wang, Q. (2021). A n-dimensional chemotaxis system with signal-dependent motility and generalized logistic source: Global existence and asymptotic stabilization. *Proceedings of the Royal Society of Edinburgh Section A, 151*, 821–841.

Lv, W., & Wang, Z. A. (2022). Global classical solutions for a class of reaction-diffusion system with density-suppressed motility. *Electronic Research Archive 3*(30), 995–1015.

Ma, M., Ou, C. H., & Wang, Z. A. (2012). Stationary solutions of a volume filling chemotaxis model with logistic growth and their stability. *SIAM Journal on Applied Mathematics, 72*, 740–766.

Ma, M., Peng R., & Wang, Z. A. (2020). Stationary and non-stationary patterns of the density-suppressed motility model. *Journal of Physics D, 402*(132559).

Maini, P. K., Myerscough, M. R., Winters, K. H., & Murray, J. (1991). Bifurcating spatially heterogeneous solutions in a chemotaxis model for biological pattern generation. *Bulletin of Mathematical Biology, 53,* 701–719.

Marciniak-Czochra, A., & Ptashnyk, M. (2010). Boundedness of solutions of a haptotaxis model. *Mathematical Models and Methods in Applied Sciences, 20,* 449–476.

Méndez, V., Campos, D., Pagonabarraga, I., & Fedotov, S. (2012). Density-dependent dispersal and population aggregation patterns. *Journal of Theoretical Biology, 309,* 113–120.

Meral, G., Stinner, C., & Surulescu, C. (2015). A multiscale model for acid-mediated tumor invasion: Therapy approaches. *IMA Journal of Applied Mathematics, 80,* 1300–1321.

Miller, R. L. (1979). Sperm chemotaxis in hydromedusae. I. Species specifity and sperm behavior. *Marine Biology, 53,* 99–114.

Miller, R. L. (1985). Demonstration of sperm chemotaxis in Echinodermata: Asteroidea, holothuroidea, ophiuroidea. *Journal of Experimental Zoology, 234,* 383–414.

Mizoguchi, N., & Souplet, P. (2014). Nondegeneracy of blow-up points for the parabolic Keller-Segel system. *Annales de l'Institut Henri Poincaré Analyse Non Linéaire, 31,* 851–875.

Mollison, D. (1977). Spatial contact models for ecological and epidemic spread. *Journal of the Royal Statistical Society: Series B, 39,* 283–326.

Morales-Rodrigo, C. (2008). Local existence and uniqueness of regular solutions in a model of tissue invasion by solid tumours. *Mathematical and Computer Modelling, 47,* 604–613.

Morales-Rodrigo, C., & Tello, J. (2014). Global existence and asymptotic behavior of a tumor angiogenesis model with chemotaxis and haptotaxis. *Mathematical Models and Methods in Applied Sciences, 24,* 427–464.

Msaouel, P., Opyrchal, M., Musibay, E. D., & Galanis, E. (2013). Oncolytic measles virus strains as novel anticancer agents. *Expert Opinion on Biological Therapy, 13,* 483–502.

Murray, J. D. (2001). *Mathematical biology.* New York: Springer.

Myowin, H., Pang, P. Y. H., & Wang, Y. (2020). Asymptotic behavior of classical solutions of a three-dimensional Keller-Segel-Navier-Stokes system modeling coral fertilization. *Zeitschrift für Angewandte Mathematik und Physik, 71(90).*

Nadin, G., Perthame, B., & Ryzhik, L. (2008). Traveling waves for the Keller-Segel system with Fisher birth terms. *Interfaces Free Bound, 10,* 517–538.

Nagai, T., Senba, T., & Yoshida, K. (1997). Applications of the Trudinger-Moser inequality to a parabolic system of chemotaxis. *Funkcialaj Ekvacioj, 40,* 411–433.

Nagai, T. (2001). Blow-up of nonradial solutions to parabolic-elliptic systems modeling chemotaxis in two-dimensional domains. *Journal of Inequalities and Applications, 6,* 37–55.

Osaki, K., Tsujikawa, T., Yagi, A., & Mimura, M. (2002). Exponential attractor for a chemotaxis growth system of equations. *Nonlinear Analysis, 51,* 119–144.

Ou, C., & Yuan, W. (2009). Traveling wavefronts in a volume-filling chemotaxis model. *SIAM Journal on Applied Dynamical Systems, 8,* 390–416.

Painter, K., Maini, P. K., & Othmer, H. G. (2000). Development and applications of a model for cellular response to multiple chemotactic cues. *Journal of Mathematical Biology, 41,* 285–314.

Painter, K., & Hillen, T. (2002). Volume-filling and quorum-sensing in models for chemosensitive movement. *Canadian Applied Mathematics Quarterly, 10,* 501–543.

Painter, K., & Hillen, T. (2011). Spatio-temporal chaos in a chemotaxis model. *Journal of Physics D, 240,* 363–375.

Pang, P. Y. H., & Wang, Y. (2017). Global existence of a two-dimensional chemotaxis-haptotaxis model with remodeling of non-diffusible attractant. *Journal of Differential Equations, 263,* 1269–1292.

Pang, P. Y. H., & Wang, Y. (2018). Global boundedness of solutions to a chemotaxis-haptotaxis model with tissue remodeling. *Mathematical Models and Methods in Applied Sciences, 28,* 2211–2235.

Pang, P. Y. H., & Wang, Y. (2019). Asymptotic behavior of solutions to a tumor angiogenesis model with chemotaxis-haptotaxis. *Mathematical Models and Methods in Applied Sciences, 29,* 1387–1412.

Pang, P. Y. H., Wang, Y., & Yin, J. (2021). Asymptotic profile of a two-dimensional chemotaxis-Navier-Stokes system with singular sensitivity and logistic source. *Mathematical Models and Methods in Applied Sciences, 31*, 577–618.

Patlak, C. S. (1953). Random walk with persistence and external bias. *Bulletin of Mathematical Biology, 15*, 311–338.

Paweletz, N., & Knierim, M. (1989). Tumor related angiogenesis. *Critical Reviews in Oncology/Hematology, 9*, 197–242.

Perumpanani, A. J., & Byrne, H. M. (1999). Extracellular matrix concentration exerts selection pressure on invasive cells. *European Journal of Cancer, 8*, 1274–1280.

Porzio, M. M., & Vespri, V. (1993). Hölder estimates for local solutions of some doubly nonlinear degenerate parabolic equations. *Journal of Differential Equations, 103*, 146–178.

Quittner, P., & Souplet, P. (2007). *Superlinear parabolic problems*. Birkhäuser Advanced Texts: Basler Lehrbücher, Birkhäuser Verlag, Basel.

Reyes, G., & Vázquez, J. L. (2006). A weighted symmetrization for nonlinear elliptic and parabolic equations in inhomogeneous media. *Journal of the European Mathematical Society (JEMS), 8*, 531–554.

Rosen, G. (1978). Steady-state distribution of bacteria chemotactic toward oxygen. *Bulletin of Mathematical Biology, 40*, 671–674.

Rothe, F. (1984). *Global solutions of reaction-diffusion systems lecture notes in mathematics* (Vol. 1072). Berlin-Heidelberg-New York-Tokyo: Springer.

Salako, R. B., & Shen, W. (2017). Spreading speeds and traveling waves of a parabolic-elliptic chemotaxis system with logistic source on $\mathbb{R}^N$. Discrete and Continuous Dynamical Systems, 37, 6189–6225.

Salako, R. B., & Shen, W. (2017). Global existence and asymptotic behavior of classical solutions to a parabolic-elliptic chemotaxis system with logistic source on $\mathbb{R}^N$. Journal of Differential Equations, 262, 5635–5690.

Salako, R. B., & Shen, W. (2018). Existence of traveling wave solutions of parabolic-parabolic chemotaxis systems. *Nonlinear Analysis: Real World Applications, 42*, 93–119.

Salako, R. B., Shen, W., & Xue, S. (2019). Can chemotaxis speed up or slow down the spatial spreading in parabolic-elliptic Keller-Segel systems with logistic source? *Journal of Mathematical Biology, 79*, 1455–1490.

Salako, R. B., & Shen, W. (2020). Traveling wave solutions for fully parabolic Keller-Segel chemotaxis systems with a logistic source. *Electronic Journal of Differential Equations, 53*.

Schwetlick, H. (2003). Traveling waves for chemotaxis-systems. *PAMM Proceedings in Applied Mathematics and Mechanics, 3*, 476–478.

Short, M. B., D'Orsogna, M. R., Pasour, V. B., Tita, G. E., Brantingham, P. J., Bertozzi, A. L., & Chayes, L. B. (2008). A statistical model of criminal behavior. *Mathematical Models and Methods in Applied Sciences, 18*, 1249–1267.

Short, M. B., Bertozzi, A. L., & Brantingham, P. J. (2010). Nonlinear patterns in urban crime: Hotspots, bifurcations, and suppression. *SIAM Journal on Applied Dynamical Systems, 9*, 462–483.

Simon, J. (1986). Compact sets in the space $L^p(O, T; B)$. Annali di Matematica Pura ed Applicata, 146, 65–96.

Sleeman, B. D. (1997). Mathematical modelling of tumor growth and angiogenesis. *Advances in Experimental Medicine and Biology, 428*, 671–677.

Sohr, H. (2001). *The Navier-Stokes equations. An elementary functional analytic approach*. Basel: Birkhäuser Verlag.

Stinner, C., & Winkler, M. (2011). Global weak solutions in a chemotaxis system with large singular sensitivity. *Nonlinear Analysis: Real World Applications, 12*, 3727–3740.

Stinner, C., Surulescu, C., & Winkler, M. (2014). Global weak solutions in a PDE-ODE system modeling multiscale cancer cell invasion. *SIAM Journal on Mathematical Analysis, 46*, 1969–2007.

Stinner, C., Surulescu, C., & Meral, G. (2015). A multiscale model for pH-tactic invasion with time-varying carrying capacities. *IMA Journal of Applied Mathematics, 80*, 1300–1321.

Stinner, C., Surulescu, C., & Uatay, A. (2016). Global existence for a go-or-grow multiscale model for tumor invasion with therapy. *Mathematical Models and Methods in Applied Sciences, 26*, 2163–2201.

Szymańska, Z., Morales-Rodrigo, C., Lachowicz, M., & Chaplain, M. (2009). Mathematical modelling of cancer invasion of tissue: The role and effect of nonlocal interactions. *Mathematical Models and Methods in Applied Sciences, 19*, 257–281.

Tania, N., Vanderlei, B., Heath, J. P., & Edelstein-Keshet, L. (2012). Role of social interactions in dynamic patterns of resource patches and forager aggregation. *Proceedings of the National academy of Sciences of the United States of America, 109*, 11228–11233.

Tao, X. (2021) Global classical solutions to an oncolytic viral therapy model with triply haptotactic terms. *Acta Applicandae Mathematicae, 171*(5).

Tao, Y., & Wang, M. (2008). Global solution for a chemotactic-haptotactic model of cancer invasion. *Nonlinearity, 21*, 2221–2238.

Tao, Y., & Wang, M. (2009). A combined chemotaxis-haptotaxis system: The role of logistic source. *SIAM Journal on Mathematical Analysis, 41*, 1533–1558.

Tao, Y. (2011). Global existence for a haptotaxis model of cancer invasion with tissue remodeling. *Nonlinear Analysis: Real World Applications, 12*, 418–435.

Tao, Y., & Winkler, M. (2011). A chemotaxis-haptotaxis model: The roles of porous medium diffusion and logistic source. *SIAM Journal on Mathematical Analysis, 43*, 685–704.

Tao, Y., & Winkler, M. (2012). Boundedness in a quasilinear parabolic-parabolic Keller-Segel system with subcritical sensitivity. *Journal of Differential Equations, 252*, 692–715.

Tao, Y., & Winkler, M. (2012). Global existence and boundedness in a Keller-Segel-Stokes model with arbitrary porous medium diffusion. *Discrete and Continuous Dynamical Systems, 32*, 1901–1914.

Tao, Y., & Winkler, M. (2012). Eventual smoothness and stabilization of large-data solutions in a three-dimensional chemotaxis system with consumption of chemoattractant. *Journal of Differential Equations, 252*, 2520–2543.

Tao, Y., & Winkler, M. (2013). Locally bounded global solutions in a three-dimensional chemotaxis-Stokes system with nonlinear diffusion. *Annales de l'Institut Henri Poincaré Analyse Non Linéaire, 30*, 157–178.

Tao, Y. (2014). Boundedness in a two-dimensional chemotaxis-haptotaxis system. *Journal of Oceanography, 70*, 165–174.

Tao, Y., & Winkler, M. (2014). Boundedness and stabilization in a multi-dimensional chemotaxis-haptotaxis model. *Proceedings of the Royal Society of Edinburgh, 144*, 1067–1084.

Tao, Y., & Winkler, M. (2014). Energy-type estimates and global solvability in a two-dimensional chemotaxis-haptotaxis model with remodeling of non-diffusible attractant. *Journal of Differential Equations, 257*, 784–815.

Tao, Y., & Winkler, M. (2015). Large time behavior in a multidimensional chemotaxis-haptotaxis model with slow signal diffusion. *SIAM Journal on Mathematical Analysis, 47*, 4229–4250.

Tao, Y., & Winkler, M. (2015). Boundedness and decay enforced by quadratic degradation in a three-dimensional chemotaxis-fluid system. *Zeitschrift für Angewandte Mathematik und Physik, 66*, 2555–2573.

Tao, Y., & Winkler, M. (2016). Blow-up prevention by quadratic degradation in a two-dimensional Keller-Segel-Navier-Stokes system. *Zeitschrift für Angewandte Mathematik und Physik, 67*(138).

Tao, Y., & Winkler, M. (2017). Effects of signal-dependent motilities in a Keller-Segel-type reaction-diffusion system. *Mathematical Models and Methods in Applied Sciences, 27*, 1645–1683.

Tao, Y., & Winkler, M. (2017). Critical mass for infinite-time aggregation in a chemotaxis model with indirect signal production. *Journal of the European Mathematical Society (JEMS), 19*, 3641–3678.

Tao, Y., & Winkler, M. (2019). A chemotaxis-haptotaxis system with haptoattractant remodeling: Boundedness enforced by mild saturation of signal production. *Communications on Pure and Applied Analysis, 18*, 2047–2067.

Tao, Y., & Winkler, M. (2019). Large time behavior in a forager-exploiter model with different taxis strategies for two groups in search of food. *Mathematical Models and Methods in Applied Sciences, 29*, 2151–2182.

Tao, Y., & Winkler, M. (2020). A critical virus production rate for blow-up suppression in a haptotaxis model for oncolytic virotherapy. *Nonlinear Analysis, 198*, 111870.

Tao, Y., & Winkler, M. (2020). Global classical solutions to a doubly haptotactic cross-diffusion system modeling oncolytic virotherapy. *Journal of Differential Equations, 268*, 4973–4997.

Tao, Y., & Winkler, M. (2020). A critical virus production rate for efficiency of oncolytic virotherapy. *European Journal of Applied Mathematics, 32*, 1–16.

Tao, Y., & Winkler, M. (2021). Critical mass for infinite-time blow-up in a haptotaxis system with nonlinear zero-order interaction. *Discrete and Continuous Dynamical Systems, 41*, 439–454.

Tao, Y., & Winkler, M. (2022). Asymptotic stability of spatial homogeneity in a haptotaxis model for oncolytic virotherapy. *Proceedings of the Royal Society of Edinburgh Section A, 152*, 81–101.

Tuval, I., Cisneros, L., Dombrowski, C., Wolgemuth, C. W., Kessler, J. O., & Goldstein, R. E. (2005). Bacterial swimming and oxygen transport near contact lines. *Proceedings of the National academy of Sciences of the United States of America, 102*, 2277–2282.

Vázquez, J. L. (2007). *the porous medium equations*. Oxford Mathematical Monographs: Oxford University Press, Oxford.

Viglialoro, G. (2017). Boundedness properties of very weak solutions to a fully parabolic chemotaxis-system with logistic source. *Nonlinear Analysis: Real World Applications, 34*, 520–535.

Viglialoro, G. (2019). Global existence in a two-dimensional chemotaxis-consumption model with weakly singular sensitivity. *Applied Mathematics Letters, 91*, 121–127.

Volpert, A. I., Volpert, V. A., & Volpert, V. A. (1994). *Traveling wave solutions of parabolic systems*. (English summary) Translated from the Russian manuscript by James F. Heyda. Translations of mathematical monographs (Vol. 140). American Mathematical Society: Providence, RI.

Walker, C., & Webb, G. F. (2007). Global existence of classical solutions for a haptotaxis model. *SIAM Journal on Mathematical Analysis, 38*, 1694–1713.

Wang, J., & Wang, M. (2019). Boundedness in the higher-dimensional Keller-Segel model with signal-dependent motility and logistic growth. *Journal of Mathematical Physics, 60*, 011507.

Wang, J., & Wang, M. (2019). Global solution of a diffusive predator-prey model with prey-taxis. *Computers and Mathematics with Applications, 77*, 2676–2694.

Wang, J., & Wang, M. (2020). Global bounded solution of the higher-dimensional forager-exploiter model with/without growth sources. *Mathematical Models and Methods in Applied Sciences, 30*, 1297–1323.

Wang, L., Li, Y., & Mu, C. (2014). Boundedness in a parabolic-parabolic quasilinear chemotaxis system with logistic source. *Discrete and Continuous Dynamical Systems, 34*, 789–802.

Wang, W. (2019). The logistic chemotaxis system with singular sensitivity and signal absorption in dimension two. *Nonlinear Analysis: Real World Applications, 50*, 532–561.

Wang, W. (2020). Global boundedness of weak solutions for a three-dimensional chemotaxis-Stokes system with nonlinear diffusion and rotation. *Journal of Differential Equations, 268*, 7047–7091.

Wang, X., Wang, W., & Zhang, G. (2015). Global bifurcation of solutions for a predator-prey model with prey-taxis. *Mathematicsl Methods in the Applied Sciences, 38*, 431–443.

Wang, Y., & Ke, Y. (2016). Large time behavior of solution to a fully parabolic chemotaxis-haptotaxis model in higher dimensions. *Journal of Differential Equations, 260*, 6960–6988.

Wang, Y., & Liu, J. (2022). Large time behavior in a chemotaxis-Stokes system modeling coral fertilization with arbitrarily slow porous medium diffusion. *Journal of Mathematical Analysis and Applications, 506*(125538).

Wang, Y., & Xiang, Z. (2015). Global existence and boundedness in a Keller-Segel-Stokes system involving a tensor-valued sensitivity with saturation. *Journal of Differential Equations, 259*, 7578–7609.

Wang, Y., & Xiang, Z. (2016). Global existence and boundedness in a Keller-Segel-Stokes system involving a tensor-valued sensitivity with saturation: the 3D case. *Journal of Differential Equations, 261*, 4944–4973.

Wang, Y. (2016). Boundedness in the higher-dimensional chemotaxis-haptotaxis model with nonlinear diffusion. *Journal of Differential Equations, 260*, 1975–1989.

Wang, Y. (2017). Global weak solutions in a three-dimensional Keller-Segel-Navier-Stokes system with subcritical sensitivity. *Mathematical Models and Methods in Applied Sciences, 27*, 2745–2780.

Wang, Y., & Li, X. (2017). Boundedness for a 3D chemotaxis-Stokes system with porous medium diffusion and tensor-valued chemotactic sensitivity. *Zeitschrift für Angewandte Mathematik und Physik, 68*(29).

Wang, Y., Winkler, M., & Xiang, Z. (2018). Global classical solutions in a two-dimensional chemotaxis-Navier-Stokes system with subcritical sensitivity. *The Annali della Scuola Normale Superiore di Pisa, Classe di Scienze, 18*(5), 421–466.

Wang, Z. A., & Hillen, T. (2007). Classical solutions and pattern formation for a volume filling chemotaxis model. *Chaos, 17*, 037108.

Wang, Z. A. (2013). Mathematics of traveling waves in chemotaxis-review paper. *Discrete and Continuous Dynamical Systems - Series B, 18*, 601–641.

Wang, Z. A., Xiang, Z., & Yu, P. (2016). Asymptotic dynamics on a singular chemotaxis system modeling onset of tumor angiogenesis. *Journal of Differential Equations, 260*, 2225–2258.

Wei, Y., Wang, Y., & Li, J. (2022). Asymptotic behavior for solutions to an oncolytic virotherapy model involving triply haptotactic terms. *Zeitschrift für Angewandte Mathematik und Physik, 73*(55).

Wiegner, M. (1999). The Navier-Stokes equations-a neverending challenge? *Jahresbericht der Deutschen Mathematiker-Vereinigung, 101*, 1–25.

Winkler, M. (2008). Chemotaxis with logistic source: Very weak global solutions and their boundedness properties. *Journal of Mathematical Analysis and Applications, 348*, 708–729.

Winkler, M. (2010). Boundedness in the higher-dimensional parabolic-parabolic chemotaxis system with logistic source. *Communications in Partial Differential Equations, 35*, 1516–1537.

Winkler, M. (2010). Does a volume-filling effect always prevent chemotactic collapse. *Mathematicsl Methods in the Applied Sciences, 33*, 12–24.

Winkler, M. (2010). Aggregation versus global diffusive behavior in the higher-dimensional Keller-Segel model. *The Journal of Differential Equations, 248*, 2889–2905.

Winkler, M. (2011). Global solutions in a fully parabolic chemotaxis system with singular sensitivity. *Mathematicsl Methods in the Applied Sciences, 34*, 176–190.

Winkler, M. (2011). Blow-up in a higher-dimensional chemotaxis system despite logistic growth restriction. *Journal of Mathematical Analysis and Applications, 384*, 261–272.

Winkler, M. (2012). Global large-data solutions in a chemotaxis-(Navier-)Stokes system modeling cellular swimming in fluid drops. *Communications in Partial Differential Equations, 37*, 319–351.

Winkler, M. (2013). Finite-time blow-up in the higher-dimensional parabolic-parabolic Keller-Segel system. *Journal de Mathématiques Pures et Appliquées, 9*(100), 748–767.

Winkler, M. (2014). Global asymptotic stability of constant equilibria in a fully parabolic chemotaxis system with strong logistic dampening. *Journal of Differential Equations, 257*, 1056–1077.

Winkler, M. (2014). Stabilization in a two-dimensional chemotaxis-Navier-Stokes system. *Archive for Rational Mechanics and Analysis, 211*, 455–487.

Winkler, M. (2014). How far can chemotactic cross-diffusion enforce exceeding carrying capacities? *Journal of Nonlinear Science, 24*, 809–855.

Winkler, M. (2015). Large-data global generalized solutions in a chemotaxis system with tensor-valued sensitivities. *SIAM Journal on Mathematical Analysis, 47*, 3092–3115.

Winkler, M. (2015). Boundedness and large time behavior in a three-dimensional chemotaxis-Stokes system with nonlinear diffusion and general sensitivity. *Calculus of Variations and Partial Differential Equations, 54*, 3789–3828.

Winkler, M. (2016). The two-dimensional Keller-Segel system with singular sensitivity and signal absorption: Global large-data solutions and their relaxation properties. *Mathematical Models and Methods in Applied Sciences, 26*, 987–1024.

Winkler, M. (2016). Global weak solutions in a three-dimensional chemotaxis-Navier-Stokes system. *Annales de l'Institut Henri Poincaré Analyse Non Linéaire, 33*, 1329–1352.

Winkler, M. (2016). The two-dimensional Keller-Segel system with singular sensitivity and signal absorption: Eventual smoothness and equilibration of small-mass solutions. Preprint.

Winkler, M. (2017). Emergence of large population densities despite logistic growth restrictions in fully parabolic chemotaxis systems. *Discrete and Continuous Dynamical Systems - Series B, 22*, 2777–2793.

Winkler, M. (2017). How far do chemotaxis-driven forces influence regularity in the Navier-Stokes system? *Transactions of the American Mathematical Society, 369*, 3067–3125.

Winkler, M. (2017). Global existence and slow grow-up in a quasilinear Keller-Segel system with exponentially decaying diffusivity. *Nonlinearity, 30*, 735–764.

Winkler, M. (2018). Renormalized radial large-data solutions to the higher-dimensional Keller-Segel system with singular sensitivity and signal absorption. *Journal of Differential Equations, 264*, 2310–2350.

Winkler, M. (2018). Singular structure formation in a degenerate haptotaxis model involving myopic diffusion. *Journal de Mathématiques Pures et Appliquées, 9*(112), 118–169.

Winkler, M. (2018). Global existence and stabilization in a degenerate chemotaxis-Stokes system with mildly strong diffusion enhancement. *Journal of Differential Equations, 264*, 6109–6151.

Winkler, M. (2018). Global mass-preserving solutions in a two-dimensional chemotaxis-Stokes system with rotational flux components. *Journal of Evolution Equations, 18*, 1267–1289.

Winkler, M. (2018). Does Fluid Interaction Affect Regularity in the Three-Dimensional Keller-Segel System with Saturated Sensitivity? *Journal of Mathematical Fluid Mechanics, 20*, 1889–1909.

Winkler, M. (2019). A three-dimensional Keller-Segel-Navier-Stokes system with logistic source: Global weak solutions and asymptotic stablization. *Journal of Functional Analysis, 276*, 1339–1401.

Winkler, M. (2019). Global solvability and stabliztion in a two-dmensional cross-diffusion system modeling urban crime. *Annales de l'Institut Henri Poincaré Analyse Non Linéaire, 36*, 1747–1790.

Winkler, M. (2019). Global generalized solutions to a multi-dimensional doubly tactic resource consumption model accounting for social interactions. *Mathematical Models and Methods in Applied Sciences, 29*, 373–418.

Winkler, M. (2019). Boundedness in a chemotaxis-May-Nowak model for virus dynamics with mildly saturated chemotactic sensitivity. *Acta Applicandae Mathematicae, 163*, 1–17.

Winkler, M. (2019). Global classical solvability and generic infinite-time blow-up in quasilinear Keller-Segel systems with bounded sensitivities. *Journal of Differential Equations, 266*, 8034–8066.

Winkler, M. (2020). Can simultaneous density-determined enhancement of diffusion and cross-diffusion foster boundedness in keller-Segel type systems involving signal-dependent motilities? *Nonlinearity, 33*, 6590–6632.

Winkler, M. (2021). Can rotational fluxes impede the tendency toward spatial homogeneity in nutrient taxis(-Stokes) systems? *International Mathematics Research Notices, 11*, 8106–8152.

Winkler, M. (2021). Conditional estimates in three-dimensional chemotaxis-Stokes systems and application to a Keller-Segel-fluid model accounting for gradient-dependent flux limitation. *Journal of Differential Equations, 281*, 33–57.

Winkler, M. Approaching logarithmic singularities in quasilinear chemotaxis-consumption systems with signal-dependent sensitivities. *Discrete and Continuous Dynamical Systems—Series B.* https://doi.org/10.3934/dcdsb.2022009.

Wu, S., Wang, J., & Shi, J. (2018). Dynamics and pattern formation of a diffusive predator-prey model with predator-taxis. *Mathematical Models and Methods in Applied Sciences, 28*, 2275–2312.

Xiang, T. (2018). Global dynamics for a diffusive predator-prey model with prey-taxis and classical Lotka-Volterra kinetics. *Nonlinear Analysis: Real World Applications, 39*, 278–299.

Xu, C., & Wang, Y. (2021) Asymptotic behavior of a quasilinear Keller-Segel system with signal-suppressed motility. *Calculus of Variations and Partial Differential Equations, 60*(183).

Xue, C., & Othmer, H. G. (2009). Multiscale models of taxis-driven patterning in bacterial population. *SIAM Journal on Applied Mathematics, 70*, 133–167.

Xue, C. (2015). Macroscopic equations for bacterial chemotaxis: Integration of detailed biochemistry of cell signaling. *Journal of Mathematical Biology, 70*, 1–44.

Yang, C., Cao, X., Jiang, Z., & Zheng, S. (2015). Boundedness in a quasilinear fully parabolic Keller-Segel system of higher dimension with logistic source. *Journal of Mathematical Analysis and Applications, 430*, 585–591.

Yoon, C., & Kim, Y. J. (2017). Global existence and aggregation in a Keller-Segel model with Fokker-Planck diffusion. *Acta Applicandae Mathematicae, 149*, 101–123.

Yu, H., Wang, W., & Zheng, S. (2018). Global classical solutions to the Keller-Segel-(Navier-)Stokes system with matrix valueed sensitivity. *Journal of Mathematical Analysis and Applications, 461*, 1748–1770.

Zhao, X., & Zheng, S. (2017). Global boundedness to a chemotaxis system with singular sensitivity and logistic source. *Zeitschrift für Angewandte Mathematik und Physik, 68*(2).

Zhao, X., & Zheng, S. (2018). Global existence and asymptotic behavior to a chemotaxis-consumption system with singular sensitivity and logistic source. *Nonlinear Analysis: Real World Applications, 42*, 120–139.

Zhao, X., & Zheng, S. (2019). Global existence and boundedness of solutions to a chemotaxis system with singular sensitivity and logistic-type source. *Journal of Differential Equations, 267*, 826–865.

Zhang, Q., & Li, Y. (2015). Global weak solutions for the three-dimensional chemotaxis-Navier-Stokes system with nonlinear diffusion. *Journal of Differential Equations, 259*, 3730–3754.

Zhang, Q., & Li, Y. (2015). Global boundedness of solutions to a two-species chemotaxis system. *Zeitschrift fur Angewandte Mathematik und Physik, 66*, 83–93.

Zheng, J. (2015). Optimal controls of multidimensional modified Swift-Hohenberg equation. *International Journal of Control, 88*, 1–18.

Zheng, J. (2016). Boundedness in a three-dimensional chemotaxis-fluid system involving tensor-valued sensitivity with saturation. *Journal of Mathematical Analysis and Applications, 442*, 353–375.

Zheng, J. (2017). Boundedness of solution of a higher-dimensional parabolic-ODE-parabolic chemotaxis-haptotaxis model with generalized logistic source. *Nonlinearity, 30*, 1987–2009.

Zheng, J. (2017). Boundedness of solutions to a quasilinear higher-dimensional chemotaxis-haptotaxis model with nonlinear diffusion. *Discrete and Continuous Dynamical Systems, 37*, 627–643.

Zheng, J. (2017). Boundedness and global asymptotic stability of constant equilibria in a fully parabolic chemotaxis system with nonlinear logistic source. *Journal of Mathematical Analysis and Applications, 450*, 1047–1061.

Zheng, J. (2019). An optimal result for global existence and boundedness in a three-dimensional Keller-Segel-Stokes system with nonlinear diffusion. *Journal of Differential Equations, 267*, 2385–2415.

Zheng, J. (2021). A new result for the global existence (and boundedness) and regularity of a three-dimensional Keller-Segel-Navier-Stokes system modeling coral fertilization. *Journal of Differential Equations, 272*, 164–202.

Zheng, J. Global existence and boundedness in a three-dimensional chemotaxis-Stokes system with nonlinear diffusion and general sensitivity. Preprint.

Zheng, J., & Ke, Y. (2020). Blow-up prevention by nonlinear diffusion in a 2D Keller-Segel-Navier-Stokes system with rotational flux. *Journal of Differential Equations, 268,* 7092–7120.

Zheng, J., & Ke, Y. (2021). Global bounded weak solutions for a chemotaxis-Stokes system with nonlinear diffusion and rotation. *Journal of Differential Equations, 289,* 182–235.

Zheng, J., & Wang, Y. (2017). A note on global existence to a higher-dimensional quasilinear chemotaxis system with consumption of chemoattractant. *Discrete and Continuous Dynamical Systems-Series B, 22,* 669–686.

Zheng, P., Mu, C., & Willie, R. (2018). Global asymptotic stability of steady states in a chemotaxis-growth system with singular sensitivity. *Computers and Mathematics with Applications, 5,* 1667–1675.

Zhigun, A., Surulescu, C., & Uatay, A.: Global existence for a degenerate haptotaxis model of cancer invasion. *Zeitschrift für Angewandte Mathematik und Physik, 67*(146).

Printed in the United States
by Baker & Taylor Publisher Services